BMW 2002
Gold Portfolio
1968~1976

Compiled by
R.M. Clarke

ISBN 1 85520 2204

Booklands Books Ltd.
PO Box 146, Cobham, KT11 1LG
Surrey, England

BROOKLANDS BOOKS

BROOKLANDS ROAD TEST SERIES

Abarth Gold Portfolio 1950-1971
AC Ace & Aceca 1953-1983
Alfa Romeo Giulietta Gold Portfolio 1954-1965
Alfa Romeo Giulia Berlinas 1962-1976
Alfa Romeo Giulia Coupés 1963-1976
Alfa Romeo Giulia Coupés Gold P. 1963-1976
Alfa Romeo Spider 1966-1990
Alfa Romeo Spider Gold Portfolio 1966-1991
Alfa Romeo Alfasud 1972-1984
Alfa Romeo Alfetta Coupés & Saloons Gold Portfolio 1974-1987
Alfa Romeo Alfetta GTV6 1980-1987
Allard Gold Portfolio 1937-1959
Alvis Gold Portfolio 1919-1967
American Motors Muscle Cars 1966-1970
Armstrong Siddeley Gold Portfolio 1945-1960
Aston Martin Gold Portfolio 1972-1985
Austin Seven 1922-1982
Austin A30 & A35 1951-1962
Austin Healey 100 & 100/6 Gold P. 1952-1959
Austin Healey 3000 Gold Portfolio 1959-1967
Austin Healey Sprite 1958-1971
BMW Six Cyl. Coupés 1969-1975
BMW 1600 Collection No.1 1966-1981
BMW 2002 Gold Portfolio1968-1976
BMW 316, 318, 320 (4 cyl.) Gold P. 1975-1990
BMW 320, 323, 325 (6 cyl.) Gold P .1977-1990
BMW 5 Series Gold Portfolio1981-1987
BMW M Series Performance Portfolio1976-1993
Bristol Cars Gold Portfolio 1946-1992
Buick Automobiles 1947-1960
Buick Muscle Cars 1965-1970
Cadillac Automobiles 1949-1959
Cadillac Automobiles 1960-1969
Chevrolet 1955-1957
Chevrolet Impala & SS 1958-1971
Chevrolet Corvair 1959-1969
Chevy El Camino & SS 1959-1987
Chevy II Nova & SS 1962-1973
Chevelle & SS Muscle Portfolio 1964-1972
Chevrolet Muscle Cars 1966-1971
Chevy Blazer 1969-1981
Chevrolet Corvette Gold Portfolio 1953-1962
Chevrolet Corvette Sting Ray Gold P. 1963-1967
Chevrolet Corvette Gold Portfolio 1968-1977
High Performance Corvettes 1983-1989
Camaro Muscle Portfolio 1967-1973
Chevrolet Camaro Z28 & SS 1966-1973
Chevrolet Camaro & Z28 1973-1981
High Performance Camaros 1982-1988
Chrysler 300 Gold Portfolio 1955-1970
Chrysler Valiant 1960-1962
Citroen Traction Avant Gold Portfolio 1934-1957
Citroen 2CV 1948-1988
Citroen DS & ID 1955-1975
Citroen SM 1970-1975
Cobras & Replicas 1962-1983
Shelby Cobra Gold Portfolio 1962-1969
Cobras & Cobra Replicas Gold P. 1962-1989
Cunningham Automobiles 1951-1955
Daimler SP250 Sports & V-8 250 Saloon Gold Portfolio 1959-1969
Datsun Roadsters 1962-1971
Datsun 240Z 1970-1973
Datsun 280Z & ZX 1975-1983
De Tomaso Collection No. 1 1962-1981
Dodge Charger 1966-1974
Dodge Muscle Cars 1967-1970
Dodge Viper on the Road
Excalibur Collection No. 1 1952-1981
Facel Vega 1954-1964
Ferrari Cars 1946-1956
Ferrari Collection No. 1 1960-1970
Ferrari Dino 1965-1974
Ferrari Dino 308 1974-1979
Ferrari 308 & Mondial 1980-1984
Motor & T&CC Ferrari 1966-1976
Motor & T&CC Ferrari 1976-1984
Fiat Pininfarina 124 & 2000 Spider 1968-1985
Fiat-Bertone X1/9 1973-1988
Ford Consul, Zephyr, Zodiac Mk.I & II 1950-1962
Ford Zephyr, Zodiac, Executive, Mk.III & Mk.IV 1962-1971
Ford Cortina 1600E & GT 1967-1970
High Performance Capris Gold P. 1969-1987
Capri Muscle Portfolio 1974-1987
High Performance Fiestas 1979-1991
High Performance Escorts Mk.I 1968-1974
High Performance Escorts Mk.II 1975-1980
High Performance Escorts 1980-1985
High Performance Escorts 1985-1990
High Performance Sierras & Merkurs Gold Portfolio 1983-1990
Ford Automobiles 1949-1959
Ford Fairlane 1955-1970
Ford Ranchero 1957-1959
Thunderbird 1955-1957
Thunderbird 1958-1963
Thunderbird 1964-1976
Ford Falcon 1960-1970
Ford GT40 Gold Portfolio 1964-1987
Ford Bronco 1966-1977
Ford Bronco 1978-1988
Holden 1948-1962
Honda CRX 1983-1987
Hudson & Railton 1936-1940
Jaguar and SS Gold Portfolio 1931-1951
Jaguar XK120, 140, 150 Gold P. 1948-1960
Jaguar Mk.VII, VIII, IX, X, 420 Gold P.1950-1970
Jaguar 1957-1961
Jaguar Mk.2 1959-1969
Jaguar Cars 1961-1964
Jaguar E-Type Gold Portfolio 1961-1971
Jaguar E-Type 1966-1971
Jaguar E-Type V-12 1971-1975
Jaguar XJ12, XJ5.3, V12 Gold P. 1972-1990
Jaguar XJ6 Series II 1973-1979
Jaguar XJ6 Series III 1979-1986
Jaguar XJS Gold Portfolio 1975-1990
Jeep CJ5 & CJ6 1960-1976
Jeep CJ5 & CJ7 1976-1986
Jensen Cars 1946-1967
Jensen Cars 1967-1979
Jensen Interceptor Gold Portfolio 1966-1986
Jensen Healey 1972-1976
Lagonda Gold Portfolio 1919-1964
Lamborghini Cars 1964-1970
Lamborghini Countach & Urraco 1974-1980
Lamborghini Countach & Jalpa 1980-1985
Lancia Fulvia Gold Portfolio 1963-1976
Lancia Stratos 1972-1985
Land Rover Series I 1948-1958
Land Rover Series II & IIa 1958-1971
Land Rover Series III 1971-1985
Land Rover 90 & 110 1983-1989
Lincoln Gold Portfolio 1949-1960
Lincoln Continental 1961-1969
Lincoln Continental 1969-1976
Lotus & Caterham Seven Gold P. 1957-1989
Lotus Elite 1957-1964
Lotus Elite & Eclat 1974-1982
Lotus Elan Gold Portfolio 1962-1974
Lotus Elan Collection No. 2 1963-1972
Lotus Cortina Gold Portfolio 1963-1970
Lotus Europa Gold Portfolio 1966-1975
Lotus Turbo Esprit 1980-1986
Motor & T&CC on Lotus 1979-1983
Marcos Cars 1960-1988
Maserati 1965-1970
Maserati 1970-1975
Mazda RX-7 Collection No. 1 1978-1981
Mercedes Benz Cars 1949-1954
Mercedes Benz Competition Cars 1950-1957
Mercedes Benz Cars 1954-1957
Mercedes Benz Cars 1957-1961
Mercedes 190 & 300 SL 1954-1963
Mercedes 230/250/280SL 1963-1971
Mercedes Benz SLs & SLCs Gold P. 1971-1989
Mercedes S & 600 1965-1972
Mercedes S Class 1972-1979
Mercury Muscle Cars 1966-1971
Metropolitan 1954-1962
MG Gold Portfolio 1929-1939
MG TC 1945-1949
MG TD 1949-1953
MG TF 1953-1955
MG Cars 1959-1962
MGA & Twin Cam Gold Portfolio 1955-1962
MG Midget 1961-1980
MGB Roadsters 1962-1980
MGB MGC & V8 Gold Portfolio 1962-1980
MGB GT 1965-1980
Mini Cooper Gold Portfolio 1961-1971
Mini Muscle Cars 1961-1979
Mini Moke 1964-1989
Mopar Muscle Cars 1964-1967
Morgan Three-Wheeler Gold Portfolio 1910-1952
Morgan Plus 4 & Four 4 Gold P. 1936-1967
Morgan Cars 1960-1970
Morgan Cars Gold Portfolio 1968-1989
Morris Minor Collection No. 1 1948-1980
Shelby Mustang Muscle Portfolio 1965-1970
Mustang Muscle Cars 1967-1971
High Performance Mustang IIs 1974-1978
High Performance Mustangs 1982-1988
Oldsmobile Automobiles 1955-1963
Oldsmobile Cutlass & 4-4-2 1964-1972
Oldsmobile Muscle Cars 1964-1971
Oldsmobile Toronado 1966-1978
Opel GT 1968-1973
Packard Gold Portfolio 1946-1958
Pantera Gold Portfolio 1970-1989
Panther Gold Portfolio 1972-1990
Plymouth Barracuda 1964-1974
Plymouth Muscle Cars 1966-1971
Pontiac Tempest & GTO 1961-1965
Pontiac Muscle Cars 1966-1972
Pontiac Firebird & Trans-Am 1973-1981
High Performance Firebirds 1982-1988
Pontiac Fiero 1984-1988
Porsche 356 1952-1965
Porsche Cars in the 60's
Porsche Cars 1960-1964
Porsche Cars 1964-1968
Porsche Cars 1968-1972
Porsche Cars 1972-1975
Porsche 911 1965-1969
Porsche 911 1970-1972
Porsche 911 1973-1977
Porsche 911 Carrera 1973-1977
Porsche 911 Turbo 1975-1984
Porsche 911 SC 1978-1983
Porsche 914 Collection No. 1 1969-1983
Porsche 914 Gold Portfolio 1969-1976
Porsche 924 Gold Portfolio 1975-1988
Porsche 928 1977-1989
Porsche 944 1981-1985
Range Rover Gold Portfolio 1970-1992
Reliant Scimitar 1964-1986
Riley Gold Portfolio 1924-1939
Riley 1.5 & 2.5 Litre Gold Portfolio 1945-1955
Rolls Royce Silver Cloud & Bentley 'S' Series Gold Portfolio 1955-1965
Rolls Royce Silver Shadow 1965-1981
Rover P4 1949-1959
Rover P4 1955-1964
Rover 3 & 3.5 Litre Gold Portfolio 1958-1973
Rover 2000 & 2200 1963-1977
Rover 3500 1968-1977
Rover 3500 & Vitesse 1976-1986
Saab Sonett Collection No.1 1966-1974
Saab Turbo 1976-1983
Studebaker Gold Portfolio 1947-1966
Studebaker Hawks & Larks 1956-1963
Avanti 1962-1990
Sunbeam Tiger & Alpine Gold P. 1959-1967
Toyota MR2 1984-1988
Toyota Land Cruiser 1956-1984
Triumph TR2 & TR3 1952-1960
Triumph TR4, TR5, TR250 1961-1968
Triumph TR6 Gold Portfolio 1969-1976
Triumph TR7 & TR8 1975-1982
Triumph Herald 1959-1971
Triumph Vitesse 1962-1971
Triumph Spitfire Gold Portfolio 1962-1980
Triumph 2000, 2.5, 2500 1963-1977
Triumph GT6 1966-1974
Triumph Stag 1970-1980
TVR Gold Portfolio 1959-1990
VW Beetle Gold Portfolio1935-1967
VW Beetle Gold Portfolio1968-1991
VW Beetle Collection No.1 1970-1982
VW Karmann Ghia 1955-1982
VW Bus, Camper, Van 1954-1967
VW Bus, Camper, Van 1968-1979
VW Bus, Camper, Van 1979-1989
VW Scirocco 1974-1981
VW Golf GTI 1976-1986
Volvo PV444 & PV544 1945-1965
Volvo Amazon-120 Gold Portfolio 1956-1970
Volvo 1800 Gold Portfolio 1960-1973

BROOKLANDS ROAD & TRACK SERIES

Road & Track on Alfa Romeo 1949-1963
Road & Track on Alfa Romeo 1964-1970
Road & Track on Alfa Romeo 1971-1976
Road & Track on Alfa Romeo 1977-1989
Road & Track on Aston Martin 1962-1990
Road & Track on Auburn Cord and Duesenburg 1952-1984
Road & Track on Audi & Auto Union 1952-1980
Road & Track on Audi & Auto Union 1980-1986
Road & Track on Austin Healey 1953-1970
Road & Track on BMW Cars 1966-1974
Road & Track on BMW Cars 1975-1978
Road & Track on BMW Cars 1979-1983
Road & Track on Cobra, Shelby & Ford GT40 1962-1992
Road & Track on Corvette 1953-1967
Road & Track on Corvette 1968-1982
Road & Track on Corvette 1982-1986
Road & Track on Corvette 1986-1990
Road & Track on Datsun Z 1970-1983
Road & Track on Ferrari 1975-1981
Road & Track on Ferrari 1981-1984
Road & Track on Ferrari 1984-1988
Road & Track on Fiat Sports Cars 1968-1987
Road & Track on Jaguar 1950-1960
Road & Track on Jaguar 1961-1968
Road & Track on Jaguar 1968-1974
Road & Track on Jaguar 1974-1982
Road & Track on Jaguar 1983-1989
Road & Track on Lamborghini 1964-1985
Road & Track on Lotus 1972-1981
Road & Track on Maserati 1952-1974
Road & Track on Maserati 1975-1983
R& T on Mazda RX7 & MX5 Miata 1986-1991
Road & Track on Mercedes 1952-1962
Road & Track on Mercedes 1963-1970
Road & Track on Mercedes 1971-1979
Road & Track on Mercedes 1980-1987
Road & Track on MG Sports Cars 1949-1961
Road & Track on MG Sports Cars 1962-1980
Road & Track on Mustang 1964-1977
R&T on Nissan 300-ZX & Turbo 1984-1989
Road & Track on Peugeot 1955-1986
Road & Track on Pontiac 1960-1983
Road & Track on Porsche 1951-1967
Road & Track on Porsche 1968-1971
Road & Track on Porsche 1972-1975
Road & Track on Porsche 1975-1978
Road & Track on Porsche 1979-1982
Road & Track on Porsche 1982-1985
Road & Track on Porsche 1985-1988
R&T on Rolls Royce & Bentley 1950-1965
R&Ton Rolls Royce & Bentley 1966-1984
Road & Track on Saab 1972-1992
R&T on Toyota Sports & GT Cars 1966-1984
R&T on Triumph Sports Cars 1953-1967
R&T on Triumph Sports Cars 1967-1974
R&T on Triumph Sports Cars 1974-1982
Road & Track on Volkswagen 1951-1968
Road & Track on Volkswagen 1968-1978
Road & Track on Volkswagen 1978-1985
Road & Track on Volvo 1957-1974
Road & Track on Volvo 1975-1985
Road & Track - Henry Manney at Large and Abroad

BROOKLANDS CAR AND DRIVER SERIES

Car and Driver on BMW 1955-1977
Car and Driver on BMW 1977-1985
Car and Driver on Cobra, Shelby & Ford GT40 1963-1984
Car and Driver on Corvette 1956-1967
Car and Driver on Corvette 1968-1977
Car and Driver on Corvette 1978-1982
Car and Driver on Corvette 1983-1988
Cand D on Datsun Z 1600 & 2000 1966-1984
Car and Driver on Ferrari 1955-1962
Car and Driver on Ferrari 1963-1975
Car and Driver on Ferrari 1976-1983
Car and Driver on Mopar 1956-1967
Car and Driver on Mopar 1968-1975
Car and Driver on Mustang 1964-1972
Car and Driver on Pontiac 1961-1975
Car and Driver on Porsche 1955-1962
Car and Driver on Porsche 1963-1970
Car and Driver on Porsche 1970-1976
Car and Driver on Porsche 1977-1981
Car and Driver on Porsche 1982-1986
Car and Driver on Saab 1956-1985
Car and Driver on Volvo 1955-1986

BROOKLANDS PRACTICAL CLASSICS SERIES

PC on Austin A40 Restoration
PC on Land Rover Restoration
PC on Metalworking in Restoration
PC on Midget/Sprite Restoration
PC on Mini Cooper Restoration
PC on MGB Restoration
PC on Morris Minor Restoration
PC on Sunbeam Rapier Restoration
PC on Triumph Herald/Vitesse
PC on Spitfire Restoration
PC on Beetle Restoration
PC on 1930s Car Restoration

BROOKLANDS HOT ROD 'MUSCLECAR & HI-PO ENGINES' SERIES

Chevy 265 & 283
Chevy 302 & 327
Chevy 348 & 409
Chevy 350 & 400
Chevy 396 & 427
Chevy 454 thru 512
Chrysler Hemi
Chrysler 273, 318, 340 & 360
Chrysler 361, 383, 400, 413, 426, 440
Ford 289, 302, Boss 302 & 351W
Ford 351C & Boss 351
Ford Big Block

BROOKLANDS RESTORATION SERIES

Auto Restoration Tips & Techniques
Basic Bodywork Tips & Techniques
Basic Painting Tips & Techniques
Camaro Restoration Tips & Techniques
Chevrolet High Performance Tips & Techniques
Chevy Engine Swapping Tips & Techniques
Chevy-GMC Pickup Repair
Chrysler Engine Swapping Tips & Techniques
Custom Painting Tips & Techniques
Engine Swapping Tips & Techniques
Ford Pickup Repair
How to Build a Street Rod
Land Rover Restoration Tips & Techniques
Mustang Restoration Tips & Techniques
Performance Tuning - Chevrolets of the '60s
Performance Tuning - Pontiacs of the '60s

BROOKLANDS MILITARY VEHICLES SERIES

Allied Military Vehicles No.1 1942-1945
Allied Military Vehicles No.2 1941-1946
Complete WW2 Military Jeep Manual
Dodge Military Vehicles No.1 1940-1945
Hail To The Jeep
Land Rovers in Military Service
Off Road Jeeps: Civ. & Mil. 1944-1971
US Military Vehicles 1941-1945
US Army Military Vehicles WW2-TM9-2800
VW Kubelwagen Military Portfolio1940-1975
WW2 Jeep Military Portfolio 1940-1990

2743

CONTENTS

Page	Title	Publication	Date	
5	BMW 2002	*Road & Track*	Apr.	1968
7	BMW 2002 Road Test	*Autosport*	May 17	1968
9	BMW 2002	*Sports Car Graphic*	July	1968
10	BMW 2002 Road Impressions	*Autocar*	May 16	1968
12	Turn Your Hymnals to 2002	*Car and Driver*	Apr.	1968
16	Long Distance Sprinter - BMW 2002 Road Test	*Motor*	June 8	1968
18	In a class of its Own - BMW 2002 Road Test	*Motor Racing & Sportscar*	July	1968
20	BMW 2002 Road Test	*Motor Trend*	July	1968
23	BMW 2002 15,000 Mile Report	*Autocar*	Dec. 12	1968
27	BMW 2002 ... a Controlable Urge Road Test	*Motor Manual*	Dec.	1968
30	BMW 2002 vs. Sunbeam Rapier H120	*Car and Car Conversions*	Apr.	1969
32	BMW 2002 Automatic Road Test	*Road & Track*	Nov.	1969
34	BMW 2002 Road Test	*Car and Driver*	Mar.	1970
39	Fun in a BMW 2002 Ti	*Foreign Car Guide*	Jan.	1969
41	BMW 2002 vs. Fiat 124 vs. Volvo 164 - Comparison Test	*Car Life*	Sept.	1969
48	BMW 2002 Automatic	*World Car Guide*	Feb.	1970
50	BMW 2002 Road Test	*Autosport*	July 30	1970
52	BMW 2002 Alpina	*Motor*	Jan. 30	1971
54	BMW 2002 vs. Fiat 124 Coupé Comparison Test	*Autocar*	Feb. 4	1971
60	BMW 2002 vs. Triumph 2500PI vs. Rover 2000TC Comparison Test	*Car*	Apr.	1971
65	BMW 2002 Road Test	*Road Test*	May	1971
70	BMW 2002 Super-Modified Road Test	*Modern Motor*	July	1971
75	BMW 2002 Tii	*Autosport*	July 22	1971
80	BMW 2002 Tii - 2 Litre with Quick Acceleration Road Test	*Autosport*	Nov. 4	1971
82	BMW 2002 Tii Road Test	*Autocar*	Dec. 30	1971
87	BMW 2002 - The Mountain Marvel	*Road Test*	Nov.	1971
90	BMW 2002 Tii Road Test	*Motor*	Dec. 18	1971
94	BMW 2002 Tii Driving Impression	*Motor Trend*	Jan.	1972
96	BMW 2002 Automatic - 10,000 Mile Long Term Test	*Autocar*	Jan. 13	1972
100	Turbocharged BMW 2002	*Road & Track*	Dec.	1972
102	New BMW 520 and 2002 Cabriolet	*Autosport*	Mar. 15	1973
103	Four of the Best - BMW 2002 Tii, Rover 3500S, Alfa Romeo 2000 GTV, Audi 100 Coupé S Comparison Test	*Motor*	Jan. 13	1973
108	One Satisfied Customer	*Competition Car*	Oct.	1973
110	A Roof for All Seasons - BMW 2002	*Wheels*	Feb.	1974
113	Lost Ground Regained Brief Test	*Road & Track*	Jan.	1973
114	Remodelled BMW 2002 Tii Road Test	*Autosport*	Mar. 14	1974
116	Well Blow Me Down - BMW 2002 Turbo	*Motor*	Apr. 27	1974
118	Suck it and See (It Go!)	*Motor*	May 25	1974
120	Turbo to Nivelles - BMW 2002 Turbo	*Competition Car*	June	1974
122	BMW 2002 Turbo Road Test	*Road & Track*	July	1974
125	BMW 2002 Tii + Five	*Modern Motor*	June	1974
129	How to Blow £4,229 Fast!	*Fast Car*	Aug.	1974
132	BMW 2002 - Hot Stuff from Bravaria Road Test	*Asian Auto*	Feb.	1974
138	BMW 2002 Road Test	*Driving*	Oct.	1974
142	BMW Turbo Road Test	*Autosport*	Oct. 17	1974
144	Bavarian Motor Wonder Road Test	*Road Test*	Jan.	1975
149	BMW 2002 Turbo Road Test	*Autocar*	Nov. 2	1974
155	BMW 2002 - Boost for a New Generation?	*What Car?*	Jan.	1975
156	BMW 2002 Tii - The Ten Tenths Touring Car Road Test	*Modern Motor*	Sept.	1975
161	BMW 1600 & 2002 1967-1976 Used Car Classic	*Road & Track*	Mar.	1981
165	Charger! - BMW 2002 Turbo Road Test	*Classic & Sportscar*	Feb.	1984
166	BMW 2002	*Car - Australia*	Dec.	1989
169	Bargain BMW Buying Used	*Practical Classics*	Mar.	1990

ACKNOWLEDGEMENTS

Our first book on BMWs sucessful 2002 model was a slim volume of only 70 pages - BMW 2002 Collection No.1 which we published in 1981. When it went out of print in 1986 we added 30 more pages and reissued it as - BMW 2002 1968-1976 - demand was so great that it had to be reprinted again in 1989. Since 1986 many more articles on the marque have come to hand and so we are again increasing our coverage by adding a further 70 pages and issuing it as one of our 'Gold Portfolio' series.

Brooklands Books primary aim is to inform and help second and subsequent owners of interesting cars by making available once again the best contemporary articles that were written about their vehicles. This new volume now contains the views of writers from four continents. I am sure BMW 2002 enthusiasts will join me in thanking the publishers of Asian Auto, Autocar, Autosport, Car, Car and Car Conversions, Car and Driver, Car - Australia, Car Life, Foreign Car Guide, Classic & Sportscar, Competition Car, Driving, Fast Car, Modern Motor, Motor, Motor Manual, Motor Racing & Sportscar, Motor Trend, Practical Classics, Road Test, Road & Track, Sports Car Graphic, What Car?, Wheels and the World Car Guide for allowing their copyright material to be reproduced here.

R M Clarke

The 2002 began as the brainchild of Max Hoffmann, BMW's US importer, who wanted a car to follow up the 1600-2, which had been very well received in the USA. The logical car for him to import would have been the hot 1600ti, but its engine could not be made to meet the new exhaust emissions regulations. So he suggested that BMW should substitute the 2-litre engine from its larger saloons, which would give excellent performance even in emissions-controlled form.

The BMW engineers did as Hoffmann suggested, and the result took the motoring world by storm on its announcement in 1968. Not only did it provide the performance-oriented model which Hoffmann wanted for the USA, but in European guise without the power-sapping emissions-control gear, it offered affordable high performance in a compact and agile package. So popular did the 2002 become that it went on to sell more than 300,000 examples over the seven seasons of its production.

It was only to be expected that BMW would capitalise on this success. The original 2002 was complemented by an automatic version in 1969, and in due course by a whole range of other derivatives: 2002ti, 2002tii, 2002 convertible, three hatchback touring models, and finally the legendary 2002 Turbo. When the last 2002 variant rolled off the Munich assembly lines in 1975, the grand total for all models came to just under 418,000.

Today, the qualities which made the 2002 range so appealing in the late 1960s and early 1970s continue to endear them to motoring enthusiasts. The 2002 models are widely and rightly recognised as classics of their period, and enjoy a large and committed following. This book will be welcomed by enthusiasts everywhere, and will also form an invaluable reference work for those who earn their living by writing about interesting motor cars.

James Taylor

New from Bayerische Motoren Werke
BMW 2002

AT THE 1967 Frankfurt Show BMW introduced a new variation on the popular 1600 2-door sedan, the 1600 TI. Like previous TI BMWs it has dual carburetors (instead of one), a higher compression ratio and hotter cam timing than its plain counterpart, plus stiffened suspension and appropriate identifying marks here and there. But BMW has discovered that its dual-carburetor engines don't lend themselves to the control of exhaust emissions. So, for the American market, BMW engineers substituted the existing single-carburetor 2000 engine and christened the resulting car 2002.

In this case the emission control regulations have precipitated a car that is eminently more suitable for American taste than the car we would have had otherwise. The 1600 TI and 2002 develop about the same maximum power—118 bhp @ 6200 rpm and 113 @ 5800 respectively—but there's a big difference in torque: the 2002 produces 116-ft @ 3000 rpm, against the 1600 TI's 97 @ 4500. The generous torque curve of the 2002 allows use of a 3.64:1 final drive ratio, vs. 4.11:1 for the 1600 and 1600 TI; again American tastes are

better served with fewer revs/mile and hence less engine fuss on the freeway.

The success of the 1600 2-door is well known but just to refresh the memory, we found it to be one of the best automotive values to be had from any country—brimming with good handling, good ride, good finish, refinement of running and even good style. It was only natural that BMW should expand on the new 2-door line, and if the price is not too much extra over the reasonable $2600-plus of the 1600 it should give BMW another healthy increase in sales for 1968—sales for the make were more than tripled in the U.S. in 1967.

The engine is exactly the same unit as used in the standard 2000: a 1990-cc, 4-cyl inline sohc unit with a single Solex carburetor. Clutch and gearbox likewise are the familiar and excellent units. In the chassis department, all elements are basically the same as in other BMWs—MacPherson-geometry front suspension and semi-trailing-arm independent rear suspension. Spring and shock-absorber rates, however, are tightened up in anticipation of more vigorous driving and radial tires are standard equipment on the 2002. In other words, the 2002 won't have the impressively smooth ride of the 1600—but it should be a lot more fun to drive on a winding road, and undoubtedly its wheels will be firmly planted on the ground with the excellent independent suspension.

Like the 1600, the 2002 is on a 98.4-in. wheelbase and has an overall length of 166.5 in. Track is unchanged, front and rear, at 52.4 in., and the car is 62.6 in. wide. BMW gives a figure of 10.5 sec for 0-60 mph and a top speed of 105 mph; we drove the car briefly in Germany and actually attained 108 mph on the autobahn. We'll have a full road test next month.

BMW 2002 SPECIFICATIONS

Engine type..........4 cyl inline, sohc
Bore x stroke, mm..............89 x 80
Displacement, cc/cu in.......1990/121.4
Compression ratio................8.5:1
Bhp @ rpm.................113 @ 5800
Torque @ rpm, lb-ft........116 @ 3000
Transmission.......4-speed manual, all-synchromesh
Gear ratios: 4th (1.00)...........3.64:1
  3rd (1.35)....................4.91:1
  2nd (2.05)....................7.46:1
  1st (3.84)...................13.95:1
Brakes.....................disc/drum
Tires......................165SR-13
Wheels......................4½J-13
Front suspension: MacPherson struts, lower A-arms, coil springs, tube shocks, anti-roll bar.
Rear suspension: semi-trailing arms, coil springs, tube shocks.
Curb weight, lb....................2070
Wheelbase, in.....................98.4
Track, front/rear............52.4/52.4
Overall length....................166.5
  width.........................62.6
  height........................54.0

ROAD TEST
by John Bolster
BMW 2002

Acute acceleration, remarkable roadholding

MORE than half a century ago, BMW were manufacturing some very advanced aero-engines. The still-famous opposed-twin shaft-driven motorcycles followed, and then the Austin Seven car was built under licence. Soon, a small six-cylinder car that was totally BMW replaced it, and in the latter nineteen-thirties the 328 sports car electrified us, from which the immortal 2-litre Bristol engine was derived.

After the war, BMW at first built very expensive eight-cylinder cars, and then went to the other extreme with tiny air-cooled vehicles. Suddenly, they hit the jackpot with a range of medium-sized overhead-camshaft cars, and now the factory is having to be greatly enlarged to cope with the demand. Production at present is 475 cars per day, of which over half are for export, and it is intended that this shall rise to 700 per day during the next two years. This would be a good output if the cars were "popular" models, but as BMW make only fairly costly high-quality cars, such a demand must prove the exceptional merit of the products.

The subject of the present test is the 2002, which has been on sale for several months on the Continent and this week is announced on the British market in right hand drive form. The compact body shell of the 1600 is used, with brakes and suspension suitably modified for continuous high-speed use. It is called a coupé, to distinguish it from the much more bulky bodies of the 1800 and 2000 series, but in fact it could equally well be described as a 2-door 4-seater saloon, and though it is so much smaller and lighter than the larger cars, it is by no means short of space inside.

In place of the 1573 cc engine of the 1600, a larger four-cylinder engine, also with an overhead camshaft, is installed. This is an 89 mm x 80 mm (1990 cc) unit with a five-bearing crankshaft, and it produces a net output of 100 bhp at 5500 rpm. Much more important, however, is the torque of 115.7 ft/lb at 3000 rpm, compared with 91.1 ft/lb at the same speed for the 1573 cc unit which normally powers this chassis. Put another way, the 2-litre engine is now having to push along about 4.5 cwt less than before, so a marked increase in performance is assured.

MacPherson struts in front and semi-trailing arms behind are the standard suspension arrangement for BMW cars. Good as the roadholding is of the larger models, the 2002 feels altogether more crisp and responsive to handle, which is saying a great deal. Perhaps the more compact body is also more rigid, and certainly the weight reduction helps, but the most likely explanation is simply that this is a later design, which takes advantage of all the experience which has been accumulated.

In accepting the extra cost of independent rear suspension, BMW have faced the fact that you can have comfort or roadholding with a rigid axle, but not both. Since this is a car which is exported to many different countries, it must ride well over bad roads as well as on the billiard table surfaces of England. As part of my road test, I collected a BMW in Munich and drove it across Germany, Belgium and France. I can therefore state with some authority that the riding comfort and stability are absolutely outstanding.

The cornering power, with German Dunlop radial ply tyres, is very high indeed. The machine is beautifully balanced, having just enough understeer to give high speed straight running, but when driven reasonably hard the characteristic is neutral. Ultimately, the rear end can be made to hang out a little on power, but it does not fully break away. A very fine sense of control is imparted and there always seems to be a reserve of roadholding to cope with the unexpected. The servo-assisted brakes are

SPECIFICATION

Car tested: BMW 2002 coupé, price £1597 8s 2d including PT.
Engine: Four cylinders, 89 mm x 80 mm (1990 cc). Single chain-driven overhead camshaft. Compression ratio 8.5:1. 113 bhp (gross) or 100 bhp (net) at 5500 rpm. Solex twin-choke downdraught carburetter. Bosch coil and distributor.
Transmission: Single dry plate clutch. 4-speed all-synchromesh gearbox with short central lever, ratios 1.0, 1.345, 2.053, and 3.835:1. Open propeller shaft. Chassis-mounted hypoid final drive, ratio 3.64:1.
Chassis: Combined steel body and chassis. Independent front suspension by MacPherson struts containing telescopic dampers with helical springs plus bottom wishbones. Worm and roller steering with 3-piece track rod. Independent rear suspension by semi-trailing arms and helical springs with telescopic dampers. Anti-roll torsion bars front and rear. Disc front and drum rear brakes with vacuum servo. Bolt-on disc wheels fitted Dunlop 165 SR 13 radial ply tyres.
Equipment: 12-volt lighting and starting with alternator. Speedometer. Rev counter. Fuel and water temperature gauges. Cigar lighter. Heating, demisting and ventilation system. Flashing direction indicators. Reversing lights.
Dimensions: Wheelbase, 8 ft 2.5 ins; track, 4 ft 4 ins; overall length, 14 ft 1 in; width, 5 ft 4 ins; weight, 18 cwt 52 lbs.
Performance: Maximum speed, 108 mph. Speeds in gears: third, 84 mph; second, 64 mph; first, 34 mph; standing quarter-mile, 17.8 s. Acceleration: 0-30 mph, 2.9 s; 0-50 mph, 6.8 s; 0-60 mph, 9.9 s; 0-80 mph, 20 s.
Fuel consumption: 24 to 30 mpg.

PERFORMANCE GRAPH

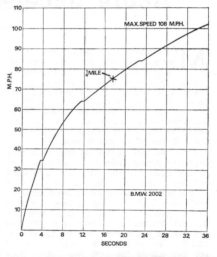

light in action and do not tend to fade, the car remaining steady during panic stops.

The engine is delightful. No doubt a TI version will be offered for those requiring to motor at over 120 mph, but for all normal purposes this power unit is ideal. The combination of exceptional medium-speed torque with a car of light weight produces quite remarkable flexibility, and the top-gear performance is reminiscent of a V8. Really rapid overtaking is possible in the medium ranges without touching the gearlever, and though there is no reason to neglect the gearbox, such driving does make for silent and effortless progress, with marked fuel economy.

Both the clutch and gearbox are in fact very pleasant to handle. My only criticism concerns the pedals, which are too high above the floor and not well placed for heel and toe. The choice of gear ratios is not critical because the engine has such a wide range, but those chosen appear to suit the car and its power unit very well.

The silence of the engine and transmission are matched by the absence of wind noise. The tyres are quiet on most road surfaces, too, although they are of the radial ply type. Adequate ventilation is provided, as was proved in an unexpected heat wave, and the rear windows can be used as extractors without creating a noise or draughts. When the heater is in operation, air outlets are arranged to demist the rear and side windows, but there are no independent fresh air face-level ventilators, which one would expect in such a car.

A typically BMW interior treatment is used. This firm do not consider that thick pile carpets and elaborate walnut veneer panelling are necessary adjuncts of the car of today. The trimming is well carried out but tends to be plain, with no suggestion of spurious luxury. The seats are very comfortable, giving plenty of support to the small of the back, and the upholstery breathes well in hot weather. The external appearance is exactly right, being strictly functional but neat and attractive without calling unnecessary attention to the speed capabilities of the 2002. The new wheel trim adds to the purposeful look of the car.

On Continental roads, this unobtrusive car slips past the other traffic without exciting remark. The cruising speed may be an easy 80 or 90 mph, or more likely an equally effortless 100 mph, the choice of speed simply depending on the driver's mood and on whether he wants to do 24 or 30 mpg. In addition to my drive across the Continent, I subsequently borrowed a second 2002, which I took to Snetterton with the object of testing its handling safely to the limit. When the circuit was wet and slippery, the roadholding was phenomenally good, cornering either as on rails or in controlled skids, according to the driver's choice. The quick gearchange and the high third gear were exploited to the full, especially in recording the excellent acceleration figures which accompany this report.

The BMW 2002 gives a remarkable combination of performance, roadholding, and riding comfort. It is also exceptionally refined, and whereas most 2-litre engines broadcast the fact that they have only four cylinders, this overhead-camshaft power unit could have almost any number of cylinders, as far as the passengers can tell, once it is on the move. Like all the best modern cars, this one has compact overall dimensions but seems big inside. It will suit so many people that an even further enlargement of the factory may soon be required.

Photos/Randy Holt, Jr.

BMW 2002

Fine cars have traditionally been a trademark of BMW, but there was always one drawback — namely, price. The average American buyer would look at the $4000 price tag, then the size of the automobile, and frankly admit to himself that for $1500 less he could buy a big American sedan. But, then, that's the American buyer for you: he's just not interested in durability, economy, and good total performance. Either that, or he's not aware of the performance BMW has to offer.

When the German company announced its 1600 TI, a smaller version of the 2000 sedan, the $2500 sticker price was low enough to stir up considerable interest. Now BMW has plunked a 1990-cc, four-cylinder power plant under the hood of the same car and named it the 2002. And if there's anything that comes close to being a perfect compact sedan, it's the 2002.

There's plenty of muscle under the hood now, with 113 horsepower pulling 2210 pounds of automobile. Performance-wise, the 2002 will reach 60 miles an hour in about 11 seconds and achieve a top speed of a little more than 100 mph. Moreover, the factory claims that with some fine tuning, the 0-60 performance figure can be improved to a little under 10 seconds.

The 2002 sells for $2988 in the United States, which is a pretty good price if you consider the car's excellent comfort and roominess. In fact, there is a lot more space inside the 2002 than you'll find in most Detroit cars.

Marketing in the United States is becoming increasingly important to BMW, and 2002 sales are expected to give the company a new high for 1968. If it doesn't, then there are more ignorant car buyers out there than we thought!

ROAD IMPRESSIONS
BMW 2002

BY GEOFFREY HOWARD

It was announced in time for the Geneva Salon last March that BMW would be offering the plain 1,990 c.c. engine in their little two-door saloon as an optional addition to the range. On the Continent, the two-door 1600 is called the 1600-2 to prevent confusion with the earlier four-door 1600, which was never imported here. It was therefore decided to name the new car the 2002. In many ways it is even nicer to drive than all the other BMWs.

Basically the 1600 is a lighter and much more manageable car than both the 1800 and 2000. Lighter both in weight (by some 3cwt) and in general airiness brought about by the low waistline and generous glass areas. Even as a 1600 it shows a surprising turn of speed, with 0 to 60 mph in 12.5sec and a mean maximum of 102 mph (full AUTOCAR road test, 26 October, 1967). The 1600 develops 85 bhp at 5,700 rpm, so the effect of putting the larger, 2000 engine with 100 bhp at 5,500 rpm is pretty exciting. On a car slightly suspect due to intermittent misfiring, we recorded 0 to 60 mph in only 10.6sec and a mean maximum of 107 mph.

Our experience of the new model so far stems from two different cars. An early right-hand-drive example was used for performance measurements and general assessment, while another brand new one was picked up from the factory in Munich a few weeks ago and driven home. This one we are keeping for a long-term assessment.

Really, therefore, our story of the 2002 starts on the production line in Bavaria, where there is still a kind of leisurely, let's-get-it-right attitude and none of the mass-produced fervour predominant with larger manufacturers. Currently nearly 500 cars per day are being built.

Speedometers go into the cars reading 99,969, the object being that zero starts from when the customer takes delivery. A 10 per cent sample are given a road test, while the remainder make do with roller running; all engines are run on the bed for six hours, before being installed.

We had chosen Granada red for our long-term car, and it really looked resplendent in the delivery car park. Obviously this one had become one of the 10 per cent, as the mile-ometer read 99,979. Running-in instructions were generous: up to 80 mph for the first 250 miles or so, and then up to 100 mph if all seems well. The BMW delivery centre is on the end of the main Nürnberg-Frankfurt *autobahn* so the engine was able to run at steady revs on a light load, ideal for breaking the friction of initial assembly and bedding-in bearings, right from the start.

All was quiet and smooth as we bowled along at about 75 mph for the first two hours. Despite the pillarless sides and lack of window frames on the doors there is virtually no wind noise and engine and transmission were both inaudible. The only noticeable noise was a general kind of faint commotion (camshaft chain whirr, tyre roar and the like), but it was still easy to talk without raising our voices.

After a quick "plastic" service area lunch we continued towards Cologne, averaging an easy 70 mph. The water temperature never quivered from the middle of its gauge (and I had checked the fan belt at the lunch stop) so we increased cruising speed into the 90-mph band once we had passed Frankfurt. It was a very hot, sticky day and it was better to keep the rear quarter panes open to improve the through-flow ventilation. We missed having any cool face-level outlets.

Without straining ourselves or the car we reached Brussels before 7 p.m., having left Munich at 10 a.m., giving a journey time of 8½ hours and a distance of 510 miles (an average of 60 mph). It impressed me to travel in a new car at 100 mph for a short burst almost the day it was made, and it was a relief to get what running-in was necessary over and done with in only a day.

The next afternoon we arrived back in England, the BMW going better than ever. It was only then that we began to use the gearbox, which was a bit stiff; engine and axle were freeing nicely, but the brakes still needed bedding.

Several subtle styling changes have been made to distinguish the 2002 from the 1600; these also apply to the 1600 TI, which is not imported here. At the front, most of the grille is painted matt black, just the surrounding flange and two horizontal bars remaining as polished aluminium. Wheel trims are satin finished, and 165-13in. radial tyres are standard; our car was fitted with German SP Sport tyres, others had Continental radials. Inside there is a woodrim steering wheel with matt black spokes, and all cars for Britain have the optional rev counter instead of a clock as the third dial in front of the driver. Seat backs recline.

The early car we tried had the original high pendant pedal layout (like the 1600 we tested last year) with both brake and clutch 7½in. off the floor. This made it impossible to heel and toe, and one had to lift one's left foot right off the floor to work the clutch. On the new car, the brake has been lowered an inch, which helps, but I managed to get both pads about 2in. lower still by some crafty cutting and welding. It transforms the driving position.

Being based on the 1600 TI, the 2002 has front and rear anti-roll bars and handles beautifully on twisty roads. It corners flat and securely, with neutral characteristics basically, instantly convertible to power oversteer if one wants to steer with the throttle. Compared with other BMWs the engine seems to have a much easier time, and it never sounds quite as busy at speed. Flexibility is excellent and it pulls like a lion from low speed.

After driving the 2002 for about 1,300 miles I handed it over very reluctantly to Graham Robson, who will be running it for the next few months. The way things are going we will certainly clock 10,000 miles before the end of the year and our long-term report will appear then. Anyone wanting to follow in our footsteps should inquire about discount available for taking delivery at the factory—I can recommend it.

1. The 2000 engine has a single carburettor and an alternator. On the far side are the screen washer reservoir and summer/winter air intake

2. Radial-ply tyres are standard and the wheel trims have a satin-chrome finish. There are towing hooks back and front

3. Front seat backs can be reclined, and they also lock to prevent them tipping forward. Rear quarter windows hinge out for added ventilation

4. Spare wheel and tools live under a trapdoor in the floor of the boot (removed here)

5. A full-size rev counter matches the speedometer and the steering wheel has a wooden rim. This picture was taken before we lowered the pedals

PERFORMANCE DATA

Figures in brackets are for the BMW 1600 tested in AUTOCAR of 26 October 1967

Acceleration Times (mean): Speed range, gear ratios and time in seconds.

mph	Top	3rd	2nd	1st
	3.64 (4.11)	4.89 (5.73)	7.46 (8.43)	13.95 (15.80)
10-30	— (11.3)	6.4 (7.5)	3.8 (4.6)	2.5
20-40	8.1 (10.1)	5.6 (6.7)	3.5 (4.2)	—
30-50	7.6 (10.1)	5.5 (6.6)	3.7 (4.7)	—
40-60	8.4 (10.6)	5.8 (7.1)	5.1 (—)	—
50-70	8.5 (12.5)	6.2 (8.4)	—	—
60-80	10.2 (16.6)	8.6 (12.1)	—	—
70-90	13.2 (—)	—	—	—
80-100	17.9 (—)	—	—	—

FROM REST THROUGH GEARS TO:

30 mph	3.5 sec (4.1 sec)
40 mph	5.3 sec (6.1 sec)
50 mph	7.6 sec (8.8 sec)
60 mph	10.6 sec (12.5 sec)
70 mph	14.2 sec (17.5 sec)
80 mph	18.2 sec (23.9 sec)
90 mph	26.4 sec (—)
100 mph	37.2 sec (—)

Standing quarter-mile 17.4 sec 77 mph (18.5 sec 72 mph)
Standing kilometre 32.8 sec 96 mph (34.3 sec 86 mph)

MAXIMUM SPEEDS IN GEARS:

	mph	kph	rpm
Top (mean)	107 (102)	172 (164)	5,800 (6,250)
(best)	110 (106)	177 (171)	5,960 (6,500)
3rd	89 (90)	143 (145)	6,500 (7,500)
2nd	59 (60)	95 (97)	6,500 (7,500)
1st	31 (32)	50 (52)	6,500 (7,500)

OVERALL FUEL CONSUMPTION FOR 1,800 MILES: 25.5 mpg; 11.1 litres/100km
(26.3 mpg; 10.7 litres/100 km)

Turn Your Hymnals to 2002– David E. Davis, Jr. Blows His Mind on the Latest from BMW

As I sit here, fresh from the elegant embrace of BMW's new 2002, it occurs to me that something between nine and ten million Americans are going to make a terrible mistake this year. Like dutiful little robots they will march out of their identical split-level boxes and buy the wrong kind of car. Fools, fools! Terrible, terrible, I say. Why are you blowing your money on this year's too-new-to-be-true facelift of the Continental/Countess Mara/Sprite/Sprint Status Symbol/Sting Ray/Sex Substitute/ Mainliner / Belair / Newport / Overkill / Electra / Eldorado / Javelin / Toad / GTO / GTA/GTB/GTS/GTX/Reality Blaster/ Variant/Park Lane/Park Ward/Ward-Heeler/XK-E/Dino/Dud car when you should be buying a BMW 2002, I ask.

Down at the club, Piggy Tremalion and Bucko Penoyer and all their twit friends buy shrieking little 2-seaters with rag tops and skinny wire wheels, unaware that somewhere, someday, some guy in a BMW 2002 is going to blow them off so bad that they'll henceforth leave every stoplight in second gear and never drive on a winding road again as long as they live.

In the suburbs, Biff Everykid and Kevin Acne and Marvin Sweatsock will press their fathers to buy HO Firebirds with tachometers mounted out near the horizon somewhere and enough power to light the city of Seattle, totally indifferent to the fact that they could fit more friends into a BMW in greater comfort and stop better and go around corners better and get about 29 times better gas mileage.

Mr. and Mrs. America will paste a "Support Your Local Police" sticker on the back bumper of their new T-Bird and run Old Glory up the radio antenna and never know that for about 2500 bucks less they could have gotten a car with more leg room, more head room, more luggage space, good brakes, decent tires, independent rear suspension, a glove box finished like the inside of an expensive overcoat and an ashtray that slides out like it was on the end of a butler's arm—not to mention a lot of other good stuff they didn't even know they could *get* on an automobile, like doors that fit and seats that don't make you tired when you sit in them.

So far as I'm concerned, to hell with all of 'em. If they're content to remain in the automotive dark, let them. I know about the BMW 2002, and I suspect enthusiasts will buy as many as those pink-cheeked Bavarians in their leather pants and mountain-climbing shoes would like to build and ship over here. Something between nine and ten million squares will miss out on this neat little 2-door sedan with all the *cojones* and *brio* and *elan* of cars twice its size and four times its price, but some ten thousand keen types will buy them in 1968, so the majority loses for once.

The 2002 is BMW's way of coping with the smog problem. They couldn't import their little 1600 TI, because their smog device won't work on its multi-carbureted engine. So they stuffed in the smooth, quiet 2-liter (single carburetor) engine from the larger 2000 sedan and—SHAZAM—instant winner!

To my way of thinking, the 2002 is one of modern civilization's all-time best ways to get somewhere sitting down. It grabs you. You sit in magnificently-adjustable seats with great, tall windows all around you. You are comfortable and you can see in every direction. You start it. Willing and un-lumpy is how it feels. No rough idle, no zappy noises to indicate that the task you propose might be anything more than child's play for all those 114 Bavarian superhorses.

Depress the clutch. Easy. Like there was no spring. *Snick*. First gear. Remove weight of left foot from clutch. Place weight of right foot on accelerator. The minute it starts moving, you know that Fangio and Moss and Tony Brooks and all those other

PHOTOGRAPHY: J. BARRY O'ROURKE

big racing studs retired only because they feared that someday you'd have one of these, and when that day came, you'd be indomitable. They were right. You are indomitable.

First stoplight. I blow off aging Plymouth sedan and 6-cylinder Mustang. Not worthy of my steel. Too easy. Next time. Big old 6-banger Healey and '65 GTO. GTO can't believe I'm serious, lets me get away before he opens all the holes and comes smoking past with pain and outrage all over his stricken countenance. Nearly hits rear-end of truck in panicky attempt to reaffirm virility. Austin-Healey a different matter. Tries for all he's worth, but British engineering know-how and quality-craftsmanship not up to the job. I don't even shift fast from third to fourth, just to let him feel my utter contempt.

Nobody believes it, until I suck their headlights out. But nobody doubts it, once that nearly-silent, unobtrusive little car has disappeared down the road and around the next bend, still accelerating without a sign of the brake lights. I learn not to tangle with the kids in their big hot Mothers with the 500 horsepower engines unless I can get them into a tight place demanding agility, brakes, and the raw courage that is built into the BMW driver's seat as a no-cost extra.

What you like to look for are Triumphs and Porsches and such. Them you can slaughter, no matter how hard they try. And they always try. They really believe all that jazz about their highly-tuned, super-sophisticated sports machines, and the first couple of drubbings at the hands of the 2002 make them think they're off on a bad trip or something. But then they learn the awful truth, and they begin to hang back at traffic signals, pretending that they weren't really racing and all. Ha! Grovel, Morgan. Slink home with your tail between your legs, MG-B. Hide in the garage when you see a BMW coming. If you have to race with something, pick a sick kid on an old bicycle.

But I don't want you to get the notion that this is nothing more than a pocket street racer. The BMW 2002 may be the first car in history to successfully bridge the gap between the diametrically-opposed automotive requirements of the wildly romantic car nut, on one hand, and the hyperpragmatic people at *Consumer Reports*, on the other. Enthusiasts' cars invariably come off second-best in a CU evaluation, because such high-spirited steeds often tend to be all desire and no protein—more Magdalen than Mom.

CU used to like the VW a lot, back when it was being hailed as the thinking man's answer to the excesses of Detroit, but now that the Beetle has joined Chevrolet at the pinnacle of establishment-acceptance, it's falling from CU's favor. But the BMW 2002 is quite another matter. It is still obscure enough to have made no inroads at all with the right-thinking squares of the establishment. It rides like a dream. It has a surprising amount of room inside. It gets great gas mileage. It's finished, inside and out, like a Mercedes-Benz, but it doesn't cost very much. All those qualifications are designed to earn the BMW a permanent place in the *Consumer* hall of fame. But for the enthusiasts—at the same time, and without even stepping into a phone booth to change costume—it goes like bloody hell and handles like the original bear. No doubt about it, the BMW 2002 is bound to get Germany back into the CU charts, to borrow a phrase from the pop vernacular.

If it wasn't already German, I'd be tempted to say it could be as American as Mom's apple pie or Rapp Brown's carbine. Not American in the same sense as the contemporary domestic car, with all its vast complexity and *nouveau riche* self-consciousness, but American in the sense of Thomas Edison and a-penny-saved-is-a-penny-earned and Henry Ford I (before his ego overloaded all the fuses and short-circuited his mind and conscience). The 2002 mirrors faithfully all those basic tenets of the Puritan ethic on which our Republic was supposedly based. It does everything it's supposed to do, and it does it with ingenuity, style, and verve.

In its unique ability to blend fun-and-games with no-nonsense virtue, this newest BMW also reflects another traditional American article of faith—our unshakable belief that we can find and marry a pretty girl who will expertly cook, scrub floors, change diapers, keep the books, and still be the greatest thing since the San Francisco Earthquake in bed. It's a dream to which we cling eternally, in spite of the fact that nobody can recall it ever having come true. But, as if to erase our doubts, along comes an inexpensive little machine from Bavaria that really can perform the automotive equivalent of all those diverse domestic and erotic responsibilities, and hope springs anew.

I'll be interested to see who those 10,000 owners of the 1968 BMW 2002 actually turn out to be. The twits won't buy it, because it's too sensible, too comfortable, too easy to live with. The kids won't buy it because it doesn't look like something on its way to a soft moon-landing and it doesn't have three-billion horsepower. BMW buyers will—I suspect—have to be pretty well-adjusted enthusiasts who want a good car, people with the sense of humor to enjoy its giant-killing performance and the taste to appreciate its mechanical excellence.

They will not be the kind who buy invisible middle-of-the-line 4-door sedans because that's what their friends and neighbors buy, nor will they be those pitiful men/boys who buy cars and use them as falsies for fleshing out baggy jockstraps. Good horses don't like bad riders, and it's doubtful if the 2002 will attract too many of the timid or confused fantasy-buyers. It's too real.

That last phrase is kind of a key to the whole BMW bag. It is too real. For a couple of years now, "unreal" has been a big word with the semi-literate savages of hot rodding. It's supposed to be a high compliment, but it turns out to be an unwittingly incisive comment on the whole metalflake-angel hair-Batmobile scene. LSD is a drag, not a drug, for that group. Gurus like George Barris and Ed Roth were blowing their minds on fiberglass and tuck-and-roll upholstery while the Indians still thought peyote nuts were something you put on chocolate sundaes.

Let me tell you there's nothing unreal about the 2002. Give it a coat of pearlescent orange paint and surround the pedals with lavender angel hair and it would just naturally die of shame. Like a good sheep dog, it is ill-suited for show competition, only becoming beautiful when it's doing its job. It is a devoted servant of man, delighted with its lot in life, asking only that it be treated with the respect it deserves. You can't knock that...

The Germans have a word for it. The German paper *Auto Bild* called the 2002 *Flüstern Bombe* which means "Whispering Bomb," and you should bear in mind that the German press speaks of bombs, whispering and otherwise, with unique authority. They, too, saw something American in the car's design concept, but only insofar as BMW had elected to stuff a larger, smoother engine into their smallest vehicle.

But that's really pure BMW, when you think about it. The current 2000 series started life in 1962 as a 1500, then it became an 1800 and finally a full two liters—going from 94 to 114 horsepower in the process. The current 1600 was introduced about a year-and-a-half ago, and BMW-ophiles everywhere began to think of that glorious day in the future when the factory would decide to put in the 2-liter engine. Well, sports fans, the glorious day has arrived, and the resulting automobile is everything the faithful could have been hoping for.

The engine cranks out 114 hp at 5800 rpm, and the way it's geared it just seems to wind forever—it'll actually turn 60 mph in second, and an easy 80 mph in third. Top

CONTINUED ON PAGE 38

Supplement to MOTOR ROAD TEST No. 20/67 ● BMW 2002

GERMAN CARS

Long distance sprinter
Motorway mile eater, sports car and town potterer all rolled into one well engineered package

SO BMW have now joined the shoehorn bandwagon, the fashionable habit of putting your larger engine into a smaller chassis; and what admirable transport this makes the 2002, the cross between the 1600/2 coupé body and the 2000 engine. You pay £1,600 for what is essentially a two-door four-seater, but the one design embodies so many different features which are better than single features of more specialised examples; the performance for most of the range, and economy are better than the very fast Cortina Lotus, ride and general comfort are in the Triumph 2000 class, and the roadholding and general behaviour on all dry surfaces are matched by very few other saloons or mass production sports cars.

But despite this build-up which suggests quite correctly that the 2002 is very much a driver's car, it is also very much a touring car, quiet and docile to drive, its controls smooth and light, and it is just as happy in the suburban shopping centre as on one's favourite highway. Drawbacks lie in details only; not everyone liked the driving position with the high steering wheel and the pedals are not only too far from the floor but with the brake pedal in good adjustment it is difficult to heel and toe. On one car we cured this fault with a block of wood on the accelerator, but we understand that later cars will have pedals arranged so that you shouldn't need to lift heels off the floor. The only other complaint is the door seals; they seem to be so strong that it requires a lot of effort to shut the doors—too much, in fact, for more than one inside doorhandle—a change is on the way again. But at least they keep the noise out.

When the 2-litre engine was first rumoured for the smaller body, many expected that the 2000TI unit would be used as a basis for competition, but in fact the basic 2-litre engine gives quite enough performance for normal use and the essential competition bits are all homologated extras. As with all BMW units starting is instantaneous with full choke and is accompanied by the characteristic whine of the chain driven overhead camshaft; it takes about two miles of gentle driving for running temperature to come up and the engine to pull with its customary strength; top gear performance is particularly good and doesn't begin to flag until nearly the 100 mark. The power/weight ratio is around 100 b.h.p. per ton with one occupant and the 2002 makes particularly good use of it; helped by a flat torque curve and very suitable gear ratios, getaway is very impressive. With a squeak from the tyres and a minor graunchy protest from the rear suspension it leaps away to reach 50 m.p.h. in under 7 seconds and by holding 2nd gear to an unstrained 6,700 r.p.m. 60 m.p.h. in 9.2 seconds against the 8.3 and 11.8 seconds for the 1600. By 100 m.p.h. it is beginning to drop behind the more powerful Cortina Lotus and the best lap speed on our banked circuit was 106 m.p.h. A mean

Price: £1,249 plus £348 8s. 2d. PT equals £1,597 8s. 2d.
INSURANCE: AOA Group rating, 6; Lloyds, 7

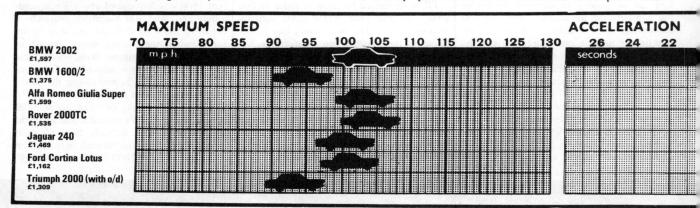

Performance

Performance tests carried out by *Motor's* staff at the Motor Industry Research Association proving ground, Lindley.
Test Data: World copyright reserved; no unauthorised reproduction in whole or in part.

Conditions
Weather: Dry, light winds up to 5 m.p.h.
Temperature 44°-48°F. Barometer 29.70 in. Hg.
Surface: Dry concrete and tarmacadam.
Fuel: Premium 98 octane (RM) 4 star rating.

Maximum speeds

	m.p.h.
Mean maximum speed	107.4
Mean lap banked circuit	106.0
Best one-way ¼-mile	111.1
3rd gear	93.0
2nd gear } at 6,800 r.p.m.	61.0
1st gear	33.0

"Maximile" speed: (Timed quarter mile after 1 mile accelerating from rest)
Mean 105.4
Best 107.1

Acceleration times

m.p.h.	sec.
0-30	3.1
0-40	4.8
0-50	6.8
0-60	9.2
0-70	13.3
0-80	17.6
0-90	24.7
0-100	37.4
Standing quarter mile	17.3

m.p.h.	Top sec.	3rd sec.
10-30	—	5.7
20-40	7.6	5.3
30-50	7.4	5.3
40-60	7.7	5.6
50-70	8.8	6.5
60-80	9.8	7.9
70-90	12.5	11.3
80-100	20.3	—

Fuel consumption

Touring (consumption midway between 30 m.p.h. and maximum less 5% allowance for acceleration) .... 27.5 m.p.g.
Overall .... 24.0 m.p.g.
(= 11.8 litres/100 km.)
Total test figure .... 1,680 miles

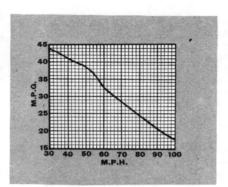

Specification

Engine
Cylinders	4
Bore and stroke	80 mm. x 89 mm.
Cubic capacity	1,990 c.c.
Valves	single o.h.c. and rockers
Compression ratio	8.5 : 1
Carburetter	Solex 40 PDSI twin choke
Fuel pump	Solex mechanical
Oil filter	Fram full flow
Max. power (net)	100 b.h.p. at 5,500 r.p.m.
Max. torque (net)	116 lb.ft. at 3,000 r.p.m.

Transmission
Clutch	Fichtel and Sachs
Top gear (s/m)	1.000
3rd gear (s/m)	1.345
2nd gear (s/m)	2.053
1st gear (s/m)	3.835
Reverse	4.180
Final drive	Hypoid bevel 3.64 : 1
M.p.h. at 1,000 r.p.m. in:—	
Top gear	18.5
3rd gear	13.7
2nd gear	9.0
1st gear	4.8

Chassis
Construction	Unitary

Brakes
Type	Disc/drum servo assisted
Dimensions	9½ in. disc, 8.0 in. drums

Suspension and steering
Front	Independent by Macpherson struts, lower wishbones, coil springs and anti-roll bar
Rear	Independent by semi-trailing arms, coil springs and anti-roll bar
Tyres	165—13 radials (German Dunlop SP's on test car)
Rim size	4½J—13

Coachwork and equipment
Battery	12v neg. earth, 44 a/h cap. Otherwise as 1600 specification.

---

speed observed on MIRA's three straights averaged 107.4 m.p.h. so the true maximum on straight roads should approach 110 m.p.h. which, like our standing start figures, is rather better than BMW claim.

We brought one 2002 back from Munich as part of its running-in; despite cruising easily at a (permitted) 80-90 m.p.h. we still returned 30 m.p.g. over the 750 odd miles to London via Luxembourg. Back in this country with a different fully run-in car we used much more throttle and the consumption dropped to 23 m.p.g. during the period which included testing, rising to 26 m.p.g. for more normal but still spirited use. On the car tested the touring consumption was only 27.5 m.p.g. but the car we first used would certainly have returned over 30.

With the extra power, BMW have added a higher final drive ratio than with the 1600 which gives pleasantly long legged touring and really useful maxima in the gears; the excellent gearbox encourages constant use as the lever slides around with silky precision through well chosen ratios and with a smooth clutch it was difficult to get a jerky change however fast. Two of our staff developed a pet hate against the rough surfaced gear lever knob but that is a very personal and minor detail.

Roadholding on the 1600 is of a pretty high order but on the 2002 it is even better; the addition of anti-roll bars at both ends and minor spring rate adjustments have stopped the inside rear wheel lifting rather prematurely, and the car is generally better balanced, moving smoothly through neutral steer characteristics at speeds many saloons wouldn't even look at, to a gentle final oversteer which is easily countered on well geared and not too heavy steering. This gives good feel in the wet, but the German SP tyres fitted to our test car were not good in these conditions; cornering grip was low and it was rather too easy to get wheelspin under power which of course tended to provoke the rear end prematurely, if still safely.

Good damping control keeps the car on a pretty even keel even on the limit with no lurching or sudden movements as you change from one lock to another in an S-bend. This also keeps the ride on the firm side, but still pitch free and with a degree of comfort which is enhanced by the seats. You can sense the car moving on irregular surfaces but occupants don't seem to mind and four adults can be quite comfortable for long distances—the large fuel tank range allows around 300 miles between stops which on a German autobahn might be covered in 3½ hours non-stop.

That is another side of the 2002's character—the long distance runner. Few cars manage to have quite such a character variation but play each part so well; we thought it an outstanding car.

M

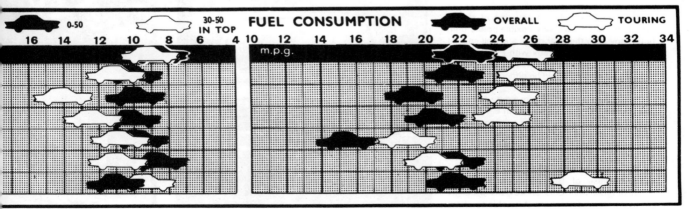

In a class of its own – the BMW 2002

If your ideal high performance saloon car has a matt black bonnet, go-faster stripes and mock alloy wheels you will be disappointed by the BMW 2002. The only concessions it makes to 'performance styling' are discreet black grille bars and a very luxurious highly polished wood rim steering wheel. Otherwise it is outwardly identical to the BMW 1600, and that hardly looks like a racing car. And yet the 2002 can see off virtually all the stripes-and-noise brigade and all but the very best 2 litre sports cars. At a price, of course. Actually £1,597 tax paid.

BMWs have a reputation for solid, fast, *very German* quality saloons. The four cylinder single overhead camshaft engine which powers their complete range started life five years ago as a 1,500 cc unit in the now familiar four-door bodyshell. It has since grown to 1,800 and 2,000 cc, and last year a 1,600 cc version was introduced in a smaller, lighter two-door body. The performance market has been catered for by the 1800 and 2000 TI models which have been available for some time; a 1600 TI is also marketed in Germany. But with basically the same unit in every model it was not surprising that BMW should follow the current trend of popping the largest engine into the smallest body. Like the 1600, the resulting 2002 is called a Coupe in Britain, although it clearly isn't one in the accepted sense of the word. In Germany there's a proper 1600 GT Coupe (derived from the Glas which BMW bought up a while back) so it's all very confusing.

The 1,990 cc engine has a single down-draught twin-choke Solex when fitted to the 2002 and gives 113 gross bhp, about 100 net (for comparison, the 1600 gives 85 bhp). According to the handbook the maximum permitted revs are 6,200 but the test car was not fitted with a rev counter (production models are) and we found that speeds well in excess of the little red marks on the speedo could be attained without any sign of mechanical unrest. This is a very flexible power unit, but in any case the gear ratios are well spaced and the four speed gearbox with Porsche-type synchromesh all round proved delightfully smooth, although it could be difficult to engage first gear from rest.

The acceleration figures are pretty fantastic, an indicated 50 mph coming up in 6.4 seconds from a standing start, making this one of the fastest saloons of any size or price. Even the 0-100 time makes a lot of sports cars look pretty silly. For once the manufacturer's figures are quite a lot down on ours. This is partly because of German laws, which are much tougher than ours on advertised claims (so every car in a batch must be able to attain

Performance

Acceleration:

0-30 mph	2.9 sec
0-40 mph	4.5 sec
0-50 mph	6.4 sec
0-60 mph	8.5 sec
0-70 mph	11.5 sec
0-80 mph	15.1 sec
0-90 mph	20.0 sec
0-100 mph	26.9 sec

Speeds in the gears:

1st	33 mph
2nd	71 mph
3rd	94 mph
4th	110 mph

Overall fuel consumption on test: 26.1 mpg.
Anticipated average consumption in normal use: 27.5 mpg.

Specification

Body-chassis: All-steel unitary construction.
Engine: Four-cylinder, in-line, water-cooled. Bore 89 mm, stroke 80 mm, displacement 1,990 cc. Compression ratio 8.5 to 1. Maximum power 113 bhp (SAE) at 5,800 rpm, 100 bhp (DIN) at 5,500 rpm. Maximum torque 115.7 lb-ft at 3,000 rpm. Single overhead camshaft. Single Solex 40 PDSI carburettor fed by mechanical pump.
Transmission: Mechanically operated single-dry-plate, 7.9 inch clutch. Four speed, all-synchro gearbox with floor shift. Ratios: 1st 3.83 to 1; 2nd 2.05 to 1; 3rd 1.35 to 1; 4th 1.00 to 1; reverse 4.18 to 1. Hypoid-bevel final drive, 3.64 to 1. (Limited slip differential available.)
Suspension: Front, independent with McPherson struts, lower wishbones, coil springs, telescopic dampers and anti-roll bar. Rear, independent with semi-trailing arms, coil springs with rubber auxiliary springs, telescopic dampers and anti-roll bar.
Steering: Worm and roller, 3.5 turns lock to lock. Turning circle 34 ft 2 in.
Brakes: Front, 9.45 inch discs. Rear, 7.87 inch drums, power-assisted. Total frictional area 243 sq in.
Wheels and tyres: 13 x 4½J pressed steel disc wheels with German Dunlop SP 165 x 13 tyres.
Weights and measures: Approx kerb weight 18 cwt; length 166.5 in; width 62.5 in; height 55.5 in; wheelbase 98.4 in; ground clearance 6.3 in; fuel tank capacity 10.1 gallons.

any published figures) and partly because BMW are anxious not to upset owners of the slower 1600 TI which isn't sold here.

The suspension, like the 1600, is by struts and wishbones at the front and semi-trailing arms with coil springs at the back, both springs and shock absorbers having been reset for the additional urge. Anti-roll bars are fitted front and rear. The car is quite lightly sprung and well damped, and understeers mildly, allowing fast, safe cornering which is not drastically upset by changes in road surface. The transition from understeer to oversteer is progressive and occurs at relatively high speeds. When pushed to the limit the inside back wheel lifts, and violent take-offs, such as during our acceleration tests, generated a good deal of wheel hop. This also showed up in the wet, but even so its cornering speeds under these conditions on the standard German Dunlop SP radials were still impressive. The servo-assisted brakes proved powerful and the steering precise, with just the right amount of 'feel'. All in all it is a very well balanced and 'forgiving' motor car.

The 2002 has the sort of styling that looks neat and right without being particularly fashionable or particularly dated. The test car was silver, a colour which always seems to suit BMWs well, and had a completely black interior. Visibility is excellent with deep screen and side windows and very thin pillars.

The purposeful theme is carried through to the driving compartment. The front seats are generously large, shaped and fully adjustable for rake. The dished wood rim steering wheel, with its oversize padded horn boss, is placed quite high but most people had no trouble finding a comfortable driving position.

The facia is somewhat Rover 2000-like with a full length open shelf on which the instrument binnacle rests. The instruments consist of three matching dials—a central speedometer, a combined fuel and water temperature gauge which incorporates all the usual warning lights, and on our test car, an enormous clock where the rev counter is to be ('12.45 in 3rd, old boy'). Double push-pull switches for lights, wipers and fan blower are arranged round the instruments; these proved difficult to reach when strapped in. One column stalk operates the dipswitch and flasher, the other, indicators and washers (which automatically operate the wipers for a few strokes).

BMW have paid a lot of attention to safety. Everything seems to be made of rubber—the switches, window winders, even the gear lever knob is moulded rubber. The ventilation and heating system is quite sophisticated, with demister ducts not only to the screen but also to the side windows, and extractor slots along the back window. There are no face-level vents though, and this is an annoying omission since the considerable wind noise when the quarter lights are open spoils what is otherwise a very quiet car.

It is difficult to know how to classify this car. Although it is a full four seater, with the front seats well back there isn't a lot of rear leg room. The boot is spacious, and with an over 100 mph cruising speed (where it's allowed) it makes a tremendous touring car for two people with a lot of luggage. It is compact, economical and quiet. Whether you think of it as an over-expensive hot saloon or a poor man's Porsche depends on the value you put on those matt black bonnets and go-faster stripes. The 2002 may be difficult to classify but it forms a very good class of its own.

BMW 2002

Model Number Or Target Date For Detroit?

by Bill Sanders

With sublime adroitness and a straight face, BMW deftly puts down Detroit by masquerading their newest bomb as a "Family Sedan." It's got "Family Sedan" comfort, luxury and convenience — but the similarity ends in the first corner.

Attention Messrs. Virgil Boyd, Roy Chapin, Ed Cole and Bunkie Knudsen: if the inner sanctums in that citadel of Great American Industry, Detroit, Mich., are astir with tension over rising import sales, that tension may soon turn to outright panic. An audacious new protagonist has just been unleashed on the automotive scene by the Bayerische Motoren Werke AG, of Munich, Germany.

Latest offering from the Bavarian craftsmen is the BMW 2002, and it's such a rugged contender, in any class, rumor has it even Ring Magazine and The Sporting News dig it for the image rub off. Esthetically, it's rather ungainly, and, from the outside it will be dismissed by many of our countrymen as just another compact. Lucky for us, gentlemen. Because, everyone who gets behind the wheel, even for a few blocks, asks the same question: why can't they build an American car like this? In performance, handling, ride, comfort and — the list grows too long — every other automotive characteristic, the BMW 2002 deals from the top of the deck, and, in most instances, outshuffles cars costing much more bread.

Imported car sales in 1967 were up 17% over 1966. So far this year they are cornering approximately 10% of the total U.S. new car market, compared to 9% in 1967, and time for concern is upon us. GM and Ford are now both working on compact cars in an attempt to stem the tide, and that brings us full cycle to our original thesis: will they, can they, turn out a street machine engineered as well as the BMW 2002 and compete in the 3 grand range? What about it, gentlemen?

Powertrain & Performance

BMW arrived at the 2002 as a result of several coincidental occurrences. Their 1600 was, and is, immensely successful in the U.S. With the advent of the 1600-2 (2-door), and the 2-carburetor TI option, BMW was moving into the rarified atmosphere of high-performance, superb handling machinery, while remaining in a competitive price market. Americans were discovering this phenomenon and digging it. Then: Smog Control. The 2-jug mill couldn't pass inspection so the next best thing became a substitute, the 2-liter (1990cc) powerplant with one carburetor. Coupled to a 3.64:1 rear end, performance remains typically BMW; vigorous and potent. No complaints are ever heard from the 2000 engine, even when subjected to the roughest treatment. Lug it down in high gear or down shift at high rpms to 3rd or 2nd; you'll never get a whine or growl from engine or drivetrain. We found we could practically run in any gear or down shift at high rpms to 3rd wind it out in 2nd or 3rd to some far out reaches of engine endurance. Still with no outcry. Our test car wasn't equipped with a tach, an option we highly recommend with the 2002. An engine so vehemently responsive invites overrevving, and when no protest is felt or heard, that tendency easily becomes habitual. Plenty of power always seems to be waiting in reserve, in any gear, except when lugging down in high.

A test of the reclining bucket seats and who wants to read further? Trunk is large with spare hidden under floor panel. Under hood, new air induction system is visible. Large diameter steering wheel is positioned at awkward angle for efficient handling in tight spots.

PERFORMANCE
Acceleration (2 aboard)
- 0-30 mph .... 3.2 secs.
- 0-45 mph .... 6.8 secs.
- 0-60 mph .... 11.0 secs.
- 0-75 mph .... 18.1 secs.

Passing Speeds
- 3rd gear 40-60 mph .... 7.2 secs. 527.04 ft.
- 4th gear 40-60 mph .... 9.2 secs. 673.44 ft.
- 3rd gear 50-70 mph .... 7.1 secs. 624.80 ft.
- 4th gear 50-70 mph .... 9.0 secs. 792.0 ft.

Standing Start ¼-mile:
75 mph .... 18.8 secs.

Speeds in Gears:
- 1st .... 27 mph @ 6000 rpm
- 2nd .... 50 mph @ 6000 rpm
- 3rd .... 77 mph @ 6000 rpm
- 4th .... 101 mph @ 6000 rpm

MPH per 1000 RPM: 16.9 mph

Stopping Distances:
- from 30 mph .... 34 ft.
- from 60 mph .... 121 ft.

Mileage Range .... 17.7-26.8 mpg
Average Mileage .... 23.5 mpg

SPECIFICATIONS
Engine: 4 cyl in-line sohc. **Bore & Stroke:** 3.50 x 3.15 ins. **Displacement:** 121.5 cu. in. **Horsepower:** 113 @ 5800 rpm. **Torque:** 116 lbs. ft. @ 3000 rpm. **Compression Ratio:** 8.5:1. **Carburetion:** 1 Solex 40 PDSI. **Transmission:** 4-speed synchromesh. **Final Drive Ratio:** 3.64:1. **Steering:** worm and roller. **Steering Ratio:** 17.6:1. **Turning Diameter:** 34.1 ft. curb-to-curb, 3.5 turns, lock-to-lock. **Tires:** Michelin XAS 165SR-13. **Brakes:** 9.4-in. disc front, 7.9-in. drum rear. **Suspension:** Front: MacPherson struts, lower A-arms, coil springs, tube shocks. Rear: Semi trailing arms, coil springs, tube shocks. **Body/Frame:** unit steel construction. **Capacities:** Overall Length: 166.5 ins. Overall Width: 62.6 ins. Overall Height: 54.0 ins. Wheelbase: 98.4 ins. Front Track: 52.4 ins. Rear Track: 52.4 ins. Curb Weight: 2210 lbs. Fuel Capacity: 12.1 gals.

Whether cornering on a road course (far left) or making panic brake stops (above), the sports car that masquerades as a family sedan unabashedly displays its noble heredity. Credit for such superb handling must be given to costly independent rear suspension that features diagonal trailing arms. Wheels are sloped to negate constant track and camber changes, so common to cross-shaft, rigid axles. Coil springs and tube shocks (see above) aid in giving comfortable ride.

BMW 2002 continued

BMW meets smog emission controls by using an air injection system on the 2002. It is set up in an ordinary way with adjustments to carburetor and distributor. Engine configuration is a basic sohc BMW design, with 5-main bearings, but with 8 balance weights compared to 4 on the 1600. And, the oil pump is driven off the crankshaft by a chain drive, as on other BMWs, rather than conventionally off the camshaft.

Handling, Steering & Stopping

You keep telling yourself: The 2002 is basically a family sedan, not a sports car. If you can remember that fact when it comes to handling, the credibility gap is a little easier to understand. Cornering is fantastic. On a winding, tight, rugged mountain road the car has no peer in its class. On an actual road racing course it performs with some race machines — up to a point. You still have to remember it's a sedan! One of the two bad features (that's all we could find) on the 2002 is the steering wheel. It's much too large for good handling, and is positioned at a bad angle, a situation one never fully gets used to. In actuality, the 2002 has a slight understeer when engaged in the usual, spirited driving it was designed for, but the large wheel often creates the sensation of oversteering. On a mountain road with many tight turns, we noticed a slight, temporary oversteer when entering an extremely sharp turn in 4th gear at high speed — 60 to 70 mph — but this can be corrected by downshifting and judicious use of the throttle. On the road course, excessively hard cornering brought the inside rear wheel slightly off the road, but that situation never develops even under the most rugged highway or street driving. Some wheel hop is also evident during extremely fast acceleration. Our test car was equipped with optional Michelin XAS radial tires and we recommend them for the additional price as they make a tremendous difference in ride and handling. Independent suspension at all four wheels adds immeasurably to the handling characteristics of this car, and we ask repeatedly, in vain it seems, why can't our own companies, with their vast resources, develop an independent suspension like this for the family sedan?

Shift lever on the 2002 is in an ideal spot for quick, easy-to-manipulate shifts. BMW pedals still go through the floor rather than being suspended, as is the case on most new cars. This seems to necessitate less pedal pressure and makes operation easier. Pedals are an adequate size although the accelerator could be a little larger and at a less steep angle. Brake and accelerator are positioned for easy heel and toeing. Braking is also an excellent characteristic of the car. Some swerve was noticed when stopping from 60 to 0 mph, with only a slight tendency to lock up. Front discs pull down quite evenly and straight. Brake swept area to weight ratio is quite good.

Comfort, Convenience & Ride

Even with 4-wheel independent suspension and sports car handling, the 2002 lays a comfortable, though pleasingly firm, ride on you. Big, bucket seats are well designed and, with an extensive front/rear adjustment, give an excellent seating location for the shortest chick or tallest guy. Front/rear adjustment, coupled with multiple reclining position backs, seem to give the two front 2002 seats more positions than a 6-way power seat system. Visibility is expansive all around, with low window sills and high windows. Dimmer switch and high beams are incorporated into a lever located on the left side of the steering column. Turn indicator is on the right. The only other bad feature of the 2002 has the turn indicator and windshield washer/wiper combined on the same lever. Often, while attempting to flick the turn indicator without removing the right hand from the wheel, you manage to have the windshield washed and wiped also — unintentionally.

An extremely quiet ride is an attribute not usually found on a small car. Disturbances from wind and road are at an absolute minimum with windows up. Vent window handles are actually knobs and easy to operate for precise window adjustment. Ventilator fan is effective, although rear windows need to be opened slightly on inordinately hot days. Opting for the well designed, useful sunroof will compensate for any lack of cooling and we recommend that extra also. With sunroof open at high speeds, wind disturbance and buffeting was insignificant.

Ignition switch is recessed and located at a spot on the steering column that makes it difficult to reach. If you happen to be a heavy smoker, you may also find the ashtray too small. This BMW body is endowed with good trunk space and tire and tool kit location under floor panels makes packing easy and keeps luggage clean. High liftover makes loading somewhat difficult, especially for women. Quality control is ultra fastidious for a car in this price range, giving one more gold star to a faultless machine.

Americans who accidentally stumble into a BMW dealer and test drive the 2002 are in for a surprise. Keep your fingers crossed, fellas—maybe the word won't get around right away, say until about the year 2002. /MT

photos by George Foon, Gerry Stiles

OPTIONS & PRICES
Price of car as tested: $3477 (P.O.E. West Coast base price $2908). Standard Accessories and Prices (included automatically): Leatherette Upholstery: $45. Power Brakes: $45. Safety Belts: $25. Bumper Guards: $8. Chrome Tip Exhaust Pipe: $2. Michelin XAS Radial Tires: $59. Reclining Seats: $48. Torsion Bar Stabilizers: $20. **Optional Accessories and Prices:** Manually Operated Sunroof: $135. Polaris Paint (Silver Gray): $42. Blaupunkt AM/FM Radio: $140. Tachometer: $40. Velour Floor Mats: $33. Trunk Mats: $15. Right Hand Outside Rear View Mirror: $8.50. Locking Gas Cap: $4.

BMW 2002—15,000 miles

A smooth, superbly engineered road-burner

By Graham Robson

● LONG TERM ASSESSMENT

In the author's opinion, this is probably the finest dual-purpose car in Europe. Smaller 1600 body combined with 2000 engine gives four-seat family comfort and splendid performance. Famous BMW ohc engine, fine engineering and sports-car-like road holding: 14,000 miles in six months with complete reliability, the car still as good as new. No flashy styling, but what much-travelled businessman could ask for more? Even faster 120 bhp version now available—the 2002TI.

JUST a year ago, if anyone had said that much of my 1968 motoring would be in a fast roadworthy sports saloon costing over £1,600, I would have smiled cynically and replied that such good things never happened to me. If he had gone farther and prophesied that there would be a choice of two such cars I would have doubted his sanity! Yet this all came to pass in April, when Geoffrey Howard and I had to decide between an Alfa Giulia Super and a BMW 2002; this was soon settled, as Geoff had already "owned" two BMW's, so he took the Alfa instead.

In my two previously published assessments, on the Renault 16 and the Vauxhall Victor 2000, I longed for more performance in each case, both for safety and enjoyment. The BMW provided more than I could have hoped, and completely captivated me from the start. It's not going too far to say that this is the most satisfying car I have ever possessed, and is likely to be unbeaten in the near future. My own business needs call for a long-legged car which will take me many hundreds of miles up and down motorways, across country, or even scooting about Europe to sporting assignments, and the 2002 filled the bill admirably. If I had to pay all the expenses it would also be nice to have the creditable fuel economy (about 25 mpg), complete reliability, and to note the heavy demand for second-hand versions when the time came to sell. Mine has covered over 15,000 miles in only eight months, and is still on top of its form; we hope to keep it for some time yet and experience BMW's legendary long-life qualities.

The new thinking in BMW saloons began in 1961, with the Michelotti-styled 1500. This basic body shell is still in production for the current 1800 and 2000. The engine was then a brand-new ohc four-cyl design, canted over at 30deg in the car, bore and stroke being 82mm and 71mm. Later the engine grew to 1600, then 1800 and finally to a 2000; actually 1,990 c.c. with bore and stroke of 89mm and 80mm. Only two years ago BMW announced a smaller, two-door body shell, fitted with the 1600 engine and known locally as the 1600-2. The obvious then happened, but not until this spring, when BMW married the 1600 body shell to the 2000 engine, added front and rear suspension anti-roll bars, and dubbed the new road-burner a 2002. For some months this was definitely *the* BMW to buy, but BMW have now achieved the ultimate by slotting in the higher-tuned 2000TI engine to evolve the 2002TI. Timo Makinen is due to drive a works 2002TI on the Monte next month; if the weather is not too snowy the results could be worth watching!

Road-burner extraordinary; the BMW 2002, distinguished by its black radiator grille

The 2002, then, is a true sports saloon, with 100 bhp from its lightly-tuned 2-litre engine, and a weight of only 2,060 lb—which compares closely with 110 bhp and 2,027lb for the Cortina-Lotus; unladen power-weight ratios are 92 bhp/ton (BMW) and 99 bhp/ton (Cortina-Lotus). Like the Ford it is a two-door, four seater saloon (wrongly named a coupé by its importers) though not quite as roomy in spite of being a bit longer. There is a really nice bonus in the form of fully-independent rear suspension, but a big penalty in the high price tag of £1,632, because of the high import duties from the Common Market. In Germany, of course, it is a very different story, as the 2002 costs under £1,000.

Nice things about the BMW's engineering include the well-laid-out independent suspension and fine handling, the sweetly smooth ohc engine, and the many inclusive "extras" such as reclining front seats, wooden steering wheel and combined wiper-washer controls.

My car was delivered by road all the way from BMW in Munich, so that the running-in could be complete on arrival. In fact running in is a bit of a joke for this car; we were allowed 80 mph (4,300 rpm) for the first 250 miles, then up to 100 mph (not much below maximum speed!) for the rest of the way. For normal running-in, 75 mph in top gear is allowed from the start. Steady speed *autobahn* cruising kept stresses down, but may also have delayed the general loosening-up process a bit. As collected, its speedometer read 99,979 (usually they read 99,969 so that pre-delivery running will bring up the zero for customer-collection). Soon after the car got back to England, with 1,474 miles on the clock, I took delivery and was allowed to use full throttle from then on. The car kept its German export plates for a while, which was very good indeed when I needed parking immunity, but bad in traffic, where we British tend to lean on foreigners; it did not collect its British registration until 3,475 miles. Our experience in the next 13,000 miles has been almost undiluted pleasure and satisfaction; literally the only two things to go wrong have been a broken driver's door pull—this happened very early—and a broken throttle pedal, of which more later. The car has been off the road only for routine service—even with the throttle problem I could keep going—and has not missed a beat for the whole eight months.

We have made three modifications. The first was to re-position the brake and clutch pedals (by re-shaping and welding the shafts) so that they were nearer the floor; right-hand-drive 2002s have pendant pedals, their pads an excessive $7\frac{1}{2}$in. off the floor, but left-hand-drive cars have them sprouting from the floor—inexplicable. Fitting my Radiomobile was complicated, at the time, by there being no fitting kit for this particular bodyshell. My local specialist—Gordon March of Coventry—solved the problem neatly by using the complete pod from my Renault 16, which slotted into the central storage bin as if the two had been designed together. Even now I hear that there's no kit, but Gordon March tells me he is using parts of a BMC 1100/1300 kit which fit even more easily. Because both the bonnet and boot lids are full-length, getting an aerial on to the car is nearly impossible. My car has the gutter fitting which is really too small for perfect reception, but this seems to be the only solution.

Front seat belts are compulsory, of course; mine are Britax with BMW decals. Finally we fitted a wooden gear lever knob to replace the horrid rubber variety which is common to the 1600. My car was shod with German Dunlop SP tyres, rather coarse-treaded and known as doggie-bones in the trade; I always kept them well blown up to 30 psi for a reasonable balance between ride, comfort and stability.

I fell in love with the car almost at once. Even in traffic it was obvious that there was lots of performance, superb handling, and that inde-

The bonnet, heavily sound proofed with polyurethane foam, hinges well out of the way. The box on the end of the air cleaner trunking selects hot or cold air, depending on the season

A plywood floor covers the spare wheel well; the single-skin boot sides are very prone to damage from loose items sliding about inside. The "porthole" by the right-hand hinge is part of the ventilation exhaust system

BMW LONG TERM...

finable "something" that is an aura of fine engineering where price takes a back seat in development—this all added up to a most pleasurable whole. I'm a long-distance motorist, possibly covering more miles in a year than anyone else on the staff, and need this *rapprochement* with a car to make me content. The Renault 16 had character but lacked performance, while with the Victor it was the other way round.

At first, the car wasn't as loose-limbed as expected, probably because the *autobahn* delivery service had retarded running-in, but a sample set of figures taken against the road test 2002 (our issue of 16 May 1968) showed there wasn't a lot of difference. By 4,000 miles all was well, and the car has continued to feel good ever since. The fact that our performance check, carried out only last week, shows performance to be slightly down at 16,000 miles, merely proves that BMW make cars to production tolerances, which do deteriorate slightly with age and that our car wasn't "fiddled" in any way for long term press use. I had occasion, some weeks ago, to make a visual check against another 2002, when I was "jumped" by AY 1 just getting on to M1 at Hendon. For the next 60-odd miles I confirmed that his car seemed to have more snap from medium speeds, but that there was precious little in it. The driver of AY 1 seemed to enjoy himself, and so did I.

On one occasion, the car did lose quite a lot of its zest, and in the most unusual way. The throttle set-up on these cars has a true organ-pedal control, the usual type with a steel blade encased in rubber. Perhaps the combination of over-enthusiastic foot-to-the-floor and a sub-standard pedal helped, but after a time the pedal began to bend, just below the roller fixing to the rest of the linkage. I had to leave for France and the Alpine Rally with the pedal in this condition, one that worsened as we swept down French *autoroutes* and caused maximum cruising speeds to drop back. Before Marseille the metal blade had broken inside the rubber, leaving me with only an uncomfortable half-throttle to use, a maximum speed of 90 mph and a splendid overall fuel consumption for the trip of 28 mpg! A new pedal cured the problem on return, performance is back to normal and fuel consumption to a more believable 25 mpg.

Cynics who might believe I think the car is perfect may now learn of minor gripes. On a car of this price I would have expected the wiper arcs to be matched to r.h.d., but they're not and they leave a nasty blind spot above and to the right of my eye-line. I find minor switches difficult to reach when wearing belts, and they have sharply uncomfortable edges. It's a constant bore to have to lock both doors from the outside with a key as neither will slam-lock, and I don't like the unprotected flanks of the boot compartment, easily damaged by sharp objects from within. It would be nice to have a bigger fuel tank than 10 gallons, but I must have been spoiled by 12 in the Victor.

The car is now on its second set of Dunlop tyres, the "doggie bones" having been exchanged for a brand new set of Dunlop SP "aquaspurts" at 11,868 miles. The German Dunlops were not swopped round at all—the spare being unused—and calculations of life rendered meaningless by a very expensive tyre burning session in North Wales when I got the car stuck facing the wrong way down a very steep hill with unmade surface. When removed, the front tyres still had 5mm of their 8mm unused—this indicating a probable 28,000 miles to the legal limit; but only 2mm remained at the rear, quite a bit of the spent rubber being not too far from Nefyn in North Wales! Let's make an estimate of 20,000 miles if they'd all been swopped round and cared for. It's quite certain that the rears would wear out first anyway; there is certainly enough power to squeak the tyres on a full-blooded take off, and

Above: The radio fitting kit came straight from the Renault 16 and fits neatly in the central cubby. The wooden gear lever knob replaces the standard rubber one

Left: The paler parts of the brake and clutch pedals show where they have been heat treated to give better location. This type of pedal will be fitted to all 2002s next month

Below: As both boot and bonnet "decks" are hinged, a gutter-mounted aerial was fitted. BMW usually fit theirs either on the roof or in the centre of the rear boot lid

the inside rear wheel sometimes lifts and spins when I take advantage of the formidable cornering power. The "aquaspurts" grip even better than the German tyres, especially in the wet when the renowned jets in the tread remove phenomenal amounts of water from the road. They show few signs of wear just yet and will certainly last out until we have to sell the car next year.

One of the things that appeals strongly is the feeling of having a real Q-car. Only the *cognoscenti* smile knowingly when they see the BMW and 2002 badges, for even in its rich red colour with the fashionable black grille, the car looks like a very modest and sedate family saloon. One can only suppose that BMW's "Move Over" advertising campaign for the car is to educate everyone who might unwittingly get in the way....

As with Alfas and most expensive GT cars, I feel that BMW *must* employ a "sensations" engineer, one who can build in the right noises and responses after development engineers have done their job. Nothing else could really explain that subdued but characteristic whine from the ohc camshaft drive, the silky gearchange that feels as if it isn't connected to vulgar gear wheels at all, and the responsive and delicate handling that flatters a driver so much.

Speaking now as a road tester, I am suspicious of an engine tune which doesn't need any choke even for cold starts (though mpg figures are well up to scratch) and I think the steering wheel is much too big. It is, of course, a very fast car on our crowded roads, whether for thrusting across country, or for cruising along motorways. There's absolutely no temperament from the docile 2-litre engine, which always gives the impression of being throttled down rather than tuned up, and I get very idle over gear changing at times. Speed-limited cruising at 70 mph is ludicrously slow for the 2002—I know that on the Continent one is regularly passed by other 2002s wafting along at speeds up to 100 mph. It's the most high-geared of all BMW's (with an axle ratio of 3.64) so 70 mph is only 3,800 rpm. Helped by the fat Dunlops, and the stiffened suspension, the handling is truly superb, especially as the steering is so light and precise, and with ultimate grip several degrees better than my personal bravery index I'm hardly likely to run it out of road. Understeer or over-steer means little with this sort of car; it's finely enough balanced usually, and I suppose power-on oversteer with a spinning rear wheel would represent the limit.

It is nice to have the ultra-solid BMW bonnet locking handle by the passenger's legs. This must be closed positively to lock the bonnet, which hinges at the front to reveal all when service is needed. Also nice is the combined, typically German, wiper-washer stalk on the steering column which is sadly missing from domestic products. One would like better heating and ventilation, especially in the methods of getting used air out of the car, and for my money there should be more instruments, particularly an oil pressure gauge and a clock. I suppose one can't have everything, for a clock on German 2002s is replaced by the so-useful rev-counter for British-market cars. The front seats give ample lounging room, and enough location for me without excessive bucketing. The car is frankly too small to be considered a true four-seater, and with my driving seat in its usual way-back position I wouldn't like to inflict a long journey on any adult passenger behind me.

This assessment reads like a saga of contentment, and so it should. When I bade a sad goodbye to my Renault 16 at the end of last year, my closing words were, "almost any new staff car will be less comfortable, less cleverly sprung and less distinctive...". Let's just say that the BMW has equalled it and beaten it by really spirited road manners that I will never fail to appreciate. Most motoring writers would vote the 2002 as one of 1968's "in" cars; count me among them.

BMW 2002 LONG TERM...

PERFORMANCE CHECK

Maximum speeds

Gear	mph		kph		rpm	
	R/T	Staff	R/T	Staff	R/T	Staff
Top (mean)	107	103	172	166	5,800	5,600
(best)	110	103	177	166	5,960	5,600
3rd:	89	89	143	143	6,500	6,500
2nd:	59	59	95	95	6,500	6,500
1st:	31	31	50	50	6,500	6,500

Standing ¼-mile R/T: 17.4 sec 77 mph
Staff: 18.7 sec 73 mph
Standing kilometre, R/T: 32.8 sec 96 mph
Staff: 34.9 sec 89 mph

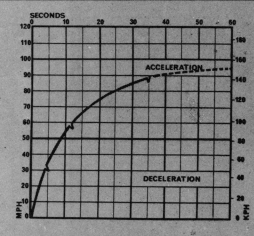

ACCELERATION

Time in seconds R/T:	3.5	5.3	7.6	10.6	14.2	18.2	26.4
Staff:	4.0	6.1	8.8	12.3	16.8	23.6	35.9
True speed mph	30	40	50	60	70	80	90

SPEED RANGE, GEAR RATIOS AND TIME IN SECONDS

mph	Top (3.64)		3rd (4.89)		2nd (7.46)		1st (13.95)	
	R/T	Staff	R/T	Staff	R/T	Staff	R/T	Staff
1—30	—	—	6.4	6.9	3.8	4.0	2.5	2.9
20—40	8.1	8.7	5.6	6.0	3.5	4.0	—	—
30—50	7.6	8.5	5.5	5.9	3.7	4.6	—	—
40—60	8.4	9.0	5.8	6.7	5.1	—	—	—
50—70	8.5	10.4	6.2	8.8	—	—	—	—
60—80	10.2	12.9	8.6	10.6	—	—	—	—
70—90	13.2	20.2	—	—	—	—	—	—

FUEL CONSUMPTION
Overall mpg R/T: 25.5 (11.1 litres/100km)
Staff: 25.0 (11.3 litres/100km)

NOTE: "R/T" denotes performance figures for the BMW 2002 tested in Autocar of 16 May 1968.

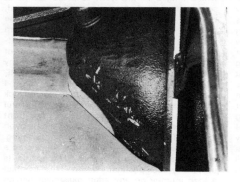

Damage to the unprotected wheel arches in the BMWs boot, caused by sharp-edged luggage crashing about. Some form of trim protection is needed

Brief Specification Resumé
FOUR-CYLINDER front-mounted ohc engine of 1,990 c.c. (89 x 80 mm). Power output 100 bhp (net) at 5,500 rpm.; max. torque 116 lb.ft. at 3,000 rpm. Four-speed, all-synchromesh gearbox; 18.5 mph per 1,000 rpm in top gear. ATE brakes, with 9.5in. front discs and 8 x 1.5in. rear drums. All independent suspension by coil springs, with anti-roll bars at front and rear. Fuel tank 10gal. Kerb weight 2,060 lb. Four-seater saloon body with two doors.

COST AND LIFE OF EXPENDABLE ITEMS

Item	Life in Miles	Cost per 10,000 Miles
		£ s d
One gallon of 4-star fuel, average cost today 6s 4d	25	126 13 4
One pint of top-up oil, average cost today 3s 0d	5,000	6 0
Front disc brake pads (set of 4)	10,000	4 14 8
Rear brake linings (set of 4)	21,000	18 2
Dunlop SP tyres (front and rear)	20,000	17 10 0
Service (main interval and actual costs incurred to date)	4,000	20 14 8
Total		**170 16 10**
Approx. standing charges per year		
Depreciation		200 0 0
Insurance		38 0 0
Tax		25 0 0
Total		**433 16 10**

Approx. cost per mile = 10½d
The BMW 2002 has been available in Britain only since the spring, so depreciation is an estimated figure based on current market prices and on a basis of 10,000 miles per year. The figure quoted for insurance is the sum due after deduction of a 60 per cent no-claims bonus; the writer would also have to agree, for this type of car, to a £25 excess and a "named driver" clause

BMW 2002:

..... a controllable urge

BMW's re-entry in 1965 into automobile manufacture must go on record as one of the most successful comebacks at all time.

The impact of the re-appearance of this legendary name was felt almost immediately by many European manufacturers, and now, ever-increasing market acceptance has placed BMW as one of the sales leaders in Germany.

In the last financial year, BMW carved a substantial slice out of the Mercedes Benz share of the home market.

This fact alone gives an insight into what type of car BMW is producing and what segment of the market it is aiming for.

BMW started with a virtually clean sheet when first planning their new range of cars, and every aspect of every vehicle they produce shows evidence of this.

The aim was to produce a car which offered fast, comfortable transport with a strong accent on safety — identical to the Mercedes Benz philosophy.

And not suprisingly, the vehicles which evolved are very similar in many respects to Mercedes Benz — although they fall into a slightly smaller size category.

Limited availability and high prices have kept BMW sales down in Australia, but recent substantial price reductions and a growing band of enthusiastic owners who are constantly singing the marque's praises are doing much to rectify the situation.

Once you have driven a BMW, you begin to understand why the car is making such a mark in European countries, where purchase prices are low enough to bring it within the reach of the average individual.

Any BMW is mechanical evidence of the fact that a car with excellent road holding power need not have firm, harsh suspension, and that a car which is capable of riding smoothly and sure-footedly over roughly corrugated roads need not have a suspension which causes it to lean frighteningly on fast corners.

In the short space of time since their introduction, these cars have shown that the word "compromise" does not exist at Bayerische Motoren Werke.

Initially, the BMW range comprised two basic models sharing the same body shell — the "1800" and the "2000".

Shortly afterwards, the range was expanded to include the "1600", with a smaller, entirely new body and a single overhead camshaft 1600 cc engine, developing 96 bhp SAE at 5800 rpm.

And the latest BMW to reach our shores is the "2002", which can virtually be described as a 1600 with the engine from the 2000 installed under the bonnet.

With this two litre, 113 bhp engine pushing along a body weighing just over 2000 lbs., it can be understood that the 2002 is something of a performer — although it's not quite the rubber burner that some would expect.

Gearbox ratios remain the same as those of the 1600, but final drive ratio has been altered to give economical and relaxed high speed cruising rather than neck-snapping acceleration.

Perhaps the main benefit of the two litre engine is the improved torque which allows the car to accelerate from low rpm strongly and without fuss.

Fuel economy, too, is a big feature of the 2002, and we recorded a figure of 25.5 mpg during testing. Under normal conditions, it would not be hard to exceed 28 mpg.

Apart from the differences already mentioned, there are few other mechanical changes from the 1600, which gives some idea of the reserves that are built into BMW cars.

Externally, the only way to pick the 2002 from the 1600 is the black-painted grille, anodized hubcaps, and the 2002 motif at the rear.

Inside, there are few obvious differences, apart from the new three spoke steering wheel with its huge padded hub and the addition of a tachometer to the instrument panel.

The safety theme is evident everywhere you look, even down to the heater control levers which on first glance look to be of hard unyielding design, but which do in fact break away on impact.

All-round visibility is excellent, yet the car doesn't make passengers feel conspicuous like some of the mobile glass houses on the road today.

Front seats are designed to make the driver and his companion as comfortable as possible, and fore and aft adjustment is so generous that any driver, regardless of the length of his legs, can achieve a very satisfactory driving position. Those in the back, however, are not as well catered for and if the front seats are pushed right back there just isn't *any* legroom.

As the 2002 is fitted with lap-style safety belts, controls can be reached fairly easily, but I felt that the windscreen wiper switch could be located at a more convenient distance — perhaps it could be incorporated in the column-mounted stalk which operates the washers?

Another small gripe concerns the location of the floor pedals, which are arranged so that operation of the brake pedal requires a definite lifting of the right foot rather than pivoting on the heel. Likewise, the clutch pedal is suspended high off the floor and the driver is forced to adopt a "Knees Up Mother Brown" attitude when gear changing.

Apart from these shortcomings, con-

2000 engine is a snug fit in 1600 engine bay

trols are well placed for relaxed driving.

Gear lever placement is beyond reproach, operation is smooth and positive, and I didn't beat the synchromesh once during 300 miles of testing.

At high cruising speeds, the 2002 is an extremely quiet car and engine noise intruded only under hard acceleration during testing. Suspension noise was hardly discernible over rough roads and even severe corrugations couldn't provoke the tail to hop off line — all credit to the well designed semi-trailing link rear suspension.

The suspension is one of the 2002's biggest features, and part of the secret is the extensive use of rubber to absorb shocks that would normally transmit themselves to the body.

As mentioned before, this suspension is unchanged from the 1600, yet it is capable of containing the extra performance of the 2000 motor with enormous reserves of safety.

Top recorded speed was 111 mph with the engine spinning at 6000 rpm. Even at this high velocity (which betters the claimed factory figure by 5 mph) the car felt well balanced and noise level was low.

This long legged character is one of the 2002's main charms, and there are few cars that can eat up the miles more effortlessly — and none that I can think of with more safety.

The theory at BMW is that to be made as safe as is humanly possible, a car should be designed to avoid accidents in the first place through good braking and handling. In a BMW, if an accident can't be avoided, the front and rear sections will collapse progressively to absorb impact and any protuberances inside the specially strengthened passenger compartment are designed so as not to inflict injury.

All this attention to safety design, and the development and testing that is involved, naturally costs money and the 2002 is priced beyond the reach of many motorists at $4360.

But if safety, comfort and performance count for anything, it's worth every penny of it.

Body design is clean and attractive from any angle. Note good glass area.

MOTOR MANUAL ROAD TEST
NEW SERIES
DRIVER COMMENTS

BMW 2002

CAR FROM: Grand Prix Motors Pty. Ltd. 208 Riversdale Road, Hawthorn, Vic.
PRICE AS TESTED: $4360 (plus radio, stereo tape)
OPTIONS FITTED: Radio, stereo tape recorder.

ENGINE:
- Type .... 4 cylinder SOHC
- Bore and stroke 89x80
- Capacity .... 1990 cc
- Compression ratio 8.5:1
- Power (gross) 113 bhp @ 5800 rpm
- Torque .. 115.7 lbs. ft. @ 3000 rpm

TRANSMISSION: 4-speed, all-synchro.

CHASSIS:
- Wheelbase .... 98½ inches
- Length .... 166½ inches
- Track F .... 52⅜ inches
- Track R .... 52⅜ inches
- Width .... 62¾ inches
- Clearance (minimum) 6¼ inches
- Test weight .... 2068 lbs.
- Fuel capacity .... 10.1 gallons

SUSPENSION:
Front: McPherson struts and rubber mounted lower wishbones, anti-roll bar.
Rear: Rubber-mounted semi trailing arms, coil springs, telescopic shock absorbers, anti-roll bar.

BRAKES: Power assisted.
Front: 9.45 inch discs.
Rear: 7.9 inch drums.

STEERING:
- Type: Worm and roller.
- Turns lock to lock: 3.5
- Turning circle: 31 ft. 6 in.

WHEELS/TYRES: 4½ inch steel rims with 165 SR 13 radial ply tyres.

PERFORMANCE:
Zero to
- 30 mph .... 3.2 seconds
- 40 mph .... 5.2 seconds
- 50 mph .... 7.0 seconds
- 60 mph .... 9.9 seconds
- 70 mph .... 12.8 seconds
- 80 mph .... 16.6 seconds
- 90 mph .... 22.0 seconds
- 100 mph .... NA

Standing quarter mile 17.5 seconds.
Fuel consumption on test 25.5 mpg on S fuel.
Fuel consumption (expected) 28 mpg.
Cruising range 283 miles.

SPEEDOMETER ERROR:
Indicated 30 40 50 60 70 80 90 100
Actual   28 37 46 55 64 74 84 NA

MAXIMUM SPEEDS IN GEARS:
- First .... 29 mph
- Second .... 54 mph
- Third .... 82 mph
- Fourth .... 111 mph

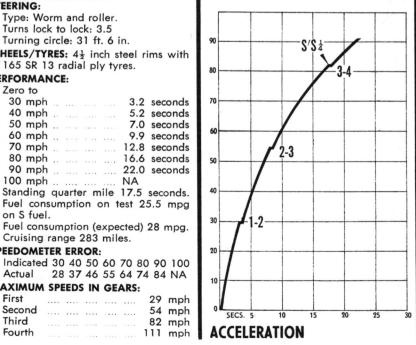

ACCELERATION

ENGINE:
- Starting .... Good
- Response .... Good
- Vibration .... Low
- Noise .... Low

DRIVE TRAIN:
- Shift linkage .... Very good
- Synchro action .... Very good
- Clutch action .... Good
- Noise .... Low

STEERING:
- Effort .... Moderate
- Response .... Excellent
- Road feel .... Excellent
- Kickback .... Low

SUSPENSION:
- Ride comfort .... Very good
- Roll resistance .... Very good
- Pitch control .... Very good

HANDLING:
- Directional control .... Very good
- Predictability .... Very good
- Resistance to sidewind .... Good

BRAKES:
- Pedal pressure .... Low
- Response .... Very good
- Fade resistance .... Very good
- Directional stability .... Very good

CONTROLS:
- Wheel position .... Very good
- Pedal position .... Good
- Gearshift position .... Very good
- Panel controls .... Good

INTERIOR:
- Ease of entry and exit .... Good
- Noise level .... Low
- Front seat comfort .... Very good
- Front head room .... Good
- Rear seat comfort .... Good
- Rear leg room .... Poor
- Rear head room .... Good
- Instrument comprehensiveness .... Good
- Instrument legibility .... Good

VISION:
- Forward .... Very good
- Front quarter .... Very good
- Side .... Very good
- Rear quarter .... Very good
- Rear .... Very good

CONSTRUCTION:
- Sheet metal .... Very good
- Paint .... Very good
- Chrome .... Very good
- Upholstery .... Very good
- Trim .... Very good

GENERAL:
- Headlights - highbeam .... Good
- Headlights - lowbeam .... Good
- Wiper coverage .... Fair
- Wipers at speed .... Very good
- Maintenance accessibility .... Good
- Luggage space .... Very good

Hair on Their Chests

BMW 2002 AND SUNBEAM RAPIER H 120

ON THE face of it there is nothing to connect these two cars whatsoever except for the price and that they are both, loosely, two-door saloons. And yet both have a good deal in common — at least, that's how it struck us after driving the two of 'em in fairly short succession. Both, of course, are hair-on-the-chest versions of existing models and both are good examples of the cars which manufacturers add a bit of muscle to so that they can point them at a slightly wider market. The BMW is an even more sporting version of an already sporting motor-car, and in 1968 did what you might call Rather Well in Group 5 racing; the Rapier is a slightly more elaborate version of the same thing, although in neither normal nor H120 form has it achieved any kind of competition record. Where BMW have used a smallish, lightish shell and dumped in a bigger engine, Rootes have taken the other tack and have got someone — in this case Holbay (whose racing engines have a considerable competition record) — to squeeze a power bonus out of the engine the car is usually fitted with. Rootes also go a little more wild on the finish — with the H120 you get a go-faster paint trim, with zonking great stripes down the sides, plus sexy-looking wheels, without even asking, while the more conservative BMW gives you a matt-black grille, the words 2002 on the blunt end and relies on pretty shattering performance to do the rest.

Looking at them in more detail, the BMW comprises the 1600 two-door body shell, giving the car four seats with rear-seat legroom strictly controlled by the height of the driver. Into this is slotted the oversquare (89 m.m. x 80 m.m.) four-cylinder engine of 1990 c.c. Hemispherical combustion chambers, inclined overhead-valves in an inverted "V" arrangement and operated by a single overhead camshaft, plus a moderate compression ratio of only 8.5 to 1 and a Solex 40 PDSI carb help it to produce 100 b.h.p. at five-five and yet retain commendable flexibility and economy. This engine is mated to a four-speed gearbox — all-synchro, of course — which in turn drives through a high-geared back axle: at 3.64 to 1, the final drive gives 18.5 m.p.h./1000 r.p.m. in top: in terms of British speed limits, the sickening seventy is equivalent to under 4000 r.p.m. The suspension is similar to that of the 1600 — in other words, you get i.f.s. with struts and wishbones, plus an anti-roll bar which is one of the things BMW added for the 2002 version. Independence at the blunt end is by semi-trailing arms, coil springs and, again, an anti-roll bar; standard equipment includes 4½ in. rims (13 in. wheels, by the way) and the list of optional equipment includes a five-speed box and a limited-slip diff. Brakes are disc at the front, drums astern; an alternator looks after the sparks and the body shell includes extractor slots above the rear window with ducts in roof pillars and waist for the BMW version of Aeroflow. Total weight is 2070 lb., unladen, which gives an unladen power/weight ratio (for those too idle to do their own sums) of over 90 b.h.p./ton. Which suggest it oughter go a bit, and by Gawd it does. Inside, you get full carpeting, reclining seats, and instrumentation by means of rev-counter, speedo, water temperature and fuel gauges, plus odd warning lights and things.

There's nothing much to show on the outside; B.M.W. spent all the money on pretty shattering performance and handling.

The H 120 has plenty of exterior goodies, enough urge to be interesting, and excellent finish throughout.

Holbay Racing Engines have never been particularly famous for road conversions, preferring to win their reputation from the successes of their single-seater racing engines. However, the Rootes Arrow power unit didn't give them much trouble, and they have come up with an engine which, again, has a good deal of top-end urge and yet is still tractable enough for traffic use. The normal Rapier 1725 c.c. "four" (81.5 m.m. x 82.5 m.m.) gets a compression increase on its standard alloy head up to 9.6 to 1 and the ports are modified. A pair of 40 DCOE Webers, a free-flow exhaust manifold and a modified cam with more valve lift completes the deal, and all this gives 105 b.h.p. at five-two. It also gets wider section tyres, a noticeably higher final drive ratio (up from 4.22 to 3.89 to 1, giving 17.3 m.p.h./1000 r.p.m. in direct top, and 21.6 in overdrive top), the stripes and so on we mentioned before and a spoiler tail shape. Inside the car is as beautifully and fully equipped as are most Rootes products in the upper price bracket, and you get full instrumentation — including an oil temperature gauge — plus reclining seats and an adjustable steering column: in other respects it is pure Rapier, just as we tested in our September, 1968 issue.

It manages to convey a feeling of luxury which the BMW somehow misses: instinctively we felt the German car had no carpets — which it has, of course: it just feels a little more spartan, somehow. Wind noise has a good deal to do with this, probably; on the Rapier, it is commendably low, even at 90 m.p.h., whereas the same speed in the BMW makes conversation a frustrating thing to attempt. On the other hand, the BMW's steering is outstandingly light and precise, whereas on the Rapier, the change from 155 to 165 section tyres seems to have improved the roadholding but only at the expense of the steering, which is now definitely on the heavy side. And even the Rapier's improved handling is not up to BMW standards: the German car is undoubtedly one of the best in its class — or even out of it — from this point of view. Balance is near perfect, and the very effective i.r.s. provides extremely good traction out of bends although, as you might expect, full urge out of a greasy hairpin will get the inside wheel really spinning. The Rapier, however, needs to be cornered smoothly, and you can't fling it about as you can the BMW, which will just go round corners, as simply as that; the only real indication that you are getting near the limit is an uncomfortable angle of heel as the body rolls further and further. The Rapier is controllable, but inspires less confidence, and the fact that in the wet the front goes first while, on dry surfaces, oversteer is the dominant effect can be a shade off-putting until you get used to it. It's only fair to add, of course, that in the wet anything over four thousand on the clock means you can boot the tail round to overcome the understeer, and that in the dry the final oversteer is something you don't reach until you're cornering ruddy 'ard, Kipling. The difference between the two cars in a nutshell is that the BMW is highly chuckable, while the Rapier just ain't.

So far as performance is concerned, the BMW has the higher top speed, by quite a margin, reaching a best one-way of 112 m.p.h. and a mean of 110, compared with the Rapier's best figure of 107 and mean of 106, both in overdrive top: in direct top you can get 104 or so. There isn't much to choose between 'em when it comes to the spacing of gear ratios: the BMW gives you 32, 60 and 90 in first, second and third; the Rapier lets you go to 33, 54 and 80, with the ton available in overdrive third. Acceleration figures are comparable: 0-60 wants 10.7 seconds for the Rapier, 9.3 for the BMW; 0-80 figures are 18.1 and 16.8 seconds respectively. Both cars will potter along at thirty in top, but neither is very keen on the idea — the Rapier's handbook, in fact, advises against running at less than two thousand revs in top, and the car doesn't really start happening until the rev-counter is showing around 4000. Similarly, the BMW isn't really happy until you're doing about three-five, while top gear performance is slightly poorer than that of the Rapier: the 70-90 figure in top for the BMW was 13.0 seconds, and for the Rapier — in overdrive top, yet — was 11.4

Fuel consumption is a point on which the BMW scores marginally: we got 25 m.p.g. on this one, compared with 23 for the Rapier. But the Rapier carries more of the go-juice, with a fifteen-gallon tank compared with the ten-gallon effort fitted to the BMW, and this in fact gives the Rapier a slight advantage when it comes to a question of range between fuel stops. And actually buying the cars? Well, your friendly neighbourhood Rootes man will let you have an H120 Rapier for £1,634; the BMW man across the street will sell you a 2002 for exactly forty shillings less. But back home in Germany, Wolfgang can, we're told, buy a 2002 for under a thousand quid — around the price, over here, of, say, a C*rt*n* GT. Makes you think, doesn't it?

Martyn Watkins

PERFORMANCE DATA

	Rapier H120	BMW 2002
Mean maximum speed	106	110
Acceleration— 0-30	3.0	2.7
0-40	5.2	4.2
0-50	7.3	6.6
0-60	10.7	9.3
0-70	13.8	12.6
0-80	18.1	16.8
Fuel consumption	23 m.p.g.	25 m.p.g.

BMW 2002 AUTOMATIC

*Cars tell you how they want to be driven—
and this one tells you it wants to be driven hard*

CARS TELL YOU how they want to be driven. If you're in a big family type car with featherbed suspension, numb steering and have only the vaguest idea where the extremities are, you drive one way. If you're in a tautly suspended small sedan with sharp steering and can see where everything is, you drive another way. In the BMW 2002 Automatic, the car tells you it wants to be driven hard. You can see where you're going, it goes where you direct it, it responds eagerly to your impulses and it stops when it's told to stop.

If you'll permit another generalization, we'll also say that drivers reflect the characteristics of their cars. To most drivers of American cars, BMW drivers rank as impertinent and impatient. Because the car is eager and aggressive, the driver is eager and aggressive. Even our kindly old editor found himself impatient with the great passenger barges basking around in traffic. Ordinarily the gentlest of men, he had to resist the temptation to intrude himself on their normal state of somnolence by flashing his lights and nipping in and out around them.

BMWs want to go. Around corners (because the suspension is excellent), accelerate briskly (because the engine is eager and the gearbox works right), cruise as fast as the road will permit (because it's comfortable going fast) and stop (because the brakes are fully up to the job). BMWs, it seems to us, do everything a car should—and do it better.

Our test car is the BMW 2002 Automatic which is the 2002 equipped with the 3-speed ZF automatic already available in the 2000. This transmission works equally well in the 2002 and while the shifts aren't as silky as with a great U.S.-designed automatic, the engine does have sufficient torque to avoid the ski-slope feeling common to many small-engined automatics where you build up a great stack of revs only to fall into the next gear (and off the torque curve) with a lurch and take forever to get back up again. There's also a safety override on the automatic that prevents over-revving on manually selected downshifts.

You give away a little performance with the automatic, as you would expect. The 4-speed manual we tested earlier covered the standing quarter in 17.6 sec and our automatic took 17.9. It is still plenty lively, of course, and goes about the job with a delightful eagerness.

Other refinements that have been made in the 2002 Automatic include an automatic choke, a worthwhile improvement in convenience. Our test car also had the optional anti-roll bars front and rear, which reduce body lean and give a crisper handling feel. There's still considerable body lean in hard cornering but this is well controlled and upsetting to neither driver nor passengers.

The test car also had the optional Michelin XAS 165-13 tires which increase the sure-footedness but also result in the steering being noticeably heavier.

For some strange reason, the already disgraceful quantity of speedometer error in the 2002 has been added to in the Automatic. In our test car, when the speedometer was reading 60, the actual speed was only 53 and you had to have the needle close to 80 to keep up with 70-mph traffic. The odometer error is also ridiculous in that only 92 actual miles are covered for every 100 recorded on the dial. When BMW engineering does so many things so well, it is disillusioning to find such details ignored.

So far as the driver's comfort is concerned, the ventilation system has good flow but is poorly directed as there is no face-level venting and the fresh air that does come in is directed at the inboard calf. The window winders are tiresomely low-geared—as are the wind wings—and the windows roll down only far enough to leave an uncomfortable ridge for the arm to rest on.

The instruments are well displayed in an attractive black setting and in addition to a drop-down glove box there is a useful parcel tray molded into the padding in front of the passenger's front seat. The seats are very comfortable, fully adjustable, and the buckles of the 3-point seat belts pull down beside the hip where they should be. There isn't much room left in the rear when the front seats are in their normal position but there's adequate space for four when it is necessary to carry that number. The trunk is of a practical size and shape (the spare fits under the floor mat) and the compartment is fully finished off, a nice touch.

The pricing structure of the BMW 2002 is of the variety we find rather distasteful. The basic list of the 2002 Automatic (West Coast POE) is $3340, a bargain figure. But once you go into the showroom, you find that they don't have any cars to sell for this price as the importer orders all of them equipped with radial tires ($59 extra), vinyl upholstery ($45), chrome exhaust tip ($2), reclining seats ($48) and power-assisted brakes ($45). These things, plus the anti-roll bars ($20), tachometer in place of clock ($40) and the dealer preparation charge ($80) result in your basic 2002 Automatic at $3340 having a real price of $3679. And if you decide not to have all these extras, which aren't really extras, you must place your order and await delivery. Most importers long ago abandoned this type of almost-misleading pricing because of the ill-will it created among potential customers. The BMW 2002 Automatic isn't overpriced at its "real" price, so why aggravate the buyer?

BMW 2002 AUTOMATIC ROAD TEST RESULTS

PRICE
List price.................$3340
Price as tested............$3687

ENGINE & DRIVE TRAIN
Engine...........4-cyl inline, sohc
Bore x stroke, mm.....89.0 x 80.0
Displacement, cc/cu in.1990/121.5
Compression ratio...........8.5:1
Bhp @ rpm............113 @ 5800
  Equivalent mph..............98
Torque @ rpm, lb-ft..116 @ 3000
  Equivalent mph..............52
Transmission type: 3-spd automatic
Gear ratios, 3rd (1.00).....3.64:1
  2nd (1.52)................5.53:1
  1st (2.56)................9.32:1
Final drive ratio..........3.64:1

GENERAL
Curb weight, lb............2270
Weight distribution (with
  driver), front/rear, %.....55/45
Wheelbase, in...............98.4
Track, front/rear.....52.4/52.4
Overall length............166.5
  Width....................62.6
  Height...................54.0
Steering type......worm & roller
Turns, lock-to-lock..........3.5
Brakes......disc front/drum rear
Swept area, sq in..........243

ACCOMMODATION
Seating capacity, persons......4
Seat width,
  front/rear........2 x 20.5/51.0
Head room, front/rear....40.0/38.0
Seat back adjustment. degrees..90
Driver comfort rating (scale of 100):
  For driver 69 in. tall........90
  For driver 72 in. tall........85
  For driver 75 in. tall........85

PERFORMANCE
Top speed, high gear, mph....102
Acceleration, time to distance, sec:
  0-100 ft....................4.0
  0-250 ft....................6.8
  0-500 ft...................10.3
  0-750 ft...................13.2
  0-1000 ft..................15.5
  0-1320 ft (¼ mile)........17.9
  Speed at end, mph..........70
Time to speed, sec:
  0-30 mph....................4.9
  0-40 mph....................7.1
  0-50 mph....................9.8
  0-60 mph...................13.3
  0-80 mph...................25.7
  0-90 mph...................39.5

BRAKE TESTS
Panic stop from 80 mph:
  Distance, ft...............380
  Max. deceleration rate, %g...84
  Control..............excellent
Fade test: percent of increase in
  pedal effort required to maintain
  50%-g deceleration rate in six
  stops from 60 mph..........33
Overall brake rating.....very good

SPEEDOMETER ERROR
30 mph indicated......actual 28.0
40 mph.....................36.0
60 mph.....................53.0
80 mph.....................71.0

CALCULATED DATA
Lb/hp (test weight)..........23.7
Cu ft/ton mi................91.8
Mph/1000 rpm (high gear)... 17.2
Engine revs/mi.............3500
Engine speed @ 70 mph.....4120
Piston travel, ft/mi.......1840
R&T wear index..............61
R&T steering index..........1.20
Brake swept area, sq in/ton...182

FUEL
Type fuel required......premium
Fuel tank size, gal.........12.1
Normal consumption, mpg....19.5

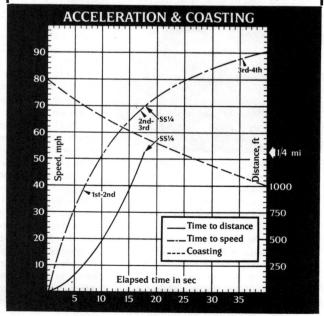

ACCELERATION & COASTING

YOU'RE BETTER OFF NOT THINKING OF IT AS A SEDAN

CAR and DRIVER ROAD TEST

BMW 2002

A lot of deutschemarks have gone over the exchange counter since we tagged the BMW 1600 "the world's best $2500 sedan." (February, 1967.) That was back in the days when you could buy all the deutschemarks you wanted for a quarter a piece. Now, after last fall's revaluation, you can't touch a mark for 27 cents, and the same car, well, OK, the same car with 400 more ccs and a handful of options that any "best" car should have, will set you back a solid $3500—if you insist on a radio and an automatic transmission you'll have to unroll about four grand.

As a $3500-sedan the BMW 2002 isn't looking too tall. For all of the things that you seek in a sedan—comfort, room and general convenience of operation—you could do better, for less money, with a Peugeot 504 or a Saab 99 or any number of American intermediates like an Olds Cutlass or a Chevelle. Competition is tough, both in price and in design, in the sedan business, particularly now in a time when smaller, more efficient automobiles are the only high fliers in the marketplace. Considering the delicacies available in the $3000-3500 sedan range, the BMW 2002 is hard pressed to keep its nose above water.

But there are shallower depths in which the 2002 can operate quite happily. Forget about the sedan body and pretend that it's a sports car—a transformation that's almost automatic in your mind anyway after you've driven it a mile or two. With the possible

exception of the new Datsun 240Z (which is not yet available for testing), the BMW will run the wheels off *any* of the under-$4000 sports cars without half trying. It is more powerful and it handles better.

Of course, forgetting about the BMW's sedan body is pure fantasy—the upright, block-like shape is too definite to be conjured or contemplated into something more Italianesque. It is, and always will be, a sedan with all of the attendant styling minuses and convenience pluses that that entails. But, spiritually if not stylistically, it is a sedan with a difference, a sedan that can beat the sports cars at the game they invented. That is the secret of the BMW's uniqueness, and if it continues to flourish in the market at its newly inflated price, it will be for that reason alone.

Although its price is far more jarring than its visual impact, the 2002 is a plainly honest machine, a kind of Bavarian Road Runner without the humorous overlay. It's the absolute hot setup in Germany where over-30 fat-cats bully their way through schools of VWs on the autobahnen with much flashing of lights and cold, sideways glances. Opels and Fords are driven by those who have renounced solid German technology in favor of vogueish, Detroit-inspired sheet metal, and the young man seeking to leave his mark in the well-understood games of traffic saves his money for a BMW 2002, or more likely the vitamin-enriched 2002 TI. From the outside the two seem identical, but the TI generates 22 more horsepower with the aid of a pair of 2-bbl side-draft Solexes and more compression, and its bigger brakes, stronger spindles and wider wheels strongly suggest the purpose of its existence. With the TI, maybe an orange one that sticks out of the somber greys, steel blues and whites of Teutonic traffic like one of the Weathermen at a Police Ball, you are the man to be reckoned with in anything but a top-end dash down the autobahn. And as if the TI wasn't enough, those overkill specialists down in BMW's dyno rooms have just put the finishing touches on the TII The extra "I" denotes fuel injection, with bigger exhaust valves and more compression included as frosting. When last we were in Munich the only TII engine available for testing was not in a 2002 but in the larger 2000 sedan. With 500 pounds extra weight its acceleration wasn't much better than a 2002 TI, but it would crank right past the redline in fourth gear with nonchalance. In the smaller car it will be devastating.

Meanwhile, back in the States, where exhaust emission controls are the law of the land, you have to settle for the plain vanilla 2002—no "T"s or "I"s are permitted. BMW engineers consider the TI a hopeless case when it comes to meeting the emission specs but are optimistic about the injected model —optimistic but non-committal when you ask when. Those same engineers shake their heads in mild disbelief at the thought of Americans classifying the "de-toxed" 1600 and 2002 as high performance automobiles. They feel that the necessary air pump and

BMW 2002

SPIRITUALLY, IF NOT STYLISTICALLY, THE BMW 2002, EVEN WITH AN AUTOMATIC TRANSMISSION, IS A SEDAN WITH A DIFFERENCE—A *SEDAN* THAT CAN RUN THE WHEELS OFF MOST OF THE UNDER-$4000 SPORTS CARS WITHOUT HALF TRYING

restrictions in ignition timing have hobbled the 2002 almost to the level of the European 1600 and rendered the 1600 a total invalid.

The engineers, of course, have numbers to back up their pessimism about the power available in the models imported to the U.S., but the 2002 is still very definitely a performance car, simply because its capabilities are not based solely on engine output. It's an agility specialist with controls that accurately telegraph back to the driver enough about what's going on so that he can comfortably operate near the limits. Its compact dimensions, more than a foot shorter than a Maverick, allow you to squeeze through bottlenecks that no American car could even consider, and the driver's fantastic view of the world around takes the guesswork out of the squeezing operation.

The 2002 is happiest in point A to point B dashes, and the more trying the circumstances the better. The ride is not soft but admirably controlled. The car is sure-footed on rough roads and you'll find yourself upshifting, downshifting, keeping the revs up and angling through corners at speeds that will make passengers wish they had taken a bus and left the driving to anybody but you. If you are at all susceptible to fantasizing, the 2002 will have you believing that every little outing is a special stage of international rally and prestige of the entire factory rests upon your shoulders. And before you've spent very many miles behind the wheel you'll discover that you can heel-and-toe like you've been doing it all your life. Yes, heel-and-toe, that anatomically impossible operation that has resulted in flat spotted tires and graunched gears every time you've tried it, is as easy as closing a door in the 2002. Such is not the stuff sedans are made of, but then we've already established that the 2002 is not a sedan.

For this test we have a pair of 2002 non-sedans—one with the recently introduced automatic transmission and the other with a standard 4-speed manual. By testing both cars together we could see exactly what effect the automatic has on performance and on the BMW's dashing personality.

In both cases the effect is pronounced. BMW buys the automatic from ZF. It's a 3-speed device, with a torque converter and it performs with noteworthy distinction. Part-throttle upshifts are very smooth, full-throttle upshifts happen at about 6200 rpm, just before the red line, and if you stand on the gas pedal it's smart enough to downshift if it's operating in a speed range that would make that operation at all helpful. And the shift linkage, though not inspired, is simple and accurate enough so that you are not likely to inadvertently make a wrong selection. But all of this favorable comment notwithstanding, the automatic is little more pleasant than having one foot in quicksand. It changes the 2002's nature from aggressive to passive; changes it from a sports car to a sedan, and that is where the water gets deep. The automatic probably hurts performance less than it hurts the performance "feel." Quarter-mile times were slower by 0.7 seconds with the automatic, most of which is lost at the start because there is not enough torque to spin the wheels, and the trap speed was slower by exactly 2 mph. The manual transmission 2002, even with all of its parasitic emission control devices working away, accelerates through the quarter in 17.1 seconds at 78.6 mph which makes it the top eliminator, by a slight margin, over all of the imported cars in its price class.

The BMW has about the same edge in handling. The test cars both had the optional front and rear anti-sway bars to compliment the traditional MacPherson strut front and semi-trailing arm rear suspension. Peugeot uses a similar arrangement in the 504 and it's the only other sedan, domestic or imported, in this price class that can match the BMW's handling on poorly surfaced roads. When you corner hard on smooth roads the 2002 assumes a mild understeering attitude which is subject to change without notice. With only a hint of its intentions it will hang its tail out in the kind of fat drift angle that dirt trackers live by. To the timid this probably sounds treacherous, and it would be if the BMW was intent on continuing this tail-wag into a spin, but it doesn't. In fact, after a couple of tries you discover you can stay hard on the throttle and use the steering to keep everything pointed in the right direction. That's when it all gets to be fun. The steering has enough feel to be very helpful during these maneuvers, but its slow (3.75 turns lock-to-lock) ratio means that you will be called upon for some occasional arm-winding corrections.

Only one situation can embarrass the BMW's suspension and that is spinning the rear tires. Every BMW we've ever tried, and that includes the 2800 coupe, suffers wheel hop if you lose traction during acceleration. Never mind that the text books say it can't happen—the Germans have found a way.

They've also found a way to make braking performance consistent. Both cars pulled slightly to the left under braking, locked up their front wheels first and required 277 feet (0.77G) to stop from 80 mph. Even though the stopping distances set no records the brakes do merit high marks because of the accuracy with which you can apply them. The power booster is particularly impressive because of its lightning response time which, among other advantages, allows you to pump the brakes very rapidly should a situation demand it.

CONTINUED ON PAGE 38

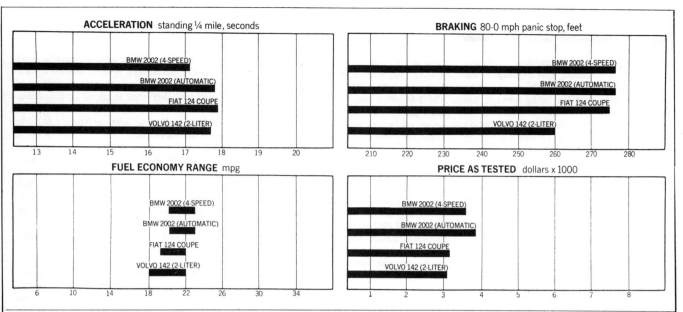

BMW 2002 4-SPEED (AND AUTOMATIC)
Importer: Hoffman Motors Corporation
375 Park Avenue
New York, N.Y.

Vehicle type: Front engine, rear-wheel-drive, 4-passenger, 2-door sedan

Price as tested: $3599.80 ($3845.80) (Manufacturer's suggested retail price, including all options listed below, Federal excise tax, dealer preparation and delivery charges, does not include state and local taxes, license or freight charges)

Options on test car: Base car, $3159.00; dealer preparation, $35.00; deluxe upholstery, $45.00; power brakes, $45.00; anti-freeze, $3.80; bumperettes, $8.00; Michelin XAS tires, $59.00; sun roof, $135.00

(2002 Automatic, $3475.00; dealer preparation, $35.00; power brakes, $45.00; deluxe upholstery, $45.00; anti-freeze, $3.80; bumperettes, $8.00; exhaust tip $2.00; reclining seats, $48.00; tachometer, $40.00; anti-sway bar, $20.00; Michelin XAS tires, $59.00; special paint, $65.00)

ENGINE
Type: 4-in-line, water-cooled, cast iron block and aluminum head, 5 main bearings
Bore x stroke . . 3.50 x 3.15 in, 89.0 x 80.0 mm
Displacement . . . . 121.4 cu in, 1990 cc
Compression ratio . . . . 8.5 to one
Carburetion . . . . 1 x 1-bbl Solex 40 PDSI
Valve gear . . chain-driven single overhead cam
Power (SAE) . . . . 113 bhp @ 5800 rpm
Torque (SAE) . . . . 115 lbs/ft @ 3000 rpm
Specific power output . . . . 0.93 bhp/cu in, 56.8 bhp/liter
Max recommended engine speed . . . 6400 rpm

DRIVE TRAIN
Transmission . . . . 4-speed, all-synchro (3-speed auto)
(Max. torque converter ratio . . . . . . 2.1 to one)
Final drive ratio . . . . 3.64 to one

Gear	Ratio	Mph/1000 rpm	Max test speed
I	3.84	4.7	30 mph (6400 rpm)
II	2.05	8.9	57 mph (6400 rpm)
III	1.35	13.7	88 mph (6400 rpm)
IV	1.00	18.3	111 mph (6150 rpm)
I	2.56	7.1	46 mph (6400 rpm)
II	1.52	12.0	77 mph (6400 rpm)
III	1.00	18.3	107 mph (5850 rpm)

DIMENSIONS AND CAPACITIES
Wheelbase . . . . 98.4 in
Track, F/R . . . . 52.4/52.4 in
Length . . . . 166.5 in
Width . . . . 62.6 in
Height . . . . 55.5 in
Ground clearance . . . . 6.3 in
Curb weight . . . . 2260 lbs
Weight distribution, F/R 55.1/44.9%
Battery capacity . . . . 12 volts, 55 amp/hr
Alternator capacity . . . . 420 watts
Fuel capacity . . . . 12.1 gal
Oil capacity . . . . 4.3 qts
Water capacity . . . . 7.0 qts

SUSPENSION
F: Ind., Macpherson strut, coil springs, anti-sway bar
R: Ind., semi-trailing arms, coil springs, anti-sway bar

STEERING
Type . . . . Worm & roller
Turns lock-to-lock 3.75
Turning circle curb-to-curb . . . . 32.6 ft

BRAKES
F: . . . . 9.45-in disc, power assist
R: . . . . 9.05 x 1.57-in drum, power assist

WHEELS AND TIRES
Wheel size . . . . 13 x 4.5-in
Wheel type . . . . stamped steel, 4-bolt
Tire make and size . . . Michelin 165 HR 13 XAS
Tire type . . . . Tube type, radial ply
Test inflation pressures, F/R . . . . 26/26 psi
Tire load rating . . . . . . 1010 lbs per tire @ 32 psi

PERFORMANCE
Zero to	Seconds
30 mph	2.5 (3.2)
40 mph	4.2 (4.8)
50 mph	6.5 (7.2)
60 mph	9.6 (10.3)
70 mph	13.2 (14.5)
80 mph	17.7 (20.2)
Standing ¼-mile	17.1 (17.8) sec @ 78.6 (76.6) mph
Top speed (observed)	111 (107) mph
80-0 mph	277 ft (0.77 G)
Fuel mileage	20-23 mpg on premium fuel
Cruising range	242-278 mi

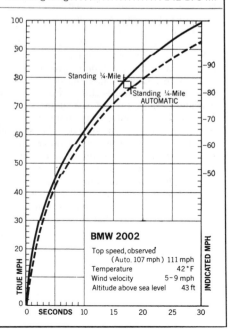

BMW 2002
Top speed, observed . . . . 111 mph
(Auto. 107 mph)
Temperature . . . . 42°F
Wind velocity . . . . 5-9 mph
Altitude above sea level . . . . 43 ft

BMW 2002

Continued from page 15

speed (which doubles as cruising speed) is a shade over a hundred, and nothing in the chassis, running gear, or engine ever gives the impression that it's being worked too hard. It's like effortless, no kidding. It couldn't come down the side of a mountain any more gracefully if Gower Champion choreographed the whole trip.

Maybe the neatest part of the whole deal is the fact that the 2002 was originally proposed as a kind of second-choice, American anti-smog version of the wailing 1600 TI they were selling in Germany, but the second-choice version turns out to be better than the original. The 2002 is faster 0-60, and faster at the top end as well. Not to mention the fact that it's a whole lot smoother and quieter.

How they can do all that good stuff and then screw it up with one of those incredible Blaupunkt radios is a little hard to imagine, but that's what they did. The rule with Blaupunkt and Becker seems to be, "The Bigger and More Complicated and Expensive Our Radios Are, The Lousier The Reception." The 2002 had a lovely-looking AM/FM affair neatly slipped into its console—easily a hundred-and-fifty bucks worth of radio—and I couldn't pick up a Manhattan station from the far end of the Brooklyn Bridge. Honestly. It was maybe the dumbest radio anybody ever stuck in an automobile, like all Blaupunkt and Becker radios, yet the German car makers—for reasons unknown—continue to use them.

It's a great mystery. Motorola, Bendix, Delco, and Philco can all sell you foolproof, first-class radios for about 75 bones—the Japanese can knock one off for about 98 cents—but the best German car radio you can buy throws up its hands in despair if you expect it to pull in a station more than three-quarters of a mile away.

Fortunately, the BMW is fast enough that you can keep picking up new stations as the old ones fade away. What you really want to do in this case, though, is install a good domestic stereo tape system. Maybe a little kitchen, too. The car is nice enough that you'll probably want to spend an occasional weekend in it—especially when you're fighting with your wife, or there's nothing good on television.

A final word of advice. The crazy-mad little BMW 2002 is every bit as good as I say it is—maybe better. If the 1600 was the best $2500 sedan *C/D* ever tested, the 2002 is most certainly the best $2850 sedan in the whole cotton-picking world. Besides the model-number was increased by 25%, but the price increase for the larger engine only amounted to 14%, and if *that* ain't a fair deal . . .

Feel free to test-drive one, but please don't tell any of those ten million squares who are planning to buy something else. They deserve whatever they get. Now turn your hymnals to Number 2002 and we'll sing two choruses of Whispering Bomb. . . ●

BMW 2002

Continued from page 36

All of this is to say that, as a sports car, the 2002 is more than just agreeable. As a sedan it's not disagreeable but it's not as comfortable as its price would indicate. We are not enamored with the driving position. The large diameter steering wheel is positioned more toward horizontal than conventional Detroit practice, and if you position the seat far enough forward so that the top of the wheel is within easy reach, you find that the control pedals are too close, particularly the accelerator. Since the seat is high you are forced into maintaining an unnatural and tiring angle between the gas-pedal foot and its attached leg. Also, the seat backs do not support you in such a way that would be comfortable for long distances. The seats themselves are mildly contoured buckets and do a remarkably good job of restraining you laterally, partially because the deeply textured insert discourages sliding around and partially because the softness of the cushions make them effectively more bucket-like than they would appear. The seat/shoulder belt arrangement, however, reflects German contempt for fussy attempts at self-preservation. Both straps join, with an unfathomable adjusting system, at a single buckle, and the fixed end of the shoulder strap mounts, below the rear window in such a way that is sure to cause high compressive loads on your spine in the event of a crash. We wouldn't wear the shoulder strap on a bet.

While the front seats have plenty of space in every direction for two adults, leg room in the rear is a dear commodity. We have no complaints about head and shoulder space, and because the side windows swing out rather than roll down, the side padding in the elbow area has been moved out toward the exterior sheetmetal to give a surprising amount of lateral passenger space. In general, the interior gives the impression of good materials and workmanship applied to a conservative design.

It's this quality, fortified with sophisticated engineering, that makes the premium German cars, Porsche, Mercedes-Benz and BMW, highly desirable to the discerning buyer. Considering the German's reputation for attention to detail, we find it unthinkable that the 2002, after two years in production, still has an overheating problem. Both of the test cars, if allowed to idle for 10 minutes, even in 30°F weather, would have their temperature gauges registering in the red zone. If it's that bad in the winter, we suspect summer driving in slow traffic would be intolerable. The importer had even installed a higher capacity cooling fan in the automatic transmission car, but its most significant contribution was an unpleasant roar at cruising speeds.

Overheating isn't the only unpleasantry associated with owning a 2002 as you will probably find out at your dealer's parts counter. Because we've been hearing comments about the high price of BMW parts (Porsche and Mercedes-Benz as well) we checked the prices of a few—a muffler, head gasket, clutch assembly, taillight lens, front fender and front bumper—that we felt an average owner would most likely have to replace. Then we compared this list with that of other similarly priced cars: Fiat 124 coupe, Triumph TR6, Volvo 142, Peugeot 504 and a Chevelle V-8. The total for the BMW was significantly higher than for any other car. Because of differences in design—for example, the 2002 has a 3-piece front bumper and you might get by with replacing only part of it—you can't project the entire parts price structure from this short list. Still, we would certainly caution against abusing your clutch (over $100) or breaking a taillight lens (almost $17).

To speak further of prices would be to belabor the point. In the things that the 2002 does best—harassing Triumphs, petrifying your passengers and awakening every dormant Fangio trait in your psyche—it can't be beat. Moreover, it's our firm belief that those things are cheap at any price. And since all of this subversion is camouflaged by the most Puritan of sedan bodies your wife won't have a clue until it's too late. ●

High-glassed as it may seem, the BMW 2002 TI is more sure of foot than most so-called sports cars. A timid Sloniger realized he could have got through this corner twice as fast.

NEW CARS FOR '69

FUN IN A BMW 2002 TI

When you want a car to beat a BMW the best choice is almost always a yet-newer model from that Bavarian stronghold of hot happenings. As of right now, ultimate performance for the money around sedanville is called a BMW 2002 TI, latest of their two-door line.

Mind you, I would be very tempted to stick with a "tame" 2002 for every-day family use. BMW swears they have sorted all four throats of those two dual sidedraft Solex 40 carbs for the TI version but I'd have to try them on a cold morning first. We do know that pickup, even when you jam the throttle down, is clean in normal weather.

No matter how hot BMW makes a sedan some soul always wants a meaner one so it came as no surprise, considering the firm, to find this 135 hp TI following a year after their 2002 of a mere 113 hp. Both, of course, are engine additions to the 1.6 two-door.

The latest is a road-registered, ultra-compact sedan (sold in England as a coupe) with competition credentials. And they plan to turn these loose on the public for something like $2750 ex-works or roughly 10% over the 2002 base price. The brain boggles: 135 (SAE) hp at a reasonable 5500 rpm from a mere 2 liters and 123 lbs.-ft. of torque down around 3600 revs with a sedan price tag. Such healthy torque comes from 89 x 80-mm dimensions. In short this is the familiar 2000 TI engine from their 4-door sedan range stuck into a car weighing only 2070 lbs (the 4-door is nearly 500 heavier).

Small wonder the baby bomb accelerates like a track machine. Would you believe 9.3 seconds to 60 mph? And if that won't do the job, buy a 5-speed option and get there in 9.1 seconds. This roadable rod will turn the quarter in mid-sixteens and that from 1990 cc, using road rubber.

In their Munich manner BMW gives the top speed as "over 115 mph" which it is without strain. Consumption naturally rears its head here with more carburetion than engine but you can hope for 16 mph using all the potential on an average and far better on the Interstates. The performance weight, just to wrap up our figures, is 17.2 lbs/hp.

It sure goes to show what they can pack into a 98-inch wheelbase when the stodgy board of directors isn't looking. What's more, BMW carefully avoided any risk of power catching up with chassis when urge was boosted 20%. They reworked the underpinnings too, not that many worried greatly with what I consider the best handling sedan known.

You get thicker (to 1/2-inch) front discs of greater diameter plus 5-inch wheel rims mounting fatter radial tires and since that widens the track a hair, they add stronger wheel bearings and wrap up the game with beefier front wishbones and rear trailing arms as well

39

as bigger shocks. Underchassised, maybe, but not until they Yankee V-8 power.

Speaking of brakes the 2002 TI comes with 4-piston front calipers and fail-safe lines arranged so one part of the front hydraulics can fall out and leave you 75% stopping power. No combination of brake failures can reduce front-brake effect below that of the rears, either.

Before this sounds too much like a BMW paean, I might add that they still haven't entirely solved the problem of pedal angles for normal ankle joints. The seats should be far more dished (always a weak feature of their two-door line) and while I like a leather wheel, this one is too thin. For that matter, 12-gallons of fuel is ridiculous with such thirst in a ground eater. And finally, shifts may be just fine when your precision is beyond reproach but snap down-shifts from third to second can easily hang up in the gate.

Ratios, even with the 4-speed box, seem fine largely because the engine is so elastic despite the state of tune. The 2002 TI is damn near as fast in third as most family iron flogging along in top so you can imagine how easy overtaking can be. Third and second could have been made closer but the loss isn't obvious unless you drive a 5-speed and really try.

Major lack at the moment is a limited slip differential, at least on the option list. Cornered even normally hard the 2002 TI will smoke its inside rear wheel when you feed all that torque through for rapid corner exits. BMW is paying here for stickiness so outstanding you can get the car sideways and unload the inside wheels without being either a race driver or even particularly white around the mouth afterwards.

This new TI doesn't so much corner on rails as flip around bends nearly sideways like a plumb bob at the end of a very safe string. You progress from neutral handling to power oversteer by simply twitching it to ridiculous (by normal sedan standards) angles and continuing on your merry way. Then in the process, you realize about at the apex of most bends that a BMW could do it a lot faster with equal ease if you weren't such a tortoise.

This exactly is the deep secret of BMW fun. Just about anybody feels like a great driver with one, yet the non-racers will be driving so far within the car's limits that it would take a man asleep to run out of road. You can only wonder what BMW will do for an encore — now if we shoehorned their new six. . . . .   *by Sloniger*

The bit at the top is the four cylinders. Toward you is a carburetor for each one, feeding out of an air cleaner that could easily service 10 BMW's.

In a car like the 2002 TI, one wishes that the space in front of the driver could be more fully occupied with instrumentation without, of course, upsetting the padding to idiot light ratio.

THREE SPORTING IMPORTS

The Fiat 124, BMW 2002 and Volvo 164 are small, and they're fun to drive. Are you paying attention, Detroit?

COMPARISON TESTING AGAIN. But before the alert reader quickly points out that these three cars are not really comparable, let us point out that we are comparing them to *domestics*. Detroit has decided (again) that it is time to build small cars and here are three the automakers ought to take a look at. Each has evolved from a different segment of the automotive spectrum. The Volvo is an example of a large (for imports) sedan that combines good interior room with compact external dimensions. The Fiat Sport Coupe is a sports car expanded for the family man—sort of a European Ponycar. The BMW represents a sedan with sporting character. None of the three can be classified as an economy car, but all are small cars, with good performance, adequate room and are a joy to drive. Except for the late Corvair, Detroit's erstwhile compacts never seemed to combine all of these qualities.

Take the Volvo 164 for example. Here is a car that roughly corresponds to the average U.S. compact, with a 106-in. wheelbase, 6-cyl. engine, five-passenger accommodation and large

BMW 2002 is a sedan body with sophisticated sports car chassis. Independent rear suspension, good power make this the fastest lapper of the three on any given road course.

FOUR-WHEEL disc brakes actually improved after CAR LIFE's strenuous fade test.

THREE IMPORTS

NEW VOLVO 6-cyl. is essentially the old four with two cylinders added. The 182-cid engine gives sparkling performance to 2800-lb. car.

trunk. Usable space is as good as any American car, yet it weighs only 2800 lb. And because it is light, performance is good. The engine is relatively small, 182 cid, yet it will run rings around any of our standard compacts —in a straight line. Its 17-sec. quarter-mile puts it right up there with the V-8 family cars. Volvo's secret, other than its light weight, is a new 6-cyl. engine. Essentially one and a half B20 4-cyl. engines, it is completely conventional in design, with overhead valves and short stroke. It gets good power because Volvo believes in letting an engine breathe. Dual carbs and free flowing exhaust are part of the standard package, allowing the engine a good top end. Because the engine is slightly peaky (for a family sedan) a four-speed gearbox is standard. This has a beautiful ratio spread and nice synchros, but one of the largest gear levers in

1969 VOLVO
164 FOUR-DOOR SEDAN

DIMENSIONS
Wheelbase, in. 106.3
Track, f/r, in. 53.1/53.1
Overall length, in. 185.6
 width. 68.3
 height. 56.7
Front seat hip room, in. ...21 x 2
 shoulder room 54
 head room 37.4
 pedal-seatback, max. 43
Rear seat hip room, in. 55
 shoulder room 55
 leg room 37
 head room 35
Door opening width, in. 31
Trunk liftover height, in. 36

PRICES
List, FOB factory $4160
Equipped as tested $4340
Options included: AM/FM radio.

CAPACITIES
No. of passengers ....4+1
Luggage space, cu. ft. 23.2
Fuel tank, gal. 15.5
Crankcase, qt. 6.3
Transmission/dif., pt. ....n.a.
Radiator coolant, qt. 13

CHASSIS/SUSPENSION
Frame type: Unit steel.
Front suspension type: Independent by s.l.a., coil springs, antiroll bar.
 ride rate at wheel, lb./in. ...n.a.
 antiroll bar dia., in. ....n.a.
Rear suspension type: Live axle with trailing arms and track bar, coil springs.
 ride rate at wheel, lb./in. ...n.a.
Steering system: Power assisted cam and roller.
 overall ratio 15.7:1
 turns, lock to lock 3.7
 turning circle, ft. curb-curb. .31.5
Curb weight, lb. 2860
Test weight 3170
Distribution (driver),
 % f/r 53.6/46.4

BRAKES
Type: Power assisted discs front and rear.
Front rotor, dia., in. 10.7
Rear rotor, dia. 11.6
 total swept area, sq. in. 433
Power assist
 line psi at 100 lb. pedal. ...n.a.

WHEELS/TIRES
Wheel rim size ....15 x 4.5
 optional size ....none
 bolt no./circle dia. in. 5/4
Tires: Goodyear Power Cushion.
 size. 6.85-15

ENGINE
Type, no. of cyl. IL-6
Bore x stroke, in. ......3.50 x 3.15
Displacement, cu. in. 182.0
Compression ratio 9.2:1
Fuel required ....premium
Rated bhp @ rpm ....145 @ 5500
 equivalent mph 107
Rated torque @ rpm ...163 @ 3300
 equivalent mph 64
Carburetion: Zenith-Stromberg CDSE 2x1.
 throttle dia. 1.75
Valve train: Overhead rocker arms, pushrods, mechanical lifters.
cam timing
 deg., int./exh. ....n.a.
 duration, int./exh. ....n.a.
Exhaust system: Single, reverse-flow muffler.
 pipe dia., exh./tail 1.5/1.5
Normal oil press. @ rpm ....n.a.
Electrical supply, V./amp. 12/35
Battery, plates/amp. hr. 37/60

DRIVE TRAIN
Clutch type: Single dry disc plate.
 dia., in. 9.0
Transmission type: Four-speed manual, fully synchronized.
Gear ratio 4th (1.00:1) overall. .3.73:1
 3rd (1.34:1) 4.90:1
 2nd (1.97:1) 7.35:1
 1st (3.14:1) 11.70:1
Shift lever location: Floor.
Differential type: Hypoid.
 axle ratio 3.73:1

the industry. An automatic transmission is optional and will probably be popular in this country.

When the going gets curvey, the Volvo really gets into its element. Handling was good, comparable to an optionized Swinger or Chevy Nova. It has the predominant understeer characteristic of a long wheelbase, front engine sedan, but this decreases as the car approaches the limit. For any hairy mountain work, it is for all intents and purposes a neutral handling car, yet safely understeering enough for your wife to drive.

Brakes—four-wheel disc—are good, could have been better with better tires. They actually faded *up,* increasing in effectiveness to 27 ft./sec./sec. after eight panic stops from 80. Power steering, rare on any import, is like a few of the better ones we've tested the past year. First time behind the wheel, the driver doesn't notice that it has power. When he does, he notices that it combines just the right amount of boost with a comparatively quick (15:1) steering ratio.

Volvo combines the best features of both performance and utility, something Detroit says it wants, but seldom pulls off in a compact.

Compared to the Fiat 124 Sport Coupe, the Volvo was dull. It's hard to really categorize the Fiat. It is a stretched version of the 124 sports car, but passenger space is as good as any small sedan. It's a small car, very light, with a small engine, yet its performance is surprising. It cannot really be called a four-passenger sedan because it is very low and rakish, having definite sports car lines. Probably it can best be called an Italian Ponycar. But most of all, it's Italian.

Which means it does everything with emotion. The engine, all 89 cubic inches of it, will rev its little heart out, just to go fast. Its five-speed gearbox is there to help it out, and give the driver an opportunity to display his driving prowess. The steering is quick, handling is sporting, and the four-wheel disc brakes do what is expected (31 ft./sec./sec.). Driving to the corner drugstore isn't a chore, it's fun. No matter how uninclined a driver might feel, before too many miles the car has lured him into enjoying himself; revving to the 6900-rpm red line, making snap shifts, heel and toe downshifting, leaving the braking to the last minute, taking corners near the limit, etc. Many a time, after an extremely hard day, we would stumble into the car wanting

LUXURIOUS Volvo interior had domestic-like dash, excellent finish throughout.

CAVERNOUS trunk competes with anything from Detroit, but has high liftover.

CAR LIFE ROAD TEST

CALCULATED DATA
Lb./bhp (test weight)..........21.9
Cu. ft./ton mile...............102.5
Mph/1000 rpm (high gear).....19.4
Engine revs/mile (60 mph)....3085
Piston travel, ft./mile 1620
CAR LIFE wear index............50

SPEEDOMETER ERROR
Indicated	Actual
30 mph	33.2
40 mph	43.1
50 mph	53.0
60 mph	62.8
70 mph	72.6
80 mph	82.3
90 mph	92.0

MAINTENANCE
Engine oil, miles..............6000
 oil filter, miles.............6000
Chassis lubrication, miles.....6000
Antismog servicing, type/miles .....
 clean PCV/12,000
Air cleaner, miles............24,000
Spark plugs: Bosch W175-T35.
 gap, (in.)..................0.030
Basic timing, deg./rpm.10BTDC/700
 max. cent. adv., deg./rpm.....n.a.
 max. vac. adv., deg./in. Hg...n.a.
Ignition point gap, in.........0.010
 cam dwell angle, deg..........40
 arm tension, oz...............20
Tappet clearance, int./exh.....0.020
Fuel pressure at idle, psi........3
Radiator cap relief press., psi....10

PERFORMANCE
Top speed (5600), mph..........109
Test shift points (rpm) @ mph
 3rd to 4th (6000)..............80
 2nd to 3rd (6000)..............54
 1st to 2nd (6000)..............34

ACCELERATION
0-30 mph, sec..................3.6
0-40 mph......................5.0
0-50 mph......................6.8
0-60 mph......................9.6
0-70 mph.....................12.5
0-80 mph.....................16.6
0-90 mph.....................24.5
0-100 mph....................38.9
Standing ¼-mile, sec........17.63
 speed at end, mph.............82
Passing, 30-70 mph, sec........8.9

BRAKING
Max. deceleration rate from 80 mph
 ft./sec./sec..................27
No. of stops from 80 mph (60-sec.
 intervals) before 20% loss in
 deceleration rate..8 stops—no loss
Control loss? None.
Overall brake performance.very good

FUEL CONSUMPTION
Test conditions, mpg..........14.3
Normal cond., mpg...........16-18
Cruising range, miles......250-280

(Acceleration graph: MPH vs Elapsed Time in Seconds, showing 1st, 2nd, 3rd, 4th gears and Quarter Mile point)

THREE IMPORTS
continued

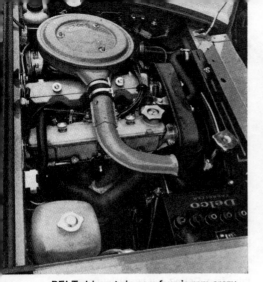

BELT driven twin cam four is rpm crazy, produced 96 bhp at 6500.

SURPRISINGLY roomy trunk is as good as any Ponycar (luggage for two).

PHOTOS BY DARRYL NORENBERG

only to get home and have that first martini. We always wound up driving eight-tenths, getting home sooner than we wanted and no longer thirsty.

It's hard to say just what it was that made the Fiat such a wicked little temptress. It was probably the combination of the high-winding engine, five speeds to play with and the handling. The handling stood out. When we got the car it wasn't yet fully

FIAT cockpit proved extremely comfortable, even for long trips.

FOUR-WHEEL DISC brakes gave a good peak deceleration rate (31 ft./sec/sec.), but stopping distance from 80 mph was high and they faded quickly.

1969 FIAT 124 COUPE

DIMENSIONS
Wheelbase, in.................95.3
Track, f/r, in............53.0/51.8
Overall length, in............162.0
  width........................65.8
  height.......................52.8
Front seat hip room, in....21 x 2
  shoulder room................54
  head room....................40
  pedal-seatback, max..........41
Rear seat hip room, in.......54.5
  shoulder room................55
  leg room.....................37
  head room....................37
Door opening width, in........41
Trunk liftover height, in......32

PRICES
List, FOB factory............$2981
Equipped as tested...........$3150
Options included: Five-speed trans., AM radio.

CAPACITIES
No. of passengers..............4
Luggage space, cu. ft........9.6
Fuel tank, gal..............11.8
Crankcase, qt.................4
Transmission, pt..............3
Radiator coolant, qt..........8

CHASSIS/SUSPENSION
Frame type: Unit body frame.
Front suspension type: Independent by s.l.a., coil springs, antiroll bar.
  ride rate at wheel, lb./in......n.a.
  antiroll bar dia., in...........n.a.
Rear suspension type: Live axle, trailing arms, track rod, coil springs, antiroll bar.
  ride rate at wheel, lb./in......n.a.
Steering system: Worm and Roller.
  overall ratio..............16.4:1
  turns, lock to lock.........2.75
  turning circle, ft. curb-curb....36.1
Curb weight, lb..............2130
Test weight..................2530
Distribution (driver),
  % f/r..................54.3/45.7

BRAKES
Type: Disc front and rear with vacuum assist, suspension actuated proportioning valve.
Front rotor, dia., in...........8.9
Rear rotor, dia., in............8.9
  total swept area, sq. in......297
Power assist
  line psi at 100 lb. pedal.......n.a.

WHEELS/TIRES
Wheel rim size............13 x 5K
  optional size................none
  bolt no./circle dia. in........4/4
Tires: Pirelli Cinturato radials.
  size........................165-13

ENGINE
Type, no. of cyl............IL-4 dohc
Bore x stroke, in........3.15 x 2.81
Displacement, cu. in...........87.8
Compression ratio.............8.9:1
Fuel required..............premium
Rated bhp @ rpm.........96 @ 6500
  equivalent mph...............116
Rated torque @ rpm....82.5 @ 5000
  equivalent mph................89
Carburetion: 1x2 Weber 34 DCF 2V.
  throttle dia., pri./sec....0.95/1.02
Valve train: Two-belt driven overhead cams, bucket tappets.
  cam timing
    deg., int./exh........26-66/66-26
    duration, int./exh.......272/272
Exhaust system: Dual cast headers, dual head pipes, single muffler and resonator.
  pipe dia., exh./tail.........1.1/1.1
Normal oil press. @ rpm...40 @ 3000
Electrical supply, V./amp.....12/53
Battery, plates/amp. hr......37/48

DRIVE TRAIN
Clutch type: Single dry disc plate.
  dia., in......................7.9
Transmission type: Five-speed manual, fully synchronized.
Gear ratio 5th (0.913:1) overall.3.74:1
  4th (1.00:1).............4.10:1
  3rd (1.41:1).............5.88:1
  2nd (2.18:1).............8.95:1
  1st (3.80:1)............15.53:1
Shift lever location: Floor.
Differential type: Hypoid.
  axle ratio................4.10:1

broken in. One staff member was going on a weekend jaunt, so he was elected to put on the miles. The day before, he had driven one of the 510 Datsun sedans with the very good independent rear suspension. After 200 miles in the Fiat he was convinced that it, too, must have had independent rear suspension. Peering under the car to see what kind, he discovered that the rear axle was live, and well located: four trailing radius rods, a track bar, an anti-roll bar and coil springs.

There is much to be said for independent rear suspension, but there's also much to be said for well executed live axles. Springing was very soft as sporting imports go, with very little harshness. Body roll while cornering on the limit wasn't very apparent to the driver, but subsequent pictures showed a substantial amount. It was a basically understeering car, and, having little torque, throttle oversteer could not be provoked. It did have a very strange trailing throttle oversteer characteristic in right-hand turns when pitched into the corners right on the limit. We suspect it was the result of a small jacking effect caused by the track bar. It wasn't dangerous, however, being completely controllable and since it happens so near the limit of adhesion, it is doubtful that many drivers will ever discover it.

The Fiat was the most underpowered of the group, although there was little feeling of having to row it around town with the gearbox. Fifth gear is actually an overdrive, to be used for open road cruising, but only offers a 9% rpm drop from fourth, which hardly changes the engine note. We were bothered by the low power in passing at highway speeds. Acceleration from 50 or 60 mph in *any* gear left a lot to be desired. If in fifth at the time of initiating the pass, a shift to fourth would not be enough. By the time the driver finds third in the notchy shift gate, his chance may have passed.

The engine itself is a delightful little 4-cyl. with double overhead camshafts driven by a toothed belt (similar to the Pontiac ohc six), putting out 96 bhp from its 89 cubic inches. It is incredibly revvy, sounding very comfortable even when buzzing along hours on end at 4500. It did have a strange vibration/noise between 3000 and 5000 rpm that had the effect of making the driver nervous enough to keep his revs above 5000, where things smoothed out and sounded normal. Engine noise was not objectionable at the high rpm, due partly to the fan clutch, partly to the belt cam drive.

Just as the Fiat displayed its Italian breeding, the BMW 2002 seemed to have a characteristic German personality. Efficient, precise, functional, tight, sturdy. The 2002 is a sedan in the purest sense, being relatively boxy with lots of interior room, much of it vertical. However, the sedan aura stops at the functional aspect, and sports breeding takes over. The 2002 is actually the 1600 sedan hot rodded with the larger 121-cid engine slipped in. Suspension system is a well engineered MacPherson strut in front, and semi-trailing arm independent rear. A smooth shifting, fully synchronized four-speed adds the finishing touches to the sporting image.

Most noteworthy item about the car is the engine, a very modern 4-cyl. with chain driven single overhead cam. It pulls smoothly and strongly up to and right past its 6300-rpm red line, sounding happy whether it's being lugged or revved mercilessly. There is a reason it sounds relaxed. It was designed for European driving, which is usually flat out all day. The BMW is built to take that kind of use,

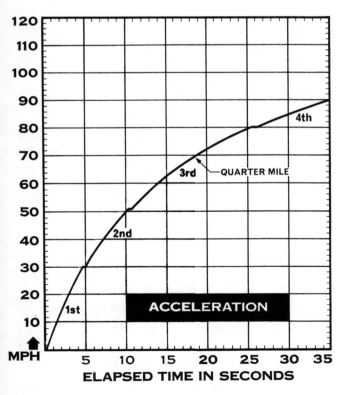

CAR LIFE ROAD TEST

CALCULATED DATA
Lb./bhp (test weight)............26.4
Cu. ft./ton mile.................67.7
Mph/1000 rpm (high gear)......17.8
Engine revs/mile (60 mph)......3375
Piston travel, ft./mile..........1580
CAR LIFE wear index...........53.3

SPEEDOMETER ERROR
Indicated	Actual
30 mph	29.4
40 mph	39.2
50 mph	49.4
60 mph	59.9
70 mph	69.1
80 mph	77.9
90 mph	85.5

MAINTENANCE
Engine oil, miles/days.........6000
oil filter, miles/days.........6000
Chassis lubrication, miles.....12,000
Antismog servicing, type/miles......
  clean PCV system/12,000
Air cleaner, miles.......clean/3000,
  replace/6000
Spark plugs: Champion N6Y.
  gap, (in.)..................0.022
Basic timing, deg./rpm.......10/1000
  max. cent. adv., deg./rpm..24/n.a.
  max. vac. adv., deg./in. Hg....n.a.
Ignition point gap, in..........0.018
  cam dwell angle, deg...........n.a.
  arm tension, oz................n.a.
Tappet clearance,
  int./exh..............0.018/0.020
Fuel pressure at idle, psi..........3
Radiator cap relief press., psi....14

PERFORMANCE
Top speed (6400), mph.....104 (4th)
  (5800)...............103 (5th)
Test shift points (rpm) @ mph
  4th to 5th (6400)..............104
  3rd to 4th (6900)...............80
  2nd to 3rd (6900)...............51
  1st to 2nd (6900)...............30

ACCELERATION
0-30 mph, sec...................4.7
0-40 mph........................7.0
0-50 mph.......................10.0
0-60 mph.......................14.2
0-70 mph.......................18.7
0-80 mph.......................25.5
0-90 mph.......................34.9
0-100 mph......................55.0
Standing ¼-mile, sec..........18.6
  speed at end, mph...........69.8
Passing, 30-70 mph, sec........14.0

BRAKING
Max. deceleration rate and stopping distance from 80 mph
  ft./sec./sec..........31/325 ft.
No. of stops from 80 mph (60-sec. intervals) before 20% loss in deceleration rate..................6
Control loss? None.
Overall brake performance. very good

FUEL CONSUMPTION
Test conditions, mpg..........20.4
Normal cond., mpg...........22-26
Cruising range, miles.......270-320

BMW has high C.G. and body lean characteristic of European sedans.

THREE IMPORTS

GOOD position for wheel and shifter, but BMW seats lacked side support.

EXCELLENT sohc engine is smooth, powerful, revvy, and very sturdy, for flat-out Continental driving.

SPARE tire stowage beneath flat trunk floor, like tool kit, is a good feature.

PHOTOS BY DARRYL NORENBERG

1969 BMW
2002 TWO-DOOR SEDAN

DIMENSIONS
Wheelbase, in. 98.4
Track, f/r, in. 52.4/52.4
Overall length, in. 166.5
  width. 62.6
  height. 54.0
Front seat hip room, in. ......2 x 24
  shoulder room 51
  head room 40
  pedal-seatback, max. 43
Rear seat hip room, in. 50
  shoulder room 50
  leg room 32
  head room 38
Door opening width, in. 40
Trunk liftover height, in. 32

PRICES
List, FOB factory. $3053
Equipped as tested. $3400
Options included: Anti-roll bars, radial tires, power brakes, reclining seats, tachometer.

CAPACITIES
No. of passengers ....4+1
Luggage space, cu. ft. 15.9
Fuel tank, gal. 12.1
Crankcase, qt. 4.2
Transmission/dif., pt. 2.1/1.9
Radiator coolant, qt. 7.4

CHASSIS/SUSPENSION
Frame type: Unit steel.
Front suspension type: Independent, MacPherson struts, lower A-arms, coil springs.
  ride rate at wheel, lb./in. .....n.a.
  antiroll bar dia., in. ....n.a.
Rear suspension type: Independent, semi-trailing arms, coil springs, anti-roll bar.
  ride rate at wheel, lb./in. .....n.a.
Steering system: Worm and roller.
  overall ratio. 17.6:1
  turns, lock to lock. 3.5
  turning circle, ft. curb-curb. .34.1
Curb weight, lb. 2220
Test weight. 2550
Distribution (driver)
  % f/r. 53.6/46.4

BRAKES
Type: Disc front/drum rear with power assist.
Front rotor, dia., in. 9.4
Rear drum, dia. 7.9
  total swept area, sq. in. 243
Power assist
  line psi at 100 lb. pedal. .....n.a.

WHEELS/TIRES
Wheel rim size. ....13 x 4.5J
  optional size. ....none
  bolt no./circle dia. in. 4/4
Tires: Michelin XAS Radial.
  size. 165SR-13

ENGINE
Type, no. of cyl. IL-4
Bore x stroke, in. ....3.50 x 3.15
Displacement, cu. in. 121.5
Compression ratio. 8.5:1
Fuel required. ....premium
Rated bhp @ rpm. ....113 @ 5800
  equivalent mph. 105
Rated torque @ rpm. ......116 @ 3000
  equivalent mph. 54
Carburetion: Solex 40 PDSI.
  throttle dia., pri./sec. ....n.a.
Valve train: Single overhead cam, rocker arms.
  cam timing
    deg., int./exh. 5-52/52-4
    duration, int./exh. 237/236
Exhaust system: Single reverse-flow muffler.
  pipe dia., exh./tail. 1.5/1.5
Normal oil press. @ rpm. ....n.a.
Electrical supply, V./amp. 12/35
Battery, plates/amp. hr. 37/44

DRIVE TRAIN
Clutch type: Hydraulically operated single dry disc.
  dia., in. 7.9
Transmission type: Four-speed manual with Porsche-type synchromesh.
Gear ratio 4th (1.00:1) overall. .3.64:1
  3rd (1.34:1). 4.91:1
  2nd (2.05:1). 7.46:1
  1st (3.84:1). 13.97:1
Shift lever location: Floor.
Differential type: Hypoid bevel.
  axle ratio. 3.64:1

and do it for 100,000 miles. Of the three cars, this seemed the least fussy about which gear it was in. It could be lugged, although most drivers responded to its performance image and automatically kept the revs up. Acceleration was comparable to the Volvo, and easily up to any American driving situation. Passing maneuvers never presented a problem, as fourth usually had enough torque to accomplish them, or, if things did get dicey, third gear and the high rev capability were always there.

Handling was not as good as we had thought it would be. Not that it was bad—we just were expecting a lot more than we had any right to. The independent rear suspension is a joy on rough roads, giving what seems like infinite compliancy. Like the Fiat, the BMW seemed to bring out the sporting driver in a person, inducing him to corner at higher speeds than he normally might in a lesser car. It's another case of the car instilling confidence in the driver, letting him know he can drive it safely out there, making driving to work an outlet instead of a chore. Understeer was the predominate characteristic, though right at the limit, the inside rear wheel would lift off, and the car would switch quickly over to a oversteer.

Curiously, the BMW had the best brakes of the three, yet it was the only one that didn't have four-wheel discs. The rear drums exhibited the usual early lock-up tendency common on disc/drum combinations. Our eight stops from 80 fade test failed to decrease the rate or increase the stopping distance.

A good descriptive label for the BMW is a sedan body on a sports car chassis and drive train. It proves one more time that utility need not be synonymous with dullness.

Don't get us wrong. These cars are not without their faults. Several things were annoying. The dash controls were very inconsistent, not just because of the different European standards, but their placement, operation, and even reliability were sometimes questionable. The BMW had rather uncomfortable seats (very poor buckets—you're not alone Detroit) with poor pedal, steering wheel and shift lever relationship. The Volvo's inside door handles were difficult to operate and if the Fiat's seat adjustment had crushed the engineering editor's finger one more time....

Where they all fell down was in the ventilation system and the seat belt arrangement. At first we couldn't figure out why they all had interior vents *and* front quarter windows. Now we know. The vent systems are not adequate, and have to be supplemented with the window vents. The seat belt/shoulder strap arrangement was also inferior to what is available on most domestic cars. Even the Volvo system, which is credited with the introduction of the diagonal strap, had a confusing and difficult buckle-up sequence. The result is that people seldom bother with them at all. With a simpler system, they at least use the seat belt.

But the biggest thing of all, about all of them, is the price. They are all expensive for what the buyer gets. For any of these prices, a corresponding American car is available that offers at least as much, sometimes more performance, convenience, and luxury. It's just that you get it in a larger size car, that almost by definition is less "fun to drive." It just seems that the American auto industry, for the moment at least, cannot scale down the enthusiast cars to a more economical size. These three cars prove that it can be done and we think, if Detroit wanted to really try, they could make cars that had performance, economy, and still be exciting to drive. And do a better job for the price. Are you listening Detroit? ∎

CAR LIFE ROAD TEST

CALCULATED DATA
Lb./bhp (test weight)..........22.6
Cu. ft /ton mile................90.5
Mph/1000 rpm (high gear)......18.3
Engine revs/mile (60 mph)....3280
Piston travel, ft./mile........1722
CAR LIFE wear index..........56.5

SPEEDOMETER ERROR
Indicated	Actual
30 mph	28.4
40 mph	38.1
50 mph	47.7
60 mph	56.8
70 mph	65.4
80 mph	74.2
90 mph	83.0

MAINTENANCE
Engine oil, miles/days.........8000
oil filter, miles/days.........8000
Chassis lubrication, miles......8000
Antismog servicing, type/miles..none
Air cleaner, miles.............8000
Spark plugs: Champion N9Y.
gap, (in.)...................0.024
Basic timing, deg./rpm...align timing marks/2000
max. cent. adv., deg./rpm.37/2400
max. vac. adv., deg./in. Hg.10/n.a.
Ignition point gap, in..........0.016
cam dwell angle, deg...........60
arm tension, oz................n.a.
Tappet clearance, int./exh....0.008/0.008
Fuel pressure at idle, psi..........3
Radiator cap relief press., psi......14

PERFORMANCE
Top speed (5700), mph..........104
Test shift points (rpm) @ mph
3rd to 4th (6400).............86
2nd to 3rd (6400).............57
1st to 2nd (6400).............30

ACCELERATION
0-30 mph, sec...................3.8
0-40 mph.......................5.3
0-50 mph.......................7.1
0-60 mph......................10.0
0-70 mph......................13.5
0-80 mph......................18.2
0-90 mph......................28.0
0-100 mph.....................49.0
Standing ¼-mile, sec..........17.4
speed at end, mph...........78.9
Passing, 30-70 mph, sec........9.7

BRAKING
Max. deceleration rate and stopping distance from 80 mph
ft./sec./sec./..........29 ft./sec./sec., 287 ft.
No. of stops from 80 mph (60-sec. intervals) before 20% loss in deceleration rate...8 stops, no loss
Control loss? Slight.
Overall brake performance..excellent

FUEL CONSUMPTION
Test conditions, mpg..........17.4
Normal cond., mpg...........22-27
Cruising range, miles........265-325

Acceleration graph: elapsed time in seconds (0-35) vs MPH (0-120), showing 1st, 2nd, 3rd, 4th gear shifts and quarter mile point.

NEW CARS FOR 1970

BMW 2002 Automatic is a spirited performer off-road and on. It's easily capable of 100-mph cruising at 17 miles to the gallon.

BMW 2002 AUTOMATIC

Text and photos by Sloniger

One of the world's finest small sedans gets a ZF automatic with a penalty in performance more imaginary than real.

The BMW 2002 Automatic tale can very nearly be compressed into a heroic couplet: all the outstanding pep and handling of the finest small BMW *plus* a 3-speed automatic box for those ever more inevitable traffic jams.

I've lauded the basic BMW often enough here to skip over its essentials and get down to what's new, which is Munich's addition of two-pedal action in their smallest sedan. Still, it wouldn't hurt to note that the finest cornering known to sedan drivers and a punchy 2-liter engine putting out 100 hp still work wonders in a 1-ton 2+2. A 105-mph top and 115 lbs.-ft. of torque for instant passing have just got to swing!

Sadly enough this kind of sports car engineering in the small family class costs cash. You wouldn't really settle for a basic 2002, nice though it may be, because the extras list carries heated rear window, reclining seats, full carpets and trunk trim and, of course, the automatic box which adds 10% to the price by itself. Even in Germany the "as delivered" bite is roughly 20% over base price. Maybe that's too much — unless you really love to drive.

And if you love to drive why buy an Automatic BMW 2002? The reasons are mostly because their ZF box doesn't steal enough performance to bother any but track addicts, it has low-cog holds for mountain work and will please the wife on her way to market. For that matter he-men will be caught smiling around town, too.

In theory an Automatic 2002 should be about 3-mph slower than its manual cousin and accelerate to 80 at about the same rate as their smaller 1.6 liter two-door with manual box. Above 80, the bigger engine more than balances the automatic.

In practice this WCG test machine was obviously at the top end of production tolerances, being actually faster than any manual 1-carb 2002 I've driven. They admit you can get from 96 to 104 horsepower and this red bomb must have been broken in right as well. Zero to 60 in 13 seconds and 110 mph tend to pop the eyes.

By way of comparison the 2002 A is slightly slower from zero to 60 than say a Renault R16 TS or Fiat 124 Coupe, but it's faster than the 1.7-liter Capri. It also costs about 10% more than the larger 2.3-liter V-6 Capri.

Just as amazing as the poke was overall consumption over 1,850 miles run flat out every second that traffic allowed. We still got 17.5 mpg. Mind you, any manual BMW driver in speed-limitless Europe is tempted, nay certain, to use full singing revs in every gear, right up to the 6,400 red line. The automatic, though, shifts up at 5,000 if you leave it in D and won't overreach 6,000 in top no matter how long the straight.

This boxy shape is familier to Americans but the shiftless feature will make it a little easier to explain the purchase to their wives.

Relatively small brake pedal precludes left foot usage and the T-handle shift lever is vaguely stopped and lettere

Somebody ought to explain to the Germans that automatic transmissions are taken pretty much for granted in other parts of the world.

Left to itself the 3-speed box shifts at 35 and 65 mph, while pulling the T-handle down a notch or two allows you to pull full revs any time. That last technique is hardly worth its slight margin.

At one point we covered two-thirds of France on those 2-lane roads with no big-city bypasses and averaged 70 mph with nary a tire squeal. Very few sport cars will do likewise and I never took the lever out of D position.

The chief reason for that was trouble in finding a notch with their tunnel-top selector. The detents are soft so you can easily overshoot and even if you wanted to look down each time, actual gear positions don't match the letters alongside anyway.

Another drawback is a very high creep factor, though the tach insisted the car wasn't idling over 800 rpm or so. It takes a lot of brake to hold it at red lights unless you shove the lever out of D, and the brake pedal is much too small and off to the right for left-foot action.

There is a fairly long pause between forward and reverse or vice versa when you're maneuvering in tight places, and there's a certain wait on cold mornings before it will take up the drive smoothly. Fortunately the car warms very quickly. The jerk on brutal kickdown is noticeable but not a bother unless you are going slowly enough for the box to drop two cogs at once, which is a real head snapper. Normal upshifts, even on strong throttle, come softly and normal downshifts in city traffic pass unnoticed.

In any case, no 2-door BMW is really comfortable for more than two adults on long hauls. The back seat is a city standby at best.

The car itself, with options noted, is highway comfortable and fitted with many small thoughts like shelf strips to keep your smokes from sliding, plenty of pockets and bins and big round dials, plus a leather-covered wheel. It is also cursed with frameless windows like all this series and they whistle like a traffic cops' convention. Besides, they need 5.5 turns to raise or lower, a teutonic design syndrome dating back to armored Grosser Mercedes.

But balancing likes and dislikes, I've always considered the 2002 an ultimate driver's sedan if you cover a lot of miles on mixed roads and can exploit both its balance and speed potential. Now, with the automatic, the wife likes it even better than I, which could make quite a parlay in the sales of a marque normally tagged for men only. ●

ROAD TEST / John Bolster — BMW 2002 automatic

The conservative lines of the BMW 2002 belie its performance.

Automatic with a sporting character

It is perhaps not generally realised that the role of the automatic gearbox is changing. The majority of the large cars are now sold in automatic form but, to the uninitiated, the sports car is inseperable from do-it-yourself gear selection. The pioneer work of Chaparral and McLaren made it clear that, far from being confined to non-performing cars for aunties, the automatic gearbox will eventually become a valuable performance-increasing adjunct of racing cars. BMW are aware of this trend, and they are now offering the ZF automatic transmission on the cars of their range which appeal to the more sporting driver.

By the yardstick of 1970, a sports car does not have to be a stark open two-seater. The BMW 2002 consists of the engine from the 2000 series in the compact two-door body shell of the 1600. Set up for fast driving with firm suspension, the resulting car handles in a most sporting manner, the reduced weight also giving the efficient overhead-camshaft 2-litre engine a chance to show what it can really do. Perhaps the most marked characteristic of this unit is its phenomenal torque at low and medium revs, in which respect it can certainly out-perfom all of its competitors. This is just what is required with automatic transmission.

It is unfortunately true that many engines which are used with automatic gearboxes are totally unsuited to the work. Gutless, non-pulling engines, or those lacking flexibility, are a complete waste of time. The transmission has to be set to change up and down all the time, the result being a noisy, fussy car that exploits none of the real advantages of automatic driving.

The ZF box on the BMW takes advantage of the sports-type engine. It has an instant kick-down which allows the unit to run up to peak revs, and the full-throttle up-changes give a considerable increase in performance. Let us compare a manual 2002 with the automatic model which I have just tested.

I do not change up on full throttle during road tests with a manual box, as this is a somewhat destructive manoeuvre which should be confined to Shelsey and Prescott and which no private owner would habitually employ on the road. The automatic box, on the other hand, is built to do it all the time and that is why it can beat a manual box over certain parts of the acceleration range, in spite of its slightly greater weight. I do use controlled wheelspin at standing starts, which makes the manual car a little quicker off the mark, but the automatic gains by its power-sustained change-up at just over 40 mph, so its 0 to 50 mph is only a tenth of a second longer than that of the hand-change car. At traffic lights the automatic would beat the manual 2002 unless the driver risked the noise and smoke of a wheelspin start.

By staying in third the manual driver can beat the automatic man in the 0-80 mph range, who has changed up at 75 mph, but the automatic can beat the manual over the standing quarter-mile. As for timed maximum speed, it's 108 mph for the manual 2002 and 107 mph for the automatic (when, incidentally, the speedometer cheerfully alleges 115 mph). So one can say that the very small loss of performance with the ZF box may actually amount to an appreciable gain under typical road conditions.

The slightly heavier gearbox has no effect on the handling of the car. It has a neutral characteristic under normal conditions, changing to moderate oversteer when really pressed. There is remarkably little roll, with a delightfully responsive feel to the steering, and the independent rear suspension pays dividends on bumpy corners. Compared with

The car rolls little under hard cornering, the rear suspension paying dividends on bumpy surfaces.

less responsive cars that understeer markedly, the BMW needs more correction in gusty winds at speeds over 100 mph.

The ride is distinctly firm at low speeds, and the car always has a taut feel about it. For fast cruising over bad roads, however, the standard of comfort is higher than would be expected. The brakes are very powerful indeed and the car does not tend to deviate during emergency stops. The hand brake is effective, and there is a parking lock on the transmission quadrant.

The engine is quiet at high cruising speeds, but it tends to be noisy when accelerated up to full revs. The gearbox has an audible whine when the car is going slowly in low gear, but it is silent thereafter. I let the car choose its own change-up speeds when recording acceleration figures, but manual changes can be made and this is useful for holding a lower gear during fast cornering. A safety device prevents bottom gear from being accidentally engaged at speeds which would entail turning the engine at destructively high revolutions, but its influence is not normally felt.

The 2002 feels very refined in normal driving, the flexible engine doing much of its work in top gear, and the machinery only becomes obtrusive when peak revs are required. There is a welcome absence of road noise, and only a little wind noise at maximum speed. It is curious that no eyeball-type fresh air vents are provided on this modern car. The interior is less austere than on previous BMWs, with attractive upholstery and trim; the seats are very comfortable and give quite reasonable lateral location for fast cornering.

Both on the road and on the circuits, the BMW 2002 is a highly respected car. With the collaboration of ZF it is one of the first medium-sized cars to have an automatic transmission that causes no appreciable loss in performance or increase in fuel consumption. I have been a die-hard upholder of manual gearboxes, with a secret leaning towards unsynchronised vintage types, but I enjoyed the automatic BMW so much that I might even be converted!

SPECIFICATION AND PERFORMANCE DATA

Car tested: BMW 2002 two-door saloon with automatic transmission. Price: £1,954 including tax.
Engine: Four cylinders, 89 mm x 80 mm, 1990 cc. Single chain driven overhead camshaft operating inclined valves in light alloy cylinder head. Compression ratio 8.5 to 1. 100 bhp net at 5500 rpm. Downdraught Solex carburetter.
Transmission: ZF automatic gearbox with torque converter, ratio 0 to 2.11 to 1, and three-speed epicyclic train, ratios 1.0, 1.52 and 2.56 to 1. Divided propeller shaft to hypoid final drive, ratio 3.64 to 1. Driveshafts with constant velocity joints to rear hubs.
Chassis: Combined steel body and chassis. Independent front suspension by MacPherson struts and lower wishbones with anti-roll bar, ZF-Gemmer worm and roller steering gear. Independent rear suspension by semi-trailing arms and anti-roll bar. Coil springs and telescopic dampers all round. Twin-circuit servo-assisted disc front and drum rear brakes. Bolt-on disc wheels fitted 165-SR13 radial-ply tyres.
Equipment: 12-volt lighting and starting with alternator. Speedometer, rev counter, water temperature and fuel gauges. Cigar lighter, heating, demisting and ventilation system. Windscreen wipers and washers. Flashing direction indicators.
Dimensions: Wheelbase, 8 ft 2½ in; track, 4 ft 4 in; overall length, 13 ft 10½ in; width, 5 ft 2½ in; weight, 18 cwt 2 qrs.
Performance: Maximum speed, 107 mph; standing quarter-mile, 17.4 s. Acceleration: 0-30 mph, 3.9 s; 0-50 mph, 6.9 s; 0.60 mph, 10.4 s; 0-80, 21.4 s.
Fuel consumption: 23 to 27 mpg.

The automatic 2002's performance is scarcely inferior to that of the manual version.

The interior is plain and simple, but well-finished (above). The familiar single-cam 2-litre engine develops 100 bhp net (below).

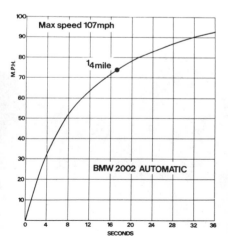

MOTORING PLUS

BY JIM TOSEN

BMW 2002 ALPINA

One of the advantages of having a roving Continental editor like Paul Frère is that he has the opportunity to drive and examine cars which are often out of bounds for practical or economic reasons to home-based scribes. It can also be tantalizing as most of the creations recorded in Continental Diaries are just not obtainable in Britain. This used to apply to the Alpina-modified BMWs, an example of which so impressed Frère that he bought for himself a highly modified 2002TI with 163 bhp engine, five speed gearbox, lowered suspension and limited slip diff. This car recorded a top speed of 126.4 mph, good enough for a Belgian National record incidentally, and last August had proved to be completely reliable over nearly 12,000 miles. But then virtually the same bottom end has withstood 280 bhp in the turbocharged factory Group 5 cars!

Having read and listened to Paul's description of this fabulous car we were immediately interested when Crayford Engineering announced that they were to import Alpina equipment. Director David MacMullan offered us his own 2002 which had just been kitted out with some of the latest goodies. He pointed out that they were still getting established on the Alpina front, that the car was getting on a bit and not a particularly good example and that we should forget any ideas about matching the Frère car's top speed or acceleration times of from rest to 60 mph in 6.8sec. and 100 mph in 18.7sec. But it made an already potent and excellent road car even better and in effect makes up for the lack of a TI version of the 2002 in the British BMW range, not that the standard car is exactly sluggardly. And our test car, from which the comparison figures are taken, seems to have been a particularly swift example.

The Frère car incidentally (described in detail on August 22, 1970) started life as a TI with the 9.3 to 1 compression pistons of that car which the Alpina mods raised to 11.5:1. The Alpina head raises the cr of the standard engine to 10.2:1.

The cylinder head comes with special valve springs, modified guides, combustion chambers reworked and balanced and both inlet and exhaust ports opened up and flowed in position. This costs £250 complete with the second wildest camshaft giving 300° valve opening, or if you really want the ultimate at the top end at the expense of flexibility and a really depressing fuel consumption, there is the full race 324° profile. The standard Solex carburetter is replaced by a pair of Webers with specially opened up chokes to match the head costing £67 for the pair.

Other Alpina modifications not carried out on the MacMullan car include what is called the sports chassis. This consists of reinforcing the rear suspension by boxing in the semi-trailing arms, uprated rear

Alpina Price List

Cylinder head complete with modified cam, valve springs, guides, flowed ports and chambers.
Compression raised to 10.2 to 1   £250
Cam separately 300 degree type   £38
324 degree racing cam   £40
Gaskets etc to suit   £10
Twin Weber carburetters modified to suit head   £67 (pair)

Crayford 2002 with Alpina modified cylinder head and 300° camshaft. The neat filter box for the Weber carburetters was added after tests were completed.

dampers, stronger front hubs and bearings and ventilated front discs, all for £139. Tyres up to 175 section can be fitted to the 2002 provided the front wheel arches are slightly flared. The test car was shod with Goodyear G800s on alloy wheels. Crayford hope to bring in a few of the Alpina full race engines with forged pistons and an output of 180 bhp which cost £850 on an exchange basis, provided the customer's engine has covered less than 5000 miles, and for the ultimate there is a Kugelfischer fuel injection kit which costs £600.

Apart from an air scoop let into the bonnet, the big GT wheels and Alpina flashes on the back, the MacMullan car looks like any other 2002 and retains most of that car's good manners and freedom from temperament. It started easily after a night out in the cold, warmed up smartly and though the figures show there is some loss in top gear acceleration up to 60 mph the engine gives the impression of having an exceptionally wide power band with no flat spots. Sometimes it showed a tendency to fluff after a long spell in dense traffic, but a short squirt on the open road soon cleared this.

The power really comes in around 2700 rpm which with the standard 3.64 to 1 final drive is equivalent to 50 mph in top. The last modified BMW we tested had the lower 4.1 axle which it seems would be better suited to this car since the engine spins freely and safely to an easy 7000 rpm and must develop peak power rather higher than the standard unit's 5500 rpm. On this gearing at our maximum speed, around 112 mph, the engine is still barely pulling 6000 rpm. A lower final drive would of course also benefit acceleration; not that 0 to 60 mph in 8.4sec: and 0 to 100 mph in 27.3sec. is exactly slow, but the former is less than a second quicker than our standard figure and viewed quite clinically

Performance

	BMW Alpina 2002	Standard BMW 2002
Maximum speed		
Mean	111.2 mph	107.4 mph
Best ¼ mile	112.5	111.1
Maximile	110.4	105.4
Acceleration		
0–30 mph	2.7 sec	3.1 sec
0–40	4.5	4.8
0–50	6.3	6.8
0–60	8.4	9.2
0–70	11.6	13.3
0–80	15.4	17.6
0–90	19.1	24.7
0–100	27.3	37.4
Standing ¼ mile in top	16.4	17.3
20–40 mph	9.6	7.6
30–50	8.9	7.4
40–60	8.9	7.7
50–70	9.3	8.8
60–80	9.8	9.8
70–90	10.8	12.5
80–100	13.2	20.3
In third		
10–30 mph	6.3	5.7
20–40	5.6	5.3
30–50	5.8	5.3
40–60	6.0	5.6
50–70	6.1	6.5
60–80	7.6	7.9
Fuel consumption		
Constant 30 mph	25.1 mpg	43.8 mpg
40	26.1	41.0
50	27.9	38.5
60	27.9	32.3
70	25.6	28.5
80	22.0	24.3
90	18.5	20.6
100	15.2	16.8
Overall	17.9	24.0
Touring	24.1	27.5

hardly seems a fair return for an investment of around £300.

With the lower axle we have got a tuned 2002 up to 60 mph in under 8sec. Another deterrent to tyre-smoking starts is shudder in the transmission which seems to apply to all 2002s to some extent. It would appear that the body mounted differential transmits tramp disturbances up the drive line to the engine, through the mounts into the front suspension cross member and up the steering column where the effect on the driver can be quite alarming. The more wear in the rubbers at the various attachment points the worse it gets, and as I have said this one was a fairly well used car. It was even noticeable accelerating hard out of a sharp corner when the steering sometimes developed a mind of its own. This completely kills its response and self-centring effect, never one of the best points on a BMW. Otherwise, within the limits of the 2002's tendency to lift the inside wheel when cornering hard, the car's roadholding and traction on the big Goodyears were very good. Paul Frère found that with the full conversion the optional BMW differential with 25 per cent locking factor was not sufficient to prevent the inside wheel spinning under power on sharp corners and opted for a diff with 75 per cent factor. Another favourite gripe of ours, directed at all older BMWs, is the uncomfortable pedal layout which makes heel and toe gear-changes almost impossible. Apparently the makers have at last paid some attention to this in the latest cars.

Though the car was obviously more raucous than standard, nobody who drove it objected to the noise. There is some harshness in the middle of the range and a lot of intake roar from the gaping throats of the Webers under power. The air cleaners shown in the illustrations were not fitted at the time of our test so we cannot comment on their effectiveness, but they must make some difference.

Steady speed fuel consumption figures show that the inlet tracts are rather too wide for economy at low speeds and the car is actually slightly more economical at constant 70 mph than it is at 30 mph. With much of our test conducted on icy or congested roads this is reflected in the rather heavy overall consumption of just under 18 mpg, though the touring figure, based on a steady speed of 71 mph, is not so far behind the standard figure and we had no difficuty in exceeding 20 mpg on a long run out of town.

Obviously when assessing the value of this Alpina kit it must be borne in mind that it is always much more difficult to make a significant improvement on a very good engine like the 2 litre BMW than a more ordinary mill, and as Crayford have found even BMWs, originating from a country with very strict trade description regulations, do show some considerable variation in standard power output. Alpina have done enough to convince me that they know their way around the BMW engine and a determined effort to promote their wares in this country is good news for BMW enthusiasts.

2 CAR TEST

Fiat 124 Coupe
BMW 2002

SOMETIMES, a particular type of car becomes fashionable. At the moment, fashion favours the sporting four-seater. From the multitude of cars in this class, two stand out as direct competitors, of great interest (judging by letters) to many of our readers. These are the BMW 2002 and the Fiat 124 Coupé 1600. Recent price increases on both sides have still left them well within striking distance of each other—the BMW is the more expensive by £77, a difference of just over 4 per cent. They are both more expensive than the Ford Capri–Opel Manta–Sunbeam Rapier group which offers the same sort of image. The Alfa GTV, on the other hand, costs a good deal more than the Fiat or BMW and in any case doesn't give as much back seat room.

We are therefore left with the BMW, which is no more or less than the little 1600 saloon with a bigger engine and stiffer suspension; and the Fiat, based on the floorpan and suspension of the ordinary 124 saloon but with a new coupé body shell, twin-cam engine and five-speed gearbox. What separates these two from their lower-priced equivalents could be broadly described as driver appeal; engine response, steering and handling. Even more than usual, therefore, our two-car test technique of convoy driving, turn and turn about, is the most likely way of finding which is the better car.

Description—BMW 2002

Fitting a larger (but not more heavily tuned) engine into an originally smaller capacity car has nearly always made a pleasant ideal. Several manufacturers have lately tried their hands at it, but few have succeeded as well as BMW with the 2002. It first appeared at the 1968 Geneva Show, being in effect the two-year-old 1600 two-door close-coupled four-seater with the standard 2-litre engine from the bigger, squarer-bodied BMW 2000.

The engine follows classic BMW design—a chain-driven single ohc aluminium cross-flow head topping notably over-square cylinders (89 x 80 mm, giving 1,990 c.c. capacity). With the comparatively modest 8.5-to-1 compression ratio and a downdraught Solex carburettor, it produces 100 bhp (DIN) at 5,500 rpm, with maximum torque of 116 lb. ft. at 3,000 rpm. Such an engine, put in a body of modest proportions weighing 18½ cwt at the kerb, means excellent performance and flexibility, and a potentially long stride; obviously deciding that acceleration matters more than anything else, BMW slightly under-gear the car overall at 18.5 mph per 1,000 rpm in top. The gearbox is a middling wide-ratio four-speed one.

Suspension is basically the same as the 1600's—on coil springs, with MacPherson struts in front, and semi-trailing arms at the back—with the addition of anti-roll bars at each end. The steering box hides worm and roller gear, and the servo-assisted brakes are as on the 1600, ATE-Dunlop 9½in. discs in front and 8 x 1½in. drums at the rear. Apart from the rev counter that replaces the 1600's clock, equipment is more sensible-family-car than sports-saloon. The price in Britain is £1,874.

Description—Fiat 124 Coupé 1600

At first sight, Fiat's contender in this comparison might not seem the obvious one,

of any make. The BMW's two-door body has been called a coupé, but it isn't really, whereas there can be no argument about the Fiat's title. The 124 coupé first appeared at Geneva a year before the Bavarian car, fitted then with a bigger bore 1,438 c.c. version of the 1,197 c.c. 124 saloon engine; instead of the original pushrod ohv light-alloy cylinder head, a tooth-belt-drive twin overhead camshaft head with single twin-choke Weber carburettor was provided. This engine gave 90 bhp (DIN).

Then last year Fiat put the 1,608 c.c. 80mm bore-and-stroke engine from the 125 saloon into the coupé as an option—though with some alterations, like two bigger Weber carburettors and a higher (9.9-to-1) compression ratio which put the power up to 110 bhp (DIN) at 6,400 rpm with 101 lb. ft. torque at 3,800 rpm. As those figures show, the Fiat power unit is obviously a more highly developed "sporting" engine than its 24 per cent larger rival. It has a 1cwt heavier car to propel via rather lower overall gearing (17.5 mph per 1,000 rpm in top), but a five-speed gearbox instead of only a four-speed.

Also on coil springs, the Fiat has wishbone independent front suspension and a live rear axle located by trailing arms and a Panhard rod, with anti-roll bars at each end. Steering is another worm-and-roller set-up, and the Fiat-Bendix brakes are all discs of just under 9in. dia; Bonaldi servo assistance is standard. Equipment inside the car is comprehensive, the only instrument lacking being an ammeter. The British price is £1,797.

Performance—BMW

Considering the undeniably sporting nature of the car, the BMW's engine and driveline are remarkably "soft". The single-carburettor engine develops peak power at 5,500 rpm, and peak torque at only 3,000 rpm, and with good basic breathing there is plenty of punch and response through the whole speed range from 1,000 rpm onwards. At the same time, the flatness of the power curve means that there is no sensation of being "over the top"; the engine pulls lustily all the way to the red line at 6,500 rpm. Nevertheless, it is as well to remember that the car *is* over its power peak at maximum speed, suggesting that slightly higher overall gearing might be of benefit. The gear ratios are excellently chosen, giving roughly 30, 60 and 90 mph in the three intermediate ratios. The gearchange itself feels slightly sloppy, yet allows very quick changes to be made with very little effort.

In absolute terms the BMW is fast but not electrifying, just failing to break the 10sec mark to 60 mph, despite the fact that a slight overspeed brings it to this point in second gear. It is easy to make a rapid getaway from rest, thanks to the nicely progressive clutch and accelerator linkages. It takes less than half a minute to reach 90 mph from rest, but from there on the acceleration starts to tail off a little. Even so, there is still enough response left when cruising at 100 mph to feel the car trying harder when the throttle is opened.

Perhaps the most impressive aspect of the BMW's performance is its flexibility. Top gear will not quite pull from 10 mph, but third gear returns such good figures from this speed that one only needs to take second gear when seeking the last edge of acceleration. From 20 mph onward, top gear is perfectly satisfactory for all normal driving. In other words, it is open to a BMW driver to be extremely lazy if he so wishes, to avoid changing gear very often.

Performance—Fiat

The Fiat's engine, aspirated by two large twin-choke carburettors and boasting a pretty high compression ratio, is inevitably a taut-feeling unit with a lot of sporting character. Peak power is developed right up at the red line—6,400 rpm—and peak torque is only reached at 3,800 rpm. To make the most of the rather restricted power curve, a five-speed gearbox is standard. This has very close ratios, but they are rather bunched towards the lower and middle speed ranges. First gear will not take the car to 30 mph, nor second gear to 50; third gear is just about good for 70 mph with a little overspeeding.

The mean maximum speed of 109 mph comes a little below the peak power speeds. But it must be remembered that fifth is an overdrive top, and it feels a little like tempting providence to cruise on the yellow line, as is perfectly possible at the moment. By British standards, in fact, the car is rather under-geared.

The Fiat's figure of 10.7sec to 60 mph is slightly misleading, since it includes two gearchanges, one of which is the awkward second-third change, where it is quite easy to

Considering they started with the floorpan of the 124 saloon, the Fiat stylists created something of a classic with their coupe. Because it is much lower than the BMW behind it, the Fiat looks longer. In fact, the German car is bigger all round

TWO CAR TEST...

overshoot into fifth and spoil the run. A more reliable indication of acceleration overall is the 27sec to 90 mph (with the car still in fourth gear). Following the change to the wider-spaced fifth, acceleration becomes much more leisurely.

The gearchange itself is well matched to the character of the engine: taut and precise, but slightly heavy. There is moderate spring-loading towards the third-fourth gate.

Although flexibility is not the Fiat's strong point, third gear will pull it away from 10 mph, and fourth from 20 mph; while it can just be persuaded to pull from 20 mph in fifth gear also. Most of the time, however, the benefits of changing down are sufficiently obvious to ensure that the gearbox gets used a good deal. The synchromesh is extremely strong, even on first gear—which tends to be used for little other than standing starts because it is so low.

Performance differences

The Fiat's engine, giving away nearly 400 c.c. to the BMW unit, has to work a good deal harder to achieve the same results. In consequence, it feels much the more sporting of the two. Against the stopwatch, the two cars are very evenly matched indeed—more so than almost any other pair of cars we have tested in this way. Their standing quarter-mile times are identical, and in the couple of cases where the Fiat appears to lose out, this is because of its over-close bottom-end gear ratios.

At the very top end, the Fiat has something of an edge and is able to pull steadily away from the BMW. This reflects the Fiat's slightly higher maximum speed, probably the result of superior aerodynamic form.

The Fiat is also at a slight advantage in circumstances where it happens to have exactly the right gear, for example when accelerating down a short straight between two corners on a twisty road. Overall, however, the BMW is better placed in this case with its massively spread power band. Our figures show how very much more flexible the German car is.

Handling and ride—BMW

Needing 3.7 turns across lock-to-lock with a compact mean turning circle diameter of 31ft 5in., the steering of the BMW is of about average gearing by modern saloon car standards. It is accordingly fairly light. The quite large steering wheel (15½in.) amplifies play in the worm-and-roller steering gear to an effective 1½in. at the rim; this seems too much if tried when standing still, but in fact is not often noticed when driving. The car responds positively and accurately to whatever you do with the wheel. There seems to be a minimum of damping in the steering itself, so that feel is excellent, though with little kickback. Straight-running stability is fairly good, although one is conscious of having to steer along motorways.

Hurry the BMW through open bends and you find it generally delightful. It doesn't roll too much, and seems pretty well balanced, having no excessive understeer like so many bigger-engined small-ish cars. Play with the accelerator in appropriate places and the tail can be broken away very controllably; on slippery surfaces the inside back wheel will spin fairly easily, but generally it is both that slide. Roadholding on the German-made Dunlop SP 57 radials, however, is good in all conditions.

As well as roll (not much, as already said) one other sort of attitude must be mentioned when talking about the 2002, and that is the degree of squat exhibited under hard acceleration. It is very much part of the likeable character of the car, fairly noticeable, but not at all objectionable. Ride is quite good. The car feels sure of its feet on a bumpy, joggly English country road without riding harshly; it is well damped and there is little feeling of float at any time. (It is the seating that disappoints here; one is not held much by the too-flat surfaces of the seats, seeming to bob about somewhat relative to the car's motion).

Handling and ride—Fiat

It is often a surprise to an experienced driver to learn that the Fiat has worm-and-roller steering. This is because it has the slightly damped movement and excellent precision and response of a good rack-and-pinion arrangement without any noticeably large amount of the slop found in other steering gear systems involving more than the two links of a rack set-up. There is plenty of feel, and a certain amount of kickback on uneven roads. At three turns of the 15in. steering wheel for a somewhat bulky 35ft 9in. turning circle, it is quite high-geared, a bit heavy when parking, and immediately responsive. Straight running is excellent, except to a small degree on bumpy roads when the car wanders very slightly.

Along winding roads, the Fiat is another car that seems to revel in being driven fast. It does however understeer a lot, more so than its predecessor; there is sometimes a tendency towards spinning the inside rear wheel on accelerating hard out of a slippery corner entered slowly. Rear-end breakaway is not often encountered except in slippery conditions, and then only with the car in the right gear for full power to be applied. Roll is more noticeable than it was before, though still is not at all excessive.

Likewise, ride is softer than before, perhaps at the expense of handling. While there is a suggestion of float over poor going, bumps do come through quite firmly at times.

Handling and ride differences

The BMW's steering is lower-geared, lighter, better-insulated against road-shocks, tighter locked and so in spite of being 4½in. longer the car is the more manoeuvrable. But though certainly good enough, the steering is not as accurate as the Fiat's and not quite so responsive. The 2002 is definitely more tail-happy than the 124, which understeers a lot more and rolls a bit more. Although the BMW's steering feels delightfully free of friction where the Fiat's feels damped in comparison, feel in both is good. But the Fiat's seats, mentioned here for the contribution they make towards handling and ride, are very much better, locating the driver properly. The Fiat apparently runs straighter and is possibly more stable in a side wind. In ride the BMW is slightly softer but better damped and seems to keep better contact with the road.

Fuel consumption

As one might expect, the BMW with its much smaller carburettor area turns out to be the more economical car—although the difference is not unduly great. Since both cars are rather under-geared, one might speculate on how much better they might be with higher overall gearing. As it is, there is undoubtedly more scope for driving the BMW for economy if its owner so wishes—the Fiat is over-choked for

Top to bottom: The Fiat has a sloping back window and a chopped-off tail; the BMW has a much more upright build and more rounded lines at the back. The BMW's bonnet rises sills and all, giving good access; accessibility to most components under the Fiat's bonnet is also good. Both bonnets are front-hinged, for extra safety.

The chunky BMW's styling is based on a very prominent waistline which continues all round the car. The 2002 has a matt-black grille to distinguish it from the smaller-engined 1600. Glass area is exceptionally generous, but visibility in the rain is spoiled for British drivers by the left-handed wiper pattern. The BMW has single headlamps of conventional continental pattern rather than the four tungsten-halogen units fitted to the Fiat

that sort of thing and mildly objects to being trickled along in a high gear for very long. One point worth bearing in mind is that neither car has a very large tank—even the BMW is pushed to make 250 miles between fillings, and the Fiat needs refuelling sufficiently often to be irksome on a really fast run. Neither car uses any oil to speak of.

Brakes

To any driver who is not going to use either car to anywhere near three-quarters of its capability, the brakes of each will seem excellent. Both require very low effort for ordinary stops, (the Fiat especially so) or even for one middling high-speed panic-stop. The Fiat starts to stop almost immediately you begin to press the pedal, and its working movement seems unusually long. If anything, the pads seem almost too willing to work at first, and one soon learns that they are pretty soft (that is, giving lessening retardation as they are heated). On this test Fiat there was a suggestion of an irregularity in a disc as one slowed gently, retardation varying rhythmically.

But attempt to drive either car fast along a give-and-take road involving straights and frequent corners and you encounter serious fade. The Fiat is slightly the worse of the two, in that the necessary increase in brake pedal pressure is the more suddenly experienced. Admittedly both recover quite quickly after fading, but the sort of vastly enjoyable rapid descent (and even ascent) of an Alpine pass which both cars are so ideally meant for would have to be markedly restrained. A surprising state of affairs in either case, the more so with the Fiat, which is built within sight of the Alps.

Noise

The Fiat is overall the noisier of the two, understandably perhaps in view of its more heavily tuned engine, for that is the main source of noise here. But to anyone who likes sporting character this will not matter; the Fiat's intake noise when working hard is not at all unpleasant. In a lengthy traffic jam one is a little conscious of the electric fan switching in and, on the test car, of some gearbox noise in neutral. Whine on drive or overrun is audible but not serious. There is a little more bump-thump on the Fiat, but it isn't bad either, nor is wind-noise.

The BMW engine is more subdued at all times, making a pleasant busy hum when working, and a very slight high-frequency vibration felt in the gearlever from around 5,500 rpm onwards. Bump-thump is quite quiet too; it is the quite loudly buffeting wind noise around the driver's and front passenger's windows that is most noticeable at the car's happy 100 mph cruise.

Fittings and furniture

The BMW and Fiat employ entirely different approaches to interior design. In the Fiat, the interior matches the sporting appeal of the body, with a comprehensive set of instruments and a massive central console. The BMW on the other hand, retains the simple interior of the 1600 saloon from which it was derived.

There are major differences in the driving positions. The relatively tall BMW accommodates a wide range of drivers in comfort. The front seats are quite high, which helps to make the visibility extremely good. Now that BMW have re-worked the pendant pedals in their right-hand drive cars, they are much easier to use. The BMW steering wheel, quite deeply dished, has a rather large-diameter rim with a plain plastic finish.

In the Fiat, the lower-set driving position which is a result of the lower build of the car (it stands nearly four inches lower than the BMW) means that even though fore-and-aft seat movement is quite generous, it is only the

TWO CAR TEST...

smaller drivers who can make themselves really comfortable. Taller drivers are also handicapped by the poor wheel-to-pedals relationship, which puts the wheel at full stretch with the pedals still too close. On the credit side, the pedals are well spaced and there is a proper rest for the left foot clear of the clutch. The steering wheel is smaller than that in the BMW and in keeping with the overall character of the car it has a wood-finish rim and matt-black, perforated spokes.

The front seats themselves follow the general trend. In the BMW they are pretty ample devices with a modicum of shaping, with fully reclining backrests. In the course of a long drive they stay pretty comfortable, but they hardly provide enough sideways support during enthusiastic cornering. The Fiat seats are smaller and very positively shaped indeed, especially the cushion. In this way they provide excellent support in all directions, but feel tight for drivers of ample width.

The BMW's back seat is a remarkably plain bench, quite high-set and well angled but, unexpectedly, lacking a centre armrest. There is noticeably less headroom in the back than there is in the front, where it is better than average. The Fiat's back seat also lacks a centre armrest, but this matters less because the seats themselves are sculptured into two individual places. They are low-set but well angled, with sufficient space for small passengers to make themselves comfortable. They, like those in front, have a very good view; in fact, both cars are notable for their generous area of glass.

Both cars place a speedometer and rev counter in front of the driver, but while the BMW otherwise has only a single combination dial with fuel contents, water temperature and warning lights, the Fiat has four separate dials placed towards the centre of the car. The Italian car has a profusion of controls, while the BMW manages with just four knobs and two column-mounted stalks. While the BMW has two-speed wipers, the Fiat has a single continuous speed with a rheostat variation, plus its very useful intermittent-wipe facility. Both cars have their washers electrically operated; the BMW still has its wipers set up for left hand drive. Where headlamps are concerned, the Fiat with its four-lamp tungsten-halogen system is much the better equipped, although the single units on the BMW are surprisingly good.

Both cars have water-valve heaters, with the BMW system being the easier to control. The Fiat (whose heater controls are on the centre console, aft of the handbrake) tends to produce either hot or cold air, but not warm. The Fiat has much the better ventilation system, with two butterfly-type inlets at the ends of the facia and two others in the centre console. Its air extraction is not very good, however, unless one or both rear windows are opened.

The BMW has a tremendous advantage in boot space, offering more length and, very important, quite a lot more height in its luggage compartment. It also has a very large, wide-opening boot lid, in contrast with the Fiat's restricted opening. Both cars have a fairly high sill over which luggage must be lifted, and both bury the spare wheel and tool kit (quite a good kit in both cases, incidentally) beneath the boot floor.

BMW 2002
Price £1,874
(total including purchase tax)

Maximum speeds

rpm	mph	
5,800	107	Top
6,500	89	3rd
6,500	59	2nd
6,500	31	1st

Acceleration

Ind. mph	sec	mph
33	3.2	0-30
44	5.1	0-40
54	7.3	0-50
64	10.1	0-60
75	14.5	0-70
85	19.5	0-80
96	27.6	0-90
107	43.2	0-100

17.8sec 77mph Standing ¼-mile

Top (3.64)	3rd (4.89)	mph
—	4.1	10-30
7.7	4.3	20-40
7.5	5.0	30-50
8.2	5.5	40-60
9.2	5.9	50-70
10.2	7.0	60-80
12.7	—	70-90
21.2	—	80-100

25.5	Overall mpg
28	Typical mpg
Negligible	Oil consumption

SPECIFICATION
FRONT ENGINE, REAR DRIVE

ENGINE
Cylinders . . . 4, in line
Main bearings . 5
Cooling system . Water; pump, fan and thermostat
Bore 89.0mm (3.50in.)
Stroke 80.0mm (3.15in.)
Displacement . 1,990 c.c. (121.4 cu. in.)
Valve gear . . Single overhead camshaft and rockers
Compression ratio 8.5-to-1. Min. octane rating: 97RM
Carburettor . . Solex 40 PDSI
Fuel pump . . Solex mechanical
Oil filter . . . Full flow, renewable element
Max. power . 100 bhp (DIN) at 5,500 rpm
Max. torque . 116 lb. ft. (DIN) at 3,000 rpm

TRANSMISSION
Clutch . . . . Fichtel and Sachs diaphragm spring, 7.9 in. dia.
Gearbox . . . Four-speed, all-synchromesh
Gear ratios . . Top 1.0; Third 1.34; Second 2.05; First 3.83; Reverse 4.18
Final drive . . Hypoid bevel, 3.64-to-1

SUSPENSION
Front . . . . Independent; MacPherson struts, lower wishbones, coil springs, telescopic dampers, anti-roll bar
Rear . . . . . Independent; semi-trailing arms, coil springs, telescopic dampers, anti-roll bar

STEERING
Type . . . . . ZF-Gemmer worm and roller
Wheel dia. . . 15.5in.

BRAKES
Make and type ATE disc front, drum rear
Servo . . . . ATE vacuum
Dimensions . F. 9.5in.; R. 9.0in. dia., 1.57in. wide shoes
Swept area . Total 299 sq. in. (260 sq. in./ton laden)

WHEELS
Type . . . . . Pressed steel, 4.5in. rim, 4-stud fixing
Tyres—make . Dunlop
—type . . SP57 radial-ply tubeless
—size . . 165-13in.

BMW: The front seats are large but not very positively shaped, but there is lots of room for a large driver; the back seat is a plain bench with no dividing armrest but with recessed elbow niches at each side; the boot opening rises complete with the tops of the rear wings. Luggage space is large for the size of car but loading must be done over a high sill

Fiat 124 Coupé 1600
Price £1,797
(total including purchase tax)

Maximum speeds

	mph	rpm
Top	109	6,200
4th	93	6,400
3rd	68	6,400
2nd	44	6,400
1st	26	6,400

Acceleration

mph	sec	Ind. mph
0-30	3.5	31
0-40	5.3	41
0-50	7.4	51
0-60	10.7	61
0-70	14.4	72
0-80	19.4	82
0-90	27.0	93
0-100	43.1	105

Standing ¼-mile 78mph 17.8sec

mph	3rd (5.85)	4th (4.30)	Top (3.79)
10-30	8.4	—	—
20-40	6.3	10.1	13.6
30-50	5.7	8.9	12.1
40-60	6.2	9.0	11.4
50-70	6.8	9.4	12.3
60-80	—	10.0	14.6
70-90	—	10.6	19.3
80-100	—	—	29.1

Overall mpg	23.6
Typical mpg	26
Oil consumption	negligible

SPECIFICATION
FRONT ENGINE, REAR DRIVE

ENGINE
- Cylinders: 4, in line
- Main bearings: 5
- Cooling system: Water; pump, electric fan and thermostat
- Bore: 80.0mm (3.15in.)
- Stroke: 80.0mm (3.15in.)
- Displacement: 1,608 c.c. (97.5 cu. in.)
- Valve gear: Twin overhead camshafts, direct-acting
- Compression ratio: 9.9-to-1. Min. octane rating: 98RM
- Carburettors: 2 Weber 40DIF 10/11
- Fuel pump: Fispa mechanical
- Oil filter: Fram full flow
- Max. power: 110 bhp (net) at 6,400 rpm
- Max. torque: 101 lb. ft. (net) at 3,800 rpm

TRANSMISSION
- Clutch: Diaphragm-spring, 7.87 in. dia.
- Gearbox: Five-speed, all-synchromesh
- Gear ratios: Top 0.83; Fourth 1.0; Third 1.36; Second 2.10; First 3.67; Reverse 3.53
- Final drive: Hypoid bevel, 4.3-to-1

SUSPENSION
- Front: Independent; double wishbones, coil springs, telescopic dampers, anti-roll bar
- Rear: Live axle, trailing arms, Panhard rod, coil springs, telescopic dampers, anti-roll bar

STEERING
- Type: Worm and roller
- Wheel dia.: 15.0in.

BRAKES
- Make and type: Fiat-Bendix disc front and rear
- Servo: Bonaldi vacuum
- Dimensions: F. 8.95in. dia.; R. 8.95in. dia.
- Swept area: Total 270 sq. in. (227 sq. in./ton laden)

WHEELS
- Type: Pressed steel, 5.0in. rim, 4-stud fixing
- Tyres—make: Pirelli
- —type: Cinturato CN53 radial-ply tubeless
- —size: 165-13in.

Personal choice

I can't recall any pair of cars which caused more agony and heart-searching than these two when it came to choosing the better. Our performance figures show them to be incredibly well matched against the stopwatch; the similarity extends to most other aspects of driving as well.

To start with the BMW I find it fits me very well, and I have a great deal of admiration for its tremendous flexibility, its supple ride and its relative quietness, although this is less so when full throttle is used. Its very large boot is also a plus point.

The Fiat offers me a far less comfortable driving position, so it starts off with a deficit. But I can't deny that it has better steering and probably better handling, too, although one eventually comes to terms with the BMW's slight sloppiness and realizes that one is talking of very small margins. The difference is, I think, that the average driver would be able to step into the Fiat and drive it fast straight away, while he would have to spend a bit of time getting used to the BMW.

I admire Fiat's doing the decent thing and providing a five-speed gearbox to go with their rather peaky engine, but I think the ratios are too squashed down towards the bottom end. The BMW's four much wider ratios work very well in practice.

In neither car am I altogether happy with the brakes. Both might give you a few worrying moments at the end of a fast downhill drive, with the Fiat worrying me more because when its brakes fade, they fade more suddenly.

In general, the Fiat has much more of that taut feel which a truly sporting car should have. This is true of the engine, the gearchange, and the steering. At the same time, it is an eminently practical car with a number of clever or helpful features. The fact that it has been given headlamps to match its performance is just one instance.

Even so, on balance, I will take the BMW. Perhaps I admire the two-facedness of the German car. It is easier to drive smoothly and quite fast, yet it takes *more* skill if it is to be driven right up to the limit. In this way, it suits both my driving moods: the relaxed (which is most of the time) and the keen. It is not as pretty a car, but then I can't see the car once I am inside it, so I don't really care.

The thing which really tips the balance for me is the driving position. It is a measure of how closely these two cars are matched that, were I *under* five feet ten (especially four inches under, instead of over as is actually the case), I would plump for the Fiat. I feel sorry for you, seventy-inchers; for you, the choice is one which you are bound to regret—half the time!

Jeffrey Daniels

Personal Opinion

With the strong rider that I'd require distinctly near-adequate braking equipment on both cars, this is a nearly impossible choice. I prefer the Fiat's looks—but the BMW is just as pretty. I prefer the Fiat's gearchange—but the BMW's, once you've bought a decent gearlever knob, is just as delightful. I thoroughly enjoy the very Italian and sporting character of the Fiat's engine—and I revel in the BMW's great spread of good torque. I prefer the Fiat's steering and its straight stability—and I prefer the BMW's handling and its relative lack of understeer. The Fiat makes a pleasant noise, but the BMW's eager yet quieter hum is just as appealing. (Here one must stop to apologise about all these I's; any opinion, for what it is worth, is personal; these two cars arouse strong personal feelings.)

As my colleague Jeffrey Daniels says, one could probably adapt the driving position of the Italian Standard Apeman's Fiat to suit the not-necessarily-so-standard Lesser Armed Longshanked British Ape, with the aid of dished steering wheels and some attention to the seat runners. Even so, I think my frequently lazy nature and stolid British faint distrust of high Italian revolutions over a long period of office would settle for the torquey 2002; not that there is much stolid about that little motor car. But instead of another steering wheel and longer seat runners, I'd have to buy another seat, an oil pressure gauge and an ammeter; assuming that they'd already done something about the brakes.

Michael Scarlett

Fiat: The boot is relatively small and shallow, with a restricted opening and a high sill. The back seats are formed into two individual spaces: there is little width for three in any case; there is no centre armrest but there are rests at the sides. The front seats are very deeply shaped, but their relationship with wheel and pedals is not right for tall drivers

TRIUMPH 2500PI v BMW 2002 v ROVER 2000TC

GIANT TEST

EVERY SERIOUS MOTORing observer knows how Munich's BMW concern came close to disaster and of its subsequent spectacular rise from the ashes over the last few years, just as he is equally familiar with the more restrained return from senility of Rover with its highly successful 2000 series. And he will be aware that Triumph is also making some pretty worthwhile cars nowadays after plugging on with exceedingly boring ones in the past.

These three names compete with one another in the lucrative executive-saloon area of the British market; indeed they have been actively exploiting this field. Their approaches in terms of design are similar but far from identical, although it is very evident just how cut-throat the market place can be when prices are mentioned.

And in the cases of the respective high-performance versions of the base models the infighting is closer still. Rover's 2000TC and Triumph's 2.5PI are identical in price at £1867, despite a £132 difference (in the Triumph's favour) when you compare the basic 2000s from which they are derived. Ranged against them—and in many ways more than their match—BMW has its 2002 priced at just £7 above the British competition at £1874. It may not convey the air of luxury that the Rover and Triumph contrive to exude but it more than compensates in the performance and handling stakes—which is the other half of the appeal of these so-called executive saloons.

Not that the BMW owes its success to being a very much later design: all three cars can trace their design history back a decade or so ago at the very least.

The Rover and the Triumph first appeared in 1963 and set the pattern for well appointed two-litre saloons. Since then the Rover has remained virtually unchanged, the TC (for twin carburettor) coming along to counter valid criticism that the SC was, and is, under powered; the only alteration of note came last autumn when the factory saw fit to tamper with one of the best facias in the business and went on to spoil the front grille and bonnet line for an encore.

The Triumph on the other hand has led a less sheltered existence. It has been through one partial restyling and now comes with an optional, enlarged, fuel injection version of the standard engine; with this its name changes to 2.5 (for litres) and PI (for petrol injection). Basically, though, and especially where the suspension, steering and brakes are concerned, it is little different from the original 2000s.

The BMW reached the British market in 1968, having been available on left-hand drive markets for some time before that. It may seem a much newer car, but in fact is an amalgam of rather older components. The single overhead cam engine, for instance, was a foundation stone of BMW's recovery while the body shell is that of the 1600 saloon, introduced to keep the firm going in that end of the market while it made hay with the 2000 and laid its plans for the 2500/2800. The 1600 used running gear similar to the 2000's and possessed handling qualities far in excess of its performance potential. It cried out for the extra muscle of the 2000 engine and, when the benefit came, turned into a car that topped even the most optimistic expectations. The 2002 is a driver's car in the true sense of the cliché, able to put up cross-country averages that it takes a good (and considerably more expensive) sports/GT car to beat.

STYLE AND ENGINEERING

Although starting to look dated (or is it just overfamiliar?) the Rover remains a prime example of just what can be achieved by the British motor industry's own stylists. The Triumph was originally Michelotti-designed so naturally enough, perhaps, BLMC turned to him when a facelift was wanted for the autumn motor show season of 1969. With an eye to the soaring costs of new body dies, he revamped the ends of the shell while leaving the middle largely untouched. The Triumph gains visually from its generous external dimensions whereas the Rover is similar in width and height—though a few inches shorter—and impresses as a well proportioned design. The BMW, conversely, does not. In width and height it is little different from the British cars, but length is a good deal less so that the overall impression is slightly ungainly. The effect is heightened by the large area of glass and the correspondingly disproportionate gap between waist and roof. Compared to the 2000s, the 2500/2800 range and the coupé shell, the 1600/2002 is very much the ugly duckling of the BMW family; albeit, a very popular ugly duckling.

But what it lacks in looks it more than makes up for in engineering. The chunky-looking four-cylinder, five-main bearing engine is the mechanical basis of all BMWs. A crossflow head carries a single chain-driven camshaft operating inclined valves through two rows of rockers. The valves open into a combustion chamber of modified hemispherical layout. Carburation is by a single Solex. Bore and stroke are oversquare at 89 by 80mm to give 1990cc.

The Rover's power unit is similar; it is a single overhead cam with five bearings, the valves are in line and bore and stroke are identical at 85.7mm to give 1978cc. Compression ratio is a highish 10 to one, and carburation is by a pair of those trusty SUs.

By comparison the Triumph's engine is unsophisticated almost to the point of vulgarity, for it is a four-bearing six with pushrod-operated valves. Designed originally for the last of the Standard Vanguards and used now for the GT6, Vitesse and 2000, it has a longer stroke in its 2.5 application, taking the measurements out to 74.7 by 95mm and capacity to 2498cc. Its saving grace lies in the use of the tried, efficient Lucas mechanical fuel injection.

Combined with an extra 500cc above the stock 2000's capacity and better breathing, the 2.5PI gives a near-50percent boost to power, taking it up to 132bhp at 5450rpm. The Rover in TC form provides a relatively smaller gain over its parent 2000. Dual carburettors and generally improved breathing raise peak power by a little over 25percent to 114bhp at 5500rpm. The BMW engine is the cooking one and produces what appears to be a modest 100bhp at 5500rpm.

If the BMW appears down on power remember that West Germany has stringently-applied trade description laws and that, judging from the results of our maximum-speed tests, some of those British horses tend to lose their way *en route* to the back wheels. They should not get diverted in the transmission, though, for all three cars come as standard with straightforward four-speed manuals. Overdrive is an extra cost option on the Triumph but not on the other two. Automatic gearboxes are available for the Triumph and BMW (the latter at a staggering £199 extra) but not the Rover.

Coil spring suspension is common to all three and the BMW and Triumph also share MacPherson strut front ends and semi-trailing arms at the rear. The Rover diverges completely, not only from its two rivals but also from practically anything else on the market. In its suspension design it is clearly the work of some remarkably level-headed engineers, untrammelled by many preconceived ideas or by the need to use existing components. This clean sheet of paper approach resulted in something approximating wishbone geometry for the front end but with the springs operating through longitudinal top links so that they could be placed horizontally and longitudinally, feeding back loads direct into the massive structure of the scuttle bulkhead. For good measure the links are angled to reduce nose dive under braking. At the back, convention is left even further behind with a de Dion axle that telescopes to allow for track variations during roll. In turn these are induced by the use of solid non-splined half-shafts, the principle being that minor track changes are preferable to the presence of splines which can bind under acceleration to lock the suspension solid, as well as being a potential source of transmission clonk. The de Dion tube is located fore and aft by Watt's linkages.

Only the Rover takes advantage of the chassis-mounted differential to fit inboard disc rear brakes. Unusually, they are larger—in diameter, but slimmer—than the front ones. Triumph and BMW stick to outboard drums at the back with discs for the front.

Power assistance is an optional extra for the Triumph's rack-and-pinion steering that is not available on the worm and roller systems of the Rover and BMW. The Triumph and BMW share normal steel unitary construction body shells, the former with four doors and the latter with two. The four-door Rover, diverging again, is welded up in the form of a skeletal steel structure to which all the mechanical components are attached before being clothed with an outer skin that forms the visible body. The result is a car that is less susceptible to major structural damage in minor accidents and is likely to incur a smaller bill at the repair shop. The bonnet and boot lid are aluminium.

USE OF SPACE

The BMW makes the best use of space available, if only because it is a substantially shorter car than either of its rivals yet can still carry four adults in comfort plus a reasonable amount of luggage. The British importer puts a *Coupé*

Instruments: 1 Speedo **2** Odometer **3** Trip Counter **4** Tacho **5** Fuel **6** Water Temperature **7** Oil Pressure **9** Clock **10** Ammeter **Warnings: 11** Ignition **12** Main Beam **13** Oil Pressure **14** Indicators **17** Handbrake **18** Fuel Low **19** Choke Control **20** Rear Window Demister **21** Hazard **Special Items: A** Fresh Air Vent **B** Ashtray **D** Radio **Controls: 26** Choke **27** Ign/Start **28** Indicators **29** Lights **30** Dip **31** Flash **32** Horn **33** Panel Lights **34** Parking Lights **35** Wipers **36** Washers **37** Heater **38** Fresh Air Vent/Control **39** Heater Blower **40** Cigar Lighter **41** Hazard Warning **42** Map Light Control **43** Interior Light Control **44** Rear Window Demister **46** Fog & Spot **49** Overdrive **50** Stereo Control **51** Trip Resetting **52** Clock Resetting **53** Handbrake **54** Gearlever **55** Bonnet Catch **56** Fuel Reserve

tag on to its title but, in fact the 2002 is no more or less than a compact saloon with two doors. Legroom is better than you might expect in a car less than 14ft long, and headroom is as generous as the inordinately high roofline leads you to anticipate. Sufficient back seat elbow room is achieved by hollowing out the panels above the armrests. The Rover is also a four-seater, but goes a step further than the German car by being an unequivocal one. The original 2000 was the first saloon of its kind to recognise that most drivers seldom have more than two or three passengers and only very rarely want to carry more than three. So the back is designed as two individual seats. The Triumph has a more normal back seat, suitable for two and a half adults. Back seat legroom is usually the first dimension to suffer when it comes to squashing people, luggage and engine into a given length. In the BMW an acceptable amount of this commodity has been retained, and naturally it abounds in the lengthy Triumph. In the Rover rear passengers feel more cramped than simple dimensions suggest unless the front seats are well forward.

Despite its overall size the Triumph seems rather lacking in boot space. The Rover gives you less room, even with the extra-cost kit that mounts the spare wheel outside the boot lid, and it does away with the rear sill that hampers loading operations in the Triumph. The BMW is best of all, though, if you can forgive its high sill. The huge lid lifts up to expose a gratifyingly large cavity —bigger than either of its rivals' —and more roomy than in most cars of similar overall size. All three allegedly have space-saving rear-suspension systems, but only the BMW takes full advantage of the possibilities.

Down at the other end it does a similarly effective job of fitting a tall and bulky engine into the space available without making things look too cramped. The engine is canted sharply to the right to leave space for carburettor and air cleaner and at the same time provide a smoothly curved induction tract. The Rover has more than enough space for its compact power unit. The unconventional location of the spring damper units leaves surplus width that goes to waste in the 2000TC (but is very welcome for the V8 installation in the 3500 edition). The Triumph needs all of its long bonnet for its slender six and width, too, is in demand for the fuel injection equipment. Instead of the neat double-carburettor set-up of the 2000 the 2.5 has six inlet stubs curving out from the head and into a large air box running down the right-hand side of the engine.

COMFORT AND SAFETY

There is little about the BMW's interior to give away the sporting side of its character. The reclining front seats, for instance, provide the acceptable minimum of lateral grip and in the Rover, too, the temptation to boy-racerise the cockpit has been resisted. The Rover seats do, however, offer more sideways support than the BMW's. And, as in the BMW, they are high mounted, making the already excellent all round visibility seem even better. The Triumph seats you lower and with less lateral grip than the Rover. In all three the seats give sufficient support over a wide area of the body to prevent aches and pains on a long day's drive. We especially liked the Rover's ultra-simple friction-clutch method of reclining the back rests although we were less enthusiastic about its slippery leather upholstery.

Each one of the trio scored highly in matters of the layout of controls. The Rover has an unfashionably large diameter steering wheel but compensates with rake adjustment controlled by a knob down on the column. The gear lever, with lift-up collar guarding reverse, is a splendidly short, rigid lever with the handbrake right next to it, while the pedals are nicely placed for heel and toeing. Pedal positioning is also satisfactory on the BMW now that some attention has been paid to it by the engineers. Mounted high and well away from the facia, we found the steering wheel odd in its location but the positions of gear and handbrake levers are beyond criticism. The BMW's control layout suits tall men best, while that of the Rover is most comfortable for those of average height and below. The Triumph manages to satisfy the entire range of stature, aided, like the Rover, by a steering wheel adjustable for rake and secured by a quick release clamp. Last autumn's alterations to the Rover's facia have done little to improve it. The plain round speedo and tachometer are clearer than the ribbon speedo and separate rev counter of yore, perhaps, but the quickly memorised, differently shaped switches have been replaced by identical ones placed in a row and therefore confusable in the dark. One area where improvement would have been welcome is the steering column stalk switches; they control so many functions that it's too easy to get a toot when you wanted to dip... The Triumph still has the minor controls about which we enthused on the 2000 Mark Two, though again most drivers would welcome a reduction in the number of column-mounted controls. Stalks abound also on the BMW—well, there are two of them—but the other minor operations have been simplified down to a bare minimum of knobs and switches. Instrumentation comprises a matched speedo and tachometer, plus another dial for a modicum of information about the other essential and not so essential services.

All three cars were commendably free of wind-excited noise, at least up to 80mph or so, though the BMW in particular tails off quickly above this figure. The Rover and BMW were prone to a limited but nonetheless occasionally audible degree of tyre thump, once a prominent bugbear of the Rover. Engine noise levels, which were inordinately high in earlier 2000TCs, have now been reduced to an acceptable level although, as in the BMW, the power unit still produces a hard, thrumming roar when on full throttle or near the top of the rev range. The Triumph is the quietest and smoothest of the three. It is also at least a match for its competitors in matters of heating and ventilation. Its systems are comprehensive and versatile, as well as being quite finely adjustable. The Rover is nearly as good, being spoiled by feeble face-level fresh air outlets. The BMW lags behind with its rather indifferent provision for ventilation.

In spite of having fractionally firmer suspension the Rover still manages to give its occupants a ride as comfortable as that of the softly sprung Triumph and does so with much less lean on corners. The BMW is even less roll prone but its ride is that much harder.

Safety has always been a strong point of the Rover 2000 range although in such matters as collapsibility of engine and luggage compartments, strength of passenger compartment, roof and screen pillars, it is today matched by an increasing number of its rivals. Even so, there is much evidence of intelligent attention to detail, as in the padded leg-protecting shin-bins and in the small (but misleadingly wide angle) rear view mirror. Yet in such matters as the use of crushable material around the cockpit, recessed or deformable protrusions, telescopic steering columns and so on a mixture of commonsense, experience, and the US safety regulations have let other manufacturers catch up.

PERFORMANCE, HANDLING, BRAKES

All three cars started easily from cold on their manual chokes. The Triumph offers dire warnings on windscreen stickers about continuing to run the engine if it persists in misfiring after firing up (ours didn't) and about leaving the ignition on with the engine not running (fuel may be pumped through an open injector to collect in the head and seep explosively down to the crankcase). The Rover, incidentally, matches its manual choke with a pull-out knob for a reserve fuel supply. The BMW also has a fuel reserve. Both the Rover and the BMW picked up and pulled strongly right away, but the Triumph was unduly sensitive to rich-mixture settings. Once warmed up, though, the 2.5PI ran with smoothness and flexibility and gave instantaneous throttle response in a way that no carburettor engine can match.

The heron-headed Rover felt harsh and rough when working hard. Partly as a result of a none-too-flat torque curve and partly because of high gearing it proved far from flexible, being at its happiest when kept on the boil with lots of gearchanging. The change action was nicer than the notchy one on the Triumph if not as slick and positive as the BMW's. As well as being a poor puller in the high gears the Rover emphasised its unwillingness to slog by being prone to transmission snatch if driven sloppily.

The BMW proved to be in a different class altogether. Its excellent engine has a generously spread range of really meaty torque and will, if you insist, make short work of accelerating in third and top from ridiculously low speeds. It performs just as well at the other end; the peak of the power curve falls away slowly and the test car was still breathing freely as it rushed into the red at 6500rpm.

Considering this, we were surprised that the acceleration times of the much lighter BMW did not look even better when compared to its rivals'. Still, from 60mph upwards it moves away quickly and it is still pulling hard as it nears its maximum whereas the British pair are labouring at anything much above 100 and take what feels like infinity to struggle up the last few mph. The BMW, thanks to better gearing and aerodynamics, you might presume, marches straight up there—in spite of having anything up to 32percent less power to propel a roughly similar front area. The generally superior efficiency of the BMW is underlined by fuel consumption figures—a little better than the Rover (which demands five star petrol) and well ahead

	BMW	ROVER	TRIUMPH
PRICES	At £1874 an inexpensive car by BMW standards—only the 1800 and 1600 cost less at £1797 and £1648 respectively. Next up in price comes the 2000, costing £325 more at £2199. So the 2002 falls neatly between the 1600 from which it draws its body, and the 2000 whence comes the engine	At £1867 the TC is £109 more than a standard 2000. The extra cost is accounted for almost entirely by the engine. The V8-engined 3500 is only £283 more than the 2000TC and this includes around £130-worth of standard-equipment automatic transmission	An exact match for the Rover at £1867, making the pair only £7 less than the BMW. The 2·5 Triumph is £241 more than the 2000 saloon. And, as with the Rover, most of the additional cost goes on the engine—in this case 25percent larger in capacity and equipped with fuel injection
ACCELERATION from standstill in seconds			
FUEL	**24** mpg overall ★★★★ 28mpg driven carefully 220–265 miles range 10 gallons capacity	**22** mpg overall ★★★★★ 26mpg driven carefully 250–285 miles range 12 gallons capacity	**20** mpg overall ★★★★ 23mpg driven carefully 260–315 miles range 14 gallons capacity
SPEEDS IN GEARS	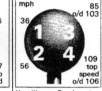		
HANDLING	Very high cornering power on Continental radials. Mild understeer easily converted to near neutral stance by judicious use of throttle and eventually changing suddenly to oversteer. Steering light and responsive but with some kickback. Basic characteristics unchanged in wet	Roadholding good—though not in the BMW's class—on Pirelli Cinturato radials. Rather strong understeer emphasised by excellent rear end adhesion, so that the tail only breaks away under extreme provocation. Overall handling excellent with an innate controllability that is retained in the wet	Handling on Goodyear G800 radials stable with initial strong understeer changing progressively to oversteer on dry and wet roads alike. Cornering power comparatively low and handicapped by excessive body roll due to soft suspension. Manual steering satisfactory, as is the power-assisted version
LUGGAGE CAPACITY cubic feet	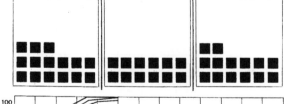		
BRAKES / FADE	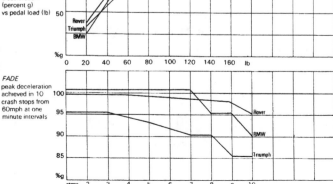		

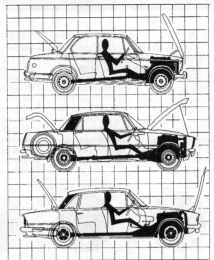

of the heavier, bigger-engined and less efficient Triumph's.

The German car tends to lead in matters of handling. Its cornering power, measured in terms of lateral acceleration, is ahead of the TC. Rover, remember, set new standards in its class when first introduced and remains a first class roadholder today. The Triumph falls a long way behind in this area, as it does in handling. The three run the gamut of understeer from the almost exaggerated behaviour of the Triumph through the still strong characteristic of the Rover to the mild understeer (easily throttle-adjusted to neutral, or even oversteer) of the BMW. All three change, as the limit is approached, to oversteer, the BMW doing so quite quickly, the others less so. Steering on the Triumph is unexceptional. The Rover's steering is light and well damped but somehow lacking in the finer degrees of road feel. We preferred the BMW's which is light, into the bargain—certainly a good deal lighter than we remember from early 2002s. Overall, the BMW has by far the best handling. The Rover is acceptably stable (unlike the BMW, which can be unsettled by a crosswind), very manageable and forgiving, but it does not present the standing invitation to fast driving that comes from the BMW. The Triumph clings to the road better than its wallowing body roll leads you to expect but its size and generally soft nature combine with soft, under-damped springing to dissuade most drivers from anything other than a staid style.

In braking tests the Rover comes out on top, being less prone to fade than the others. The Triumph was more progressive still, but it could not achieve a 1g stop during our trials. The BMW managed 1g stops at first, then began to fade rapidly when just over halfway through the test. The brakes soon recovered themselves but disturbed us by departing and reappearing very unevenly. Their action was also far from progressive, as the chart shows.

IN CONCLUSION

Cars like this trio are taking an ever increasing share of the market and after prolonged experience of them it is not difficult to see why. Even the least agile of them can match the majority of similarly priced sports cars in many respects, while a good one like the BMW will leave them standing. At the same time they provide high standards of comfort, carrying capacity, sobriety of appearance and reasonably low prices.

The most successful of these sports saloons are those like the Rover, which were designed for the job, or the BMW, which sprang from a particularly well informed, sporting-minded design team. Those like the Triumph, developed from simpler beginnings, show up less well when pushed hard.

As enthusiastic drivers we preferred the BMW but we can see that a motorist who puts more store by sumptuous looks, both inside and out, will prefer the Triumph. At best, the BMW is certainly an unexciting car with nothing to give away the other side of its character. The Rover fits neatly between the two in this as it does in most things.

We would consider the Triumph for its comfort, quietness and easy performance. Then again, the Rover matches it on most of these points and gives you greatly preferable handling and roadholding. Its peaky engine can be tiresome after a while, though.

So it really comes back to that unobtrusive, effortless, if rather plain BMW 2002. ●

DIMENSIONS:	BMW	ROVER	TRIUMPH
wheelbase	98·4	103	106
front track	52·4	53	52·5
rear track	52·4	52·5	52·8
length	166·5	180	182·3
width	62·6	66	65
height	55·5	55·25	56
ground clearance	5·5	5·5	6
front headroom	36	35	35
front legroom	26	31	25
rear headroom	34	34	35
rear legroom	8	8	9
ENGINE			
material	alloy/iron	alloy/iron	iron/iron
bearings	5	5	4
cooling	water	water	water
valve-gear	sohc	sohc	pushrod ohv
carburettors	1 Solex downdraught	2 SU variable choke	Lucas fuel injection
capacity cc	1990	1978	2498
bore mm	89	85·7	74·7
stroke mm	80	85·7	95
compression to 1	8·5	9	9·5
bhp	100	114	132
rpm	5500	5500	5450
torque lb ft	115·7	126	153
rpm	3000	3500	2000
TRANSMISSION			
control	floor lever	floor lever	floor lever
synchromesh	1-2-3-4	1-2-3-4	1-2-3-4
ratios to 1 1st	3·83	3·62	3·28
2nd	2·05	2·13	2·10
3rd	1·34	1·39	1·39
o/d 3rd	—	—	1·14
4th	1·00	1·00	1·00
o/d 4th	—	—	0·82
final drive ratio	3·64	3·54	3·45
final drive ratio (with overdrive)			
tyre size	165 × 13	165 × 14	185 × 13
rim size	4·5	5	5
SUSPENSION			
front	Macpherson strut, coil spring and telescopic damper	double links, coil spring and telescopic damper	Macpherson strut, coil spring and telescopic damper
rear	semi-trailing arm, coil spring and telescopic damper	de Dion tube, Watt's linkage, coil spring and telescopic damper	semi-trailing arm, coil spring and telescopic damper
LUBRICANT			
engine oil type SAE	10W/30	20W/50	20W/50
sump capacity, pints	7	9	8
change miles	4000	5000	6000
other lube points	1		8
lube intervals	4000	5000	6000
AIR	26psi	26psi	26psi
STEERING	worm and roller	worm and roller	rack and pinion
BRAKES	disc 9·55in	disc 10·31in	disc 9·75in
STEERING	31·5ft turning circle, ratio to 1: 17·58	31·5ft turning circle, ratio to 1: 20·3	34ft turning circle, ratio to 1: 18
AIR	26psi	28psi	26psi
BRAKES	drum 9·06in	disc 10·69in	drum 9·0in
WEIGHT	2072lb	2787lb	2632lb

BMW 2002

Better choose a color you can live with because you'll want to keep this one a long time!

The 2002 rates at the top of its class in handling capabilities and as its racing record proves, it may be considered a true sports car.

"The engineers who develop BMW cars are enthusiasts too." This simple statement in the front of the owners manual certainly sets the tone for what follows in the most comprehensive 90-page booklet we've ever seen. It is further amplified after the briefest acquaintance with any BMW automobile.

The *Bayerische Motoren Werke* is as old as the automobile engine itself and is a company that excels in the production of blue-blooded automobiles. Their 2800 series cars are among the most sought-after forms of personal transportation in Europe. Each one offers its owner a satisfying means of covering all sorts of terrain at very high speeds, and with great safety and comfort. Needless to say, they are not cheap automobiles, but worth every penny of their $7,000 plus price tag. Fortunately, BMW also build smaller cars with most of the features of their larger brethren at a price you can take.

The BMW 2002 is not a large car by any standards, but with its 166.5-inch overall length, it still offers ample room for four full-size adults in excellent comfort with enough rear seat space to squeeze in an extra body if need be. With the ever increasing demand for maximum utilization of road space, BMW are certainly doing their share by producing small cars with roomy interiors—effectively putting a "quart into a pint pot." The 2002 is a spacious family sedan with the performance and handling of a fast touring-type machine.

This model is only available in two-door form, but the large door makes entry simple for the average person. Both seats tilt forward for entry to the rear and the seat backs are released by the small lever located near the top of the seat. No stooping or fumbling for the release near the floor is needed, as on most two doors.

Once seated, the most impressive and immediate feeling one has is the excellent visibility. The roof pillars must be the slimmest we've seen in a long time though they are evidently extremely strong, forming as they do an integral part of the roll-cage passenger compartment. At eye level, there must be at least 95% glass.

The instrument binnacle is both simple and functional. In the center is the speedometer with trip odometer. To the right is a tachometer (a mandatory option on all 2002s), and this is matched to the left with a dial incorporating the fuel and temperature gauges plus the four warning lights for oil pressure, alternator charging, hi-lo beam and turn and flasher signal indicators.

Two knobs at each end of the binnacle are for lights and heater blower on the left and cigarette lighter and wipers on the right. The instrument lights are rheostatically controlled from the light switch for a wide range of intensity to satisfy most night driving needs.

Below the binnacle and on either side of the steering column are the heater and fresh air controls. These can be operated by touch and the system will produce enough air to suit polar bears. Air-conditioning, incidentally, is not a factory option. Defrosting, demisting and the doughty heater are all within easy reach. At the extreme ends of the dashboard are two small ducts that blow air directly on the side windows for demisting purposes.

Two stalks control a multitude of services. The left-hand one operates the hi-lo beam control and the headlight flasher. The right one is the turn signal plus windshield washer and a few sweeps of the blades when pulled toward the driver. A little practice is required in avoiding this

Though BMW's general styling theme is nearly a decade old, the lines remain contemporary because they are functional. Despite flow-through ventilation, vent panes are retained.

Instrumentation lacks gauges for oil pressure and alternator, but the 2002 is one of the few cars where all controls are centered in front of the driver.

control as the first turn one makes with the car can prove mildly traumatic if the wipers and washers start smearing the windshield at the same time.

Another function of the turn signal switch is redundant in this country. In Euope it's a requirement for the off-side parking lights to be lit when a vehicle is parked at night in the street. The U.S. 2002 retains this feature which only operates when the ignition key is removed and the lever is left in either left or right position.

A horn button in each spoke of the leather covered steering wheel completes *all* the controls and instrumentation. It's been quite a while since we've tested a car that puts *everything* in front of the driver, made it easy to see, and arranged it so you can operate all the controls without having to unfasten the safety harness. All driving controls are operated without the hands having to let go of the steering wheel. The feet are for clutch, brake and accelerator only, with no stabbing around to find a floor dip-switch.

A large open tray extends to the right of the instrument binnacle allowing plenty of room for storage of small items. To ensure that nothing slides around, there are small retaining ridges that prevent loose items from piling up in a heap during hard cornering. BMW 2002 power comes from a hot little four cylinder in-line overhead camshaft engine with Hemihead combustion chambers, a 1,990cc (121.4 cu. in.) cast iron block and an alloy head. Its power output is a healthy 113 hp at 5,800 rpm. The tachometer is redlined at 6,000 rpm with a safe maximum of 6,200 rpm. The engine is so well balanced that the power comes on with much the smoothness of a rotary or a turbine. Take it easy in 1st as the needle gets round that dial awfully fast. Maximum torque is at 3,000 rpm (115.7 lbs.-ft.) but the curve is practically flat for a good thousand or so rpms on either side of the peak point. The torque drops off towards the 5,500 rpm mark and best shifting for fast times was usually between 5,000 and 5,500 rpm.

The compression ratio is a mild 8.5 to one but this still requires the use of premium gas to satisfy emission modifications. The engine gave a solid 23.4 mpg average under very hard driving during the test and cruising at freeway speeds brought 25 mpg. The car is so much fun to drive up through the gears that most owners will probably go for their jollies by using engine and gears rather than be miserly with the throttle. This car is bought for driving. With that in mind, it is a relatively simple matter to have the engine tweaked to produce a potent ma-

provides the 2002 with very good adhesion and handling. Front suspension is by MacPherson strut with lower wishbones, the spring columns incorporating double-acting hydraulic shock absorbers, coil springs and auxiliary rubber springs. Wheel travel is a healthy seven inches, which is ideal on rough roads. Rear suspension is by trailing arms located by a delta-shaped box section support beam which also locates the final drive assembly. Anti-sway bars front and rear complete the package.

Steering is ZF worm and roller with a ratio of 17.58 to one. Add a set of Michelin XAS radials and the whole becomes a handling machine that will not have many equals on any road and fewer peers, at any price. From the radial tires to the leather-covered steering wheel, an empathetic driver can be completely aware of the car's behaviour at all times. Hard driving inspires confidence and it is exceptionally difficult to get completely out

BMW 1600 and 2002 (pictured) are outwardly identical but the latter is available only as a two-door sedan. Unusual visibility is one of BMW's many plus features.

chine for rallying. The 2002 engine can be stretched to over 200 horsepower for racing purposes with very little work being required on the bottom end.

The overhead cam is Duplex roller chain driven with the usual tensioner and backlash controls. The valves are inclined at a narrow vee angle and with the cross-flow head, there is excellent swirl action in the combustion chambers. Fuel supply is maintained by an engine driven pump at 0.3 psi and the oil pump is chain-driven off the bottom of the crankshaft. A full flow oil filter is standard.

The whole engine is tilted over 30 degrees to the vertical over the front axle assembly and is mounted close to its center of gravity. Two side mounted rubber cushions on the block attach it directly to the front axle cross-member. The rear of the engine bolts directly to the gearbox.

Power is transmitted to the gear box by a hydraulically operated single dry-plate diaphragm-spring clutch, complete with an automatic wear compensator. The four-speed transmission has Porsche synchromesh on all gears and is one of the smoothest units on the road. Speed shifting is a joy because it is impossible to beat the synchromesh.

The ratios in the box are ideally matched with the engine torque. First gear has a 3.835:1, second is 2.053:1, third 1.345:1 and fourth, 1.0:1. Second will turn 60 mph and third will bring up 80 mph on the dial. Varying the shift points slightly in first and second and taking off below 2,500 rpm could produce slight

Anti-roll bars front and rear are a mandatory $40 option on 2002s equipped with a 4-speed transmission. These and the tach ($40 also) can be skipped when an automatic is specified.

variances in the times either way so the above figures should be taken as being on the conservative side. Very possibly, one driver could pick up a second on another in the same car.

Passing with the BMW 2002 is a safe operation, even if only fourth gear is used. Gearing down to third at 50 mph and stepping on it will bring up 70 mph in just under 10 seconds.

That the BMW 2002 is well endowed with power and a great transmission is only part of the picture. The engineers have certainly excelled themselves in the roadholding department. Total performance is a complete entity. You don't add bits at a time. You have to design the whole car from the wheels up, especially if it is to be assessed as a package.

Four-wheel independent suspension of shape in this car. We never came close.

Having given the owner a going machine, BMW have wisely added a compatible set of anchors to bring it smoothly to a stop in a hurry, and without wobbling all over the highway. The 9.45-inch diameter discs up front provide most of the stopping power and are backed up with 9-inch drums on the rear. The dual-circuit safety system has been standard on BMWs long before the Federal requirements became mandatory. With a weight division of 1,250 lbs. front to 970 lbs. rear, there is a slight tendency to nose down under heavy braking, but the rear end always stayed firmly on the ground, which is good indication of the suspension working well.

While we have stressed the performance of the 2002, remember that this is also a family sedan. For mundane transportation needs, there is a large 15.5-cu.-ft. trunk that will haul a month's worth of groceries. The spare wheel is stowed under the flat floor of the trunk on the left, the right side being the gas tank area. The tool kit is rather spartan compared with the exotic

1. MacPherson strut tower
2. Overhead cam
3. Solex 40 PDSI carburetion
4. Brake fluid reservoir
5. Well-ventilated battery
6. 500-watt alternator
7. Forward opening hood
8. Emission control

BMW 2002 2-DOOR SEDAN

PERFORMANCE AND MAINTENANCE

Acceleration: Gears:
0-30 mph .... 3.0 secs.— I
0-45 mph .... 6.0 secs.— I, II
0-60 mph .... 9.6 secs.— I, II
0-75 mph .... 13.9 secs.—I, II, III
0-1/4 mile .... 16.9 secs. @ 83 mph
Ideal cruise .... 70-80 mph
Top speed (est) .... 108 mph
Stop from 60 mph .... 142 ft.
Average economy (city) .... 22.9 mpg
Average economy (country) .... 25.0 mpg
Fuel required .... Premium
Oil change (mos./miles) .... —/6000
Lubrication (mos./miles) .... —/6000
Warranty (mos./miles) 12/12,000
Type tools required .... Metric
U.S. dealers .... 225 total

SPECIFICATIONS AS TESTED

Engine .... 121.3 cu. in., OHC 4
Bore & stroke .... 3.50 x 3.15 ins.
Compression ratio .... 8.5 to one
Horsepower ..... 113 (SAE gross) @ 5800 rpms
Torque .... 115.7 lbs.-ft. @ 3000 rpms
Transmission .... 4-speed, manual
Steering .... 3.5 turns, lock to lock
                  34.2 ft., curb to curb
*Brakes .... Disc front, drum rear
Suspension .... Coil front, coil rear
Tires .... 165 SR x 13, tube-type radial
Dimensions (ins.):
   Wheelbase .... 98.5   Front track 52.5
   Length ...... 166.5   Rear track 52.5
   Width .... 62.5   Ground clearance 6.3
   Height .... 55.5   Weight .... 2080 lbs.
Capacities: Fuel ... 12.2 gals.   Oil ... 4.25 qts.
   Coolant ..... 7.3 qts.   Trunk ... 15.5 cu. in.

*Power assisted as tested

RATING

	Excellent (91-100)	Good (81-90)	Fair (71-80)	Poor (60-70)
Brakes	95			
Comfort		89		
Cornering		90		
Details	91			
Finish	93			
Instruments		88		
Luggage		89		
Performance	91			
Quietness		85		
Ride		90		
Room	91			
Steering	95			
Visibility	98			
Overall	91			

BASE PRICE OF CAR

(Excludes state and local taxes, license, dealer preparation and domestic transportation): $3346 at West Coast P.O.E.

Plus desirable options:
$ 160 AM/FM radio
$ 48 Reclining bucket seats
$ 40 Tachometer (replaces clock)
$ 59 Michelin XA5 radial tires
$ 45 Tinted glass
$ 45 Vinyl upholstery
$3753 TOTAL
$1.48 per lb. (base price).

ANTICIPATED DEPRECIATION

(Based on current Kelley Blue Book, previous equivalent model: $201 1st yr. + $252 2nd yr.

Braking by servo-assisted front discs is always controllable. Two types of optional radials are offered, tube-type at $40 and Michelin XAS for $19 more.

Rear window and windshield are almost identical in size and despite the thin pillars, the body structure may be accurately described as containing an integral roll cage.

one in the BMW 2800, but is quite satisfactory for roadside work.

Flow through ventilation is standard and the rear windows open outwards to assist the system. Good nylon carpeting and high-grade vinyl upholstery (optional extra) look like they will stand the hard wear-and-tear of family life.

The 2002 meets all Federal safety and smog requirements and exceeds quite a few. The front and rear body sections are of the crushable variety, and the passenger compartment follows standard European practice in cars of this type by having a rigid-type cage construction. Minimum crash shock is transmitted into the passenger area. Properly strapped in, the odds are against severe injury in city-speed crashes. The door locks have extra safety locks to prevent them opening under impact, helping reduce the risk of ejection for passengers who may not have fastened their seat belts.

We returned the 2002 to its makers reluctantly. It is difficult to find much fault with this car. It's a sports sedan by any interpretation. Criticism degenerates to nit-picking, such as asking for a slightly larger outside mirror. The car abounds with lots of small technical features which are practical and not there purely for gadgetry's sake. Such an item is the inability to depress the inside door locking button down with the door open, making it impossible to lock the keys in the car.

We could find little fault with the quality of the paint, trim or general workmanship. It's rare to find this standard on cars nowadays, and particularly in this price bracket.

Although the car tested ran out at $3,737.00, it still represents excellent value for the money. AM/FM radio is not a must but as this has got to be one of the quieter running cars on the road, one might as well indulge a little and enjoy the benefits of the better radio.

The dealer network is expanding and at the present time, there are 225 nationwide. However, with the overall reliability of the automobile, a long trip should not be cause for concern. The owners manual is so complete that any mechanic worthy of the name should be able to perform quite complicated repairs if required. In fact, the manual is without doubt the best we have ever seen for any automobile and we would highly recommend it as mandatory reading for student drivers because it is so well stocked with excellent driving tips.

If you are contemplating the purchase of a BMW 2002, we would suggest you choose a color you can live with because you'll want to keep this one a long, long time!.

Ken Wright

BELTING fire into the BMW bomb is simple — but this private conversion turns on the heat without robbing the car of its gentle nature. ROB LUCK reports on one of the toughest private car tests we've ever done...

BMW 2002 super-modified

THROUGH most of his life his friends have described him as "that mad German bastard." But there is nothing even slightly insane about his taste in motor cars. For the man with the almost unpronounceable name, Hans Szulczyk (he calls it Shul-zig) only the best will do.

His interpretation of the ultimate in motoring is a BMW 2002 — with extra "go", plus some suitable safety, performance and comfort dress-up items.

His end-product is a beautifully balanced Porsche-Alfa beater that is alternatively docile and savage — and all for little over $6000.

The car's specifications are practically an enthusiast's dream — two dual-choke sidedraft 45DCOE Webers outside a near-stock mill apart from factory camshaft, wide based magnesium/steel wheels, cockpit equipment including two full-harness front seat belts, and beefed headlamps with added spots.

And the performance against the stopwatch is brain-snapping for a car of this size and price.

I took the car for an extensive 250 mile no-holds-barred test this month and put down some performance figures to shatter the morale of sporting GT owners in the E-Type-Volvo-Porsche-Alfa class.

Standing quarters were slashed to 15.2, a top speed of 119 mph (and still winding) and 0-50, 0-60 times of 5.5 and 7.6 seconds.

With that goes tractability as good as

COLOR PAGE: Restrained exterior modifications ran to lights, and wheels. Interior got Momo wheel, full harnesses, and comfort-option lambswool seat covers:
RIGHT: Much-modified 2002 gets off the line sideways with a little wheelspin — but put down good figures.

BMW 2002

the best in the business. The owner does all his own tuning — and it has paid off with great smoothness, progressive throttle control and good low down torque-response for a multi-Weber carburetted car.

And Mr. Szulczyk has proved that such a car can be more economical in terms of fuel consumption, registration and insurance than a big V8 thumper.

Fuel consumption in brutal testing went down to as low as 12-14 mpg with the throats fully opened for miles of acceleration figures. But our cruising figure was a respectable 26 mpg for a 75 mph average — chiefly top gear work.

Now add-in the bonus economy factors — registration cost $83.50 all-inclusive and insurance on the basic car worked out at $187 because of an accident-free record.

All that, and it'll still swallow the really expensive high performance stuff. This BMW would race a Porsche or Volvo 1800E to the end of the quarter and clean-up both. The 911T for $10,000 can only muster 16.4 while the Volvo at $7000-plus is a fraction faster at 16.0. An Alfa 1750 GTV will do the distance in 16.0 against the BMW's 15.2.

Top speed-wise it lies between the Alfa at 117 mph and the Volvo (122) and Porsche (128). Down the strip, the BMW clearly puts away all three, knocking down the 0-60 mph time in 7.6 secs against the Porsche's 8.6, Alfa's 10.0 and Volvo's 9.0.

But the big Weber taps turn on too much performance for upper gear

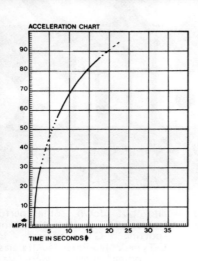

TOP RIGHT: Extra horses were chiefly bolt-on — big 45 mm Weber jugs match factory-modded cam and lift power at back wheels to 103 bhp.
FAR RIGHT: Spur mag steel wheels were the only type that fitted BMW at the time. They were bolted-up with Vredestein rubber.
ABOVE: Booted hard off the line, the little coupe sits quite flat, just blasts out great acceleration times.

ROAD TEST DATA — SPECIFICATIONS

Manufacturer: Bayerische Motoren Werke, Munchen, Germany.
Make/Model: BMW 2002. Body Type: 2 dr. Coupe.
Pricing: basic: $4846; options: 2 full harness Hemco front seat belts, 2 rear lapsash, Spur wheels, AWA Pressmatic radio, Cibie Q14 Filaments, Hella parking lights, 7 in Q 15, Momo wheels.
Test car supplied by: H. Szulcyzk.
Mileage start/finish: 12,272-12,443

ENGINE

Cylinders: Water cooled, four-in-line
Bore x stroke: 3.5 in. x 3.4 in. (89 mm x 80 mm)
Capacity: 121.4 cu.in (1990 cc)
Compression: 9 to 1 (8.5 to 1 Standard)
Aspiration: 2 45 DCOE Dual Choke Sidedraft Type 13
Fuel pump: Mechanical
Fuel recommended: 100 Octane
Valve gear: SOHC
Max. power (gross): N.A. (103 bhp back wheels; Standard (77.8 bhp back wheels). Actual engine rating, standard car 113 bhp at 5800 rpm.

TRANSMISSION

Type/locations: Four speed manual all syncro
Clutch type: s.d.p. diaphragm

Gear	Direct Ratio	Overall Ratio	MPH/1000	(KPH)
1st	3.835	13.959	4.6	(7.4)
2nd	2.053	7.472	8.6	(13.7)
3rd	1.345	4.895	14.1	(22.5)
4th	1.000	3.64	18.8	(30.5)

Final drive: 3.64 to 1

CHASSIS AND BODY

Type: Unitary
Distribution front/rear: 54/46 percent
Kerb weight: 18 cwt 90 lb (1411.8 kg)

SUSPENSION

Front: Independent by McPherson struts, coils and lower wishbones, Torsion bar stabiliser.
Rear: Independent by semi-trailing arms, coil springs, torsion bar stabilisers.
Shock absorbers: Telescopic

torque work. The modified 2002 drops seconds in the third and fourth gear acceleration ranges compared with the standard car, although there is rarely a trace of surge, roughness or uneven acceleration. It is interesting to note the figures on the chart which show the car still pulls from 20 mph in fourth (a little shudder low down).

But the owner is not impressed with mid-range upper-gear performance figures. Reflecting the attitudes of the true performance driver he says: "that's for drivers who aren't in the right gear at the right time."

In the right gear, the modified BMW certainly shifts. Response for tight mountain sections with varied inclines is brilliant — provided you work the gearbox. Full use of the rpm limit will give you 35-66—104-119 mph spacings — but a more cautious hand may wish to "cool it" at 6000 rpm. Even so, performance is still impressive.

The gearing is excellent for high-speed overtaking and the sort of driving the BMW enthusiast expects to do on a mountain road.

And the driver gets added confidence with the brilliant suspension and brakes. The stoppers are among the best the world has seen on a production car so far — on our test of the standard car we crashed it to rest from 60 mph in 3 seconds recording almost a full g from medium 51 lb pedal effort.

The modified BMW turned in the same figures — brakes are unimproved, and they are completely capable of taking the full performance punched through with the engine modifications.

Cornering was uprated to match power only slightly — with the addition of wider wheels. The owner

Wheels: . . . . Spur. 6 in. x 13 in.
Tyres: . . . . .175 SR13 Vredestein
Pressures . . . . 28 lb front/26 rear

STEERING
Type: . . . . ZF Gemmer worm and roller
Ratio . . . . .17.58 to 1
Turns lock to lock 3.5
Wheel diameter . . . . 14 in. (35.6 cm)
Turning circle, between kerbs: . . . . 34 ft 2 in. (10.3 m)
Between walls: . . . . 36 ft 1 in. (9.4 m)

BRAKES
Type: . . . . Disc front/drum rear
Dimensions . . . . 9.45 in. (24 cm)/9.05 in. (23 cm)

DIMENSIONS
Wheelbase . . . . 98.5 in. (250 cm)
Track, front: . . . . 52.5 in. (133 cm)
Rear: . . . . 52.5 in. (133 cm)
Overall length: . . . . 13 ft 10.5 in. (423 cm)
Width: . . . . 5 ft 2.5 in. (159 cm)
Height: . . . . 4 ft 7.5 in. (141 cm)
Ground Clearance: . . . . 6 in. (16 cm)
Overhang, front: . . . . 29 in. (73.6 cm)
Rear: . . . . 41 in. (104 cm)

EQUIPMENT
Battery: . . . . .12 V, 44 a/h
Alternator: . . . . 14 V, 35 A
Headlamps: . . . . 12 V
Jacking points: . . . . .4 side points

CAPACITIES
Fuel tank: . . . . 12.1 galls (46 litres)
Engine sump: . . . . 8.5 pints (4 litres)
Final drive: . . . . 1.7 pints (0.8 litre)
Gearbox: . . . . .2.1 pints (1 litre)
Water system: . . . . 1.8 galls (7 litres)

PERFORMANCE

Test conditions for performance figures; Weather: Fine, Cool; Wind: 0-4 knots; Humidity: 40 percent; Max. Temp: 56 degrees; Surfaces: Dry Hotmix.
Top speed, average: 114 mph (182.4 kph) (107.2mph/171.5 kph)
best run: . . . . . 119 mph (190.4 kph) (108.1 mph/172.9 kph)

Acceleration:
Standing Quarter Mile, average: 16.3
best run: 15.2 (16.1)
Speed at end of Standing Quarter: . . . . . . 84 mph (134 kph)
0-30 mph: 2.9 (3.0)
0-40 mph: 4.0 (4.5)
0-50 mph: 5.5 (6.3)
0-60 mph: 7.6 (8.4)
0-70 mph: 10.6 (11.2)
0-80 mph: 15.0 (14.9)
0-90 mph: 19.2 (19.9)

Speed in gears:

Gear	Max. mph 6000 rpm	7000 rpm	kph 6000 rpm	7000 rpm
1st	30	35	48	56
2nd	56	66	89.6	105.6
3rd	88	104	140.8	166.4
4th	117	—	187.2	—

Acceleration holding gears:

	2nd	3rd	4th
20-40	3.9 (—)	7.4 (4.8)	11.0 (6.9)
30-50	3.8 (—)	6.1 (4.4)	11.2 (6.4)
40-60	3.9 (—)	7.0 (4.4)	9.2 (6.7)
50-70	4.5 (—)	6.1 (5.0)	11.2 (6.9)
60-80		6.5 (6.6)	10.8 (7.2)

Fuel consumption:
Average for test: . . . . Overall — 25 mpg (10.5 kpl) for Standard
Best recorded: . . . . 30 mpg (12.7 kpl) — country
City average: . . . . .23 mpg (9.6 kpl)
Country cruising: . . . . . 26 mpg (10.9 kpl) at 75 mph average

Braking: Five crash stops from 60 mph

Stop	G	Pedal
1	.90	51 psi
2	.90	52 psi
3	.90	55 psi
4	.88	56 psi
5	.86	56 psi

30-0 mph: . . . . 1.5 secs
60-0 mph: . . . . 3.0 secs

Speedo Corrections:

30	40	50	60	70	80
29.1	39.2	48.5	58	67	77

BMW 2002

LIGHTING got particular attention. Big four-filament Cibie QI-inserts glare-out clean white light from main beams, and parkers had to be shifted to small bumper-mounted accessories. Big 9 in. Hella spots were also added. Grille from BMW symbolic vertical ovals was detached because of stone damage, difficulty in cleaning and appearance.

wasn't interested in widened steel wheels and looked for a suitable mag-type.

He could find only one wheel on the market to suit the car at the time he bought the wheels. That was the Spur mag-centre steel-rim which provide 6 in. grip with safety, strength and reliability.

They've done 8000 hard miles on the car already and there's no sign of problems. And that mileage has included some pretty tough interstate and outback trips.

The extra inch-and-a-half rim width (standard rims are only 4½ in.) keeps pace with the performance mods. The car can be hurled at corners at remarkable speeds for an ordinary road car, with delightful control. Set up with slightly higher front tyre pressures (28-26) the car understeers mildly and progressively until the nose starts to break. Approaching that point, the tiny Momo steering wheel loads up in feel, without being over-heavy to turn, and without reaction losses.

A few more pounds in the front end reduces the understeer to almost zero, and eliminates wheel loadings, but the owner likes a bit of understeer feel so that's how the car is set up.

Cornered on constant, even throttle pressures, and smooth line the car just goes round very fast with plenty of power reserve left to swing the tail around.

It is possible to throw it at a 105 mph sweeper sideways and power it through with plenty of horses to spare.

Back down at city traffic speeds the car is completely tractable. Off the mark you can select first and clutch-slip away on idle without touching the throttle. There's no shake, judder or jerk, and you can squeeze the throttle till the needle hits the big seven without any snatch or vibration in between.

For a Multiple Weber carburetted car, this type of full range smoothness is almost unique. It's only been made possible by the determination and patience of the man who owns the car — a driver capable of extracting virtually all the car's performance, but dedicated to driving a smooth, unfussy town car.

The Szulczyk (remember, Shul-zig) story started on the frightening open cycle courses of Europe, where he earned his nickname by going so fast downhill that he overtook Police escorts and arrived at controls before they were ready for him.

He "tuned" his bicycle like a car — getting down to details like dynamically balanced tyre/rims, specially fiddled tyre pressures and delicately equalised chain-links, cable-operation and balance points.

With cars he was just as fanatical — and when he acquired the BMW 2002 last year, he programmed the modifications carefully. During an overseas trip he picked up a factory BMW "mild" cam and full workshop manual. Back in Australia he acquired John Passini's book on carburettor tuning.

After a good running-in period, he handed the "stick" over to Sydney's Lynx Engineering — automotive modification experts — and gave them the go on a full Weber carburetion job.

Rolling the basic car on the chassis dynamometer turned up an output of 77.8 bhp at the back wheels, from a standard factory-claimed 113 bhp output at the flywheel. When he drove away from Lynx again, he left the dyno dial at 103 bhp.

Hans was impressed — but he felt he could do better still.

He looked at the standard Weber linkages and after a few hundred miles of jerky driving around town, decided they had to go.

Using less than $2 worth of bicycle parts he completely revised shafts, syncronisation action, springs, levers and fulcrum points. He aimed to get a smooth, progressive throttle action and he succeeded — the end result is as good as the best factory-standard throttle on any Weber-equipped car I have driven ... and that includes all comers up to Lamborghini, and Ferrari.

Operating on the principle of patience, he then set about completely re-tuning the engine. Calculating a 40 deg. overlap on the camshaft, (the standard is 8 deg) he tuned for smooth power from low down. As a basis he used timing set to standard 2000TI specifications, 190 air correction jets and 140 main jets in the Webers. The float level in the carburettor bowls was increased from 5 mm to 7 mm and the idle was set down to 700-800 rpm. Tappet clearance was bumped from 6-8 thou up to 11 thou — although standard valve springs are used.

The carburettors were completely tuned by ear — and to prove to my satisfaction that Hans could do the job, I took a screwdriver, screwed the carbs way out of sync and mixture and let him have his head. Two minutes later he had finished detail adjustments, the car sounded good, and I climbed in and recorded identical figures to the first times put down.

At the end of our tough test, the BMW sounded as sweet and smooth as ever.

Personally, I feel the quiet German has missed his vocation — I can think of literally dozens of owners of Weber-equipped vehicles who'd like to have their carburetion sorted to the same smoothness and finesse as the much-modified BMW. He's just the man for the job. Address has been withheld to prevent a queue forming at his front door. ■

NEW CARS

BMW 2002 Tii

The 2002 Tii (above) retains the small shell of the 1600 coupé but has Kügelfischer petrol injection for its 2-litre engine (below). Power output installed is 130 bhp.

2002 gets fuel-injection

It is well known that the BMW 2002 consists of a 2-litre ohc engine in the compact 1600 bodyshell. With its small size and light weight, it is a very popular model of more than lively performance. In standard, single carburetter form, it satisfies the requirements of the majority of buyers.

There is, however, a definite demand for a hot 2-litre, and lucky continental motorists have been able to buy a multi-carburetter version of the 2002. In England, unfortunately, we have had to go without for a very simple reason—the right-hand steering goes just where the carburetters should be; an insuperable obstacle. Insuperable, that is, with carburetters, but now an even hotter 2-litre BMW engine has been developed from the competition type, and there are no installation problems because the new unit has fuel-injection.

The well-tried Kügelfischer system is employed and there is no doubt that the engine can easily stand the increased stresses, as it sustains incredible outputs when turbocharged for racing. It retains the single overhead-camshaft with rockers and inclined valves that are now found on all BMW engines. The gearbox has been developed with a new type of synchromesh to give faster and smoother changes.

In view of the higher performance potential of this model, the bodyshell has been reinforced to give greater rigidity. This has permitted suspension modifications for increased cornering power, including the use of wider wheels and tyres plus more powerful dampers. Larger brakes have also been adopted.

The result of all this is an 18 cwt car with an overall length of 13 ft 10.5 in and a width of 5 ft 2.6 in, propelled by an engine with a gross output of 147 bhp that develops 130 bhp as installed. It must be emphasised, however, that this 2002 remains a tractable touring car with a four-seater body, though it is sporting rather than luxurious—BMW luxury cars have six cylinders.

Very sensibly, the makers eschew boy-racer decoration and the car has that unobtrusive appearance which pays dividends when one is in a hurry on police-infested roads. The interior is plain but functional and the driving position is first class, with a good view all round. Everything about the car seems substantial, and there is no sign that any weight saving has been attempted.

The car was placed at my disposal in France during a heatwave and though I was able to cover a useful mileage on typical *Routes Nationales*, there were no *Autoroutes* handy for sustained maximum speed tests when I was driving this particular car.

The makers claim 120 mph, which may well be possible on a long enough straight, but what is much more useful is the rapid acceleration up to 110 mph, a speed that comes up in a remarkably short distance. The car is high-geared and will cruise easily at 100 mph on a very small throttle opening.

The clutch and gearchange handle well and permit the full potential of the engine to be exploited, third gear in particular being ideal for overtaking on French roads. The fuel-injection engine has plenty of torque and there is no necessity to use the lower gears a great deal. The unit is flexible throughout its range and idles evenly, only a scarcely perceptible rhythmic surge on part-throttle acceleration betraying the presence of injection to the experienced observer.

Immediately, one notices that the suspension feels much harder than that of other BMW models. In fact, this is probably more a matter of heavier damping than stiffer springs, as is confirmed by the capacity of the car to absorb bad road surfaces at high speeds without throwing the passengers about. Some road noise is evident under these conditions but it is strictly moderate. Similarly, the engine sounds healthy and unmistakably has four cylinders, but when driven at ordinary touring speeds it is well silenced.

Though the Tii is such a satisfactory everyday touring car, there is the feel of a competition car about it. Very well balanced, it flicks through corners without effort, and it shows its breeding when forced to change its line in the middle or to take an adverse camber in its stride. The independent rear end sticks down extremely well over broken surfaces, while the absence of wheelspin on leaving sharp corners is praiseworthy. Nevertheless, an oversteering condition can be induced by suitable provocation.

The BMW 2002 Tii is not cheap, but it represents remarkable value when its performance is taken into account. You can buy more car for the money but that is not the point. The man who chooses this model will realise the amount of time that can be saved on a journey by small overall dimensions. Fierce acceleration in a ponderous package is not the answer in the traffic of today, but the same performance in a nippy little car certainly takes a lot of beating.

JOHN BOLSTER

Car described: BMW 2002 Tii 2-door saloon, price £2295 including tax.
Engine: Four cylinders 89 mm x 80 mm, 1990 cc. Chain-driven overhead camshaft. Compression ratio 10 to 1. 130 bhp (net) at 5800 rpm. Kügelfischer fuel-injection.
Transmission: Single dry-plate clutch. Four-speed, all-synchromesh gearbox with central change, ratios 1.0, 1.32, 2.015, and 3.767 to 1. Hypoid final drive, ratio 3.45 to 1.
Chassis: Combined steel body and chassis. Independent front suspension by damper struts, coil springs, and lower wishbones, with torsional anti-roll bar. Worm and roller steering gear. Independent rear suspension by semi-trailing arms, coil springs, telescopic dampers, and auxiliary rubber springs, with anti-roll bar. Dual-circuit servo-assisted brakes with front discs and rear drums. Bolt-on disc wheels fitted 165 HR13 radial-ply tubed tyres
Equipment: 12-volt lighting and starting, Speedometer, rev-counter, fuel and water temperature gauges, Cigar lighter. Heating, demisting and ventilation system. Windscreen wipers and washers. Flashing direction indicators.
Dimensions: Wheelbase 8 ft 2.4 in. Track 4 ft 4.4 in. Overall length 13 ft 10.5 in. Width 5 ft 2.6 in. Weight 18 cwt.
Performance (Maker's figures): Maximum speed 120 mph. Acceleration 0-50 mph 6.2 s.

BMW 2002tii

*First test of the fast
new fuel-injection model*

2002tii THE BMW 2002 is one of our favorite cars. After all, we did name it Best in Class in our roundup of the world's best cars earlier this year. But it has been around over three years now, and its basic machinery and body date back to early 1967 when the 1600 2-door was first introduced. So it was time for a little updating. BMW has done just that and at the same time introduced some new variations on the 2-door theme.

For the home market the 2-door line starts with a 1602—the 1600, facelifted—and goes through an 1802 (1.8-liter engine) and the 2002 (2-liter) to the top-of-line 2002tii. There had been a 2-carburetor 2002TI of 120 bhp DIN not available here; the 2002tii uses Kugelfischer mechanical fuel injection to get another 10 bhp and replaces the TI. In the U.S. from henceforth, we'll get only the 2002 and 2002 tii; Hoffman Motors is no longer importing the 1.6-liter model. But getting the hot model is something new and nice for the U.S., which has been denied the 2-carburetor versions since emission regulations went into effect.

The facelift is mild. There are new, rubber-tipped bumpers which are wider and wrap around to the wheel openings at each end of the car, and a new rubber-faced body molding running between the wheel openings at bumper level (this looks odd, conflicting esthetically with the horizontal chrome strip farther up the body side). The front seats have been reshaped, the instrument panel slightly rearranged and a few minor electrical-accessory changes made. Later in the year a new body variation, the Touring, will be available here in either 2002 or 2002tii form; this is a "hatchback" 2-door sharing the 2-door's basic body but with new roof and open-up rear end.

For now, the important news is the mechanical goodies offered in the 2002tii at approximately $400 over the regular 2002. We hadn't expected to see BMW import this engine with mechanical fuel injection as electronic injection is generally considered best for emissions, but here it is—in U.S. form somewhat detuned with its compression ratio dropped from 10.0:1 to 9.0:1 but still notably more powerful than the 2002. The Kugelfischer system is somewhat similar to

BMW 2002 tii

the Bosch injection long used on Mercedes engines, with a mechanical metering unit furnishing fuel at high pressure to port injectors. As with most such injected engines, the timed delivery of fuel to the ports makes it possible to use more radical valve timing and ram-tuned intake pipes without sacrificing low-speed tractability, and BMW does exactly that to extract the extra horses. The tii's power peak is moved up to 5800 rpm (the 2002 peaks at 5500), but more significantly the torque peak of the injection engine occurs between 4000 and 4700 rpm compared to a mere 3000 for the 2002. That torque peak is also much higher—145 lb-ft vs the 2002's 116—but it's clear that the engine needs to be wound up to get the benefit of it.

There's more than just a hotter engine. The 2002tii inherits the 2002TI's beefier chassis components, with bigger brakes front and rear and 1.4 in. additional track at both ends. Its wheels are wider—though still not very wide at 5.0 in.—and the Michelin XAS tires carry an H rating (130 mph) instead of the 2002's S rating (112 mph). In other words, the chassis has what it needs to handle the additional power; in Germany it has to be that way because people drive their cars as fast as they will go.

Transmission and gearing are also different for the tii. A new 4-speed gearbox with Borg-Warner synchromesh, similar to that used in the 6-cylinder BMWs, is standard and a 5-speed ZF box optional. The final drive ratio is 3.45:1, same as in 2800s. The 4-speed gearbox has slightly taller ratios in the indirect gears and the 3.45 final drive is taller than the 2002's 3.64:1, so the tii is longer-legged all the way.

The engine sounds and feels very much like that of a regular 2002; despite its higher output it's actually a bit quieter and smoother, a fact enhanced by the gearing when one is cruising at Interstate Highway speed. It idles just as nicely as a 2002 and pulls smoothly from low engine speeds in any gear, though of course a driver has to use the gearbox skillfully to get the most out of it. But the difference in performance doesn't show up until 4000 rpm is reached: above that there's definite "on cam" feeling that stays with you up to over 5500 rpm. BMW fans may be disappointed with our acceleration times for the standing ¼-mile and 0-60 mph, as we were; the tii is only 0.3 sec quicker in the quarter and its 0.6-sec-better 0-60 time is aided by the fact that 2nd gear takes it to 63 mph. Two explanations: one, our test car was not fully broken in, and since we journeyed to New York on a tight schedule to test the car there wasn't time to put a few hundred extra miles on it; second, that taller gearing takes its toll. There was particular difficulty getting the 2002tii "off the line" smoothly (BMW rear suspension always lets the wheels patter badly anyway) and when the wheels finally settled down the engine bogged. The real benefit comes at the high end—the tii gets to 100 mph nearly 10 sec quicker than the 2002—which means that the car is geared for top speed, not acceleration. This is a bit curious—we'd expect BMW to select "acceleration" gearing for the U.S. version and would suggest that the 2002's 3.64 gears would have been better. The 5-speed, which uses a 4.11:1 final drive, goes too far in the other direction!

The tii uses more fuel than the very economical 2002, but it still isn't what we call thirsty, doing over 22 mpg in everyday driving. In all the 2002tii offers a good combination of performance, refinement and economy, and our only serious question is about the gearing. The gearbox itself is a delight, with wonderfully smooth synchronizer action of just a slight notchiness as one moves from gear to gear offering a feel

BMW 2002 tii

of precision rather than real resistance. BMW seems to be leaning away from Porsche synchromesh, as used in the 2002 box, and toward the B-W design. We have no strong preference; some staff members prefer the smoothness of the Porsche synchros and others the notchiness of the Borg-Warner.

The throttle and brake pedals are positioned just right for blipping the throttle while downshifting, so the 2002tii is perfectly set up for vigorous driving on winding country roads. Having driven it several laps around the Lime Rock circuit and over some of the lovely roads of rural New England, we can vouch for that.

The responsive engine, nice gearbox and good controls are just part of what makes all BMWs real driver's machines. Then comes handling, and for a rather tall sedan the 2002tii is quite good. Its steering seems somewhat lighter than previous 2002s we've driven, despite the new car's greater weight; it's still not the lightest thing going but is very precise and quite quick enough. The tii comes with anti-roll bars front and rear, which keep body roll down to a moderate level, and the wider track, wider wheels and slightly stiffer tires all conspire to make the tii significantly better-handling than a 2002. Springing seems to be no stiffer. The ride is still fairly soft and smooth over large irregularities, though small ones such as tar strips bring out the harshness of the steel-belted tires. There's lots of suspension travel so that one can negotiate all kinds of evil road surfaces, even with four people aboard, at speed without fear of bottoming. Our test car's body was tight and rattle-free; some owners report that this isn't always the case, however.

The tii's stable understeer, which is moderate but stays with you even at high speeds, gave us the courage to work up our speed at Lime Rock; and this gave the brakes a good workout. They were up to it. Though they smelled strongly after a few laps, they were still effective, and in our fade test they showed no fade at all. The standard 2002's had shown a 21% increase in pedal effort in the 6-stop test, so there's a clear improvement here.

The 2002, tii or otherwise, is a very "upright" car, getting considerable passenger space into a compact overall length by virtue of its height. This means a commanding driving position, and all-around glass gives the driver a commensurate view. So it's a good traffic car. The "vertical" character carries through to an accelerator pedal that seems too upright when one holds it steady at cruising speed. BMW engineers make good use of steering-column stalks for things like headlight flashing and wipers-washers, but we've never understood why they put the directional-signal stalk on the gearshift side where the hand can be so busy. Instrumentation is minimum: BMW doesn't like oil-pressure gauges or ammeters, so we have to make do with flashing lights (many owners add supplementary gauges) for that information.

Ventilation isn't a strong feature of the 2002 and the hot, humid weather during this test emphasized that BMW might have done some facelifting in this department. The airflow through the system is not bad, though one has to open the swiveling rear side windows to maximize it (thus accepting considerable wind noise), but there are none of the dash vents that are so welcome in contemporary cars—including BMW's own 6-cyl models. Bad marks here for the 2002 and

2002tii. And no factory air conditioning is available—only an add-on kit.

Notwithstanding the weatherproofing deficiency, the BMW 2002tii is a keen sports sedan—a real blast to drive fast and yet practical enough for a small family to use for daily transportation and extended trips. The price is high and getting higher, thanks to the German currency's upward spiral, but the 2002tii is bound to give BMW's little 2-doors a popularity boost. It certainly gave our collective mood a boost—it's nice to know that even with tightening smog regulations it's possible to get more performance in a car that was already strong in that department.

COMPARISON DATA

	BMW 2002 tii	Alfa Romeo 1750 Berlina	Volvo 142E
List price, incl. prep	est. $4000	$3905	$4080
Curb weight, lb.	2310	2484	2696
0-60 mph, sec.	9.8	11.0	10.5
Standing ¼-mi, sec.	17.3	17.9	17.5
Speed at end	78.5	76	76
Stopping distance from 80 mph, ft	315	n.a.	306
Fade in 6 stops from 60 mph, %	nil	nil	20
Cornering capability, g	0.726	0.692	0.649
Fuel economy, mpg	22.7	20.4	23.7

ROAD TEST
BMW 2002 tii

PRICE
List price, east coast...est. $4000

IMPORTER
Hoffman Motors Corp., 375 Park Ave., New York, N. Y.

ENGINE
Type................sohc inline 4
Bore x stroke, mm.....89.0 x 80.0
  Equivalent in........3.50 x 3.15
Displacement, cc/cu in..1990/121.4
Compression ratio...........9.0:1
Bhp @ rpm...............140 @ 5800
  Equivalent mph............118
Torque @ rpm, lb-ft...145 @ 4500
  Equivalent mph.............91
Fuel injection........Kugelfischer mechanical
Type fuel required..regular, 91-oct
Emission control....fuel injection, engine mods

DRIVE TRAIN
Transmission.....4-speed manual
Gear ratios: 4th (1.00).....3.45:1
         3rd (1.32)...........4.56:1
         2nd (2.02)...........6.98:1
         1st (3.76)............13.0:1
Final drive ratio...........3.45:1

CHASSIS & BODY
Layout.....front engine/rear drive
Body/frame.............unit steel
Brake type....10.07-in. disc front, 9.05 x 1.57-in. drum rear; vacuum assisted
  Swept area, sq in.........306
Wheels.........cast alloy 13 x 5J
Tires.....Michelin XAS 165 HR-13
Steering type.......worm & roller
  Overall ratio.............17.6:1
  Turns, lock-to-lock..........3.7
  Turning circle, ft..........31.5
Front suspension: MacPherson struts, coil springs, tube shocks, anti-roll bar
Rear suspension: semi-trailing arms, coil springs, tube shocks, anti-roll bar

INSTRUMENTATION
Instruments: 120-mph speedometer, 8000-rpm tachometer, 99,999 odometer, 999.9 trip odometer, coolant temp, fuel level
Warning lights: oil pressure, brake system, alternator, high beam, directionals

MAINTENANCE
Service intervals, mi:
  Oil change................4000
  Filter change.............4000
  Chassis lube..............none
  Tuneup....................8000
Warranty, mo/mi.......12/12,000

RELIABILITY
From R&T Owner Surveys the average number of trouble areas for all models surveyed is 11. As owners of earlier models of BMW reported 7 trouble areas, we expect the reliability of the BMW 2002 tii to be better than average.

ACCOMMODATION
Seating capacity, persons.......4
Seat width, front/rear...2 x 20.5/51.0
Head room, front/rear...40.0/38.0
Seat back adjustment, degrees..90

GENERAL
Curb weight, lb.............2310
Test weight.................2645
Weight distribution (with driver), front/rear, %....56/44
Wheelbase, in...............98.4
Track, front/rear.....53.8/53.8
Overall length.............166.5
Width........................62.6
Height.......................55.5
Ground clearance.............6.3
Overhang, front/rear....29.0/39.1
Usable trunk space, cu ft....9.5
Fuel tank capacity, U.S. gal..12.1

CALCULATED DATA
Lb/bhp (test weight).........18.8
Mph/1000 rpm (4th gear)......19.8
Engine revs/mi (60 mph)....3020
Piston travel, ft/mi........1585
R & T steering index........1.13
Brake swept area sq in/ton...242

ROAD TEST RESULTS

ACCELERATION
Time to distance, sec:
  0–100 ft....................4.4
  0–250 ft....................7.0
  0–500 ft...................10.0
  0–750 ft...................12.6
  0–1000 ft..................14.8
  0–1320 ft (¼ mi)...........17.3
Speed at end of ¼ mi, mph...78.5
Time to speed, sec:
  0–30 mph....................3.9
  0–40 mph....................5.9
  0–50 mph....................7.7
  0–60 mph....................9.8
  0–70 mph...................13.9
  0–80 mph...................17.9
  0–100 mph..................32.0
Passing exposure time, sec:
  To pass car going 50 mph....5.3

FUEL CONSUMPTION
Normal driving, mpg........22.7
Cruising range, mi..........275

SPEEDS IN GEARS
4th gear (5600 rpm).........115
3rd (6400)...................95
2nd (6400)...................63
1st (6400)...................34

BRAKES
Panic stop from 80 mph:
  Max. deceleration rate, % g..87
  Stopping distance, ft.......315
  Control..............excellent
Pedal effort for 50%-g stop, lb..35
Fade test: percent increase in pedal effort to maintain 50%-g deceleration rate in 6 stops from 60 mph..........................nil
Parking: Hold 30% grade?.....yes
Overall brake rating.....very good

HANDLING
Speed on 100-ft radius, mph..32.9
Lateral acceleration, g......0.726

SPEEDOMETER ERROR
30 mph indicated is actually..29.5
40 mph.....................39.5
50 mph.....................50.0
60 mph.....................60.0
70 mph.....................70.0
80 mph.....................80.5
100 mph...................101.0
Odometer, 10.0 mi..........10.0

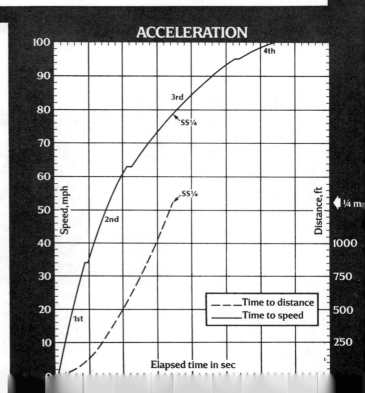

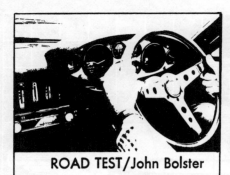

ROAD TEST/John Bolster

BMW 2002 tii: A 2-litre saloon with quick acceleration

A well-appointed and practical car, the 2002tii is a very rapid form of transport.

I always choose horses for courses, so when the French show came round I looked for something that was small enough for Paris, would idle for hours in traffic blocks, and would pass all the passionate Frenchmen on the *Autoroute du Nord*. From among the available machinery, I chose the fuel-injection BMW 2002.

The BMW, in all its forms, is very familiar to me. For the roads of Northern France, independent suspension of all four wheels is a minimum requirement and all BMWs have that. The tii has strong anti-roll bars at both ends and wide radial-ply tyres. It has the engine and gearbox in front, driving to a chassis-mounted hypoid unit at the rear, and the power unit is that well-known and almost indestructible four-cylinder, with a chain-driven overhead camshaft operating inclined valves through rockers.

Fuel injection is, admittedly, a complication as well as an expensive addition to an engine. At first glance, there might appear to be little advantage in the system over the multi-carburetter version of the same unit, for the gain in maximum power is not great. The point, however, is that the fuel injection engine gives more torque and it continues to give it over a wide band of revs. In practical terms, that means instant acceleration at any time, which is what fast road motoring is all about.

This characteristic is so marked that I wouldn't give a thank you for the optional 5-speed gearbox. With four well-spaced ratios, the fuel-injection BMW has all the gears it needs, and for this reason the acceleration is even better than would appear from the graph.

Gearchange points are not critical and if you put your foot down the car just goes, though wheelspin must be guarded against at the traffic lights.

With such vivid acceleration on tap, the little saloon cuts its way through traffic with considerable ease. It goes on accelerating up to about 112 mph, which is a speed that can be kept up indefinitely. Higher speeds take longer to reach, the shape of the body probably not attaining a particularly low drag coefficient. However, 119 mph can be reached on the level, with a little more to come under favourable conditions. The high third gear gives splendid acceleration over a very wide range.

The engine is smooth for a four-cylinder and though it sounds efficient it is by no means noisy. The level of road noise is satis-

With its independent suspension all round, the 2002tii corners and accelerates very impressively.

factorily low on most surfaces but there is a good deal of wind noise. Although the suspension is well damped and there is remarkably little roll on corners, the ride is truly excellent on the bad French roads, the wheels having sufficient travel to absorb the undulations.

The BMW has high cornering power, showing up particularly well on the faster corners. It is well balanced, having perhaps a trace of initial understeer that can soon be cancelled by spirited handling. The steering does not have quite the feel of the road that a rack and pinion can give but it is satisfactorily direct and accurate. One can pay the brakes the compliment of saying that one never thinks about them because they always behave as they should.

In spite of its performance potential, the engine never fouls its plugs in traffic blocks. However, the test car was a slow starter from cold and did not always idle regularly. No doubt this was a question of simple adjustment but I am not familiar with the injection system on this car and decided not to interfere — I hope to take lessons in this sphere as I am driving so many injection cars nowadays. Apart from this slight mal-adjustment, the engine behave impeccably throughout.

Engines with fuel injection are usually economical of petrol and the 2002 tii is no exception. It would be a reasonable guess that the hard driver will get 25 mpg and his more timid brethren 28 mpg, with 30 mpg on the distaff side. On autoroutes, with lots of 110 mph motoring, the figure falls to 20 mpg, which is still outstandingly good. There are some large-engined cars which can also cruise at 110 mph but use so much fuel in doing it that it is scarcely a practical proposition, especially at the prices ruling in France.

During the time that I was in France, I never met another car that could equal the BMW for acceleration. I was overtaken once when my speedometer was showing 125 mph, but as my rival had three times my engine capacity, this was to be expected. In any case, I would not have cared to pay his petrol bill and I could out-accelerate and out-corner him easily. Certainly, I made the right choice of mount for this particular trip.

The BMW also served me well in England. Here, there is little need for sophisticated suspension but fierce acceleration will get you out of more trouble than any other quality. The sheer unexpectedness of the British driver can place sudden demands on the motorists around him; the vivid responsiveness of the 2002 tii may save him from his own folly on such occasions. It is equally good at dodging the well-lunched Frenchman in his 2CV, who suddenly asserts his priority from a hidden side turning, for in all countries the car of instant acceleration is the safest.

The BMW 2002 tii is a well-appointed and practical car. Though arranged for through-floor ventilation, it lacks the adjustable face-level current of air which the eyeball ventilators of the bigger BMWs provide; if the quarter-lights are used for extra ventilation, they are objectionably noisy. Other faults are few indeed and the car must be regarded as a very effective and rapid piece of transportation. You could buy a bigger vehicle for the money but not one that was a greater pleasure to drive.

SPECIFICATION AND PERFORMANCE DATA

Car tested: BMW tii 2-door saloon, price £2,299 including tax.
Engine: Four cylinders, 89 mm x 80 mm (1990 cc); chain-driven overhead camshaft operating inclined valves through rockers; Kugelfischer fuel injection; compression ratio, 10 to 1; 130 bhp (net) at 5,800 rpm.
Transmission: Single dry plate clutch; 4-speed all-synchromesh gearbox with central change; ratios 1.0, 1.32, 2.015, and 3.767 to 1; hypoid final drive, ratio 3.45 to 1.
Chassis: Combined steel body and chassis; independent front suspension by coil spring damper struts and lower wishbones; worm and roller steering; independent rear suspension by trailing arms, coil springs, auxiliary rubber springs and telescopic dampers; anti-roll torsion bars front and rear; disc front and drum rear brakes; bolt-on disc wheels fitted 165HR13 radial-ply tyres.
Equipment: 12-volt lighting and starting; speedometer; rev counter; fuel and temperature gauges; clock; heating, demisting, and ventilation system; 2-speed windscreen wipers and washers; flashing direction indicators with hazard warning; reversing lights; radio (extra).
Dimensions: Wheelbase, 8 ft 2.4 in; track 4 ft 4.4 in; overall length, 13 ft 10.5 in; width, 5 ft 2.6 in; weight, 2,028 lb.
Performance: Maximum speed, 119 mph. Speeds in gears—third 91 mph, second 62 mph, first 34 mph. Standing quarter-mile, 16.9 s. Acceleration—0-30 mph, 3.4 s; 0-50 mph, 6.6 s; 0-60 mph, 8.9 s; 0-80 mph, 17.6 s; 0-100 mph, 26.8 s.
Fuel consumption: 20 to 30 mph (see text).

The fascia of the 2002tii lacks face level fresh air ventilation (above). An addition to the well-known and well-proven four-cylinder engine is fuel injection.

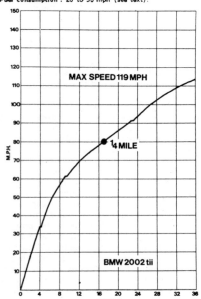

AUTO TEST

BMW 2002tii

Little to look at, great to drive

At-a-glance: High-compression, fuel-injection engine gives the smallest-bodied BMW superb performance. More advantages at top end of range, but very easy to drive in town also. Good handling except in extreme cases; generally quiet and comfortable, but better ventilation needed. Good fuel consumption.

THERE can be no doubt that the BMW 2002 has proved one of the most successful of the recent crop of "shoehorning jobs"—putting a bigger engine into an existing car. Undeniably it was helped by having the splendid BMW 1600 as a starting point, and by the fact that the new engine, while 25 per cent bigger in capacity, was hardly any larger in terms of weight or size. But the resulting 2002 was still better than one might have expected, and in some ways whetted people's appetites for more.

"More" in the first instance meant the adoption of the twin-carburettor engine from the 2000TI, but there was a snag as far as the British market was concerned. It proved impossible to engineer a right hand drive conversion of the 2002TI, because the induction sat square in the path of the steering column. As a result, British buyers have had to wait for the advent of the 2002Tii.

In the Tii, Kugelfischer indirect injection is used, along with a higher compression ratio, to give a considerable power boost. The new induction arrangement makes it possible to instal right hand drive steering.

It is notable that BMW have chosen the Kugelfischer system for their four-cylinder cars, although the six-cylinder injection models (the 3.0Si and CSi) use a Bosch system. There is no denying its efficiency in this case; compared with the standard 2002, power goes up from 100 to 130 (DIN) bhp, with the power peak coming only 300 rpm higher at 5,800 rpm. Extra power can be something of a mockery without extra torque, but here again there is an improvement, from 116 to 131 lb. ft., a considerable achievement, considering the engine is still exactly the same size. In this case, however, the peak speed goes up from 3,000 to a highish 4,500 rpm.

BMW 2002 (1,990 c.c.)

ACCELERATION

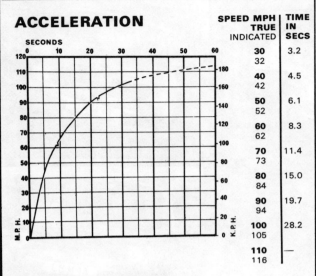

SPEED MPH TRUE	INDICATED	TIME IN SECS
30	32	3.2
40	42	4.5
50	52	6.1
60	62	8.3
70	73	11.4
80	84	15.0
90	94	19.7
100	105	28.2
110	116	—

GEAR RATIOS AND TIME IN SEC

mph	Top (3.45)	3rd (4.56)	2nd (6.97)
10-30	—	6.4	4.2
20-40	8.5	5.8	3.4
30-50	8.1	5.6	3.4
40-60	7.8	5.7	3.8
50-70	8.4	5.9	—
60-80	9.6	6.7	—
70-90	11.3	8.8	—
80-100	14.8	—	—

Standing ¼-mile 16.4 sec 83 mph
Standing Kilometre 30.3 sec 102 mph
Test distance 1,017 miles
Mileage recorder 1 per cent over-reading

PERFORMANCE

MAXIMUM SPEEDS

Gear	mph	kph	rpm
Top (mean)	116	187	5,950
(best)	118	190	6,060
3rd	93	150	6,300
2nd	61	98	6,300
1st	33	53	6,300

BRAKES

FADE
(from 70 mph in neutral)
Pedal load for 0.5g stops in lb

1	50	6 40-45
2	42	7 45-60
3	37	8 45-60
4	38	9 40-50
5	38-45	10 45-55

RESPONSE
(from 30 mph in neutral)

Load	g	Distance
20lb	0.15	201ft
40lb	0.45	67ft
60lb	0.72	42ft
80lb	0.85	35ft
100lb	0.95	32ft
120lb	0.98	30.6ft
Handbrake	0.28	107ft

Max. Gradient 1 in 3

CLUTCH
Pedal 30lb and 5in.

COMPARISONS

MAXIMUM SPEED MPH
Rover 3500S . . . . . (£1,977) 122
Alfa Romeo 1750 GTV . (£2,346) 116
BMW 2002 Tii . . . . (£2,299) 116
Triumph Stag . . . . . (£2,227) 116
Ford Escort RS 1600 . . (£1,496) 113

0-60 MPH, SEC
BMW 2002 Tii 8.3
Ford Escort RS 1600 . . . . . 8.9
Rover 3500S 9.1
Triumph Stag 9.3
Alfa Romeo 1750 GTV 11.2

STANDING ¼-MILE, SEC
BMW 2002 Tii 16.4
Ford Escort RS 1600 . . . . 16.7
Rover 3500S 16.8
Triumph Stag 17.1
Alfa Romeo 1750 GTV . . . 18.0

OVERALL MPG
BMW 2002 Tii 25.4
Alfa Romeo 1750 GTV . . . 23.9
Ford Escort RS 1600 . . . . 21.5
Triumph Stag 20.7
Rover 3500S 20.1

GEARING
(with 165—13 in. tyres)
Top . . . . . 19.5 mph per 1,000 rpm
3rd . . . . . 14.8 mph per 1,000 rpm
2nd . . . . . 9.65 mph per 1,000 rpm
1st . . . . . 5.2 mph per 1,000 rpm

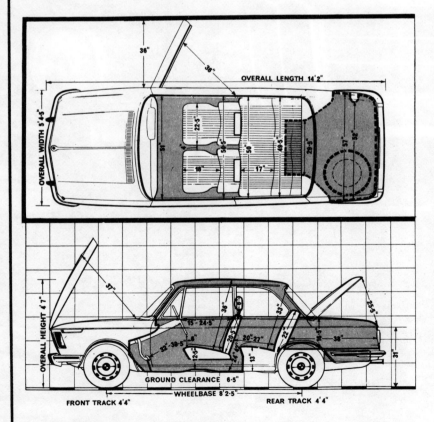

CONSUMPTION

FUEL
(At constant speed — mpg)
Fuel injection system, incompatible with flowmeter test equipment.

Typical mpg . . 28 (10.1 litres/100km)
Calculated (DIN) mpg 32.1 (8.8 litres/100km)*
Overall mpg . . 25.4 (11.1 litres/100km)
Grade of fuel . . . Super Premium, 5-star
(min. 100 RM)
*Manufacturer's figure

OIL
Consumption . . . . . . (SAE 20W/50)

TEST CONDITIONS:
Weather: Fair Wind: 5-10 mph.
Temperature: 2 deg. C. (36 deg. F)
Barometer: 29.8 in. hg. Humidity: 60 percent.
Surfaces: dry concrete and asphalt.

WEIGHT:
Kerb weight 20.6 cwt (2,312lb-1,054 kg)
(with oil, water and half full fuel tank).
Distribution, per cent F, 54.8; R, 45.2.
Laden as tested: 23.3 cwt (2,612lb-1,185kg).

TURNING CIRCLES:
Between kerbs L, 33 ft 6 in.; R, 32 ft 1 in.
Between walls L, 35 ft 0 in.; R, 33 ft 7 in.
Steering wheel turns, lock to lock 3.6.
Figures taken at 12,200 miles by our own staff at the Motor Industry Research Association proving ground at Nuneaton, and on the Continent.

SPECIFICATION

FRONT ENGINE, REAR-WHEEL DRIVE

ENGINE
Cylinders	4, in line
Main bearings	5
Cooling system	Water; pump, fan and thermostat
Bore	89.0mm (3.50in.)
Stroke	80.0mm (3.15in.)
Displacement	1,990 c.c. (121.4 cu.in.)
Valve gear	Single overhead camshaft and rockers
Compression ratio	10.0-to-1 Min. octane rating: 100RM
Induction	Kügelfisher indirect injection
Fuel pump	Kügelfisher high-pressure
Oil filter	Full flow, replaceable cartridge type
Max. power	130 bhp (DIN) at 5,800 rpm
Max. torque	131 lb.ft (DIN) at 4,500 rpm

TRANSMISSION
Clutch	Fichtel and Sachs, diaphragm-spring, 7.9in. dia.
Gearbox	4-speed, all-synchromesh
Gear ratios	Top 1.0
	Third 1.32
	Second 2.02
	First 3.77
	Reverse 4.09
Final drive	Hypoid bevel, ratio 3.45-to-1

CHASSIS AND BODY
Construction	Integral, with steel body

SUSPENSION
Front	Independent; MacPherson struts, lower wishbones, coil springs, telescopic dampers, anti-roll bar
Rear	Independent; semi-trailing arms, coil springs, telescopic dampers, anti-roll bar

STEERING
Type	ZF-Gemmer worm and roller
Wheel dia.	15.7in.

BRAKES
Make and type	ATE disc front, drum rear
Servo	ATE vacuum
Dimensions	F 10.0in. dia.
	R 9.0in. dia. 1.57in. wide shoes
Swept area	F 210.4 sq. in., R 88.6 sq. in.
	Total 299 sq. in.

WHEELS
Type	Pressed steel disc, 4-stud fixing, 5.0 in. wide rim
Tyres — make	Michelin
— type	XAS radial ply tubed
— size	165-13in.

EQUIPMENT
Battery	12 Volt 44 Ah.
Alternator	Bosch 53 amp a.c.
Headlamps	Hella, 80/90 watt (total)
Reversing lamp	Standard
Electric fuses	12
Screen wipers	Two-speed
Screen washer	Standard, electric
Interior heater	Standard, water valve type
Heated backlight	Extra
Safety belts	Extra, mounting points standard
Interior trim	Pvc seats and headlining
Floor covering	Carpet
Jack	Screw pillar type
Jacking points	2 each side under sills
Windscreen	Laminated

MAINTENANCE
Fuel tank	10.1 Imp. gallons
Cooling system	12 pints (including heater)
Engine sump	SAE 20W/50
	Change oil every 4,000 miles. Change filter element every 4,000 miles.
Gearbox	2 pints SAE 90EP. Check oil every 4,000 miles
Final drive	1.75 pints SAE 90EP. Check oil every 4,000 miles
Grease	No points
Tyre pressures	F 27; R 27 psi (normal driving)
	F 27; R 30 psi (full load)
Max. payload	750 lb (335 kg)

PERFORMANCE DATA
Top gear mph per 1,000 rpm	19.5
Mean piston speed at max. power	3,050 ft/min.
Bhp per ton laden	111.5 (DIN)

AUTOTEST
BMW 2002 TII ...

Matching changes

Relatively few changes have had to be made to the chassis to accommodate this extra output. The front disc brakes are slightly larger, and the rim width of the wheels goes up from 4.5 to 5in.; the tyres remain 165-13s, with Michelin XAs fitted to the test car.

The most important change in many ways is the raising of the final drive ratio from 3.64 to 3.45, the same as that fitted to the big 3.0S (although that, of course, runs on larger wheels). The result is a sensible overall gearing of 19.5 mph per 1,000 rpm, with 60 mph attainable in second gear and 90 mph in third. All the four-cylinder BMWs share the same gearbox, with rather closer ratios (and especially, a higher second and third) than that fitted to the six-cylinder cars.

Clearly, it is the performance which is the main point of interest where the Tii is concerned, and it fulfils all expectations. Because of the higher gearing and the higher torque peak, it feels to have lost something right at the bottom end. The standing-start time of 3.2 sec to 30 mph is no better than we achieved with our last 2002; nor is there any advantage to be seen in some of the lower speed increments in each gear. In practical terms, this is most noticed by the lazy driver. If he stays in top gear to pull away from (say) 20 mph, the engine remains perfectly smooth but things will not happen all that fast. Changing down to third helps a lot, but second is the gear he really needs. In other words, the Tii lacks some of the low-speed punch of its lower powered brother.

It doesn't take long, however, to discover the other side of the picture. From 40 mph onwards in the standing-start acceleration runs, the Tii carves greater and greater margins off the 2002 times. For instance, 60 mph comes up in 8.3 rather than 10.1 sec, turning a respectable time into one which is fast by any standard. The quarter-mile is passed in 16.4 sec, and 100 mph is reached in 28.2 — compared with 43.2 for the 2002. This point is reached just before the kilometre post, which is reached in 30.3 sec; and the half-minute kilometre is one of our yardsticks of a *really* fast car, which usually means something pretty exotic.

This sort of performance has been achieved without making the engine in any way temperamental. For starting, the driver must take note of the instruction *not* to touch the accelerator pedal until the engine is turning, and then to depress it gradually until the engine catches. This invariably takes a second or two of churning, with the accent on the invariably; even on the coldest morning of the test, with several degrees of frost, the performance was still exactly the same. The engine takes a few seconds to clear its throat, so to speak, and then pulls perfectly.

There is never any missing or spluttering in heavy traffic conditions, and the exhaust system is very quite, so that the car can be trickled along with only the most knowledgeable on-lookers aware of its performance potential. The test car had a slightly uneven tickover when warm, the revs oscillating gently between 600 and 1,000 rpm.

Good consumption

The overall consumption of our last test 2002 was 25.5; for the Tii it was 25.4 mpg. The high operating pressure of the Kügelfisher injection system makes it impossible to use our Petrometa test equipment, but BMW claim a DIN touring consumption (i.e., a steady 70 mph less an allowance for acceleration) of over 32 mpg.

In fact, the nature of the beast makes it unlikely that many owners will approach this figure. Given gentle treatment and open-road work it is certainly within reach; our best brim-to-brim figure was actually 31.1 mpg. But really hard driving, or commuting in heavy traffic, are hard on the consumption and bring it down to the low 20s.

One of the things which militates against reaching 30 mpg, oddly enough, is the spacing of the intermediate gears. With such a high third, one is tempted not merely to use it, but to stay in it along considerable stretches of winding road, and second is frequently used by an enthusiastic driver on twisting routes. By contrast, first is rather low and not often used except away from rest and in very tight bends.

At the other end of the scale, the higher final drive means that top gear matches the maximum speed quite well. Our recorded mean maximum of 116 mph actually corresponds to 5,950 rpm, which is 150 rpm over the power peak, indicating that in theory the Tii is still slightly undergeared. What matters much more, however, is the ability to cruise serenely at anything up to 100 mph. The speedometer on the test car became progressively less accurate as speed increased, and at the 116 mph maximum was

BMW stick to their usual neat instrument layout in the Tii, adding nothing to remind the driver of the injection system under the bonnet — except the extra performance. One noteworthy omission is any form of face-level ventilation

AUTOTEST
BMW 2002 TII...

showing well over 120. The rev counter, on the other hand, was commendably accurate, with a small, near-constant error on the safe side.

Handling and brakes

For a car with quite a lot of its weight over the front wheels (the split is near enough 55:45) the Tii handles well. There is not as much understeer as one might expect and the handling is relatively little affected by power. If the throttle is opened in mid-corner, the car seems to squat down and feels more stable; conversely, if the driver lifts off, it feels "on tiptoe" and wavers slightly. In neither case, however, does it vary its line by more than a fraction.

At the limit, which is naturally more easily reached in the Tii, the limitations of the semi-trailing rear suspension start to show themselves as the back end jacks-up slightly and starts to drift sideways. Full-blooded oversteer is only encountered on very fast bends, because the more usual result is for the inside rear wheel to lift and spin, slowing the car (unlike the big BMWs, the Tii does not have a limited slip differential).

Even at this end of the handling scale, which is surely beyond the aspirations of most private owners except in emergency, correction is easy and applied almost entirely by steering, while keeping on enough power to help stability. In more normal cornering, the Tii feels exceptionally reassuring, with its supple suspension keeping all four wheels on the ground even over humps and uneven surfaces.

The steering itself is distinctly middleweight, becoming heavy at low speeds. It is precise, with no lost motion about the straight-ahead position, and yet is free from all but the very worst effects of kick-back from poor surfaces. Straight-line stability is generally good, but some correction has to be made for gusty sidewinds.

In some ways, the BMW's brakes are slightly less reassuring. In normal use they respond well, except that with very gentle braking the results are less than anticipated. They also have plenty of ultimate stopping power, although it is difficult to lock the wheels in anything but a real panic stop. When the driver is pressing on, however, and using the brakes hard and frequently, they can start to feel rather "soggy", even though they never lose their ultimate effectiveness. Our fade test shows a pattern of pedal pressure reducing to begin with, then climbing and becoming speed-sensitive (with the driver having to press harder as the car slows down) and finally a tendency towards recovery. The handbrake proved no more than reasonably powerful on the level, yet held the car either way on the 1 in 3 test slope. A restart was possible on this slope, with the aid of some wheelspin.

Comfort and controls

The suppleness of the suspension, apart from helping the handling, also ensures a better than average ride. It is a good compromise between floating over rough surfaces and soaking up single large bumps or hollows, doing neither perfectly (which is very difficult) but both tolerably well. The damping is of high quality and prevents any feeling of "float". Pitching is also well controlled, but noticeable roll angles are reached in hard cornering.

The suspension has to contribute most of the ride quality, because the seats are rather hard.

Those in front are well shaped, but more (it seems) with large people in mind. The back seat is a plain bench with no centre armrest and a slight shortage of headroom; but the shortage of knee room is much more likely to worry passengers riding behind tall front seat occupants.

The controls in the small BMWs have been rearranged slightly of late. There are now two column-mounted stalks, one for dip and flash and the other for indicators, wipe and wash. Apart from a tendency to wash instead of flash, we found the arrangement a good one.

An earlier modification put right the complaints about the high-set pedals of the 2002, and there is now nothing to criticize beyond a barely noticeable offset. Nor does the steering wheel seem out of place, although it is large by modern standards. All the minor controls are within reach of a strapped-in driver, but we found the belts themselves very awkward to use.

For the most part, visibility for the driver is splendid, with large windows, slim pillars and a good view of all four corners of the car (barring the obstructive headrests). The effect is rather spoilt in wet weather by the wiper pattern, which is still set up for left hand drive, and by the persistence of misting in the interior. This is a symptom of the poor ventilation arrangements — the car badly needs face-level fresh air inlets to bring it up to date — and suggests that the optional extra heated rear window would be an investment. Indeed, it is surprising not to find it part of the standard equipment on a car of this class.

The lack of face-level ventilation means that the highly effective heater tends to make the interior dry and stuffy.

So far as noise is concerned, the extra power has not made the car any louder, inside or out. By far the most noticeable noise when cruising at really high speed is the wind noise from around the edges of the frameless windows. Road noise is kept well under control.

A worthwhile increase

The Tii is sufficiently different from the basic 2002 — much more so than at first sight — for customers to ask themselves which model would really suit them best.

The Tii is a fast car by any standards; not so much in maximum speed as in mid-range acceleration. It has few peers when it comes to hustling across country on mediocre roads; none at all, perhaps, if one considers the comfort and quietness which go with it. At the same time, drivers who do not appreciate the upper limits of performance, who are maybe less keen on gearchanging, and who do more than their fair share of town driving might be better off with the plain 2002. It is to BMW's credit that they offer the choice. □

Massive plenum chamber and inlet manifold arrangement under the bonnet seriously reduce accessibility of some items. All fluid levels can be checked readily, however, and the fuse box is neatly installed

A functional black rubber side stripe and a discreet badge on the rear panel serve to distinguish this fuel-injected 2002 from its lower-powered brother. Note the large window area and slim pillars, a great help in everyday driving

MANUFACTURER:
Bayerische Motorenwerke AG, München 13, West Germany

UK CONCESSIONAIRES:
BMW Concessionaires (GB) Ltd., BMW House, 361-365 Chiswick High Road, London W4

PRICES
Basic	£1,837.70
Purchase Tax	£461.30
Seat belts (approx.)	£14.00
Total (in G.B.)	**£2,313.00**

PRICE AS TESTED £2,313.00

The mountain marvel

A look at the BMW 2002 TII sedan with fuel injection

Anybody who knows a better all-around mountaineering sedan than the new BMW 2002 TII must have a hot line to the devil.

Following nearly 2000, mostly alpine, miles in this fuel-injected missile I would be more than willing to admit that its ventilation is laughable for instance. Or that the famous 2002 chassis is getting towards the end of its string for flat-out motoring with 130 hp in the nose.

But it remains the most nimble hairpin charger around with Kugelfischer injection which all but turns 7500 foot pass roads into sea-level freeways. The TII has more punch than a TI with less fuss. This is the engine announced nearly two years ago. They waited until it was really right.

Meanwhile BMW was lining up a few cosmetic changes in line with customer murmurs. These apply to all two-door BMWs: things like the elastic plastic (looks like rubber) stripe down each side and ditto insets in the bumpers. And the finally-reclining seats or more readable dials. In the TII you get a rev counter, heated rear window and leatherette steering wheel as well.

The body, a 320,000 best seller in five output versions since 1966, is part of the pleasure of seeking out smaller passes. Vision remains outstanding five years after it was first designed.

The English market these cars as coupes which is closer to the truth than sedan though you can carry a couple in back on short trips. Two and their luggage have really rich space, however, with the bonus of good small storage and touches like the cross-ribbed dash shelf which prevents the sunglasses sliding about.

The good-size trunk does get warmer from their exhaust than seems necessary, be it noted. And the door windows don't wind clear down despite muscle-bound handles. It takes an NFL lineman to work the knobs for their wind wings. So there are a few things left for the Munich firm to fix.

BMW also claims better air extraction in the 1971 version thanks to vents above the rear window. Forget it. This car was a sauna south of the alps, not to mention noisier than hell with the windows open as they had to be for survival.

Along with better dials and more knee padding BMW finally fitted three-speed wipers and moved their switch to a wand under the steering wheel, literally at your fingertips. That TII wheel has a thin rim and a little better grip than plastic but I still wanted gloves on hot, sticky days.

Hillclimbing by car is greatly eased by the light, direct steering and quick, precise gearbox action. Their optional five-speed box with top still direct would make it even nicer. A limited slip differential would smack of paradise. In any case a slightly longer than 2002-normal final drive ratio (3:45) gives the TII an easy freeway gait.

This is a 119 mph sedan after all, capable of 16 second quarter mile runs on a mere 120 cubes. What's more we drove it virtually flat out in all gears for two weeks, including five of the steepest (up to 16%) and highest (over 7500 feet) passes in the Alps and still got overall fuel consumption of 18.8 mpg, plus 1.5 quarts of oil in 1950 miles.

Using the car for family runs as well as high altitude work would make 25 mpg a cinch. The injection system, keyed to throttle setting, revolutions and engine temperature is very little thirstier than that single carb on the normal 2002.

With two and plenty of luggage aboard the car would still pull uphill cleanly from 2500 revs in top. It was always a cinch to start provided you remembered to stay off the throttle pedal until it fired when hot. Compared to the temperamental TI with two dual-body carbs which has been discontinued, the TII with Kugelfischer injection not only has more power and a torque band better matched to its gears, but it is far easier to drive around town as well.

Cooling is the chief remaining engine problem. Ever since the first two-liter model introduced in January of 1968 they have faced marginal cooling. When you try to use these 147 horses for any length of time, say down a hot Italian autostrada, the temperature needle will go all over red.

After all, 6000 in top is a true 115 mph and the TII just won't take that for more than 20 or 30 minutes at a time when temperatures get up around the 90 mark. This information is brought to you courtesy of the Nevada desert club.

Pushed right along up a winding road the BMW remains nearly neutral until you use all that new power to hang the tail out at any angle desired. The car shows a good deal of roll by modern standards and this serves to warn the less than ultra-expert driver that he is approaching oversteer. Those who like to charge will still find a BMW ideal.

On the other hand a chassis which first appeared with over a third less power to set the handling standards five years ago is bound to show its age by now. BMW did beef up the suspension members front and rear for this added poke. The relative softness (by sporty sedan standards, that is) only bothered us in very tight descending hairpins when you could rub a tire on the fender.

We did manage to get some brake smell on our faster downhill rushes but there was never any pedal loss and the pads continued to work without evidence of glazing.

Brake circuits are set up on a double-dual system. If one circuit fails it leaves 75% of your brake power, if the other goes out, the worst conceivable case, a driver still has 60% of his very good brakes. This is a soothing thought at the high speeds which seem so natural to a TII.

Factors like this give the new BMW its special charm on unknown, winding roads. You can hustle along, achieving amazingly good point to point averages, and yet hoard a large margin of handling safety for the unexpected.

This 2002 TII proved to be very much an automobile and engine worth waiting for.

BMW 2002 T11 2-Door Sedan
Data in Brief

PERFORMANCE AND MAINTENANCE

Acceleration:		Gears:
0-30 mph	3.5 secs.	I
0-45 mph	5.7 secs.	I, II
0-60 mph	9.0 secs.	I, II
0-75 mph	13.7 secs.	I, II, III
0-¼ mile	16.1 secs.	
Ideal cruise		90 mph
Top speed		119 mph
Average economy (city)		17.5 mpg
Average economy (country)		20 mpg
Fuel required		Super
Oil change (miles)		3500
Lubrication (miles)		3500
Type tools required		Metric

SPECIFICATIONS AS TESTED

Engine .... 121.4 cu. in.
injected inline 4
Bore & stroke .... 3.5 × 3.15 ins.
Compression ratio .... 10 to one
Horsepower .... 147 (SAE gross) @ 5800 rpms
Torque .... 150.5 lbs.-ft. @ 4500 rpms
Transmission .... 4-speed, manual
**Steering .... 3.5 turns, lock to lock
31.5 ft., curb to curb
**Brakes .... disc front, drum rear
Suspension .... spring leg front
angled wishbone coil rear
Tires .... 165 HR 13 radial
Dimensions (ins.) .... Wheelbase 98.4
length 166.6, width 62.6, height 55.5
front track 53.1, rear track 53.1
ground clearance 6.3, weight 2180 lbs.
Capacities .... Fuel 11.9 gals.
oil 4.5 qts., coolant 7.4 qts.
trunk 15.8 cu. ft.

ROAD TEST

BMW 2002 TII

Modern suspension and engine developments have blurred the distinction between sports cars and saloons, many four-seat closed cars now being quicker — round corners as well as on the straight — than ostensibly more sporting soft-topped rivals of similar capacity. This trend is admirably exemplified by the BMW 2002 range of cars which are fast as well as being sufficiently compact for sports car agility and adhesion, yet large enough to seat four adults and carry a good deal of luggage.

Even in its lowest state of tune, as fitted to the ordinary 2002, the 100 net bhp produced by the BMW 2-litre four exceeds the maximum power developed by some similarly sized sports-car engines, though 50 bhp/litre has been reliably attainable for many years. With twin Solex carburetters, however, the output goes up to 120 bhp for the 2002 Ti (Touring International) which we have not tested as it is not available in this country. But the fastest of them all, the 2002 Tii, can be bought in Britain. The extra "i" stands for injection (by Kugelfischer) which gives an extra 10 bhp, making the car 30 bhp more powerful than the 2002 we originally tested in June 1968.

It took several attempts to prove the effectiveness of this extra 30 bhp as our test car suffered from some elusive trouble. In the end we confirmed that on acceleration the Tii is usefully faster than our road test 2002 which itself was at the top end of the performance bracket — 0-60 mph is reduced from 9.2 to 8.2s. However, for maximum speed the best we could achieve was "only" 113 mph — a figure claimed for the 2002 by some journalists who can only read speedometers. In early tests of the Tii we achieved 113 mph quite easily, but by the time the acceleration had improved to a satisfactory level after various adjustments the car was unable to maintain higher speeds for a complete lap at MIRA, although the best timed quarter-mile was as high as 120 mph. On a flat road the predicted maximum is around 117 mph obtained from the mean of opposite runs. This failure to perform as BMW expect is unusual — normally we better the German company's ultra-conservative claims quite substantially.

Nevertheless the 2002 Tii is a very fast 2-litre as well as a fine sports saloon. Stiffer than the plain 2002, its handling is tauter and its roadholding marginally better, but the ride on broken surfaces is noisier. The absence of proper face-level ventilation, however, is a serious fault on such an expensive car. At nearly £2300 the Tii is £400 more than the 2002.

Performance and economy

There is a lot to be said for the BMW arrangement of a single overhead camshaft operating inclined valves via twin rocker shafts. It certainly doesn't hamper the free-revving Tii engine which is quite happy to go beyond 6000 rpm in the lower ratios without getting at all rough. The camshaft is chain driven (you can hear a sophisticated whine when the unit starts from cold). The injection system has no enrichening lever and takes care of cold starts automatically. According to the handbook you are not supposed to touch the accelerator pedal when starting but we found it helped if it was lightly touched since normal tickover — not easily adjustable — can be as low as 400 rpm, a throttle setting which is a little low to "catch" on a cold morning.

Early Tii units used to suffer from a cyclic idling beat from about 400-1000 rpm but this seems to have been cured and the Kugelfischer system is now completely untemperamental. The pull is even from low

revs although the engine isn't as happy to slog from under 10 mph in third as that of the 2002. Figures are a little misleading, however, as the more powerful engine is coupled to a final drive 5 per cent higher. The Tii was fractionally slower in top and third up to about 60 mph than the 2002 but beyond that the extra 30 bhp is evident— the 70-90 mph acceleration time in third comes down from 11.3 to 8.1s. The rev counter is red-lined at 6400 rpm but the car will reach 60 mph in second without exceeding this speed. The engine is smooth and mechanically quiet up to around 5500 rpm

but it begins to get a little harsher beyond this.

Acceleration from rest is much better than that of the 2002; it reaches 50 mph half a second sooner and 60 mph in a second less. The slightly higher bottom gear doesn't affect the ease with which the BMW takes off nor its ability to romp away from rest on a 1-in-3 hill.

The maximum speed we've quoted is in fact our maximile figure — the speed attained after one mile's acceleration from rest. Normally we build up to the full maximum over one 2.8-mile lap and time the next flyer, but the Tii started to fluff after less than a mile of the first flying lap by which time we had recorded a quarter-mile at 120 mph. By comparison with the 100 bhp 2002 the Tii should go 10 per cent faster on ideal gearing; this would give 117 mph which at 5950 rpm is just over the power peak at 5800 rpm.

It isn't possible to fit our petrol flow measuring apparatus to a fuel-injected engine but the factory claims 32 mpg at a steady 70 mph. This ties in well with our overall consumption of 24.0 mph which includes a large mileage at home and abroad. At worst it was 22 mpg, at best 28 mpg. With a full 10.1 gallon fuel tank you can just scrape 250 miles when cruising at around 80 mph and using most of the considerable acceleration. But a bigger tank would be welcome.

Transmission

Although the synchromesh has been altered in the Tii box, the gearchange feels much like that of the 2002 — it is difficult to improve on something that is already very good. The ratios are well chosen and second is very useful for overtaking. While in

Germany we tried the five-speed box that is optional in Europe and found that the ratios were unnecessarily close for the standard torquey engine. First was higher and fifth the same. On the standard box the change is very smooth and you can slice the lever around the gate confident that you will always get the right ratio and that the synchromesh will always be strong enough to allow easy engagement. The clutch was light and cushioned the drive well. We couldn't hear any noise from the body-mounted final drive unit but there was some gear whine in first.

Handling and brakes

To help cope with the extra performance a number of changes have been made to the running gear, including stiffer bodywork and dampers. The difference isn't immediately apparent (unless you've just driven a 2002), the stiffer damping being most noticeable on S bends as the car is easier to line up without uncomfortable lurching. The wheels now have 5J rims but still with 165 x 13 tyres to HR rating. On several different cars we tried Uniroyals, Continental and Michelin XAS: the Continentals were better than the Uniroyals on dry roads because they didn't squeal; the Michelins squealed but were good in the wet.

The Tii can out corner most saloons and has good steering response. There isn't too much roll and usually the handling is neutral. If you throw the car around it oversteers but in a very safe, controllable fashion. It oversteers if you ease the throttle in mid-corner, too, probably because of the semi-trailing arm rear suspension geometry. This is a safe feature and merely allows you to tighten the line without sudden steering movements.

Changes to the braking include bigger rear drums and two separately servoed systems; the pedal isn't too light, though. There was a bit of judder at the end of a long hard stop from high speed but the car passed our fade test with only a little smoke as protest. There was no increase in pedal pressure either then or after the water splash test. The centre mounted handbrake holds the car on a 1-in-3 slope and provides average emergency braking.

Comfort and controls

The seat slides through 12 reasonable positions (plus some more unreasonable forward ones) and the back rest has 18 angles, of which 10 can be used for driving. As a result most of our drivers found satisfactory settings without the high steering wheel being too uncomfortable for them to reach, though the driving position is not as good as that in the big BMWs. The pedals, too, are well placed, the organ-type throttle being easily blipped for heel and toe use. Adjustable head restraints cost extra.

For access to the rear seat there is a quick release knob on the squabs, and there is ample room for two adults in the back. The rake adjusting bar looks rather unpleasant at ankle level, though.

Stiffening the suspension hasn't improved the ride over indifferent surfaces though it is better damped for high speed undulations. The ride of the two-door BMWs isn't as good as it might be for an independent layout, as

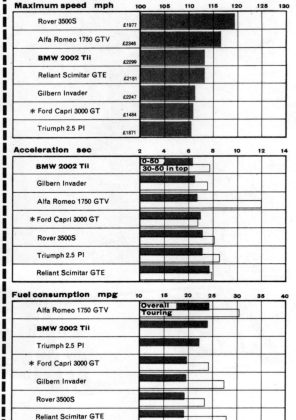

Motor Road Test No. 58/71   BMW 2002Tii

Make: BMW.
Model: 2002Tii.
Makers: BMW AG., Munich 13, W. Germany.
UK Concessionaires: BMW Concessionaires (GB) Ltd., BMW House 361-365 Chiswick High Road, London W4.
Price: £1837.70 plus £461.30 equals £2299.00.

Performance tests carried out by *Motor's* staff at the Motor Industry Research Association proving ground, Lindley.
Test Data: World copyright reserved; no unauthorised reproduction in whole or in part.

Conditions
Weather: Cloudy with winds 7-18 mph.
Temperature: 46-55°F.
Barometer: 29.8in. Hg.
Surface: Dry tarmacadam.
Fuel: Super premium 100 octane (RM) 5-star rating.

Maximum Speeds
	mph	kph
Mean maximum (see text)	113.5	183
Best one-way ¼-mile	120.0	194
3rd gear	96	155
2nd gear } at 6500rpm	63	102
1st gear	27	44

"Maximile speed: (Timed quarter mile after 1 mile accelerating from rest)
Mean   113.5
Best   118.5

Acceleration Times
mph	sec
0-30	2.7
0-40	4.4
0-50	6.3
0-60	8.2
0-70	11.6
0-80	15.0
0-90	20.2
0-100	28.8
Standing quarter mile	16.5
Standing Kilometre	30.6

mph	Top sec	3rd sec
20-40	7.7	5.4
30-50	7.8	5.4
40-60	8.0	5.6
50-70	8.4	5.7
60-80	9.6	6.2
70-90	10.8	8.1
80-100	13.1	—

Fuel Consumption
Overall consumption .. 24.0 mpg
(=11.8 litres/100km)
Total test distance ... 3182 miles

Brakes
Pedal pressure, deceleration and equivalent stopping distance from 30 mph.
lb.	g.	ft.
25	0.32	94
50	0.75	40
65	0.96	31
Handbrake	0.32	94

Fade Test
20 stops at ½g deceleration at 1min. intervals from a speed midway between 40 mph and maximum speed (=77 mph).

	lb.
Pedal force at beginning	38
Pedal force at 10th stop	42
Pedal force at 20th stop	35

Steering
	ft.
Turning circle between kerbs:	
Left	30.0
Right	29.2
Turns of steering wheel from lock to lock	3.6
Steering wheel deflection for 50ft diameter circle	1.1 turns

Clutch
Free pedal movement = ⅜in.
Additional movement to disengage clutch completely = 3¼in.
Maximum pedal load = 25lb.

Speedometer
Indicated	20	30	40	50	60	70
True	19	29	40	49	58	67½
Indicated			80	90	100	
True			77½	87	97½	

Distance recorder 2% fast.

Weight
Kerb weight (unladen with fuel for approximately 50 miles) 20.2 cwt.
Front/rear distribution 56½/43½
Weight laden as tested 23.9 cwt.

Motor Road Test No. 58/71 BMW 2002Tii

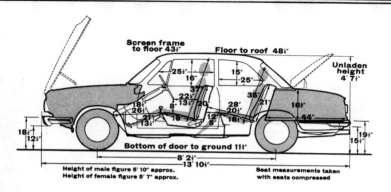

Engine
Block material	Cast iron
Head material	Aluminium alloy
Cylinders	4 in line
Cooling system	Water pump, fan and thermostat
Bore and stroke	89mm (3.50in.) 80mm (3.15in.)
Cubic capacity	1990cc. (121.4 cu.in.)
Main bearings	5
Valves	Single ohc with rockers
Compression ratio	10:1
Carburation	Kugelfischer fuel injection
Fuel pump	Mechanical
Oil Filter	Full flow
Max. power (net)	130 bhp at 5800 rpm
Max. torque (net)	131 lb.ft. at 4500 rpm

Transmission
Clutch	Diaphragm sprung sdp
Internal gear box ratios	
Top gear	1.0
3rd gear	1.32
2nd gear	2.015
1st gear	3.767
Reverse	4.096
Synchromesh	on all forward ratios
Final drive	Hypoid bevel 3.45:1
Mph at 1000 rpm in:	
Top gear	19.6
third gear	14.8
second gear	9.7
first gear	5.2

Chassis and body
Construction	Unitary

Brakes
Type	Atc disc/drum with dual servoid systems
Dimensions	9.5in. dia. discs, 9in. dia. drums

Suspension and steering
Front	Independent with Macpherson struts and lower wishbone, anti-roll bar
Rear	Independent by semi-trailing arms with coil springs and anti-roll bar
Shock absorbers:	
Front:	In strut
Rear:	Telescopic
Steering type	ZF worm and roller
Tyres	165 HR x 13 Continental
Wheels	Pressed steel disc
Rim size	5J x 13

Coachwork and equipment
Starting handle	None
Tool kit contents	Three spanners, pliers, screwdriver, plug spanner
Jack	Screw type
Jacking points	One in each side at centre
Battery	12 volt negative earth 44 amp hrs capacity
Number of electrical fuses	11 + 1 (heated rear screen)
Headlamps	Two 6.7in. 45/40W
Indicators	Self-cancelling flashers
Reversing lamp	Yes
Screen wipers	Two-speed electric
Screen washers	Two-speed
Sun visors	Two, padded
Locks:	
With ignition key	Steering
With other keys	Doors and boot
Interior heater	Fresh air
Upholstery	Skai
Floor covering	Carpet
Alternative body styles	None
Maximum load	880 lb.
Maximum roof rack load	150 lb.
Major extras available	Heated rear window, alloy wheels, head rests, rally seats, 4 QH headlights

Maintenance
Fuel tank capacity	10.1 galls
Sump	7½ pints SAE 20/50
Gearbox	2.1 pints SAE 80
Rear axle	2.6 pints SAE 90
Steering gear	0.5 pints SAE 90
Coolant	12.3 pts. (2 drain taps)
Chassis lubrication	None
Maximum service interval	8000 miles
Ignition timing	24° btdc at 2200 rpm
Contact breaker gap	0.016in.
Sparking plug gap	0.027in.
Sparking plug type	Bosch W175T30
Tappet clearance (cold)	
Inlet	0.007in.
Exhaust	0.007in.
Valve timing:	
inlet opens	4° btdc
inlet closes	52° abdc
exhaust opens	52° bbdc
exhaust closes	4° atdc
Rear wheel toe-in	0-0.12in.
Front wheel toe-in	0-0.08in.
Camber angle	—
Castor angle	4-4½°
King pin inclination	8½°
Tyre pressures:	
Front:	28 psi
Rear:	28 psi

the fairly high stance of the car requires stiffish springing for the handling that BMW owners expect. Surface breaks send a shudder through the frame as if there isn't enough compliance in the front suspension. At higher speeds it is much happier and the seats absorb road shocks as well, so passengers are never jerked around.

The interior is very airy with large areas of glass for good all-round visibility; you can also see the rear deck from the front seats which makes reversing easy. The rear bumper has a full width rubber facing and the edges and overriders are rubber faced at the front to guard against clumsy parkers.

The standard BMW lights give fair illumination with a typical Continental cut-off. One of the available extras is the system used on the 6-cylinder cars, with four QH lights and a modified front grille, at £65. These lights are extremely good.

Frameless windows are not the easiest to seal against wind noise. That on the two-door cars is not obtrusive until you are going over 80 mph. Up to that speed the car is pretty quiet as the engine seems hardly to be working. The interior air has positive outlets through the boot; opening the rear window increases the flow — and noise. The heater has controls for individual adjustment of volume to screen and interior with a third slide for temperature adjustment. Small slots at the end of the facia help keep the side windows demisted.

Fittings and furniture

Some people feel that a £2000 car should look rather more luxurious inside than does the sombre but functional black of the BMW. The three-dial instrument pod is well placed and informative and there is a clock in the centre. A shelf runs the full width of the facia with the instrument nacelle occupying part of it, leaving a small space on the right which takes a packet of cigarettes and little more.

Left of centre the tray is useful for papers and small objects but maps are best placed in the convenient console in front of the gearlever. There is a separate glove pocket in front of the passenger.

The carpeting and ambla/pvc seats are easily cleaned, as are the doors covered in the same material. The boot is very large, engulfing 10.8 cu.ft. of our Revelation cases. The spare wheel and tool kit are found under the floor.

Servicing and maintenance

The full-width bonnet is released by an over-centre lever on the passenger's side which is very easy to overlook after an oil check. But with its forward hinging action it is held down by its weight up to about 75 mph when underbonnet pressure lifts it slightly without any danger.

After initial servicing at 600 and 4000 miles, the BMW only needs garage visits every 8000 miles — an unusually long interval, so servicing is best left to the garage. Underbonnet accessibility is good enough for items that the private owner might want to check himself, like the distributor (which is just accessible against the bulkhead), the coil, plugs and reservoirs. A clear-topped fuse box is well sited on the top of the left wheel arch.

1 clock. 2 heater fan. 3 cigarette lighter. 4 screen demister. 5 fuel gauge, temp. gauge on right. 6 oil and alternator warning lights. 7 speedometer with trip and total mileage recorders. 8 trip zero. 9 rev counter. 10 lighting switch. 11 two-speed wiper control. 12 side window demister slot. 13 rear screen heater. 14 hazard warning flasher. 15 indicator flasher stalk. 16 footwell air flow. 17 indicator and main beam tell-tales. 18 horn. 19 wiper washer stalk. 20 heater temperature control

BMW 2002tii
When comparing 2002s, the "ii's" have it/By Eric Dahlquist

Impression

The traffic signals work differently in Munich than America. Instead of the amber light warning when the red is about to come on, it goes the other way 'round—like the Christmas tree at the drags. So you go down Maximilian Strasse on Friday night and choose off other guys at the stop lights just like they used to do on Detroit's Woodward Avenue back in 1966. Only the cars are different—BMWs and Alfas and Mercedes and Opels—and the traffic polizei smile because they get to see free races. The new king of Maximilian Strasse is the BMW 2002tii. But then you're down on the stretch of Autobahn between the Austrian border and Munich, flat out, hitting right around 115 mph, trying to stave off the challenge of a new BMW Bavaria 3.0. With one third more displacement and six more mph on the top end, it's only a matter of time until the Bavaria slides by but the tii makes it strain—strain hard. The BMW 2002tii isn't King of the Autobahns—but it's close.

The old 2002, the sports sedan we got and loved in the 'States didn't have it after 90, at least, not like this one. But nobody really missed it because most American 2002's only saw 90 occasionally anyway. There isn't any place for full-bore work in the U.S. except Nevada and if the truth be known, there aren't a lot of places left to do it in Germany either, so maybe that's why there is a 2002tii, a car small enough, nimble enough, and fast enough to play the old top speed game under the new restrictive rules but also be competitive in the cities. In 1969, the 2000tii was some of BMW's hot news at the Frankfurt Auto Show, a hot engine to bolster the sagging appeal of their 9-year-old intermediate. But in 1971, it's the crowning touch for their little super car, the 2002, a way to meet the emissions rap and save performance.

Someone commented earlier in the trip that the 2002tii may be wind sensitive over a hundred mph but that doesn't seem to be any great bother. You sit high in the car and the hood slope and large sweep of glass yield a kind of forward visibility available almost nowhere else except, perhaps, a bus. Arms out, grasping the pleasantly large steering wheel, the driver derives the sensation of ultimate control, of being well and truly the master of his fate. The tii feels so naturally right, the way all good automobiles ought to feel, completely comfortable through its entire speed range.

Based on the mini-racer 2002ti chassis, the tii has more robust front and rear brakes (disc/drum), generally stouter suspension components throughout (including anti-roll bars front and rear), and 1.4 inch wider track than the 2002's 52.4. At five inches, wheel rim width is broader by a half inch although the significant items are new H rated Michelin XAS 130 mph radials, superceding the 2002's 112 mph S type. Later, on a drive through the foothills of the Bavarian Alps, we measured the suspension's quality on spiralled, undulating two-lane roads and did not find it wanting.

What all of this means is that the tii driver can devastate almost everything else on the road, hauling down into corners at Porsche 911-like velocities, having the same power reserve to throttle steer as well. Handling is generally neutral with a very small amount of designed-in understeer for high speed stability. Steering is, if anything, lighter than the excellent 2002. Impressive beyond anything else is the clever way BMW's engineers have achieved this on a wide range of road surface disparities with seemingly no sacrifice to ride, a feat even mighty General Motors seems helpless to accomplish. The one thing you can do with a BMW is load in passengers and roar down almost any road without fear of bottoming; adequately damped, realistic suspension travel is the key, folks. If the tii had any riding compromise, it is a certain low speed harshness over things like tar strips and this can be laid as much to the steel tire belts as suspension.

Right here, after three days of generally flogging the tii over the autobahns and rural byways of Bavaria, we were ready to say that it was maybe the most fun BMW or anybody has ever built into a 〉〉〉

The Kugelfischer-injected 2002tii engine belts out 140 hp @ 5800 rpm with the same authority as Chrysler's 426 King Kong hemi but with a lot less fuss. High performance, low pollution.

BMW 2002tii

car, certainly the best small sedan we've tried in its class. In Munich's old center with its narrow, crowded streets and heavy traffic, the tii was even better. You need a small, agile car here, one that can carry four people yet squirt, waterbug-like, through tight gaps in the swirling traffic. On the face of it, a 121.4 CID engine pumping out 140 hp at 5800 (145 lbs. ft. of torque at 4500 rpm) might be a bit edgy at low revs but Kugelfischer mechanical fuel metering is so precise as to cancel the inherent slow-speed nervousness of the high-lift, long overlap, camshafts and tuned intake manifold runners. So, the tii surfaces as the German version of the original tri-power GTO Pontiac.

As far as quality goes, this BMW, all BMWs, are in a class by themselves. The paint flows like a competition orange mirror across the car, broken on the flanks only by inch-and-a-half wide, metal-backed plastic inserts intended to reduce parking damage. Everything fits properly. The doors shut with the solid, time-worn Teutonic authority Americans wistfully recall from Detroit twenty years ago. The seats are comfortable, fully adjustable and covered with a simulated leather material looking like it will be the last thing on the machine to wear out.

SPECIFICATIONS

Engine	OHC 4-cylinder
Displacement	1990cc (121.3 cid)
Bore and Stroke	3.50 x 3.15 ins.
Horsepower (SAE Gross)	140 hp @ 5800 rpm
Torque (SAE Gross)	145 lbs.-ft. @ 4500 rpm
Compression Ratio	9.0/1
Transmission	4-speed manual
Final Drive Ratio	3.45/1
Steering Ratio	17.58/1
Front Suspension	McPherson strut with shock coils
Rear Suspension	Semi-trailing arms with coils, tube shocks
Body	Welded all-steel unit
Brakes	Disc front/drum rear, vacuum assisted
Wheelbase	98.4 ins.
Overall Length	166.5 ins.
Height	55.5 ins.
Width	62.6 ins.
Track Front/Rear	53.1 ins.
Weight	3,065 lbs.

PERFORMANCE
Acceleration:
0-60 mph ..... 9.9 secs.
Top Speed ..... 115 mph
Gas Mileage Range ..... 22-25 mpg

If the 2002 tii has a fault, it is the conspicuous lack of a modern flow-through ventilation system incorporating adjustable face-level registers. True, by judicious manipulation of the vent wings and rear windows, a habitable temperature can be maintained, but at the expense of increased wind noise. And, with the front seats all the way back for tall drivers, rear passenger legroom suffers somewhat, though it is certainly no worse than American two-doors a third larger in size.

And there is one other thing too, though it is a suggestion rather than a complaint. On the way back from BMW's new proving ground we had a brief but spirited tussle with a brand new 2-liter Alfa Berlina. We won, happily, but it brought to mind the idea that the two cars are close in specification—2-liter engine, mechanical fuel injection, weight, size, price—except that the Alfa has a five-speed transmission as standard equipment. BMW offers an optional ZF 5-speed but it is a rare option and they are wondering whether to abandon it altogether and concentrate on other things. Certainly, the 2002 tii's flexibility is such that it doesn't really need a 5-speed and in fact they have gone up from the 2002's 3.64:1 final drive to 3.45, giving the machine an even longer-legged capability. Still, a 5-speed is a certain psychological filip for the kind of person who will buy the tii, in the same way a bicycle freak covets a 15-speed over a 10-speed.

On balance, 5-speed or no, the tii is one of the benchmarks in sedan touring car design, superbly executed and disappointing in only a few minor ways. One dark cloud, however, may loom on its horizon. With the German Deutschmark revalued and the new U.S. port-of-entry import surcharge, the price of all German cars is escalated significantly from what it was six months ago. At the time we were there, the factory was still deliberating on the tii's final U.S. sticker, somewhere said to be in the neighborhood of $4100, or around $400 more than the old 2002. Excellence doesn't come cheap. /MT

BMW's seating position and visibility are unexcelled. Note that brake and accelerator pedal are equal height for easy heel and toeing. Car could use better ventilation system.

10,000 miles Long Term Test
BMW 2002 AUTOMATIC

By David Thomas

Automatic transmission ideal for city driving, yet detracts very little from sporting appeal of this compact saloon. Rapid brake-pad wear and leaking final-drive only significant problems encountered during 10,000 hard miles. Unusually easy to maintain in pristine condition.

"THE World's Best 2-litre" proclaimed the sticker in XNJ 998J's rear window. In spite of being a BMW addict, I was ready to dismiss this as so much advertising ballyhoo. Now, having lived with the car for 10,000 miles, I'm not so sure!

From my own point of view, what is there to rival a 2002 in this capacity class? Family considerations dictate a practical four-seater, yet my personal preference is for something compact and sporting. The situation is further complicated by my insistence on reasonable standards of refinement and quietness. I can think of no other automatic which fits the bill. In manual form, the injection 2002 Tii is an attractive alternative — albeit at the disadvantage of a £400 increase in price. Having fond memories of a long-term Alfa Romeo 1750 Berlina, this (in current 2000 form) would also join my list of "possibles".

Performance

All too often, automatic transmission spells considerable penalties in terms of ultimate performance and overall economy. Not so with the 2002, whose ZF unit seems more efficient than most of its contemporaries. To underline the model's potential in automatic form, I will relate a somewhat amusing tale of an encounter which was sparked off by the aforementioned sticker. Early one morning, while waiting at an isolated set of traffic signals, a sporting vee-eight arrived alongside. The driver called my attention and, pointing at his own car, remarked "The world's best 3-litre!" On the green, he proceeded to demonstrate just what this meant in terms of performance. The temptation was too great, and the 2002 joined in the fray! To its surprising credit, it emerged the clear victor — a feat few automatic "fours" could hope to emulate.

Since *Autocar* has not previously published figures for a 2002 in automatic guise, some further comment may not be amiss. Mileage at the time of the car's visit to MIRA was around 6,000; as always, it was running extremely well.

Top speed averaged 103 mph, with 106 mph coming up in the faster direction (wind gusting to 10 mph). At the latter speed, the ludicrously optimistic speedometer showed 117 mph — all but round the clock! Why *do* BMW (along with certain other German manufacturers) always seem to fall foul of this one? In contrast, the tachometer seemed pretty accurate, indicating just a shade over the theoretical (no converter-slip) revs under steady-speed conditions.

Confirming impressions gained on the road, acceleration times were most impressive. Sufficient torque is available at the rear wheels to induce wheelspin on getaway, resulting in a 0-30 mph time of only 3.6 sec. Aided by this decidedly brisk step-off, the 2002 reaches 60 mph in 10.6 sec and the quarter-mile mark in 18.1 sec. These times are appreciably faster than those returned by *manual* versions of most rivals. The Rover 2000 TC, for example, achieved 3.9, 12.2 and 18.5 sec. With its electronic-injection engine, the Volvo 144 GL managed 4.1, 11.6 and 18.3 sec. Triumph 2000 times were 4.5, 14.9 and 19.7 sec. To be fair, these are more spacious, four-door models. A more realistic comparison is provided by the Capri 2000 GT. In manual form, this clocked 3.5, 10.6 and 18.0 sec — pretty much the same as the 2002 *automatic*. Fast through-the-gears times are meaningless unless backed by good transmission response. The 2002's ZF unit cannot be faulted on this score. Provision is made for automatic down-changing on part-throttle, with the result that the driver rarely feels it necessary to over-ride the control mechanism. In fact, two of my colleagues thought the unit over-sensitive to throttle movement, but I disagree on this point. For the record, the maximum speeds at which kickdown changes can be effected are 62 (top-to-intermediate) and 34 (intermediate-to-low). Automatic up-changes (with the pedal in the kick-down position) take place at 40 and 68 mph, with the tachometer indicating around 5,500 rpm in both cases. There is little or nothing to be gained by exceeding these revs, although another 900 rpm is available before the needle reaches the red.

In circumstances where it is beneficial to use the selector lever, such as the approach to a tightish corner, the unit's response is unusually rapid. This holds true whether one is changing down or up. One minor criticism concerns the abruptness of the over-run change into low when holding intermediate. At all other times, change-quality is good.

Petrol consumption

Unfortunately, no steady-speed fuel consumption checks were carried out. However, making due allowance for the 5.3 per cent odometer error (over-reading), the consumption averaged exactly 23 mpg over the 10,000-mile test period. Periodic checks show that it remained in the 21.5-23.5 mpg bracket for the first 9,000 miles. Thereafter, following a belated routine service by MLG of Chiswick (see later), it improved to 25.7 mpg. The latter figure included a gentle drive to MIRA and back, which resulted in a best-ever figure of 27.4 mpg. Summing up, it would seem that a briskly driven model in good tune should return 23-26 mpg.

No oil used

No matter how hard the 2002 was driven, no measurable quantity of oil was used. In addition, the power-unit remained almost spotlessly clean.

Taking delivery

Having dealt with the subject of performance and economy, let us now take a look at the day-to-day running of the car while in *Autocar's* hands.

On arrival at Dorset House, the odometer showed 87 miles. First chore was topping-up with four-star fuel as a prelude to our regular check on consumption. Thanks to the generous size of the cap and filler-neck, this involves no risk of blow-back. Total capacity is 10 gal — barely adequate for long-distance journeys.

Oil levels (engine and transmission) were checked and a quick tour of inspection made.

1. After 10,000 miles and a year in London, the car's dark blue paint was unmarked
2. One of the delightful features of the 2002 is its slim pillars and large glass area. The front grille (inset) is painted matt black
3. The engine stayed as clean as the exterior with no visible oil leaks
4. The test car had pvc seats, although there is now a cloth option for the central panels
5. It is the rather cramped back seat which justifies the name "coupé" more than the overall styling
6. The spare wheel lives in a well under the boot floor, with tools and jack neatly clamped around it

BMW 2002 AUTOMATIC LONG TERM TEST...

The only "fault" spotted was incorrect setting of the air-intake valve; this was duly switched from "S" to "W", which feeds pre-heated air to the carburettor (it was late November).

On the way home that evening, it soon became obvious that both factory and concessionaire had missed some minor faults. Worst of these was an abnormally fast tickover (about 1,200 rpm in neutral); another was misalignment of the headlamps (set much too high). A few minutes' work in a lay-by improved matters no end — thanks to the accessible thumb-screws (beneath easily removed covers) for headlamp adjustment. Other points noted were a tendency for the brakes to squeal when applied gently and failure of the indicator stalk to self-cancel. Other than this, it showed every promise of being a delightful car.

Getting acquainted

Starting from cold the following morning was a first-turn affair, but it was soon evident that the automatic choke was providing much too rich a mixture during the warm-up period. The bi-metallic spring used in this device features water and electric heating; having satisfied myself that the latter was working, I was content to leave well alone for the time being.

One of the first things that impressed me was the excellence of the ride and freedom from road noise. There was little doubt in my mind that the car represented a significant improvement on our previous long-term 2002 (a manual model). This impression was to last throughout my period of "ownership".

Less pleasing was the behaviour of the heating and ventilation system. Water-valve control is employed on this model. In addition to its characteristic sluggish response, the system failed to maintain a reasonably constant temperature within the car. The situation was further aggravated by the absence of face-level vents. Another problem was ineffective extraction, leading to misting up in humid weather. On the credit side, the two-speed blower was unobtrusive and the heated rear window (an extra-cost option) very effective.

1,000-mile service

Soon it was time for the first service, the work being entrusted to P. & L. Motors, Enterprise Garage, Upminster. In addition to the routine tasks involved, they were asked to attend to the non-cancelling indicator switch, check the headlamp and choke settings, and fit a screen-pillar radio aerial. Faults found by the garage amounted to two over-tight tappets, excessive contact-breaker dwell (74 deg instead of the specified 60 deg), and the presence of a stray piece of polythene sheet (approx 15 × 2 in) in the air-cleaner housing. Non-cancelling of the indicator was found to be caused by a faulty steering-column collar, for which a replacement had been ordered. Both headlamp alignment and choke-setting were given a clean bill of health. Cost of materials (oil, filter and cam-cover gasket) amounted to £2.70; no labour charge was made for this work. Price of the radio aerial was £3.30, a modest £1.50 being charged for the considerable work involved in fitting it.

Automatic choke still troublesome

The car had obviously benefited from the vetting it received at the 1,000-mile mark. Although quite new still, it felt very peppy.

Shortly afterwards, I delivered it to the Radiomobile fitting centre at Cricklewood in order to have a radio fitted. No tailor-made fitting kit was available, but they made a first-class job of the installation. The set

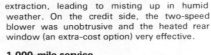

Above: Instruments include a rev counter and the horn is worked by neat push bars in the wheel spokes

performed very well, but I suspect that it would have benefited from the use of a better quality aerial (possibly a roof-mounted one).

Despite the garage's assurance that the choke was in order, warm-up was still proving troublesome. On very cold mornings, over-richness could actually bring the car to a halt. This was sufficient provocation to make me re-set the unit on a trial-and-error basis. Somewhat to my surprise, this proved quite effective. Even so, I failed to achieve that ideal setting which results in the engine being fed exactly the right mixture at all times. Oddly, there was no improvement in economy, but I was happy enough with the 22-23 mpg being returned at this time.

Although the next service was scheduled for 4,000 miles, the clock showed nearer 4,700 at the time. Again P & L Motors carried out the work. Apart from fitting the new steering-column collar under warranty, this was strictly routine. Materials (oil and air filter, plus oil) amounted to £3.51¼ and the labour to £4.00.

Selector lever problems

At around 6,000 miles, difficulty was experienced in getting the transmission selector lever to clear the neutral and park "guards" on the quadrant. As the car was urgently needed by our Editorial Director, something had to be done! On dismantling, the problem was found to have been caused by "stripping" of what clearly had been an under-sized thread on the

---

1. *Reclining backrests are standard and the tip-up arrangement does not interfere with the setting*
2. *The Radiomobile installation fills most of the central locker, but there is still room for maps underneath*
3. *This patch of wear on the carpet seems to indicate that the driver's heel mat is too short*
4. *Because of the full-width front bonnet the radio aerial must be roof or pillar mounted*

stem of the pawl. This rendered the T-handle button ineffective. As a temporary measure, the unit was reassembled without the offending part. As soon as the car was returned, a replacement (cost 50p) was fitted. In normal circumstances, of course, this work would have been covered by warranty.

Worn-out brake pads
Shortly afterwards, colleague Martin Lewis used the car to cover the Senior Service Hill Rally. On his return, he complained that the brakes were pulling to the right. Sure enough, they were. It didn't take long to find the cause — the left-hand brake pads were completely worn out, and those on the right very nearly so!

Brake-pad inspection forms part of the 4,000-mile service. The relevant instruction reads "Check *overall* (my italics) brake pad thickness and renew if less than 7 mm (0.28 in.)". This can be very misleading; what it doesn't make clear is that the pad backing plates are virtually in contact with the retaining clips at this stage — in other words, the pads are worn out.

In our case, some 8 mm had been worn away from each pad in 7,000 miles, equivalent to 4.6 mm in 4,000 miles. At this rate of wear, overall pad thickness would have to be 11.6 mm (12 mm for all practical purposes) before these items could be considered fit for a further 4,000 miles. The moral is obvious — inspect the pads yourself or make quite sure that the garage man understands the problem. In any event, you will know what to look for if the brakes start pulling!

This episode would normally have added £9.77 to the maintenance bill — £8.27 for the pads themselves, the remainder for labour. In our case, labour was our own, but the relevant charge has been added.

Change of garage
Around this time, I moved house from Upminster to Marlow. Although I had been very pleased with the treatment I'd received at the hands of P & L Motors, it now became more convenient to have the car serviced at MLG of Chiswick. Belatedly (at 9,000 miles), it was presented for its next service, another straightforward one (no extra jobs). In fact, MLG discovered that the stop-lights were inoperative — the result of a faulty switch. They also spotted a bad oil leak past the final-drive pinion seal but, at my request, left this job until a later date. Materials (stop-light switch, plugs, contact-breaker, oil and air filter, engine and EP oils) amounted to £5.55. Since this is a major service, labour charges totalled a hefty £13.75.

Performance was now better than ever and, as indicated earlier, economy had improved markedly. Some 400 miles later, I revisited MLG for renewal of the pinion seal. Materials, seal and lock-washer cost 89p, labour amounting to £7.50.

With this problem disposed of, the car was in peak condition. However, to my great dismay, my term of "ownership" was nearing its end. How I wish I could have bought it for my wife (that, at least would have been my excuse!). Paintwork was unmarked, as was the bright-work. Apart from a threadbare patch on the driver's carpet, the inside was indistinguishable from new. This was not the result of loving care and attention — it simply happens to be an extraordinarily easy car to keep in good shape.

Tyres (Continental radials) were approaching two-thirds worn at this stage. Had I kept the car, I wouldn't have been sorry to see these go. Their wet-road adhesion was never very good, and I have the feeling that a set of Dunlop SP Sports would have been very worth-while.

Summing up
What a delightful little car the 2002 had been! I loved its Jekyll and Hyde character — unobtrusive and well-mannered on the one hand, a real road-burner on the other. Its handling, being more-or-less neutral, was exactly to my taste. Its compactness and superb visibility made it a traffic car *par excellence*. Good road-insulation was a very strong point.

Nothing is perfect, of course. There was that accursed heater to contend with! Another criticism concerned the rapid brake-pad wear and tendency to fade under hard usage. The frameless windows caused a fair amount of wind noise at around maximum speed. That's just about it on the debit side!

To repeat my earlier question, what is there to rival a 2002 in this capacity class? Having re-lived the pleasures of "owning" XNJ 998J I have only one answer a 2002 Tii.

PERFORMANCE CHECK
Maximum speeds

Gear	mph R/T	mph Staff	kph R/T	kph Staff	rpm R/T	rpm Staff
Top (mean)	107	103	172	166	5,800	5,600
(best)	110	106	177	171	5,960	5,750
3rd (Inter)	89	77	143	124	6,500	6,400
2nd (Low)	59	46	95	74	6,500	6,400
1st	31	—	50	—	6,500	—

Standing ¼-mile, R/T: 17.4 sec 77 mph **Standing kilometre, R/T:** 32.8 sec 96 mph
Staff: 18.1 sec 77 mph **Staff:** 33.5 sec 92 mph

Acceleration, R/T:									
R/T:	3.5	5.3	7.6	10.6	14.2	18.2	26.4	37.2
Staff:	3.6	5.5	7.8	10.6	14.6	20.4	30.2	—
Time in seconds	0							
True speed mph	30	40	50	60	70	80	90	100
Indicated speed MPH, R/T:	33	43	53	63	73	83	93	104
Indicated speed MPH, Staff:	27	41	54	65	77	88	99	111

Speed range, Gear Ratios and Time in seconds

Mph	Top R/T	Top Staff	3rd R/T	Inter Staff	2nd R/T	Low Staff
10-30	—	—	6.4	—	3.8	2.6
20-40	8.1	—	5.6	—	3.5	3.3
30-50	7.6	—	5.5	4.9	3.7	—
40-60	8.4	7.7	5.8	5.3	5.1	—
50-70	8.5	8.2	6.2	6.9	—	—
60-80	10.2	10.5	8.6	—	—	—
70-90	13.2	16.1	—	—	—	—
80-100	17.9	—	—	—	—	—

Fuel consumption
Overall mpg, **R/T:** 25.5 mpg (11.1 litres/100km)
Staff: 23.0 mph (12.3 litres/100km)

NOTE: "R/T" denotes performance figures for BMW 2002 tested in AUTOCAR of 16 May 1968

COST and LIFE of EXPENDABLE ITEMS

Item	Life in Miles	Cost per 10,000 Miles
		£ p.
One gallon of 4-star fuel average cost today 34p	23.0	147.83
One pint of top-up oil, average cost today 18p	Service interval	Nil
Front disc brake pads (set of 4)	7,000	11.81
Rear brake linings (set of 4)	17,000	2.66
Tyres (front pair)	16,000	11.90
Tyres (rear pair)	18,000	10.58 (1)
Service (main interval and actual costs incurred)	4,000	48.32 (2)
Total		233.10
Approx standing charges per year		
Depreciation		386.00
Insurance		38.00 (3)
Tax		
Total		682.10

Approx cost per mile = 6.8p

NOTES: (1) Logical course would be to renew all four tyres at 16,000 miles, increasing total cost (for 10,000 miles) by £1.32.
(2) Includes cost of renewing brake pads at 7,000 miles and final-drive pinion seal at 9,000 miles.
(3) Insurance quotation is for a 30-year-old driver living in Marlow. No-claims bonus of 60 per cent. Excess of £25 on accidental damage.

TURBOCHARGED BMW 2002

What the factory does for an experiment a performance-minded dealer does for you

JOE RUSZ PHOTOS

A LESS LIKELY alliance for automotive progress than racing and emission controls is hard to imagine but here they are, green thumbs and grimy knuckles in cooperation to create a turbocharged BMW 2002 that meets all the laws, goes like blazes and is so unobtrusive about it that even experts don't notice until they drive the car. And if things continue to go this right, you will be able to buy one.

This worthwhile piece of work comes from Vasek Polak. That makes the racing portion of the combo perfectly clear. Polak is a former driver, known now as the owner/sponsor/entrant of the Porsche 917-10 which Milt Minter drives in the Can-Am and which R&T drove in the October issue. Polak is also a dealer for both Porsche and BMW. As a businessman he knows that he does well by supplying what the public needs. And he knows that what he sells must meet the laws, which in California prohibit the removal or modification of emission controls. (He's also a human being, that is, he breathes air and wouldn't want to produce smog even without considering the law.)

When Polak started this project he had several advantages: BMW and Porsche both had their engineers working on turbocharging for racing and for high performance production cars of the future, as witnessed by the Penske Porsche 917 and the show BMW Turbo in this issue. Enter a prime pair of sources for encouragement and advice. Polak had decided to install a turbocharged engine in his 917, which gave motivation and an in-house sort of a cross-fertilization as the techniques and procedures learned on one project could be applied to the other.

Polak's crew plucked a 2002tii out of stock, ordered an Eberspacher turbocharger—like the ones used on the 917, but smaller—and set to work.

The resulting installation looks neat and simple. Polak snorts at that, saying that it took six months of hard work and experiment to come up with such a neat and simple installation. The turbocharger is mounted on the lower right front of the engine block. The exhaust is routed from the stock manifold through the turbine wheel, then beneath the car to a small fiberglass-packed muffler just beneath the rear bumper. It's not much of a muffler, but most of the exhaust pressure is used up spinning the impeller so not much muffler is needed. The noise level goes many decibels above that of a stock 2002tii in a deep, pleasant way—noisy but not raucous and not especially noticeable most of the time.

The air intake and filter are at the right side of the radiator, with the air piped through the turbocharger and across the engine via a steel pipe to the stock air horn for the fuel injection. From that point into the engine it's all straight production 2002tii, untouched by human hands. The other changes are minor; the oil pressure gauge fitting is tapped and oil supplied from there to the impeller and drained back into the oil pan, there's a pipe from the blower pressure relief valve into the air intake upstream of the impeller and the crankcase is ventilated into the relief pipe, which is under negative pressure from the intake pipe.

Nice and simple. It can be thus because the 2002tii engine isn't simple. It's built strong from the start, designed to hold up under maximum use because that's what it gets at home. And the fuel injection is complex, as it must be. But because the injection regulates fuel volume and strength by measuring pressure in the inlet tract between throttle butterfly and valve, and because all the changes, specifically the capacity for positive pressure on the other side of the butterfly, are made away from the injection's sphere of influence, why, the injection doesn't have to be changed at all. So it isn't. Unlike turbocharged engines with carburetors there is no need for pressurized float bowls and throttle shafts, or secondary fuel pumps and electric relays, etc. And the emissions levels aren't changed, nor is the equipment. Polak fitted colder spark plugs to protect against detonation at full power. The distributor has a cutout which limits the engine to 6400 rpm because the turbocharger would otherwise push past the engine's limit, and the blow-off valve is set at 6.5 psi for the same reason. But within those limits Polak reckons the engine to be no more than just stressed,

TURBOCHARGED BMW 2002

and he hasn't changed pistons or lowered the compression ratio. In fact the test car's engine has never had the cylinder head removed.

When the work was completed, the car was put on a chassis dynometer, where it produced 124 bhp at the rear wheels at 6000 rpm. Before turbocharging, the car had 72 bhp at the same engine speed.

One is inclined to doubt this impressive figure on first acquaintance with the car. The engine starts quickly and idles and sounds just like a normal 2002tii except for the deeper exhaust tone. And in normal driving the car feels showroom stock. With a light foot on the pedal and the normal shift points, there's not as much as a hint of added power.

But the power comes on with a bang. As the figures show, using full power through the gears cuts several seconds off all the times to speeds, with the cuts getting deeper as speed goes up. The start is virtually unchanged, as there's no time for the impeller to speed up and provide boost until about 3000 rpm. Nor would brutal tricks like revving the engine and dropping the clutch work: there's no boost unless the throttle butterfly is nearly wide open.

There is no dreaded turbocharger lag as such. Rather, with wide open throttle at low rpm the car accelerates like an unblown 2002tii, briskly. When the impeller is spinning at a useful speed the needle on the gauge flicks over to six and the car surges forward; like a racing engine when it comes into its band of useful power, the difference being that there is no flat spot, no stumbling and banging, at less than the useful speed.

Once again we wish that the standard test procedure called for measurement of acceleration from a steady speed, from 30 mph to 80 mph, say, in third gear. A normal passing situation, more important than leaping from stoplight to stoplight, and a test in which the turbocharged BMW would do even better than it does on the standard test.

And speaking of standard tests, the turbocharged car went farther per gallon than did our test production 2002tii. The mpg test is a rigid procedure, calling for accelerations to set speeds, cruising at steady speeds for set distances, etc. Under these conditions the turbocharged car did better than the production car, presumably because the test conditions were more easily met by the more powerful engine. If the turbocharger was called into service often—and why have it if you don't use it?—then the fuel consumption would of course increase.

This was our third recent experience with turbocharged versions of production cars, and it was the most impressive of the three by a wide margin. The Vega truck (Jan. '72 R&T) proved to be a good performer when not hampered by derangement of its elaborate fuel pressure controls and when caught between changes of sparkplugs, for which its appetite was insatiable. We had a turbocharged Datsun 240Z on hand for the technical article about turbocharging in the May '72 issue, but it blew a head gasket before any figures could be recorded. Turbocharging works. It can also be troublesome. Polak's car has been tested and driven several thousand miles without incident and it performed without flaw during the 10 days and several hundred miles we had it. Still, the history of the technique makes durability and servicing something to think about.

What of the future? Polak has applied to have his system certified for sale in California, which requires that any change to the engine not increase the emission levels. As another step, he's concerned that the stock injection won't provide enough fuel for sustained operation at wide open throttle. Likely the fuel-power mixture will be richened or the pressure relief valve will be set at 5 psi, to decrease the amount of air being pumped into the engine. This would cut the rear wheel power down to 102 bhp at 6000 rpm but Polak thinks the gain in safety margin would be worth the lessened power.

Whichever, the cost will not be cheap. Polak estimates that the complete job, all parts and the work done in his shop by his mechanics, will be close to $1500. That's a lot and he's not eager to simply sell the kits for home conversion. It's possible that Polak and the BMW factory could arrive at some sort of franchise arrangement with certain other BMW dealers stocking and installing the turbochargers. But that's at least one step beyond the present plans and Polak can't offer any definite timetable.

Too early for a line to form, then, but when it does, the place to be in line at is Vasek Polak Motors, 199 Pacific Coast Highway, Hermosa Beach, Calif. 90245.

Comparison Data Turbocharged BMW 2002tii vs production 2002tii		
	Turbocharged	Production
Acceleration to speed:		
0-30 mph	3.5 sec	3.9
0-40	4.9	5.9
0-50	6.7	7.7
0-60	9.0	9.8
0-70	12.1	13.9
0-80	16.1	17.9
0-90	21.5	n.a.
0-100	29.6	32.0
Miles per gal.	24.5 mpg	22.7 mpg
Interior noise, dBA:		
Maximum, 1st gear	83	82
Constant 70 mph	86	76

BMW 2002 Cabriolet's detachable roof with rollover bar does not affect rigidity.

New BMW 520 and 2002 Cabriolet

I recently had an opportunity to test the two latest BMW models in the South of France. These are the 2002 Cabriolet and the 520. The Cabriolet is really an open-air version of the 2002, which we know so well, but the 520 is an entirely new four-door saloon.

The Cabriolet is identical to the existing 2002 up to the waistline. There is a hefty rollover bar incorporated in the roof styling, ahead of which is a light detachable section which fits very neatly into the lid of the boot. If rain should fall, it can be snapped back into place in a matter of seconds. Usually, the car will be driven with the front open but with the section behind the rollover bar closed, to reduce back-draughts.

However, in really hot weather one would open the back as well, which simply folds down into the rear of the car and can be concealed with a cover. The car is then completely open but of course the steel rollover bar, being an integral part of the structure, remains in place, which pleases the insurance companies. It is claimed that rigidity is not impaired by making the roof detachable and certainly the handling seems identical to that of the closed version.

This BMW is small enough to be great fun on the winding French secondary roads and is very quiet when the front of the body is open. One can enjoy the radio and, if the early spring weather is a bit nippy, the heater can be used with advantage. The scents of the countryside come wafting in through the open roof, though these are mixed with the odour of hot brakes on mountainous descents. The Cabriolet seems just as lively and flexible as the existing 2002 and its only disadvantage is the rather higher price. When closed, however, it shares an unfortunate lack of controllable fresh-air ventilation with its sister coupé.

The 520 is a much larger car, giving even more space than the old 2000, and it incorporates all the latest thinking on safety. It is a really roomy four-door saloon and weighs over 6 cwt more than the Cabriolet. To cope with this extra load, an up-rated version of the 2-litre engine is used, with a different cylinder head and twin Stromberg carburetters instead of the single Solex. These modifications make an extra 15 bhp available and the final drive ratio has been changed from 3.64 to 4.1 to 1.

Elaborate equipment is a feature, with four quartz Halogen headlights, courtesy lights for all four doors, a heated rear window, and a rev-counter as standard. In spite of the extra weight, light steering has been achieved, and great attention has been paid to silencing the exhaust. The rear passengers have ample space in all directions, for the trend nowadays is towards rear seats that can be occupied comfortably on long journeys and the day of the four-door two-seater is past.

From the driver's seat, the 520 seems quite a big car and the interior is luxuriously appointed. A more elaborate ventilation system is employed than on the other four-cylinder BMW models, with separate cold-air inlets each side of the instrument panel, though still without directional control. The seats are very comfortable and there is a good all-round view.

Though the lower gearing disguises the extra weight to some extent, the car naturally does not feel so responsive to the accelerator as the lively 2002. The twin-carburetter engine likes to rev, which the well-spaced ratios of the excellent gearbox allow it to do. Being more highly tuned than the single-carburetter version, the power unit is not quite so quiet, though it runs very smoothly.

On sharp corners, the 520 rolls a little more than the 2002 but it is beautifully balanced on fast bends with very little understeer. The car rides well on every sort of surface and the insulation of road noises from the interior is very effective; the body shape seems to be responsible for the low level of wind noise.

The gear ratio is about right for maximum performance, the rev counter needle just entering the red section of the dial on the autoroute. In spite of its larger size, the new body evidently has a lower drag factor than that of the 2002, the maximum speed of both models being about the same, in the region of 108 mph. Flat-out driving causes no sign of distress, the car swinging happily along for mile after mile at its maximum speed. Though I was unable to test the 520 in gale-force winds, it is perfectly stable in normal gusts and seems unaffected by the sudden blasts which the passing of huge continental lorries may cause.

The popular 2-litre class now embraces cars of considerable luxury, of which this new BMW is an example. Perhaps it places more emphasis on comfort and utility than on the sheer performance which has recently been the hallmark of this make. The 520 is evidently the first of a new series of BMW cars and one would expect to see a six-cylinder derivative before long.

The last BMW I tested was a lightweight 3-litre coupé. The 520 is at the opposite end of the performance spectrum but it will appeal to the professional man in search of a medium-sized car of exceptional refinement. It should suit British road conditions very well.

JOHN BOLSTER

SPECIFICATION AND PERFORMANCE DATA

Cars tested: BMW 2002 two-door Cabriolet, price £3299. BMW 520 four-door saloon, price £2999 including tax.
Engine: Four cylinders 89 mm x 80 mm (1990 cc). Inclined valves operated by single chain-driven overhead camshaft and rockers.
2002: Solex downdraught carburetter. Compression ratio 8.5 to 1. 100 bhp (net) at 5500 rpm.
520: Twin Stromberg carburetters. Compression ratio 9.0 to 1. 115 bhp at 5800 rpm.
Transmission: Single dry plate clutch. Four-speed all-syncromesh gearbox with central change, ratio 1.00, 1.32, 2.02 and 3.76 to 1. Hypoid final drive, ratio (2002) 3.64 1, (520) 4.1 to 1.
Chassis: Combined steel body and chassis. Independent front suspension by MacPherson struts and lower wishbones. Worm and roller steering gear. Independent rear suspension by semi-trailing arms and coil springs. Anti-roll bars front and rear with telescopic dampers all round. Servo-assisted disc front and drum rear brakes. Bolt-on disc wheels fitted (2002) 165 HR 13, (520) 175 SR 14 tyres.
Equipment: Twelve-volt lighting and starting. Speedometer. Rev counter. Fuel and water temperature gauges. Cigar lighter. Heating, demisting, and ventilation system. Flashing direction indicators. Reversing lights.
Dimensions: 2002: wheelbase, 8 ft 2.4 in; track, 4 ft 4.4 in; overall length, 13 ft 10.5 in; width, 5 ft 2.6 in; weight, 2028 lb. 520: wheelbase, 8 ft 8 in; track (front), 4 ft 7.4 in; track (rear), 4 ft 8.8 in; overall length, 15 ft 7 in; width, 5 ft 6.5 in; weight, 2712 lb.
Performance: (maker's figures) 2002: Maximum speed, 110 mph. Acceleration, 0-50 mph, 6.8 s. 520: Maximum speed, 107.5 mph. Acceleration, 0-100 kph (62 mph), 12.3 s. Standing 400 m (¼ mile approx.), 18.2 s. Standing kilometre, 33.7 s.

The 520 uses the latest thinking on safety. It weighs over 6 cwt more than the Cabriolet.

GROUP TEST

Four of the best

Alfa Romeo 2000 GTV
Audi 100 Coupe S
BMW 2002 Tii
Rover 3500 S

As it would be misleading, even impertinent, to divide this quartet into winners and losers, don't flip to the last paragraph to see what's best. We don't say. Each is a civilised performance car of redoubtable reputation, differing from the others more obviously in character than in merit. Debriefing after our usual four-man 800-mile return convoy to Wales didn't produce a unanimous worst or best, underlining that it's more a matter of personal taste that counts here rather than the accumulation of marks on a score sheet.

What we can say is that we enjoyed and respected each and every car, though not necessarily for the same reasons. None of them is perfect mind you, and some are a lot less perfect in certain respects than others. As four sound but different ways of spending £2500, we found it an intriguing exercise.

Three of the cars have been around, at least in body style, for a very long time, none longer than the Alfa Romeo which is a direct descendant of the Giulia Sprint coupe built in the days when a pretty shape didn't mean a sacrifice in window area and visibility. Bertone's little slim-pillared glasshouse is to our eyes still a model of how it should be done. Then, as now, Alfas were fun—but also fragile with it. Intelligent evolution has since made them not only faster—the Giulia progressed through 1300, 1600, 1750 and 2000 phases—but tougher, plusher, and a lot more refined as well, without losing that essential ingredient that inspires even the most disinterested driver. Even so, the Alfa 2000 GTV is on paper getting pretty long in the tooth. Past its prime or still a car for the Seventies?

If you prefer to sacrifice some space for a sports coupe that looks the part, the Alfa clearly has a head start on the neat but undistinguished-looking three-box BMW.

They used to call this one a coupe too, as a showroom euphemism for its two doors and sylphine build: now the 2002Tii label on the back spells performance rather more lucidly so the coupe disguise is no longer needed. In essence, this well-engineered car is just what it appears to be: a compact yet quite roomy family saloon built and finished to a standard a cut above the average for a car of its size. It's nothing special to behold, inside or out, and its appointments and amenities are far from extravagant for a £2500 car. Indeed, they differ very little from those of the basic 1600 which costs £500 less. If £500 seems a lot to pay for an extra 400cc and fuel injection perhaps we should remind you that Tiis won the first seven Group 1 production car races on the trot in Britain last year.

The Rover is something quite different. No sporting dashabout in the Alfa/BMW mould but no dignified sluggard that's too big for its boots either. With a relatively large V8 engine and manual transmission, the 3500S is a real go car. Moreover, it's beautifully finished and appointed and imparts through its opulence a feeling of luxury and wellbeing that the two more sporty imports lack. On paper, it makes the Alfa and BMW look rather expensive, yet it wasn't everyone's first choice by any means. So what are the snags?

By stretching a point, the Audi's ancestry can be traced back to the pretty DKW two-stroke which later sired the highly successful Audi range with its Mercedes-designed four-stroke engine. The 100 Coupe is the youngest car here and the only one with front-wheel drive. Although grafting a fastback tail on to a saloon base doesn't guarantee aesthetic success, it worked with the Audi to such good effect that it's by far the most striking car in the group, and quite a head-turner when there's an audience. Which is just as well because you pay pretty dearly for this bodywork which costs £580 more than the mechanically identical 100GL that has four doors, more headroom in the back and fewer blindspots to mar the view. Even so, the Audi coupe is easily the most roomy car of the four and as a thoroughly modern halfway house between the rorty Alfa and the silken Rover, appears on paper an attractive alternative to both.

PERFORMANCE

As we've said often enough before, engine capacity is a poor guide to performance. True, the Rover 3500S, with a V8 engine that's nearly double the size of the Audi's and 1½ litres greater than that of the Alfa and BMW, has the highest maximum speed of 119 mph. But what modest extra power it's got is offset by much greater weight (26.1 cwt is a lot for a car of this size) so on acceleration the Alfa and BMW, each weighing little more than a ton, have the edge on it all the way to 100 mph. In a straight drag race, the Alfa—aided by its better gear ratios—would reach three figures first in a creditable 27.2s, fractionally ahead of the BMW with the Rover not far behind. Although the 21.3 cwt Audi is no sluggard, it can't quite live with this company against a stopwatch, though it had no trouble keeping up with the other cars on the road. All four are at no more than a canter at 100 mph—an especially easy cruising gait for the long-legged Rover and Alfa Romeo.

Even the Rover's massive low-speed torque is not enough to better the lower geared BMW in top gear acceleration below 60 mph, and there's precious little in it at higher speeds. The Alfa will better them both in fourth gear and it's not far short of them in its overdrive fifth. Again, the Audi trails the field, its smaller engine having the most work to do on a power/weight basis. Although it's by no means inflexible at low revs, you have to make full use of the lower ratios to get the most from the Audi's 1871 cc's worth.

Just to underline how well the Alfa and BMW perform on 2 litres, it's worth noting that neither the Triumph Stag nor the Reliant Scimitar GTE, both 3½ litre rivals like the Rover, can do any better.

The BMW's slant four ohc engine is as sweet as they come, devoid of any mechanical thrash or boom periods. Indeed, for its combination of mechanical refinement and rewarding vigour, we cannot think of a finer four-cylinder engine, though the slightly harsher Alfa unit runs it very close indeed and that of the Audi, which is also smooth and willing, is hard to fault.

None of the fours, though, have quite the silken delivery of the Rover's V8 which is easily the smoothest and normally the quietest of the group—except at high revs when its usual remote waffle becomes a distinctly frenzied drone. Not that you need to exploit the upper reaches of the modest rev range very often.

With the biggest engine in the heaviest car the Rover, predictably enough, is by far the most thirsty. On our brisk group test run it recorded a hefty 17.7 mpg whereas the sprightlier BMW, under identical conditions, managed a respectable 24.6 mpg. Just to put things in perspective, though, remember that the Rover costs £292 less than the Tii, enough for 23,000 miles even on five-star fuel.

TRANSMISSION

Although all four cars scored well here, the Alfa is in a class of its own when it comes to swapping gears. To start with it has five of them — five perfectly stacked ratios which at one extreme will catapult you off the line with spinning wheels and at the other will waft you down a motorway at high speed without any apparent strain. In between, there's a gear for every corner and gradient. Then there's the action of the gearchange, so slick and easy — despite quite a long travel — that you find yourself changing gear for the fun of it. The clutch and throttle action are also excellent.

At extra cost, you can have a five-speed gearbox in the BMW as well but the standard item is a four-speed unit with a nice stubby lever topped by a nasty misshapen knob (that on our test car had been swapped for something more agreeable to clasp). If the action isn't quite so slippery and precise as that of the Alfa, it's nevertheless a satisfying change. The ratios are fairly wide apart, first being particularly low, but such is the engine's rev range (6500 rpm against the Alfa's lowly 5700) that it doesn't much matter.

There's little to choose between the gearchanges of the other two cars. The Rover's is perhaps the more precise and the movement of its tiny lever nice and short. But the lack of leverage accentuates the baulking of the synchromesh to make the change a bit notchy, especially if you rush it.

That of the Audi is free and easy but the gate is less positively defined and lever movement slightly clonky. Nor is the clutch action quite so soft and progressive as that of the Rover. Even so, a nice gearbox to play with. Surprisingly, the Audi has a higher first gear than the Alfa or BMW, perhaps to help minimise wheelspin which weight transfer from the driven front wheels tends to promote during a brisk getaway.

Gear noise was practically inaudible in the Audi. In the Rover it was thrown into prominence by the quietness of the engine, but was not obtrusive. You could also hear it quite plainly in the Alfa, despite the gruff note of the engine.

HANDLING AND BRAKES

Our drivers fill in a score sheet for each car during a group test run. Totting up the marks awarded for steering, handling and roadholding it may surprise you to learn that the Audi came out on top by quite a margin, despite low-geared steering that demands more twirling on sharp bends than the others. It transmits good road feel to your hands, though, and it is commendably light, even when parking. The car's high marks stem more from its impeccable road manners. It treats bumpy corners taken at speed with disdain, gliding through, rather than round them with reassuring poise and little body roll or lurching to upset the impression of rock-like stability. Even at high g forces, the car responds still more to the helm, tightening its line as the modest tyres find more grip. Ultimately, progressive understeer sets the limit but you have to be going pretty hard to find it. Nor does lifting the throttle in mid corner have any untoward effects.

Although the best behaved, the Audi is arguably not the most entertaining of the quartet, perhaps because it is so good. You can't, for instance, indulge in any extrovert power slides when no-one's looking, as you can in the Alfa and BMW to good effect. Both have shortcomings though. The Alfa, for instance, has very heavy steering — so heavy on our test car that we suspect the steering geometry may have been amiss, especially as understeer would set in strongly and very finally on sharp bends. Even so, the cornering powers are high and the limited slip differential (now included in the recently increased price) allows you to put the power on the road, even during hard cornering when there's little weight on the inside rear wheel. Despite the stodginess of the steering on sharp corners (and even more so when parking) it's otherwise alert and responsive and feeds you with strong messages from the front wheels.

The BMW's steering was also a bit like heavy pudding at low speeds, partly because the test car had a small non-standard wheel. But at anything more than a brisk trot it became lightish and direct. At speed on a winding road, the Tii has a slightly nervous feel but it's delightfully taut and controllable and, unlike the Alfa, will invariably step out of line at the back first. Without a

Cockpit comfort. Below left, Alfa Romeo; below right, Audi; bottom left, BMW; bottom right, Rover

limited slip differential, however, it can also become a little untidy at the limit as it kicks a spinning wheel into the air. Even so, fast and entertaining.

The Rover is altogether different. To start with, it is more softly suspended than the Alfa and BMW and rolls a lot more when pushed through the turns. Hurrying along some tortuous Welsh lanes, it became very untidy indeed, wallowing and lurching uncomfortably with its big understeering tyres scrubbing their sidewalls. Quite safe, mind, because the car clings on quite well long after the point when its behaviour discourages more spirited driving. While disappointing when extended, we feel that many drivers would rate the Rover as the easiest (and maybe nicest) car to handle on account of power steering that's very light *and* highgeared, the ideal combination for effortless parking and sharp bends. Moreover, on fast main roads where steering feel (which is poor) or ultimate grip don't come into the reckoning, the car feels immensely stable and reassuring even at speed on long corners. Easy, safe and effortless : yes, all these but not very rewarding for the enthusiastic driver.

COMFORT AND CONTROL

In this section, at least, we're prepared to split the quartet into two pairs, starting with the BMW and Alfa which we rated less highly than the Audi and Rover.

The BMW's ride is firm but not harsh or crashy, and body motion is well controlled on undulating secondary roads. So far so good for a taut and responsive performance car. But the rather flat, firm seats lack both resilience and support and some people found the high steering wheel and tall pedals a little awkward. Visibility, however, is superb — there's not a blind spot in sight. Some of the switchgear, mostly knobs on the facia, is hardly under fingertip control and the ventilation . . . well, there isn't any unless you open a window. While on performance the BMW is still a class leader, it lags conspicuously when it comes to the sort of creature comforts expected of a £2500 saloon. (Significantly, the new 520, which must eventually supersede the 2002 series, promises to excel in all those departments in which the Tii is weak.)

Next comes the Alfa Romeo. If anything, its ride is firmer than that of the BMW, becoming quite restless — but again not harsh or bumpy — on secondary roads. The nicely contoured seats got high marks but the high, floor-hinged pedals that are too close up to the driver prevented a relaxed and comfortable driving position for most of us : even when sitting with splayed legs (there's no other way really) tall drivers found the nice wood-rimmed steering wheel brushing their thighs. It could be so good with a different pedal set-up.

As in the BMW, you get an unimpeded panoramic view, despite the rather high window sills and low seating position. The switchgear is passable (it could be improved by making better use of the three fingertip stalks), but the instruments and warning light clusters are everything you'd expect of an Alfa which has been more successfuly modernised inside than the BMW. It has, for instance, effective fresh air ventilation to complement the excellent heater, so there's no need to relieve any stuffiness with an open window.

The Rover and Audi are in another league altogether here, both being exceptionally comfortable and well appointed cars. The Rover, as we've said, has a more resilient ride than the Alfa or BMW, floating smoothly over most well-made roads without the slight jittery motion of its two more sporty rivals. It copes less well, however, when extended on twisty secondary roads, heaving and jolting over the bumps, especially those it encounters in mid corner. These circumstances apart, however, the Rover rides very smoothly.

Deeply dished seats hold you in place well but, as we've observed in previous tests, the cushions are too flat and you tend to slide forward on them, especially under braking. As the steering wheel is adjustable for height and reach, however, everyone should be able to tailor an excellent driving position. As in the Alfa and BMW, the pedals are poorly arranged for

The remarkable 2-litre engines of the Alfa Romeo, top, and BMW, above

All the controls. Below left, Alfa Romeo : below right, Audi 100 Coupe ; bottom left, BMW 2002 Tii (with non-standard steering wheel) ; bottom right, Rover 3500S

heel and toe changes (not that most Rover owners will notice) but the major controls are otherwise well placed.

We liked the chunky easy-to-grip switchgear knobs, too, though their location and operation is rather confusing at first. Again, better use could be made of the column stalks. The heating and ventilation is superb and the seat belts are easy to clasp and comfortable to wear. Although neither wind nor road noise is especially well isolated, there is in the Rover an almost tangible feeling of refinement and isolation not evident in the Alfa or BMW.

You get the same feeling in the Audi too, despite its busier engine which is neither as smooth nor as quiet as the Rover's. Even so, this is a very comfortable car indeed. Its ride is excellent, more because of innate stability in pitch, roll and bounce rather than any extraordinary absorbing resilience: on secondary roads at speeds, it displayed none of the restless jittering of the Alfa or BMW at one extreme or the Rover's lurching at the other.

Although the steering wheel is not adjustable, we all found the driving position as near perfect as you could expect, thanks to well placed controls (the pedals are especially good) and excellent seats that locate you well. The switchgear is marginally the best of the four cars here (though there are better systems around elsewhere) and we gave it the highest marks for wind and road noise. The heating and ventilation also works well, though the latter isn't quite as good as that on the Rover as the vents are not directly ahead.

The Audi's big black mark here, as with so many modern fastback coupes, is rear threequarter visibility, made to look especially bad on the group test by the presence of the Alfa and BMW. The sloping roof also reduces headroom in the back although the Audi remains the most capacious car inside, both for passenger and luggage accommodation. Getting into the back is less easy, however, than in the Rover, the only car with four doors. Moreover, its rear seats are less well shaped than those of the Rover's individually contoured ones which we found very comfortable, despite restricted legroom.

For such a small car there's more room in the back of the BMW than you might expect, though the seat itself is a rather shapeless bench. The Alfa is even less satisfactory and barely qualifies as family transport.

FITTINGS AND FURNITURE

Although well built and finished, the Alfa and BMW are quite simply appointed, the 2000 GTV looking slightly the more opulent of the two with its cloth-covered seats, wood-veneer facia and more comprehensive instrumentation.

The Audi is particularly neat

GROUP TEST

PERFORMANCE

	Alfa	Audi	BMW	Rover
Max speed, mph	115.3	112.7	113.7	119.0
Max in 4th	99	—	—	—
3rd	73	87	96	88
2nd	50	61	63	57
1st	30	35	27	34
0-60 mph, sec	8.9	10.8	8.2	9.3
30-50 mph in top, sec	10.9	10.5	7.8	8.1
Group test mpg	21.1	23.7	24.6	17.7
Touring mpg	26.9	30.4	—	23.6
Fuel for 10,000 miles, £	180	160	154	215
50 lb on brakes, g	1.0	1.0	0.75	1.0
Turning circle, ft	35.6	31.5	29.6	44.0
Steering turns, 50 ft circle	1.1	1.25	1.1	1.0
True speed at ind 70 mph	70	65	67.5	70

PRICES

Alfa Romeo 2000 GTV
£2255 plus £471 pt equals £2726

Audio 100 Coupe S
£2130 plus £445 pt equals £2575

BMW 2002 Tii
£2067 plus £432 pt equals £2499

Rover 3500S
£1825 plus £382 pt equals £2207

EQUIPMENT

	Alfa	Audi	BMW	Rover
Adjustable steering	No	No	No	Yes
Cigar lighter	Yes	Yes	Yes	Yes
Clock	No	Yes	Yes	Yes
Electric windows	No	No	No	No
Fresh air vent	Yes	Yes	No	Yes
Hazard warning	No	Yes	Yes	Yes
Head restraints	Yes	Yes	Extra	Extra
Heated backlight	Yes	Yes	Yes	Yes
Laminated screen	No	Yes	No	No
Outside mirror	No	Yes	Yes	No
Parking lights	No	Yes	Yes	Yes
Petrol filler lock	No	Yes	No	Yes
Radio	No	No	No	No
Rev counter	Yes	Yes	Yes	Yes
Seat belts: front	No	No	No	Yes
rear	No	No	No	No
Seat recline	Yes	Yes	Yes	Yes
Seat height adj	No	n/a	No	No
Sliding roof	No	No	No	No
Wiper delay	No	Yes	No	Yes

ALFA ROMEO 2000 GTV

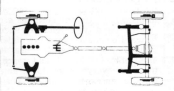

Cylinders	4 in line	Chassis con	Unitary body/chassis
Capacity	1962 cc; 120 cu in	Front sus	Ind by coil springs and wishbones
Bore/stroke	84.0/88.5 mm; 3.31/3.48 in	Rear sus	Live axle located by radius arms and A bracket. Coil springs
Valves	dohc		
Compression	9.1:1	Steering	Recirculating ball
Max power	131 bhp DIN at 5500 rpm	Brakes	All disc with twin servos, divided circuit
Max torque	134 lb ft DIN at 3000 rpm		
Gearbox	5-speed manual	Tyres	165 HR14 radials
mph/1000 rpm	22 in top	Weight	20.1 cwt; 1022 kg

AUDI 100 COUPE S

Cylinders	4 in line	Chassis con	Unitary body/chassis
Capacity	1871 cc; 114.2 cu in	Front sus	Ind by coil springs and wishbones
Bore/stroke	84.0/84.4 mm; 3.31/3.33 in	Rear sus	Tubular beam axle located by trailing arms and transverse arm
Valves	ohv pushrod		
Compression	10.2:1		
Max power	112 bhp DIN at 5600 rpm	Steering	Rack and pinion with dampers
Max torque	118 lb ft DIN at 3500 rpm	Brakes	Disc/drum with servo
Gearbox	4-speed manual	Tyres	185/70 HR 14 radials
mph/1000 rpm	19.7 in top	Weight	21.3 cwt; 1082 kg

BMW 2002 Tii

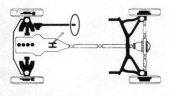

Cylinders	4 in line	Chassis con	Unitary body/chassis
Capacity	1990 cc; 121.4 cu in	Front sus	Ind by MacPherson struts, lower wishbone and coil springs
Bore/stroke	98/80 mm; 3.5/3.15 in		
Valves	sohc	Rear sus	Ind by semi-trailing arms with coil springs
Compression	10.0:1		
Max power	130 bhp DIN at 5800 rpm	Steering	Worm and roller
Max torque	131 lb ft DIN at 4500 rpm	Brakes	Disc/drum with servo assistance
Gearbox	4-speed manual	Tyres	165 HR x 13 radials
mph/1000 rpm	19.6 in top	Weight	20.2 cwt; 1026 kg

ROVER 3500S

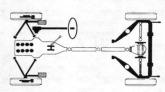

Cylinders	8 in V formation	Front sus	Ind by transverse links with longitudinal upper links pivoting on bulkhead. Coil springs
Capacity	3528 cc; 215 cu in		
Bore/stroke	88.9/71.1 mm; 3.5/2.8 in		
Valves	ohv pushrod	Rear sus	Telescopic de Dion tube located by fixed-length driveshafts and Watts linkage. Coil springs
Compression	10.5:1		
Max power	153 bhp DIN at 5000 rpm		
Max torque	203.5 lb ft DIN at 2750 rpm		
Gearbox	4-speed manual	Steering	Power-assisted worm and roller
mph/1000 rpm	23.5 in top	Brakes	Discs all round, with servo
Chassis con	Steel monocoque with unstressed outer panels	Tyres	185 HR14 radials
		Weight	26.1 cwt; 1324 kg

Alfa Romeo 2000 GTV

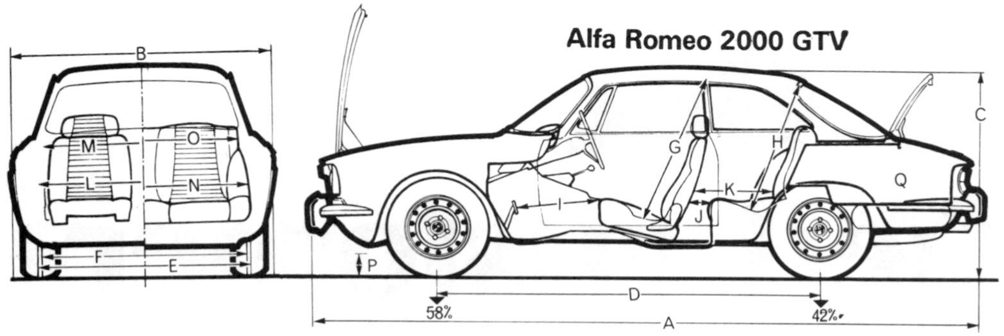

Audi 100 Coupe S

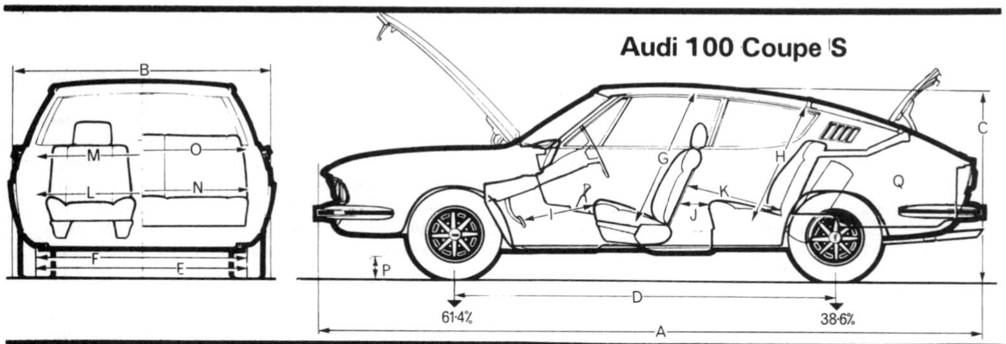

BMW 2002 Tii

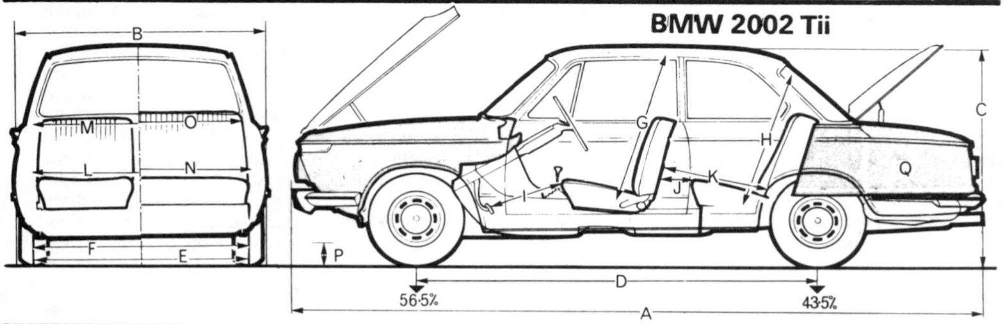

Rover 3500 S

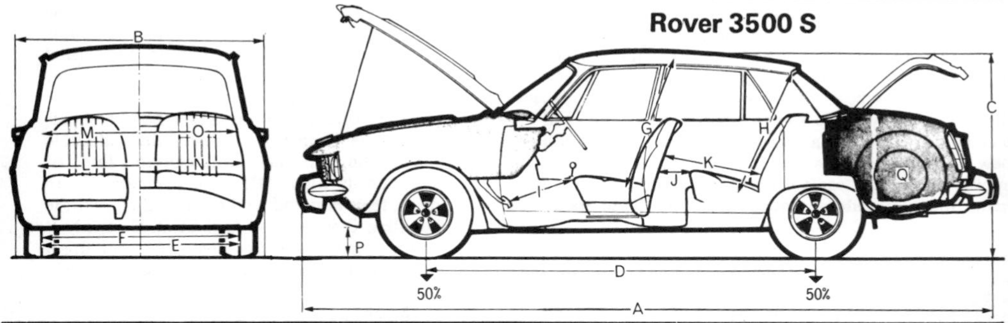

	ALFA ft in	AUDI ft in	BMW ft in	ROVER ft in
A overall length	13 5½	14 7¾	13 10½	15 0½
B overall width	5 2	5 8¾	5 1½	5 7¼
C overall height	4 4¼	4 4½	4 7½	4 8¼
D wheelbase	7 8¾	8 5	8 2½	8 7½
E front track	4 4¼	4 8½	4 4¼	4 5
F rear track	4 2½	4 8¾	4 4½	4 4
G seat* to roof	3 1	3 2½	3 1	3 4
H seat* to roof	2 10	2 11¼	2 11	3 0¾
	1 10	1 9½	1 9	1 11¼
I seat to pedals†	1 3½	11½	1 1¼	1 3¼

	ALFA ft in	AUDI ft in	BMW ft in	ROVER ft in
J legroom gap†	7½	11	1 0	10¾
	¼	1	5	2½
K seat to seat†	1 11¾	2 5	1 8½	2 4¼
	1 5	1 6¾	2 4	1 8½
L elbow width F	4 3	4 8½	4 4½	4 6
M shoulder width F	3 11½	4 8	4 3	4 3¾
N elbow width R	4 3	4 8	4 6	4 5¾
O shoulder width R	3 9	4 7½	4 2	4 3¼
P min ground clearance	5	6½	5	7¾
Q boot (cu ft)‡	6.8	12.7	10.8	10.3

* seat compressed † max and min figures ‡ measured with boxes, not suitcases

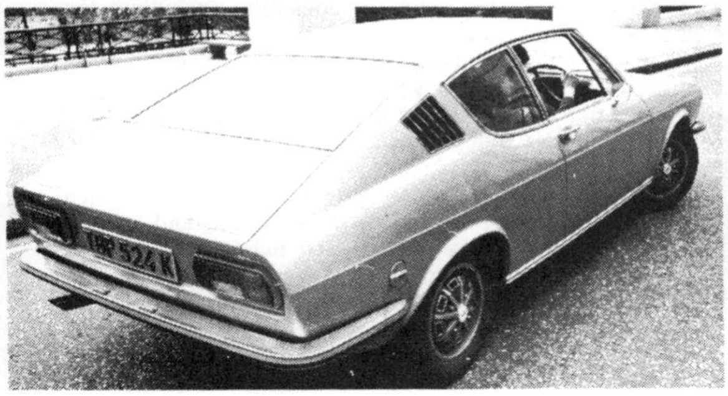

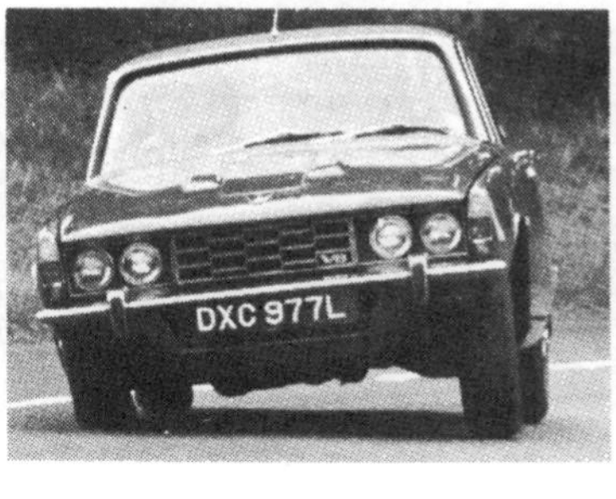

and attractive inside, the greater space and larger areas of plush upholstery creating a very real impression of superior luxury.

Although the cheapest car of the group, the Rover is the most luxurious inside and, despite the use of plastic wood on the door cappings, also the best finished we thought. The interior is well planned, too, if not as roomy as the Audi's: the facia design, for instance, with its exquisite instrument cluster, and let-down shinbins beneath a really useful facia-top tray, is still a model of how it should be done.

IN CONCLUSION

Like we said at the beginning, no best or worst. So let's do it alphabetically. At £2760, the 2000GTV seems to us to be getting a mite expensive for a car that relies heavily on driver appeal to offset its deficiencies. But it still looks the part and its superb engine and transmission alone give it an endearing character, not to say a high, effortless performance. It's the sort of car you get up to drive early on Sunday, just for the fun of it.

On sheer performance, the Audi Coupe can't quite match the Alfa. But although less sporting in its feel and handling, its cornering power is every bit as good (better on some roads) and the ride and stability so exceptional that, as we discovered, it can comfortably hold a well-driven Alfa on the road. For a slightly smaller outlay you get more car and comfort for your money, not to say striking styling that's just on the right side of being flashy. Perhaps what the Coupe does best of all, however, is to underline what an excellent car the mechanically identical 100 GL is for nearly £600 less!

The BMW is the least distinguished to look at (especially when parked alongside the Audi Coupe) though we like its neat lines and good space utilisation within a compact three-box shape. Although very well made and finished, it doesn't fare well on appointments and creature comforts for a car costing so much. However, the performance of its refined and economical fuel-injected engine is outstanding and the handling taut and entertaining. Like the Alfa, it inspires spirited driving through its willingness to go.

In some respects, the Rover is outstanding. It's fast, smooth, quiet, comfortable and very well appointed. On a long trans-continental journey, it would, by a small margin, be our first choice —provided we could afford the heavy fuel bills. But on the twisty bits the car is too soft and wallowy to endear it to the keen driver. It depends how — and where — you drive.

Far left: poor rear threequarter visibility from the Audi Coupe. Left: lots of roll on the Rover

One satisfied customer

IT'S FUNNY HOW you get attached to some cars, and not to others. Six years ago I bought a new Cooper S which I kept for two years. Its acceleration, handling and brakes were marvellous: it was also harsh, noisy, tiring, and constantly out of tune. Yet I developed an affection for that car which remains undimmed by anything I have since owned. It was traded in, at the beginning of 1969, for a new Lotus Elan S4, which had the dubious distinction of being the first Stromberg-equipped Elan in the London area. It was quite dreadful; in fact, to this day, I am convinced that it never came from Hethel at all, that it was rather the pay-off for some awful mis-deed of mine. A nightmare. It didn't go — OK, OK, we don't know much about Strombergs, but we're working on it — but, worst of all, it didn't handle. This was an Elan, remember, not an early Spitfire. And it was never sorted out. Finally, in total desperation, I swopped it for an earlier S3. A straight swop, I might add. The difference was night and day. Suddenly, all you have read about the joys of Lotus are true. This one had Webers and, like the first car, didn't notice corners. But this one went round them. Then it all went sour. An evil gentleman with a beard and a mews garage prevailed upon me to have the engine rebuilt. The car was off the road for thirteen weeks. Waiting for bits, said the man. My guess is that the bits never arrived, because the car never ran properly after that. Unfortunately, the laws of libel preclude my mentioning the name of that shark. Anyway, if you're toying with the idea of having your Elan rebuilt, talk to me first.

I don't learn quickly. When, at the end of 1970, Lotus announced a £100 bonus scheme for all those trading in old Elan for new, I fell for it. By now, Chapman was planning to launch the Elan Sprint and this scheme was intended to expedite the sale of all the Stromberg engines. "We've got Strombergs well sorted now." As I said, I don't learn quickly. They hadn't. However much you enjoy driving a car, you get tired of waiting for the AA man, and finally I divorced Lotus on grounds of mental cruelty. Not irreconcilable differences, though. An Elan is a great girl-friend, but not a wife. Of course, if you want both, that's another story.

My next car was an Alfa Romeo 1750GTV. I had always loved its looks, and the Italian firm build magnificent engines, but there were troubles. Traction in the wet was awful, and the rust was diabolical. High-speed cruising was effortless, and the brakes were fantastic, the best on any car I have ever owned. Also, the noise alone has to be worth 50 per cent of the price. My main reason for selling the car was the very high cost of spares and servicing, but I was sorry to part with it, I must say.

During the time I had the Alfa, one of the things which most disturbed me was the way I seemed to get blown off by every BMW 2002 on the road. Let's face it, the 2002 doesn't *look* quick, but after a test-drive in the fuel-injected Tii, I was convinced. I ordered a red one, and it arrived in April 1972. Options fitted were cloth upholstery (at no extra cost), tinted glass, and a heated rear window. Also I had my Blaupunkt radio and Philips cassette player transferred from the Alfa. As the car was to be used for long continental journeys, I had one of BMW's superb rally seats installed, together with a smaller steering wheel. To complete the picture, a set of Minilite sports wheels was added, and a pair of Hella quartz spot-lights. The Minilites are excellent value at £35 for a set; the looks of the car are transformed. And night driving is made so much easier by these lights: Ford used Hella lights on the Safari in 1972, when Mikkola and Palm scored that famous victory in their Escot, and I can see why Hella were chosen. They are absolutely superb. With only a few days remaining before the Spanish Grand Prix, I rushed around to get 1000 miles on the odometer, so as to have the first service done before I left. This completed, I set off for Jarama — not without a degree of worry, I might say. After all, it was a new car and, as a consequence of my eventful motoring past, I associated new cars with trouble.

Happily, I was totally wrong, and the journey down to Madrid was completed with nary a hint of drama. In fact, the only difficulty we had was finding our way around the city in search of an edible meal, which we never did find. From Spain, it was up to Spa the following weekend for the 1000 km sports car race on the magnificent Francorchamps circuit. (You're not alone, Jacky.) By now, I was thoroughly enjoying the BMW, it being fully run-in, and revving freely up to the 6500 limit. Obviously, there was no way I could resist a few laps around Spa. It's a drug. You forget that each lap is nearly nine miles, and you just want to go on and on. On my way to the German Grand Prix in 1971, I called at Spa in the Alfa, and went round fifteen or twenty times. Highly exhilarating, but there's no doubt that the BMW was significantly quicker, especially on the long up-hill sections. I frightened myself a couple of times on the down-hill section, and retired to my hotel lost in admiration for the guys who drive race cars round there...

The handling of the BMW is marvellous, far more impressive than the ultimate roadholding, in fact, although the limit of adhesion is high. It's a very secure car to drive, feels taut and well put-together, and the handling is very close to neutral. Certainly it doesn't understeer like the Alfa, and this I like. Really, the handling is lose-and-catch, rather like a Lotus Cortina, and very flattering to the driver.

From Spa, it was back down to Monte Carlo. I must say here and now that I don't go a lot on the British approach to travelling abroad — two or three hundred miles a day and then start looking for an hotel. If I have a definite destination, rather than wandering aimlessly around, I like to get the journey over and done with as soon as possible. All very philistine, I know, but there it is. Consequently, I use motorways, autobahns etc. as much as possible, including the ridiculously expensive ones in La Belle France. Tolls apart, they are a boon, tending to be far less congested than in this country, and the absence of a speed limit eliminates the 69mph queues which make M1 so lethal. After the luxury of the Alfa's fifth gear, the BMW sounds a little bit busy at very high speeds, but not obtrusively so, and the car will cruise at 105-110 indefinitely without complaint. One very big plus for the Tii's fuel injection is the fuel consumption, which is quite remarkable, I think. Under no circumstances does it ever drop below 23mpg, and

average consumption is around 28. Considering the car's medium size, and the power of the engine (130bhp nett), petrol bills are incredibly low. Similarly, the car uses virtually no oil – a pint every 1000 miles.

Thinking back to Monaco 1972, only two things come strongly to mind: Beltoise and the rain. And the rain. Jean-Pierre's virtuosity was something to see, and well worth all the hassle that characterizes the Monaco Grand Prix. And there's plenty. You have to argue with the flics to allow you to park your car, so that you can go to the press office and argue about your entitlement to a pass; having got your pass, you then go and have a punch-up with officials as to where that pass will allow you to go. There are times in life when it helps to be French.

The traffic situation was chaotic that day. Getting into Monaco on race day is always a slow business, and the rain brought things almost to a halt. Two praiseworthy features of the BMW impressed me during the long wait: there wasn't a water leak anywhere, and the engine showed no signs of overheating, or sooting its plugs.

After Monte Carlo, I returned to London, and the car went in for its second service, with over 6000 miles on the clock. No problems at all. I find the car very suitable for messing around in London, thanks to its lack of temperament, but it really comes into its own abroad. I took it to France again in June, for the Rouen F2 race and the French Grand Prix. On the French trip, for the first time, there were troubles. On the Autoroute du Nord, between Arras and Paris, the engine was clearly not pulling properly, so I stopped. A quick look at the exhaust tail-pipe told it all. It was black. The automatic choke was stuck open.

As it was now early evening, I had no option but to press on to Paris; on arrival, I got into an enormous traffic jam on the peripheral road, and now there was a real problem, for the engine would not tick over, and the fuel-gauge needle was almost keeping pace with that on the rev-counter. Mercifully, I managed to turn off and get to a garage before the tank emptied, and the following day I took the car to one of several large BMW garages in Paris, that of Charles Pozzi, who also deals in various exotica and sponsors those racing Ferrari Daytonas. Yes, they would do it that day. After all, it was an emergency – I needed the car to get to a motor race! I only hope I am never faced with the same predicament in England. The mechanics worked on the car for four hours and charged me five pounds!

That episode has really been the only disagreeable incident during my 18 months of BMW ownership. After six months there was a problem with the gear selectors, which was immediately put to rights by BMW free of charge despite the fact that the car was 10,000 miles beyond its warranty limit. There was no quibbling: the attitude was that the fault should not have arisen, and therefore there was no charge. Refreshing.

At 29,000 miles, I became aware of steadily worsening clutch-slip, particularly on fast up-changes into third and top. The car was duly taken to BMW, in Chiswick. One of the problems with owning a foreign car arises when you need spares; I hasten to add that the problem, at least in BMW's case, is not one of availability, but of price. For instance, if your clutch goes, you have to buy a complete new unit, rather than a particular part. In my case, the flywheel had to be replaced as well, and the result was a bill for over £100! I used to think Alfa prices were a bit on the high side... As a bit of a guide, a wing mirror is six and a half quid!

What else had happened to the car? Well, recently, it's been under siege from Britain's increasingly large army of vandals. One morning I found that both rear tyres had been slashed and, two weeks later, the radio aerial was ripped off and left on the bonnet. Fox-hunting has been much the news recently, despicable as it is, but I feel that vandal-hunting could be quite fun. What a pity tyres don't explode when skewered...

So there you have it. One very satisfied customer. This is one of those cars you get to love, with a great many real virtues and few faults. The standard headlights aren't anything special, the car desperately needs face-level vents, and the brakes, although adequate, are by no means in the Alfa class. But performance, handling, reliability, and finish are virtually beyond criticism, and how many cars are there about which that can be said?

The only thing that really bothers me is the question of what my next car should be. After disastrous experiences in the past, I am well and truly convinced that second-hand cars are not the answer. At the same time, a new Tii is now £500 more than when I bought mine eighteen months ago, and in these glorious days of "unparalleled growth and prosperity", salaries do not keep pace with that sort of increase! I'd love another Tii, but at present, I just don't see any way. To my mind, the car has no competition at all – for my needs, anyway. There is one challenger, now I come to think of it – the 2002 Turbo, a real smoker, which will be available here soon, hopefully. I really can't imagine how much *that* one will be!

NIGEL ROEBUCK

IT IS A common belief that the convertible automobile has had its day. Ask any US manufacturer. From its peak as the pride of Detroit and Sunset Strip, the soft-top has plummeted from the American sales charts in recent years.

The feature is becoming a comparative rarity in Europe, too. Mercedes and even MG are notably withdrawing from the canvas and bow brigade as demand diminishes. Instead, fickle weather and astute sales promotion has opened the way for the modern sun-roof. More than half of some Mercedes and BMW models come so equipped. In Britain a staggering percentage of new cars are fitted with a letterbox slit for sunshine. In fact it is more practical — you can grab every minute of sun without any complicated conversion, yet the rain never catches you out.

In Australia the demise of the MGB on our market heralded the beginning of the end. The open-air sales gap is now filled by the availability of sun-roof conversions for almost every model down to and including the Mini.

Then we come to the new BMW 2002 Cabriolet. Sidestepping all the current objections to the convertible yet offering more real wind-in-the-hair freedom than any sun roof invented, the Cabriolet has been introduced by the Bavarian maker to give a sales boost to its elderly 1602/1802/2002 small car range.

Basically it is a standard 74.5 kW (100, bhp) 2002 model with a wide and novel C pillar hooped just behind the rear of the doors.

This pillar serves a double purpose. It

BMW's VERSATILE 2002 CABRIOLET HAS...
A ROOF FOR ALL SEASONS

Traditional convertibles may be dying but, after driving it in Europe, Mike Browning found the new open-or-shut BMW is a better answer anyway. Unfortunately, we'll see little of it here due to high price and low production.

acts as an efficient roll-over bar in an accident and, connected on each side to the windscreen by the door frame, it provides a rigid framework.

BMW's previous convertible attempt on the then-1600 chassis only used additional under-body strengthening to resist body flexing and scuttle shake on poor surfaces. But the latest system is considerably more successful.

Rough cobblestoned roads in Southern Bavaria produced only slight screen shake although the test car was not as rattle free as other 1602 and 2002 models I have driven. Yet it felt considerably more rigid than most convertible models.

Filling the gap between the roll-over bar and the windscreen is a quickly removable plastic panel in Porsche Targa fashion. This is held in place by two small tongues at the back and two over-centre clamps at the front, and can be removed in only a few seconds — easier accomplished by two people than one.

A simple fold-up fabric and Perspex rear window bridges the space between the roll bar and the rear of the back seat and squab and is even easier to erect and collapse. It is clipped in place by two over-centre catches, and when folded, drops neatly into a well behind the rear seat where it is covered by a built-in tonneau.

Commendably, BMW has lost little rear seat room in the conversion, and the Cabriolet can be classified as a full four seater, although knee-room is a bit restricted. And the capacity of the large, flat-floored luggage compartment is unaltered.

Nothing of the 2002's spirited performance and good handling is sacrificed either. The Cabriolet weighs the same as its steel-roofed brother and has the same 172 km/h (107 mph) top speed in manual gearbox form — slightly slower with the automatic transmission alternative. However, the strong point is that it will very nearly reach this speed with the roof section removed, naturally at the expense of much wind noise and a little buffeting.

Acceleration is, as you would expect, identical too: Standing quarter mile is 17.4 seconds; zero to 60 mph in 10.7. As with the 2002 sedan, it is only the widely-spaced ratios of the four-speed gearbox which prevent more rapid progress.

Probably the most exhilarating way to drive the Cabriolet is as a full convertible, with the roof section removed and clipped into its special boot holder and the rear window tucked away. If a "wind" is the criterion for a convertible, then the BMW passes the test with flying colors. For high speed comfort, a cap or scarf is essential to avoid the bottle-brush look.

To reduce the buffeting, you simply pull up the back window, then you can drive at almost any speed above 80 km/h (50 mph) with little increase in through-breeze, as the interior pressurises itself to a large extent. Only problem with this format is noise. Above 120 km/h (75 mph), you have to shout; above 135 km/h (85 mph), you can forget the radio and above 150 km/h (95 mph) any conversation must be in sign language.

Of course, in really hot weather when you don't want to baste in the sun, you leave the top on and simply pop down the rear window to create a magnificent flow-through draught. This can be aided by opening the rear quarter windows in the C pillar. They also provide additional ventilation when the Cabriolet is in fully enclosed sedan form.

Finally the many-roofed BMW can imitate a snug sedan on cold winter evenings. Unlike its flop-top forerunners, the Cabriolet is utterly watertight, has an excellent heater-demister and no chilling draughts.

Probably the only major disadvantage over the sedan version is the fact that the fabric rear window makes it easier to break into when parked.

But then at its high price, any

A ROOF FOR ALL SEASONS

Cabriolets sold in Australia will spend most of their time in garages or parking lots. Because of the comparatively small production, the model is considerably more expensive than the sedan. If imported to Australia it would leave little change out of $8000-$8500.

At that price it is not the most beautiful of cars. Without its rear window installed the tail looks weirdly elongated like a small pick-up truck. And the different textures of the plastic roof, painted or vinyl covered C pillar and the fabric rear window appear a strange blend at close range.

But the Cabriolet is compact, surprisingly roomy, comfortable, spritely, easy to drive and supremely airy: a wonderful car for those whose enjoyment of motoring is an open and shut case.

RIGHT:
The Cabriolet looks a bit strange with its rear window down but the concept is highly successful.

Roof panel clips into boot, leaves plenty of luggage space.

RIGHT:
The rear window folds behind rear seat and is covered by built-in tonneau. Rear seating is satisfactory for adults.

Brief Test

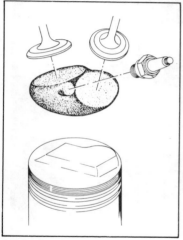

Der Dreikugelwirbelwannenbrennraum.

LOST GROUND REGAINED

*Add a new cylinder head, subtract an old air pump.
Result: a revitalized and better-running BMW 2002.*

JOE RUSZ PHOTOS

MOST CARMAKERS don't like to talk about it much, but the road performance of familiar models has dropped off badly in many cases as emission-control limits tightened and obtuse pressure from General Motors and the California government pushed down compression ratios. For most imported cars the big crunch came in 1972, when engines tailored to run on low-octane unleaded fuel became the rule.

The familiar and popular BMW 2002 was such a victim. From the beginning it had an air pump, that dreaded device that pumps fresh air into the exhaust ports so unburned hydrocarbons and carbon monoxide will be burned and converted respectively. That sounds innocent enough, but air pumps often brought with them unpleasant side effects: backfiring on deceleration, enough extra exhaust back pressure to cut power output, extreme underhood temperatures, and sometimes more. Generally, air pumps have been used on engines whose designer-engineers couldn't make them conform to the rules without one and BMW's 4-cylinder, carbureted engine was one of these. When its compression ratio was dropped from 8.5:1 to 8.3:1 for 1972 and other emission-reducing changes were made, a few horsepower were lost, and nobody ever liked the air pump anyway.

Meanwhile, at the upper end of BMW's passenger-car line, the 6-cylinder engine had a clever combustion-chamber design that was so clean burning it didn't need an air pump. BMW calls the chamber *Dreikugelwirbelwannenbrennraum,* German for tri-spherical turbulence-inducing combustion chamber. Spherical forms around the valves, with the curved side of the piston-top hump, promote much turbulence and a third, smaller sphere form around the sparkplug completes the shape (see drawing). The chamber apparently gives good emission characteristics, power output and fuel economy.

Now this feature has been extended to the 2002, restoring power to where it was before 1972 and helping the 4-cyl engine meet 1973 emission regulations without an air pump. The result is an engine with better drivability, less underhood complication, and naturally better performance.

We borrowed a fresh 2002 to see if all this really worked and we're happy to report that it does. Our test car ran with practically no symptoms of lean carburetion and went as well as the 2002 we tested for our June 1968 issue with a 10.5-sec 0-60 mph time and a 17.8-sec quarter mile. The 1968 car had done 10.4 and 17.6 respectively, and differences that small just aren't significant.

The engine retains the same character as ever: rather noisy but mechanically smooth, virile-sounding and potent. Fuel economy isn't quite as good as it was in 1968, but it needed premium fuel then; its current 23 mpg still isn't bad for a car this quick.

Otherwise the 2002 is the same likeable sports sedan, updated this year with some nice new colors and inertia-reel belts. In some ways (like ventilation) it's a dated car, and the price is high these days, but it's still one of the best compact sports sedans around. The rigid, rattlefree body, compliant suspension, roadholding, brakes and responsive engine all conspire to make it a car that just can't be driven slowly. If everyone drove one of these, traffic would move a lot more efficiently.

Road test

by John Bolster

Remodelled BMW 2002 Tii

The BMW 2002 has been manufactured for quite a number of years. As is well known, it is derived from the small and relatively light bodyshell of the 1600, fitted with the 2-litre engine. The power unit, like all BMWs, has inclined valves operated by a chain-driven overhead camshaft and rockers. The Tii model has an up-rated engine with fuel injection, developing 130 bhp.

The subject of the present test is the latest version which has been re-styled front and rear while the interior has been remodelled, with additional sound insulation. The mechanical specification remains the same, apart from a fractionally wider track. There is independent suspension front and rear, with MacPherson and semi-trailing arm geometry respectively, plus anti-roll bars at both ends. ZF-Gemmer worm and roller steering is still used and the disc front and drum rear brakes have twin servos.

Since I last tested this model, the price has risen by no less than £900. One doesn't get a lot of motor car for all that money but the compact size is an advantage and the acceleration is outstanding in any company. However, in spite of the new trim the interior is still a bit spartan for this price class and the absence of face-level eyeball ventilators is astonishing, the swivelling quarter-lights creating a great deal of wind noise. More luxurious seats would be appreciated, too.

The 2-litre BMW engine must be one of the best four-cylinder units ever built. It has none of the roughness at low speeds which plagues most of the larger fours and is outstandingly flexible. The former starting problems of the injection engine have been completely overcome, the machinery springing instantly to life on the coldest mornings, and the theoretical advantages of fuel injection in saving petrol are realised. In my previous test of the Tii, I was able to average 20 mpg when cruising at 110 mph whenever possible, in France. This time, I proved that the car will achieve well over 40 mpg at a steady 50 mph, which means that a clever driver will get 35 mpg on country journeys, perhaps

Suspension is independent all-round.

reduced to 25 mpg in thick town traffic.

A few spot performance checks showed that the capabilities of the latest car are virtually identical with those of the earlier model tested. To avoid wasting fuel, therefore, I have used some of the figures previously obtained to complete the data panel, rather than going through the repeated full-throttle tests that I normally carry out. BMW cars always have pleasant gearboxes with nicely spaced ratios, which allows the efficient power unit to give its best performance.

The suspension feels fairly hard at low speeds but is outstandingly good at coping with really bad roads. At high speeds, the car rides the bumps very well and the absence of roll on corners contributes to the comfort of the passengers. The road noise is very moderate, though the wind noise increases somewhat as the speed rises.

The BMW goes through corners in a most satisfactory manner, with moderate understeer and a tail that can be brought round under power on occasion. The steering gear ratio is about right and the spokes of the wheel are arranged to give a clear view of the instruments. The front disc brakes have 4-piston calipers and automatic pad wear compensation, in conjunction with substantial drums at the rear. The results are truly excellent, with extremely potent retardation and freedom from fading.

The engine could not be called silent, though its sound level is moderate except at maximum revs. Like all BMW engines, it has a hum that sounds efficient but it never seems to be highly stressed. With fuel injection, the already good low-speed torque becomes

Two litre engine uses Kugelfischer injection.

Latest cars have a re-styled front and rear and more comfortable interior.

Road test

Above: familiar, fascia is still rather spartan. Below: moderate understeer prevails.

Fuel injection provides good low speed torque and helps fuel consumption.

outstanding, permitting top gear to be used a great deal when conserving petrol. The torque is sufficient to promote wheelspin rather easily on wet roads, but the car remains very controllable under these conditions.

The body shape is inconspicuous, which can be quite an advantage. However, the car is unmistakably a BMW and its looks show plenty of character. The 2002 may be used as a full 4-seater, though it is best for the people in front to avoid pushing their seats right back. The luggage boot is surprisingly roomy for a car of such moderate size.

Though sheer performance takes second place at the moment, it is pleasant to know that the maximum speed is not far short of 120 mph and a reading of 125 mph is easily obtained on the slightly optimistic speedometer. More important, perhaps, is the smooth performance in top gear between 20 and 30 mph, an area where many four-cylinder cars thump and rumble until the driver is forced to change down and waste precious fuel.

There remains the question whether or not fuel injection justifies its cost in a road car. That it can give excellent fuel economy is certain and by its use a four-cylinder engine may be given much of the flexibility of a six. Though it may never be used for the more basic types of car, its virtues make it more than worth while for a machine of the calibre of the BMW.

Though this is an expensive little car, it is one of those rare dual-purpose vehicles which combines the handling and performance of a competition car with the perfect manners of a town carriage. The changes in the latest model are very slight and this is basically quite an old design, but it is still right on top for speed, acceleration, and fuel economy. To many, the racing background of the BMW adds greatly to its appeal, too.

SPECIFICATION AND PERFORMANCE DATA

Car tested: BMW 2002 Tii 2-door saloon, price £3,199 including car tax and V.A.T.
Engine: Four cylinders 89 mm x 80 mm (1990 cc). Compression ratio 9.5 to 1. 130 bhp at 5800 rpm (net). Chain-driven overhead camshaft operating inclined overhead valves through rockers. Kugelfischer mechanical fuel injection.
Transmission: Single dry plate clutch, 4-speed all-synchromesh gearbox with central change, ratios 1.0, 1.32, 2.02, and 3.764 to 1. Hypoid final drive, ratio 3.64 to 1.
Chassis: Combined steel body and chassis. Independent front suspension by coil spring damper struts and lower wishbones. Worm and roller steering. Independent rear suspension by semi-trailing arms, coil springs, auxiliary rubber springs, and telescopic dampers. Anti-roll torsion bars front and rear. Twin servo assisted disc front and drum rear brakes. Bolt-on disc wheels, fitted 165 HR 13 radial ply tyres.
Equipment: 12-volt lighting and starting. Speedometer. Rev counter. Fuel and temperature gauges. Clock. Heating, demisting and ventilation system with heated rear window. 2-speed windscreen wipers and washers. Flashing direction indicators with hazard warning. Reversing lights. Radio (extra).
Dimensions: Wheelbase 8 ft 2.4 ins. Track 4 ft 4.8 ins. Overall length 13 ft 10.5 ins. Width 5 ft 2.6 ins. Weight 2226 lbs.
Performance: Maximum speed 119 mph. Speeds in gears: Third 91 mph, second 62 mph, first 34 mph. Standing quarter-mile 16.9 s. Acceleration: 0-30 mph 3.4 s, 0-50 mph 6.6 s, 0-60 mph 8.9 s, 0-80 mph 17.6 s, 0-100 mph 26.8 s.
Fuel consumption: 20 to 35 mpg.

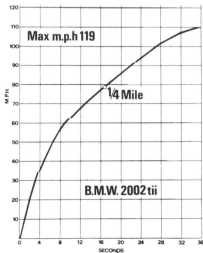

Well blow me down...

Mike McCarthy travels to Cornwall and back courtesy of a Turbocharged BMW and a Gas Turbine Herald

"Are you busy next week?" asked the Ass Ed (sorry, Asst Ed) casually.

"Yes," I said, quick as a flash (I know how his casual questions can lead to all sorts of nastiness).

"Doing what?" he asked.

As ever in times of dire emergency my mind was an absolute blank. I stood and stared at him.

"Right," he said, "You can drive down to Cornwall and back." I looked glum. "In a turbocharged BMW," he said. I looked happy.

Which is how I happened to be standing in BMW's new service centre in Brentford at 8.30 am (ghastly hour) feeling somewhat trepidatious as the efficient and bright-eyed BMW man ran over the controls (I mean, of course, pointed them out to me — what do you want, humour?) like an airline pilot.

He actually said: "It's just like any other car, but don't let the needle on that little dial in the middle of the facia climb into the red." I climbed into the king-sized hairdryer — then got out and slowly walked around the other side, because it was left-hand-drive you see. BMW man looked supercilious.

Somewhere behind and between the plastic beard in front, the cauliflower spats on the side that hide fat, fat tyres and the black plastic spoiler on the boot lid (which for some reason makes the car look like a small truck) there is a BMW 2002 that we know and love of old. Except that the technocrats at BMW (I'm sure they are bright-eyed and efficient too) have added a fairly novel option — a turbosupercharger. As everyone knows (if you don't you don't read *Motor*) this takes surplus energy from the exhaust gases and uses it to increase the pressure of the incoming air. So instead of suffering from the well-known disease manifold depression the car becomes quite Fascist and causes manifold oppression, as it were. The net result is what we of the Motoring Press (he said, full of his own self-importance) call "a vast increase in power." In fact the 2-litre engine started out life producing a reasonable 100 bhp; add the turbo and a few other bits and it kicks out an unreasonable 170 very DIN horsepower. And the only way you tell from inside that it is potent is that little dial — and a red (for danger?) facia panel.

BMW gave me a sheet of paper  with a drawing of the dial on top, and in capital letters underneath FOR MAXIMUM MOTORING. Now the dial has nothing so vulgar as numbers on it, so you can't talk about "There I was, on 5 lb boost" or whatever. Instead there are three sectors, the first white and quite large, the second tiny and green, and the third red and very large. As the sheet says, the needle stays in the white sector for normal driving. Things get more exciting when it goes on to say that when the needle is in the green sector maximum turbocharging occurs. The ultimate comes when it says, again in capitals, IF THE NEEDLE MOVES INTO THE RED SECTOR, LIFT OFF — the pressure relief valve is stuck. It's obviously intended for some sort of game, like  a pin-ball machine: you must keep it above the white but below the red. The reassurance note that a sticking relief valve is unlikely did not reassure me at all, since I am the eternal pessimist.

Anyway, I turned the key in the ignition waiting for a loud wailing and howling — the original banshee cry, beloved of aged motoring writers, and which is always mentioned whenever the Mercedes SSKL is talked about. It fired immediately, but sounded like any other BMW 2002, if anything a little quieter. Disappointing. It must be true, I thought, a turbocharger does act like an additional silencer. I gingerly executed a three-point turn in the confines of the reception area; I didn't want to disappear out through the receptionist's little glass box. It

4.5 grands worth of diminuitive Turbocharged 2002 (top); if you want one you'd better hurry, they're importing fewer than 200. Left: pity this is in monochrome. The instrument surround is bright red—a warning, perhaps, to keep your eyes open!

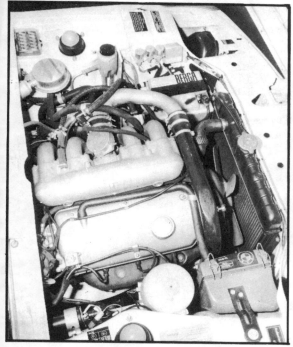

This is what 170 German horsepower looks like (left) and yes, I couldn't find the turbocharger either! The innocuous dial on the right, next to the clock, tells you (as if you wouldn't know) that it's all happening

behaved impeccably. Good grief, I thought, it's actually quite tractable. (Tractable to me means that it doesn't make me angry because of the noise/temperamental engine/ peculiar transmission/ odd controls/ sheer difficulty that a number of tuned cars display. I'm a great one for creature comforts.)

Left down the A4, past the Firestone factory. No chance to use more than 2000 or so rpm: too much traffic. Burble, burble, it went, nice and docile. No problem. I relaxed, unsticking my (by now) clammy hand from the steering wheel. I headed west, out through the dormitory suburbs to the M3. No fuss, no dramas, but also no heavy right foot. After getting lost a couple of times — why do they always put direction signs up when the best route is obvious, but never, never at critical T-junctions? — I saw the lower reaches of the M3, stretching away into the distance.

At 2000 rpm nothing much happens. The tacho needle just climbing relatively slowly to about 2700 to 2800 rpm. At that point a soft whistling begins, getting slightly louder and higher in frequency as engine revs rise. Then the needle on that little gauge rushes over to the green bit (up to now it had been hovering forlornly around the bottom end of the white bit) and suddenly you're moving. Not just quickly, but like the proverbial cat with the scalded tail. The needle on the tacho just about bends itself on the stop, you pop the clutch, slip the gearlever into third, and it all happens again, except that this time you don't drop the revs and keep the turbocharger spinning, so again there's that superb push in the backside.

By this time prudence, not to mention an almost pathological fear of the Q-car police mini-van (would you believe?) that has been known to creep around that neck of the woods dictated a return to insanity and 50 mph — well, 50 or so. Time to look around the car a bit. Ride—harsh, I suppose, but I've known worse on cooking family saloons. Steering is light and delicate (this car must have the high ratio box that Hardly referred to when he wrote about the 2002 Turbo back last September), and the right amount of feel for me. Noise is reasonable.

Off the M3 on to the dreary and depressing A303, possibly the most worked upon road in Britain. Once more no fuss, no boiling or plug fouling, no snatch or jerk. Remarkable.

Through Exeter, with a stop for petrol (only four star—hope that's all right) and coffee — or tea, I still don't know which. It was lukewarm and sickeningly sweet.

Away from Exeter and across Dartmoor on the B something-or-other. No traffic apart from the odd sheep or two — safe behind hedges, I might hastily add — and a twisty road that the well-known seaside snapshot photographer and Jaguar-owner Paul Skilleter quoth highly of. This, I say grinning inanely to myself, is where they mean you to keep that little needle firmly in the green sector. And I tried, how I tried. Superb, incredible fun, but then a few damp patches cooled my fervour and the big Pirellis. They did not like the wet.

The Hardly said that "the handling will certainly suit the Achim Warmbolds (who?) and other loose-surface German heroes with oversteer being virtually on top — (sorry) tap." And he's right. Play it wrong, like getting the turbocharger to sing at the wrong point, and whoops! It's tail-first into the flora and fauna.

Driving a turbocharged car is not at all like the pop-pop, spit-spit, backfire, hiccup then away you go bit that you get with the "cammy" overcarburettered modified cars that we sometimes drive for "Motoring Plus." It's smooth, very smooth. A hard shove is all you feel with no definite " now it's in, now it's not " point. On the other hand the turbocharger is not working very hard at low load and revs, but is spinning it's little heart out (at 100,000 rpm, would you believe) at full load and speed. So the faster you go, the faster you go, if you know what I mean. The original BRMs, with centrifugal superchargers, suffered from the same characteristic, and caused Messrs Fangio and Moss, among others, to execute some strange manoeuvres. I think BMW may have to look a little carefully at who wants to buy it . . . but they are going to change the tyres for some more appealing to the British temperament.

I was woken the next morning suffering from the late-night syndrome—the party sat up until two, telling of how brilliant and brave we were. I managed to put on a superior air to the others, because I " had driven the car before," so could sneer while they searched for the ignition switch, or reversed over the rose bushes. It was a very grey day, as they say, and pouring with rain. The idea was to go off two by two, and make our way to St Ives. A certain member of the motoring press who, up to now, I had regarded as a friend, paired up with me along with a nervous member of BMW's press corps. My friend drove.

I knew there was something I had to tell — George, shall we call him, but my brain hurt. I remembered some 10 minutes later as we clouted a bank and dropped into a ditch, all wheels locked. " These tyres aren't very good in the wet," I said. We got out, except George, who sat with episcopalian unconcern behind the steering wheel turning it this way and that as we heaved and strained. A gentleman farmer came past in gumboots, and obligingly climbing into the ditch with us to heave and strain. With a loud squelch the car came out of the ditch. The only damage was a set of muddy handprints on what had up to then been pristine silver paint. Not a scratch, otherwise. Why does it never happen to me that way?

From there on we settled down to an amicable game of " You frighten me, I'll frighten you." We stopped at Goonhilly so that George could snap away happily with his box Brownie. I had forgotten mine. We got lost many times. Neither of us could navigate very well.

After lunch in St Ives we all played musical cars: those who had driven the Turbo took over 3.3Ls, the other new car for us to thrash, I mean try. By this time my brain was working again, and I realised that really I am cut out for the life of a millionaire. It's my way of life. Wafting along with all creature comforts, being courteous to the hoi poloi even when they get in the way, and discussing lofty sentiments like the best way of cooking sole bonne femme or whether left foot braking on an automatic was good or bad.

The intention was for me to drive one of the 3.3Ls back to London. Some quick calculations showed that no way was I going to be there before midnight, so I nastily pulled rank and suggested that a seat on the plane flying back would not go amiss. BMW acquiesced. For some reason the receptionist at the airport at Newquay insisted that I had to call myself " Miss Ann Hope," which was an embarrassment but it did mean that I was travelling back to London in a Herald — which you might have gathered by now is of the flying sort.

And that is a description of the sort of hard times that we motoring journalists have to put up with. Life is not all caviar and champagne, you know. They sometimes put Guinness in as well.

MOTORING PLUS

SUCK IT AND SEE (IT GO!)

To qualify a car for saloon car racing, its manufacturer must go through a most involved process known as homologation. The most important homologation requirement is proving to the FIA that a minimum number of identical cars have been built, to ensure people race "true" production saloon cars and not tarted up prototypes. If finances allow, though, a keen manufacturer will quite naturally build the requisite number of "special" cars but because of their uncompromising nature these "homologation specials" are usually something of a waste of time or money for ordinary road use.

A welcome exception—up to a point—is the £4299 BMW 2002 Turbo. This is a turbocharged version of the 2002 Tii with corresponding modifications to the chassis. BMW admit it's an homologation special, even supplying it with screwed-on wing extensions that can be replaced easily if you wish to fit fat racing tyres. They won't manage to produce the necessary 5000 for Group 1 homologation (only 180 are being imported to the UK), so instead they'll have to settle for more doubtful benefits in Group 2 racing. All Turbos, incidentally, will be left-hand drive, as it's impossible to fit the steering box in on the right.

BMW's experience with turbocharging goes back quite a long way, for in 1969 they won the European Touring Car Championship with a 275 bhp turbocharged 2002. In the road-going 2002 Turbo, the KKK turbocharger is bolted on to a fuel injection Tii engine — a feature which gives it a few advantages over normal carburetter set-ups. To cope with the extra performance the Turbo has a strengthened gearbox, 3.36:1 limited slip differential, revised spring and damper rates, wide 5½J wheels shod with 185/70 VR 13 tyres, ventilated front disc brakes, an air dam at the front where the bumper should be and a spoiler on the boot lid. Inside there are special Rentrop rally-style seats and a boost gauge. Just to make sure that everybody knows that it is something special BMW finish it off with a lurid paint finish; we'd much prefer something that didn't catch the eye so much.

The turbocharger used by BMW is an off-the-shelf unit produced by KKK. Because the engine has a cross-flow head, it is mounted on the opposite side to that which carries all the normal induction items. A long steel pipe leads the induction air across the engine to the inlet manifold via a blow-off valve (which when the maximum boost pressure is reached dumps air unsilenced to atmosphere) and butterfly valve. The fuel injection system can compensate automatically for charge density (by measuring inlet manifold pressure) and thus should provide a near-perfect mixture at all times unlike most carburetter-turbocharger setups which supply over-rich mixture at low rpm when the turbocharger isn't blowing at full pressure. In the 2002's case this is 1.55 atmospheres or a boost of 7.8 psi, which raises the power output from 130 bhp (DIN) at 5800 rpm to 170 bhp at the same speed. The compression ratio is only 6.9:1 but premium fuel must be used because of the turbocharger's presence.

One of the most impressive features of the car is its tractability and fussless behaviour in town. You don't expect to be able to trundle along at 15 mph in top with an engine that produces 85 bhp per litre, but do it you can. No hint of the potential lying in wait at high rpm exists, it just feels like a rather slow—because of the low compresson ratio and high gearing—Tii. But plant your right foot firmly on the floor and keep it there and the effect is astonishing. Little boost exists until 3800-4000 rpm when the needle of the boost gauge climbs into the green and the power comes in with quite a rush. It's quite easy to keep the turbocharger pressure up between gearchanges if you're quick. Driving like this means that you'll see 60 mph from rest in 6.6 sec—respectable by any standards but astonishing for a production two litre saloon. Only one car of less than 2000 cc tested by *Motor* has accelerated any faster—the very special Group 6 rally Escort from 1969—and that certainly couldn't be driven through town every day. To 100 mph takes a more sane 19.1 sec.

Good as the Turbo is for out-and-out performance driving it can be a real pain under more normal circumstances. Floor the throttle at 3500 rpm to overtake a slower car and there's a good couple of seconds delay before the turbocharger gets the car really moving; by that time you're normally past the obstacle and the sudden increase in power can be an embarrassment. We'd prefer more mid-range power even if it meant a slight sacrifice in top end performance.

The changes made to the suspension make the handling extremely good under most circumstances. To all intents and purposes the car steers where you point it with no tricks such as vicious tuck-in if you lift-off. But push the car to its (admittedly high) limit of adhesion and it can become a little untidy particularly if the turbocharger cuts in half-way round a corner. On wet London streets the tail could step sideways very quickly too. Although we didn't manage it on the road the brakes faded badly after four or five laps of MIRA's road circut but did recover quickly. We thought them to be rather unprogressive and over-assisted for town use—features that are accentuated by the awkward pedal placings.

Despite its racing bias the Turbo remains an extremely civilised car. We've already mentioned the engine tractability and to complement this there are superbly comfortable seats, a firm but very resilient ride and no more noise than from the standard car. That you can make a 2-litre so quick while retaining such a high level of refinement is what makes the Turbo so impressive.

The turbo's secret laid bare. Air enters the turbocharger on the right and is forced under pressure up the long pipe over the engine. The pipe leading from the turbocharger to the front of the sump is an oil drain

	BMW 2002 Tii	BMW 2002 Turbo	Broadspeed Turbo Bullit
MAXIMUM SPEED			
Mean	113.5 mph	128 mph	138.5 mph
ACCELERATION			
mph			
0-30	2.7 sec	2.4 sec	2.9 sec
0-40	4.4	3.6	3.9
0-50	6.3	5.0	5.3
0-60	8.2	6.6	7.0
0-70	11.6	9.0	8.8
0-80	15.0	11.5	11.4
0-90	20.2	14.5	14.3
0-100	28.8	19.1	17.6
0-110	—	25.1	23.5
Standing ¼ mile	16.5	15.3	15.1
Standing km	30.6	28.3	27.4
IN TOP			
mph			
20-40	7.7 sec	10.0 sec	8.5 sec
30-50	7.8	9.2	7.8
40-60	8.0	8.6	7.5
50-70	8.4	8.2	6.3
60-80	9.6	7.3	6.4
70-90	10.8	7.2	6.3
80-100	13.1	8.5	7.1
90-110	—	—	11.2	8.0
IN THIRD			
mph			
10-30	—	7.7	6.0
20-40	5.4	6.8	5.4
30-50	5.4	6.0	5.0
40-60	5.6	5.0	4.2
50-70	5.7	4.6	3.8
60-80	6.2	4.7	4.1
70-90	8.1	5.4	5.1
FUEL CONSUMPTION			
Overall	24.0	18.5	17.2
Touring	—	—	22.3

COMPARISONS

	Capacity cc	Price £	Max mph	0-60 sec	30-50* sec	Overall mpg	Touring mpg
BMW 2002 Turbo	1990	4299	128	6.6	9.2	18.5	—
Alfa Romeo Montreal	2593	4999	135.2	8.1	8.8	13.8	—
Broadspeed Turbo Bullit	2994	3532	138.5	7.0	8.8	17.2	22.3
Jaguar E-type V12	5343	3812	146.0	6.4	6.0	14.5	16.1
Porsche 911S 2.7	2687	6993	140†	6.5	9.0	19.8	—

* In Top. † Estimated.

REX GREENSLADE

turbo to nivelles

We drive the turbocharged BMW 2002

LAST OCTOBER, at the Motor Show, an otherwise boring and monotonous day was considerably enlivened for me by my first sight of the BMW 2002 Turbo. I've always thought that the Tii was a very considerable improvement over the ordinary 2002, so the Turbo had to be something else again, a real smoker. After all, the 2002 turns out 100 bhp (NET) and the Tii 130, so I figured to myself that the Turbo with its 170 bhp (NET) must be pretty dramatic to drive. And I wasn't far wrong.

Recently I took one over to Nivelles for the Belgian Grand Prix, and I found it one of the most exciting road cars I have ever driven. BMW claim its 0-60 time to be 6.8 seconds, which is quick by any standards — quicker than a Ferrari Dino or Porsche 911, for example. I have no reason whatever to dispute their figure, for the car feels every bit as fast as that. But it is the turbocharging that makes the acceleration so dramatic. Move away from the lights, and you're a bit brought down, for the initial step-off feels just like any other 2002, nippy but hardly neck-snapping. Then, at just over 3,000 rpm, you can hear a slight whistle; on up to 4,000 the whistle becomes more apparent, and then the needle in that little gauge on the right moves out of the white section and into the green, and suddenly you know just what Mark Donohue's been talking about all these years . . . . All of a sudden, you seem to have found another hundred horsepower, and the little car simply hurtles along, the tachometer needle at 6,500, and you wonder what the hell happened. It's as if you'd been driving with the handbrake on and you'd suddenly remembered.

Chassis mods

Now all this takes acclimatisation. When the Tii was introduced, many people reckoned that the 1600/2002 chassis had been taken to its limit, that the car couldn't reasonably handle more than 130 bhp. So obviously when the Turbo road car got the green light, BMW's engineers had to give the chassis some thought. The result is remarkably successful. Everything has been beefed up; there are bigger anti-roll bars, stiffer shock-absorbers (adjustable, incidentally), and wider wheels. To accomodate the wheels, flared arches are necessary. In addition, there is a huge bib spoiler at the front, and a small 'wing' at the back. All these factors combine very successfully, for the car is extremely stable at all times, and can be flung through corners at speeds inconceivable in a Tii, for instance.

Personally, I think the new mods are practically rather than aesthetically pleasing. In short, I don't much care for the appearance of the Turbo. One of the things I've always liked about my Tii is that it is such a

tremendous Q-car; the car's appearance is sporty but not at all ostentatious. Not so with the Turbo. I accept that the air-dam is very necessary, and also the rear wing, but those wheel-arches .... They are bolted — not very elegantly — on to the car, rather like everybody had on their Cooper S when they were 19, and the side-stripes with the word 'Turbo' seem wholly unnecessary to me, quite out of character on a BMW product. I reckon that £4300 is too much money for a car with bolted on wheel-arches. It's *not* a boy-racer's car, so why make it resemble one?

Having said that, let's go on to the car itself. In a word, it's quite sensational, and every bit as impressive as I expected. In fact, all things considered, I can't recall a more enjoyable road car in a long time. It's all things to all men. In town, you can drive it just as you would any other 2002, and on the open road it's an absolute ball of fire.

Quicker and quicker . . .

I don't quite know why, but it seems that I never leave for Dover on time. I have a boat to catch, I know the time of departure, and I know roughly how long the journey will take me. But I never seem to put the three things together properly. Consequently, there is always a frantic last-minute dash round, trying to find films, tapes, press passes, sunglasses, and what have you, followed by a very rapid drive towards the White Cliffs. This time, I used a lot of back roads in an effort to avoid traffic, and the Turbo was in its element. I was hardly ever out of 2nd gear on those roads, and I can't think of another car which would have a prayer against the Turbo in these conditions. The most exciting thing about the car is the *way* in which it accelerates. Many cars have a very impressive step-off but then run out of steam just when you need it. The Turbo, on the other hand, feels fairly fast up to around 4000 rpm, and then it all starts to happen, and it goes on happening. In fact, the car seems to accelerate faster and faster all the time, and it just as well that BMW fit a rev-limiter at 6500 rpm, because the thing feels as if it would rev for ever. Another outstanding feature of this engine is its quite remarkable smoothness — once the turbo has started working.

It takes a little time to get used to the characteristics of this engine. To get the power when you need it, you have to accelerate a little earlier than normal when going round a corner. After a bit of practice, it is very satisfying, for the little car fairly catapults out of the bend and belts on down to the next one. Lovely. Having said that, I must add that it needs to be driven with a great deal of respect in the wet. If you're halfway round a corner in torrential rain, and suddenly all that extra power cuts in, the tail goes very quickly indeed. Likewise, hard acceleration in a straight line produces instant wheelspin.

Unfortunately, it is not possible to manufacture a right-hand drive Turbo, for the actual turbo-charger occupies the space needed for the steering-column. Personally, I think this is far less of a problem than many people make out, and I hate to think how much a rhd version of the car would cost, even if it were possible to make one!

The brakes of the car are well up to its performance, and everybody commented on their excellence. Likewise, the handling is marvellous, very close to neutral, and the steering is quicker than on other models in the range — which it needs to be. When you're *really* pressing on, you do have to work quite hard, but that's what makes driving satisfying, isn't it? Coming out of a corner, the turbo cuts in, the tail gets slightly out of line, and the rear wheels scrabble for grip as 170 bhp tries to put itself down on the road . . . .

The interior of the car is much like any other 2002. There are rally-type seats which are quite good but, disappointingly, not cloth-covered. A different, smaller steering wheel is fitted and, for reasons best known to BMW, a piece of red plastic has been stuck to the small fascia in which the instruments are set. Likewise, the latest versions of the 2002 and Tii have a piece of imitation wood stuck there. Horrible. Why they didn't leave the thing plain black, I can't imagine. Otherwise, the interior is plain 2002 — apart from that little boost gauge on the right, that is . . . .

Fantastic fuel consumption

In Belgium, the car was able to cope with all kinds of conditions. It crept through Brussels traffic on Saturday night, it blasted along the motorways at 130 mph, it sat in a queue at Nivelles for ages without overheating or sooting its plugs. At last there is a very high performance car without any of the traditional hang-ups. And the fuel consumption is absolutely unreal. In fact, we were quite unable to believe it at first. On the way to Nivelles, we put in twelve gallons of 4-star, and we filled up again 354 miles later! If you'd care to check it, that works out to 29.5 mpg! Amazing.

So how does one sum up the car? Well, for a 2-litre saloon, it sets new standards. Nothing else is even close. It's a beautiful car to drive, it's incredibly thrifty on gas, it seems to be reliable. It's expensive, and its external appearance does not suggest a £4300 car. I would dearly love one, but I think my first course of action would be to have the car resprayed, *without* the stripes, unbolt the wheel-arches, and have them flared professionally. Then I think you would have pretty close to the ideal road car.

NIGEL ROEBUCK

BMW 2002 TURBO

When Alex von Falkenhausen applies BMW's engine department to turbocharging, watch out!

BY PAUL FRÈRE

For those who thought the 2002tii was to be the ultimate BMW in the "02" series, the Munich factory had a surprise in store. Unveiled at the Frankfurt Motor Show last September, with 170 bhp DIN packed under its hood, the turbocharged version of the 2002tii lived up to BMW's sporting image by lifting the compact 2-door which started life with a humble 1600-cc engine of half the power into the exclusive "over 125 mph" class. Not everyone gave it a hearty welcome, however. Part of the German press severely criticized the "provoking aggressiveness" of its racy-looking front air dam bearing the inscription "Turbo" in mirror script to warn those seeing the BMW come up in their mirror of what sort of car they had to cope with. The backward script was painted over even before the show opened, but the tension between the Bavarians and part of the press remained—with the result that no BMW 2002 Turbo was available at the Motor Show test day in Hockenheim. Then came the fuel "crisis,"
which gave BMW a chance to quietly solve the problems still giving them some worry at the time the car was announced. The most important one was a tendency for the exhaust manifold feeding the turbocharger to crack under the strain of the 1650°F temperature of the exhaust gases. The solution is a manifold made of a special cast iron with high nickel content.

The turbocharging installation, applied to the basic 2002 engine with mechanical injection as used in the 2002tii, is straightforward. The turbo of KKK manufacture is fed by the full flow of the exhaust gases and delivers fresh air through a long pipe to the intake manifold on the opposite (left) side of the engine. The throttle valve is located just before the point where the pipe from the turbo joins the manifold, with the blowoff valve just ahead of the throttle. This is controlled by a mechanism sensitive to intake-manifold pressure and opens when an 8-psi boost is reached to release the excess blower output into the atmosphere. This is admittedly a rather crude way of limiting the boost pressure; an exhaust bypass would no doubt be more efficient in terms of thermodynamics and reduce the exhaust back pressure. But engine development chief Alex von Falkenhausen says there is just no room to arrange for such a layout. No provision is made to release the pressure in the pipe leading up to the intake manifold when the throttle is shut: the blower, kept spinning by its own inertia, must work against the back pressure created by the throttle valve. Experiments made with a venting valve apparently did not show up any noticeable advantage in keeping the turbo spinning fast with no exhaust gases to activate it and thus in reducing the time lag before full power again becomes available when the throttle is reopened.

Apart from a different ignition timing curve and a reduction in compression ratio to 6.9:1, the engine is identical with that of other 2002 models and uses the same camshaft, though an oil cooler is included in the circuit mainly because of the heat transferred to the lubricant by the turbocharger of which the (plain) bearings are lubricated by a derivation from the main oil circuit. Maximum bearing pressures are the same as in the aspirated 2002tii engine, as they occur at exhaust top dead center.

As the maximum boost pressure of 8 psi (reached at 3500 rpm full load) is comparatively low, I asked Von Falkenhausen why BMW had settled for this figure, giving the mild engine output of 170 bhp DIN, and if it had anything to do with the reliability of the engine or of the transmission. "No," was his answer, "the engine could easily take more and the transmission still has a small margin. We rather thought that 170 bhp was about the limit for the running gear of a car not necessarily driven by experts only."

He then handed over the keys of his personal car, which I was to take to the BMW test track. It's the only place where the Turbo could be extended in this period of blanket speed limits, though we also drove it with little respect for current legislation on some small secondary roads. Though the engine was said to be perfectly standard, this car had no air dam nor the ugly rear spoiler added to the trunk lid of production 2002 Turbos. It also had the optional 6-in.-wide (instead of 5½-in.) light alloy wheels and a small Alpina steering wheel, but in performance it was probably fully representative of the production model. Starting from cold took some (but not excessive) churning and a slight whistle was noticeable over the entire speed range, though more so when going slowly. In town the engine proved extremely smooth and tractable but slightly gutless in the lower speed ranges—not surprising in view of the low compression ratio and the comparatively high exhaust back pressure. It was not before a short stretch of *Autobahn* was reached, on the way to BMW's test track, that the car showed its Jekyll-and-Hyde character by the way it rocketed up to the provisional 100-km/h (62-mph) limit. This speed, by the way, is reached in 2nd of the four gears in the standard gearbox.

Only the test track, however, enabled the car to reveal its full personality. There was little time, so we headed straight for the 4.5-mile high-speed course of two long parallel straights with big loops at each end, banked no more than a normal road would be. Undoubtedly the most impressive thing about the car is its smooth top-end acceleration and the complete lack of fuss with which high cruising speeds can be maintained. Compared with the 165-bhp 2002 Alpina I used to own, top-end performance is very similar but very much smoother and more silent (in the Alpina intake roar was the main offender). Thus high-speed cruising—around 110–115 mph—is much more relaxed than it was in that highly tuned, unblown car which had practically the same maximum speed. The turbo car uses a higher-geared rear axle (3.36 instead of 3.64:1,) and a slight period in the Alpina between 6000 and 6200 rpm is gone in the turbocharged engine.

On the other hand, partly because of the higher gearing, the turbo 2002 has definitely less guts in the lower- and medium-speed ranges and also "feels" less lively because of the time necessary for the turbo to reach its full output when the throttle is depressed, even if the engine is still spinning fast on the overrun. Assuming the throttle has been closed for a few seconds but the engine is still running above 4000 rpm, the time necessary to feel the additional push, coinciding with the moment the manifold pressure reaches its governed maximum (easy to check on Mr von Falkenhausen's car, which was fitted with a big, non-standard manometer) is anything between two and three seconds. This second push comes in quite progressively, however, and there is no problem whatsoever with controlling the engine output in delicate conditions as when cornering at the limit or (presumably) on a slippery surface.

When the engine is made to pull hard from low speeds, the manifold pressure becomes positive at something above 2000 rpm and the maximum governed pressure is reached at 3800,

Turbocharger is low on engine's exhaust side and feeds compressed air to intake manifold through a long pipe. There's no exhaust bypass for the turbo; its output is regulated to 8.0 psi by a pressure relief valve.

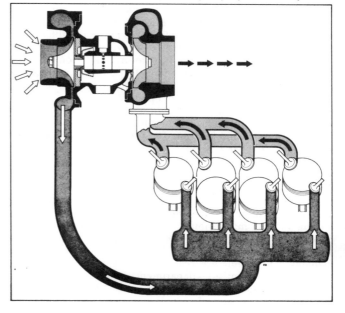

remaining steady up to the 6500 rev limit—when an ignition cutout takes care of unwary drivers. With such a difference between the power available below 4000 rpm and above, the optional 5-speed gearbox (not fitted to the car driven), in which 5th is direct but which has more closely spaced intermediates, should be a considerable advantage.

In the fast curves at the ends of the parallel straights of the track, the car felt beautifully stable and neutral with a slight tendency to tuck into the bend if the power was reduced when cornering fast and enough power to powerslide it out of the bend in 3rd when cornering near the limit. Things were easily kept under control by the precise and high-geared steering, which has strong self-centering action, excellent feel and very good high-speed stability. But the characteristic 2002 tremor around 50-60 mph is still with us, enhanced by the larger steering offset resulting from the use of ventilated front disc brakes and wider wheels. Brake fade was never a problem with that Alpina of mine, so it should not be with the Turbo either.

Unfortunately the test track was not available long enough for a full set of performance figures to be taken, nor are the straights quite long enough for the car to reach its outright maximum speed. Two runs in opposite directions gave an average of 121 mph at a tachometer reading of around 6200-6300 rpm and with the car still accelerating, so that the real maximum should be around 124 mph. It must be emphasized, however, that Mr von Falkenhausen's car uses a 1973-type body; the front air dam and small rear spoiler do nothing to improve the looks of the car but do improve aerodynamics and should raise the maximum speed by another two mph at least. But not to the 131 mph claimed by the makers, which is 'way beyond the ignition cutout point anyway.

Still, 124–125 mph is quite impressive for a full 4-seater of 2-liter capacity, and using the gearbox to the full we averaged 8 sec dead from 0 to 60 and 22.1 sec from 0 to 100 mph. The standing quarter-mile was reached in 16.2 sec elapsed time.

In the exhaust-emissions department, the Turbo is better than its unsupercharged equivalents as far as NO_x is concerned because of its low internal compression ratio, and is about the equivalent of the 2002tii for HC and CO; the emission test program is such that the blower hardly comes into its own.

Wouldn't it be nice if BMW were planning this approach to meeting the tough 1975 American emission standards?

PRICE	
List price, W. Germany	$6600

ENGINE & DRIVE TRAIN	
Type	turbocharged sohc inline 4
Bore x stroke, mm	89.0 x 80.0
Displacement, cc/cu in.	1990/121
Compression ratio	6.9:1
Bhp @ rpm, net	170 DIN @ 5800
Torque @ rpm, lb-ft	179 @ 4000
Fuel injection	Schäfer mechanical
Fuel requirement	98-oct.
Transmission	4-sp manual
(5-sp manual optional)	
Gear ratios: 4th (1.00)	3.36:1
3rd (1.28)	4.33:1
2nd (1.86)	6.23:1
1st (3.35)	11.25:1
Final drive ratio	3.36:1

CHASSIS & BODY	
Body/frame	unit steel
Brake system	10-in. vented disc front, 9.8-in. drum rear; vacuum assisted
Wheels	cast alloy, 13 x 6J
Tires	Michelin XWX, 185/70VR-13
Steering type	worm & roller
Turns, lock-to-lock	3.7
Suspension, front/rear: MacPherson struts, coil springs, lower lateral arms & compliance struts, anti-roll bar/semi-trailing arms, coil springs, anti-roll bar	

GENERAL	
Curb weight	2290
Weight distribution (with driver), front/rear, %	na
Wheelbase, in.	98.3
Track, front/rear	54.0/54.0
Length	166.0
Width	64.0
Height	55.5
Fuel capacity, U.S. gal.	18.0

CALCULATED DATA	
Lb/bhp (test weight)	15.5
Mph/1000 rpm (4th gear)	20.1
Engine revs/mi (60 mph)	2990

ACCELERATION	
Time to distance, sec:	
0-1320 ft (¼ mi)	16.2
Time to speed, sec:	
0-60 mph	8.0
0-100 mph	22.1

SPEEDS IN GEARS	
4th gear (6200 rpm)	124
3rd (6000)	95
2nd (6000)	65
1st (6000)	36

SPEEDOMETER ERROR	
100 km/h (62 mph) indicated is actually	98
160 km/h	158

TALES OF BAVARIA (CARS)

THE BEST-LOOKING best-moving 2002 yet. The 2002 TII is a comfortable, good handling cruise machine with a sleek body and refined injection system. The injection is by Kugelfischer and is one of the best sorted we've experienced. Interior comfort of the test car was enhanced by Recarro seats and a Ralla sports steering wheel.

2002 Tii + Five - dar Frugal-Fischer

FOLLOWING our road test of the BMW 520i (MM April '74) we were anxious to try another injected model from the range and jumped at the opportunity to drive a privately-owned 2002Tii fitted with a five-speed transmission. The problems we had suffered with the 520 were mostly attributable to the tuning of the car and we have been assured by BMW Australia that such poor results are certainly not the case with other injected 520s.

This being the case we arranged for an exclusive drive of the injected 2002, which belongs to Mr David Bones, proprietor of Bavariacars, a Sydney business dealing solely in BMW accessories and tuning equipment.

David's 2002Tii is the only five-speed version in Australia and if it were possible to buy one locally the price tag would read around $10,200.

The difference in performance and driving ease between the four-speed Tii and the five-speed has to be experienced. The five-speed car is much more comfortable and seems well set-up for both city driving and country touring.

The gear ratios are well spaced with fifth a direct ratio, although there wasn't much opportunity to use it around town where we had to keep the revs up. The ratios in the normal 2002 box are almost identical to the normal four-speed in the Tii, but the Tii runs a 3.45 to one rear axle ratio as against 3.64 to one for the standard 2002.

The ideal power band for the Tii appears to be between 3000rpm to 5500rpm. Maximum horsepower is 4500rpm. Maximum horsepower is 130(DIN)bhp @ 5800rpm.

The Kugelfischer injection is pure joy.

The throttle response is instantaneous and smooth. The acceleration is progressive and without jerks and shudders. The power comes on smoothly making it obvious that BMW have not only spent a lot of time matching the injection to the engine's performance, but they have made some subtle changes which have produced a flat torque curve (above about 2600rpm) which gives the Tii tremendous top-end performance.

On overun there is no throttle kickback and the idle is smooth with no lumpiness. Altogether the two-litre injected powerplant is a credit to the men who researched the design and the company who built it.

David reports that in thousands of miles of touring around Europe and Sydney he has never suffered any serious tuning problems with the car and it regularly returns about 26mpg. When the Bavarian company introduced the Tii version of the 2002 they pressed the point that the engine mods were aimed at improving acceleration rather than increasing the maximum speed.

Top speed is just over 121mph and the 0-60mph time is 8.6secs. 0-90mph comes up in 17.3secs.

The 2002Tii is obviously no slouch, but neither is it brutish. The car has an air of sophisticated refinement, much the same as the 3.2CSL coupe Editor Rob Luck drove (see report on page 43) during his time in Germany.

The seats in the local 2002Tii are the normally firm, but comfortable BMW type, but in David Bones' car they are the optional German Recaro buckets and they're another reason why we enjoyed our time in the Tii. Infinitely adjustable for rake, and movement, they are also adjustable for height.

Locking yourself into the Tii you are impressed by the incredibly tight feel of the car and at the same time aware of the 'muscle' waiting to be 'stretched'.

The 1600/1602/2002 body style has been with us a long time, but design updates to exterior trim and cabin layout give the car a timeless shape which is plain, but pleasing.

The 2002's shape in the airstream must be clean with an obviously low drag co-efficient because the car lopes along at high speed with seemingly little effort and still returns good consumption figures. The quietness of the Tii is very impressive. The handling of the car is very much the same as the other 2002 versions, but David Bones' car is helped along by a superb set of optional sports wheels made in Germany by tuning and accessory specialist Gerhard Schneider. The Tii points well and goes where its pointed with no drama. The wheels stick well, and there's very little body roll under hard cornering, but in a tight spot you could expect to see the occasional wheel lift-off.

For the Tii version of the 2002, BMW strengthened the wishbones of the front strut and also the rear trailing arms. In basic form the car feels strong, so it's no wonder they're a popular rally car in Germany and one of the reasons why the official rally cars rack up such a good record of success.

David Bones' 2002 Tii also features an electrically-operated steel sliding sunroof and this adds to the driving pleasure. It's a relatively big hole in the roof and is very comfortable on a sunny day. The spoiler at the leading edge is accurately designed to reduce wind entry into the cabin and eliminate buffeting.

The 2002 has always been a favourite of ours and after just a short run in the five-speed Tii our impression of the car is only enhanced. Significantly our test car has recorded more than 10,000 miles including touring and the daily traffic grind and the car feels as tight as on the day of delivery.

Matching impressive performance and a well-sorted injection system with great fuel economy makes the five-speed 2002Tii a real 'frugal-fischer'.

TAMPERING WITH 2002s
(or how deep breathing brings 150 bhp)

SUSPENSION mods to Dibbley's 2002 make for sure-footed handling, but occasionally a wheel lifts off in the tight spots. The ride is firm without being harsh.

IF YOUR CHEQUEBOOK doesn't quite run to $10,000-plus, allowing you to become the proud owner of a 2002 Tii, five-speed then you could look seriously at picking up a second-hand 2002 (still the best used BMW value) and having some mods carried out in the breathing department to give you Tii-type performance. Then you could look into the suspension department and pretty soon you'd have a comfortable, quick and safe touring car at nowhere near $10,000 cost.

With this exercise in mind we contacted Paul Older, driver of the rapid Craven Mild racing BMW 2002 and operator of Bavaria Cars, an exclusive tuning shop which specialises in BMW modifications using equipment and accessories developed by the recognised top BMW tuners in Germany.

Paul's autoshop had recently completed a tuning project for Sydney 2002 owner Brian Dibbley and, he invited us to drive the car which he said was one of the most docile, yet most powerful 2002s in Australia.

Remember that the new price of BMW 2002s between introduction in 1968 and last year ranged from $4400 (68) up to $7200, for the '73 car — the used prices range from $2200 up to $5400 for secondhand 2002s from '68 to '73.

The work carried out on owner Dibbley's car cost $2300. Having bought the car new and enjoyed many thousands of miles of sporty motoring Dibbley had Paul Older completely overhaul and rework the engine and suspension. The tuning program was extensive and in Brian Dibbley's opinion it was well worth it.

Firstly the engine is stripped down and the brand new bits are selected. Dibbley wanted the engine to produce 150bhp and consequently there were replacements required for the highly stressed items.

All the new parts are German pieces and the engine components are fully balanced and crack tested before re-assembly.

To obtain 150 reliable bhp from the 2002 engine a special camshaft is fitted with more than double the overlap of the stock cam, the head is extensively modified with a lot of relief around the valves and ports. Special rods and pistons are fitted plus an oil cooler. A modified fuel pump is installed to feed twin 45DCOE Weber carbs which are fitted on a special GS inlet manifold.

An interesting point about the manifold is the totally rigid fitting of the carbs to the manifold. In most instances Weber recommend using 'O' rings between the carbs and manifold for flexibility. GS's system has the carbs fitted with normal Weber trumpets and a special GS-designed air filter, with flexible rubber induction tubes which slide over the mouths of the trumpets.

The mounting for the air cleaner is very simple. As it is normally located near the front left hand suspension strut top mounting, a bracket is fitted to the cleaner which locates through one of the strut mounting bolts. Thus, when the engine moves the aircleaner remains locked in place and the movement is taken up in the rubber connecting air tubes.

INTERIOR is standard, except for additional gauges, sports steering wheel and the fabulous Recarro driver's seat. Under-bonnet shot shows the impressive twin Weber kit — the engine looks stock, but puts out an honest 150bhp!

With the abovementioned modifications to the head and pistons the compression ratio is effectively increased to 9.7 to one. The final step of the drivetrain mods is a modification to the clutch springs and a new friction plate.

Paul Older and mechanic. Peter Mabey then look at the suspension. Older says that his suspension changes aims for precise handling without undue harshness.

Once again the parts used are all-German. Bilstein gas-filled front struts replace the original equipment and at the back Bilstein gas-filled shocks are fitted, together with anti-roll bars front and rear.

The car is a little choppy, on the road, and very taut, but not uncomfortable. As an integral part of these changes, the cross members are reinforced and so are the wishbones and trailing arms.

When the engine is screwed back together it is run on a chassis dyno and perfectly tuned to give 150bhp @ 7000rpm. The dyno chart is given to the customer, so he may verify the performance of his rebuilt engine.

The parts and labour associated with the job are guaranteed, in writing, for three months. The engine was fully run before delivery to Brian Dibbley and he reports that the car ticks over like the proverbial Swiss watch, but now has the performance and reliability he wanted.

Driving the car gives you a big charge.

First impression is the tautness of the suspension and we confirm Paul Older's original plan to get positive handling without harshness. The handling is good, thanks to adjustable anti-roll bars, but the choppiness is mild compared to some stiffly sprung cars we've sampled.

The engine will easily pull 7000rpm and the performance is brainsnapping. By the same token Dibbley's car will rumble along in top gear at a mere 1500rpm and will pull away strongly without fuss.

As another aspect of the performance modifications Older has tried for flexibility and this is why, although mods can be carried out to achieve more than 170bhp, the lesser bhp figure was agreed on because of the necessity to drive the car in traffic as daily transport. During the short time we drove the car the water temperature needle remained steady as did the oil pressure and oil temperature gauges.

Dibbley's car is fitted out with Recaro seats and an optional, small diameter sports steering wheel and full harness.

Outside the only giveaway to the performance that lurks under the bonnet is an unobtrusive matt-black front spoiler and a set of light alloy ROH mag wheels.

Of course, when the car accelerates from a standing start there is further evidence of the beast under the skin. When you raise the revs and dump the clutch the rear end squats down, the tyres bite and you're away. The car pulls strongly and confidently up to 7000rpm and you snap change. If you're not careful you'll beat the syncros.

Without looking at the acceleration times on a stopwatch it's difficult to do any more than be subjective in our judgement of the car, but considering that owner Dibbley has had happy miles with the car since new and already absorbed the initial cost — for an additional $2300 he has a car which will last a long time, give great performance and certainly see the pants off most GTs.

The tuning service (Bavaria Cars) is planning several other projects with different BMW models and you can be sure you'll read about these developments here.

BMW sales in this country have never been better, the company's performance being almost totally limited by the amount of stock which is sent from Germany. Consequently we think there's a good market in the BMW tuning business and with Paul Older's attention to detail we know where most of the tuning work will probably be done.

RACING TO THE TOP

PAUL OLDER leading the pack at Amaroo Park in the Craven Mild BMW 2002. The car is subject to a continual development program and as a mark of its potential it beat a group of 'heavies' (including V8s) home at a recent Oran Park meeting.

BMW is a name which has been associated with motor cars and motor sport for a long time and their cars have never been far from the spotlight, however with the rising to prominence of the brilliant 3.5litre CSL coupes in winning last year's European Touring Car Championship it seems BMW is about to enjoy a resurgence of interest in its activities.

Locally a great deal of interest is being centred on Paul Older and his fiery BMW 2002.

Older, a 35-year-old Sydney medical specialist, has been racing since 1956, when he started out in England as an enthusiastic 17-year-old with a passion for the sophistication of race car driving. His activities have spanned road racing, rallying and other areas of motor sport, but as far as his connection with BMW goes, it all started in Sydney back in 1971 with a stock BMW 2002.

From this he progressed through a number of BMWs which were modified respectively and now he races one of the most immaculately-prepared racing 2002s in the world.

Older is a perfectionist and through his racing activities he has had the opportunity to set up an exclusive BMW tuning business in Sydney called Bavaria Cars. Drawing on tuning information from the BMW factory, and German firms like Schnitzer and Schneider, he has built-up a highly professional operation.

His own racing car is a hybrid of development ideas stemming from German sources and his own hard work and experience.

The present car is powered by a two-litre fuel-injected motor developing 270bhp at 9200rpm. The photo of the tell-tale tacho shows that it definitely will rev higher than 9200rpm and still hang together.

The head is a 16-valve Schnitzer twin-overhead-cam with chain-driven cam gears. It's a precision motor, but now seems to be strong enough to handle racetrack pounding. Older, and chief mechanic Peter Mabey have developed the car in complete co-operation with BMW in Germany and the Schnitzer works. They have changed few of the factory-recommended modifications, but what alterations they have made have worked well.

Both the front and rear suspension components were designed and made here in Australia under Mabey's supervision. The major change to the front end was a different pair of front struts which feature modified spring travel and ride-height alteration control. The rear end is also a fine piece of engineering.

To convert the car to a four-wheel disc brake setup Older and Mabey adapted a basic Schnitzer principle which involved mounting of the calipers underneath the disc, rather than at the front of the disc plate. They also shifted the lower mount of the rear shock absorbers and the trailing arm pickup — the latter mods were aimed at altering the rear camber change induced by body roll. Consequently the handling is flatter and entirely suited to local circuits.

Older has experimented with quite a number of modifications — most of them very successfully.

He altered the tappet adjusters to a different metal hardness to stop them cracking and splitting. The adjusters are, in simple terms, different thickness shims and are literally crushed by the pounding they receive from the cam follower and the valve. If they were to split there's a good chance of breaking a valve or incurring more expensive damage. With the Older system the adjusters merely crush to a smaller thickness putting the tappet adjustment out but not causing major damage. A lot of work has also been put in on the ZF limited slip differential to give it more lock-up.

The exhaust extractor is designed and constructed locally. It is a pipe-man's nightmare. The available space is like, nil, and the manifold had to be constructed in tiny pieces and welded together — the steering box and lower control arm have to be removed to take the extractor in and out.

Although BMW Australia are not involved in motorsport in Australia, Paul Older says they, and the German works, have offered a lot of technical assistance, all of which is invaluable when you're dealing with a one-off racing engine, where the designers are 12,000miles away.

The Schnitzer company, who designed and built Older's 16valve head are one of four major BMW engine tuners who have the nod of approval from the BMW company. They are situated only 70km from the BMW factory in Munich and are well known in Europe for their racing car 'packages'. They specialise in two-litre racing heads, like the 16-valve job, but they also make a wide range of tuning equipment for the 3.2CSL coupes. Schnitzer race their own 3.3CSL which develops more than 410bhp.

The other major tuning company is Gerhard Schneider (GS) and they deal more commercially with hot road cars. They have developed a two-litre 16-valve head, but are also famous for their 525 and 3.0S conversions too.

Possibly the most well known of the BMW tuners is the longest established company — Alpina. They have been involved with BMW competition for longer than any other firm and have achieved wide success. Now they are a very commercially-oriented company dealing in hot road-cars, but still with a very active motorsport department. The fourth company is the little known Koepchen concern, who mainly deal with race and rally conversions on 1600cc and two-litre engines. Then, naturally, there is BMW's own Motorsport department, which is a separate, totally autonomous company from the main BMW company.

BMW Motorsport are involved in rallying (with 2002s) and racing (with a 24-valve, 420bhp CSL and a 2002 with Schnitzer head and Weber carbs) in addition they make tuning options for road car customers.

BMW Motorsport used the Schnitzer two-litre 16-valve head in last year's ETCC, but are now developing their own. In the two-litre field Schnitzer have been way ahead of anyone else for more than three years.

The BMW factory have been mainly responsible for the Formula Two engine which powers the March-BMW driven this year by up and coming Hans Stuck. The engine is a 16-valve, fuel-injected, dry-sump powerplant putting out in excess of 275bhp and in Formula Two it is very hard to catch.

Following their capture of the Touring Car title last year racing BMW's aren't going too badly this year either. In the first two rounds of the touring car championship BMWs were first, second, third and fourth in round one at Monza. And in round two at Salzbergring last month, 3.5CSLs came home first (the factory 24-valve car), second and third (private entries) with factory Porsche Carreras fourth and fifth and a Schnitzer-engined 2002 sixth.

In Formula Two so far, BMW have won both the first two rounds with Hans Stuck behind the wheel of the March-BMW.

In touring car racing this year the only real opposition the BMWs face is the Cologne Cosworth Ford Capris (V6, 24-valve), but so far they have performed disappointingly.

Despite the fuel crisis and other short-term shockwaves the BMW company and the major engine tuning firms continue with the development of BMW engines and according to Editor Robin Luck, they just keep getting better and better. Where will it all end — 24-valve head motors for the masses?

ABOVE: Paul Older's BMW 2002 being prepared for racing. Note the rear suspension and the disc brake setup. BELOW LEFT: The business side of the Schnitzer 2002 16-valve cylinder head. BELOW RIGHT: The official dyno graph shows the BMW puts out 270bhp @ 9200rpm — after a recent race meeting the tacho tell-tale needle was stuck on 9600rpm!

HOW TO BLOW £4,299 *FAST!*

Text by Richard Hudson-Evans. Photographs by Ken Goddard.

In almost any field the first of anything invariably costs plenty—and BMW's 2002 Turbo, being the world's first turbocharged production car, is most certainly no exception.

However, the Turbo, even though the most expensive saloon car in the up to 2 litre class, is also the fastest. With its 'Batmobile' paint job and left-hooker only specification, it's just about the rarest means of conveyance with which you could possibly confront your insurance broker, not to mention the bank manager.

Needless to say, the top speed attainable, positively of Mulsanne proportions, is fast becoming sadly irrelevant in almost any country; but a figure from rest to sixty of less than seven seconds (6.8 to be precise!) gives a useful escape mechanism when a rare hole in the traffic appears.

For the record, during our far too brief tenancy, the only devices on wheels that contrived to actually stay with us were a particularly daringly conducted ▶

Suzuki 750 Triple, an RS Carrera driven by a fellow cowboy (actually an enterprising salesman who was only trying to catch up in order to persuade us to chop in the Turbo for his Porsche!), and a BMW police car, the crew of which were happily satisfied with a detailed conducted tour of the Turbo's features in the next lay-by.

The 'prettying-up' gear apart, of which more anon, obviously most of the £1300-worth of inflationary Bavarian Deutschmarks that you are paying over and above the cost of a 2002 Tii goes towards technically justifying the 'Turbo' label.

Turbocharging, for the conventionally aspirated and uninitiated lay majority, is basically a different form of supercharging. The vintage blower and driven belt are replaced with an exhaust driven turbine, which forces vast quantities of your air and mine under pressure into the bowels of the power plant. This in turn results in the plant's power being much more lusty than is possible for the same amount of your money from any amount of unpressurised carburation or the most sophisticated varieties of fuel injection.

You see, the Men of Munich tell me, at high engine revolutions, the ingress of the fuel/air mixture in any normal engine is in fact only 80% or less of the Teutonic potential. This is because the flow resistances (terribly sorry Flo!) become greater, the higher the engine speed. So, by upping the jolly old pressure of the mixture artificially from outside, the vital cylinder charge can be brought back to the sort of 100% efficiency of which any German can be proud—and mere Englishmen grateful.

Now, in the British Racing Green old days of the Bentley Boys, this pressure-upping could be achieved by a great deal of gorgeous whining from a 'blower' or engine-driven supercharger. This meant of course an extra and rather large component for the hard-worked engine to drive. By going for an exhaust-driven turbocharger, the BMW boffins have avoided such power losses by utilising the not inconsiderable energy from the exhaust gas flow.

What happens is that the exhaust energy is cunningly directed via a turbine which is shaft-connected to an impeller, which, in turn, is responsible for sucking in free air and increasing its pressure. It all means that in this way, more air can be fed to the engine than it could otherwise suck in. In the BMW Turbo's case, far from free fuel is mechanically injected by Schafer in front of each inlet valve. I can tell you that this can (and, in my case did) happen to the fairly alarming extent of a mere fourteen miles to every precious Middle Eastern gallon!

However, Israeli or Aberdonian 'Fast Car' fans fear not! You won't be swelling Arabian coffers, increasing your respective national balance of payment deficits at such a thirsty rate all the time. Besides the fuel demands of the beast being met from the North Sea long before you've cleared the last enormous H.P. payment, to be fair, my overall consumption averaged the top twenties and did, on infrequent legal occasions, even exceed thirty (albeit frustrating) miles to the gallon.

You see, the real beauty with this beast, in my opinion, is that somewhere along the Autobahn, Messrs. Jekyll and Hyde have been consulted. Apart from how much or how little fuel you're having to buy for the fifteen and a half gallon tank (incidentally, four larger than the standard 2002's), you can gauge the mood of the monster by keeping a most wary eye on a rather innocuous dial, almost an afterthought, fitted to the right of the usual—and, in this case, tastefully efficient—sports saloon instrumentation pack. For monitoring the turbocharger's pressure, this little gauge houses a needle which scans a large white, smaller green, and, if something goes dangerously wrong with the turbo, the forbidden red, sector. If you keep the needle out of the green, which in practice means keeping the revs certainly under 4000, then, as far as tractability goes, you might just as well be in a humble 2002. Yet, once the needle strays into the green area, you'll very rapidly discover that 'green' most certainly means green for plenty of Go, with a capital 'G'. The harder you wham the needle round the dial, the greater the 'G' figures achieved!

Ideally, it pays off to effect the next change of the four forward alternatives at just over 5500 rpm. By the way, the theoretical maximum power output, increased by 31% to 170 bhp, ought, one's told, to be produced at 5800 rpm; for absent-minded or totally unfeeling owners, there's an automatic electric cut-out fitted.

Now a 2002 Tii ain't exactly slow, yet your Turbo will take nearly two seconds off the zero to endorsement time, whistling you there in just under ten seconds, while the magic ton can be yours if you've just a shade under 23 seconds to spare.

The source of all the fun is basically the familiar water-cooled, 4 cylinder, 5 bearing, single overhead camshaft 1990 cc lump, front-mounted, driving the rear wheels. However, the compression has been reduced to 6.9:1, to ensure that the combined ratio resulting from the turbo blower and the motor's compression isn't too high. Just because there's this reduction in compression, apparently it's unwise to think that you can economise by slurping in lower octane fuel—you're advised to stick to at least four star. However, the fresh air gang at least have no cause to complain with the Turbo, because I'm told, even the most stringent and unrealistic West Coast exhaust emission standards are easily met. The percentage of harmful substances left in the exhaust gas, having done with driving the turbine, is lower, particularly so over long journeys where the extra available power is exploited.

Fortunately, there's so much torque available, pull-aways in top being a possibility at 20 mph, even before any turbocharging effects occur, that you really only need the four on the floor, a fifth gear not being missed all that much. Second gear, at 1.86:1, is slightly lower than other 2002s; the full set comprises a house-climbing 3.35:1 for first, then second, with third at 1.27:1, followed by direct top. As the car uses the 2002 3.36:1 final drive ratio, though this is not as high as the Tii's 3.45:1, the lack of a fifth overdrive cog isn't really a disadvantage. In any case, no licence holder is likely to complain with the 130 maximum to play with!

Suspension is all familiar stuff, the same as on the Tii, which means our old friend Mr Macpherson's struts, with wishbones, coil springs and telescopic dampers at the front, and the same springing and damping arrangements, aided by an additional rubber spring unit and some trailing arms, at the rear. Anti-roll bars—highly effective they are, too—are fitted to both ends of the cake.

For what one might assume is supposed to be a semi-luxurious sports saloon, with four entirely comfortable seats, even though it has two doors less than the Dolomite Sprint (lest we forget at half the price!), the ride turned out to be a great deal more sporting than luxurious. Roll-free, yes, and considering the power output, amazing traction,—but you're paying a great deal for a bumpy and juddery ride, which can (at best) only be described politely as 'competitive'. However, one has to admit that many roundabouts, for which one would normally need to lift off on other new cars, were absolutely flat-out in the Turbo. Full marks are due for the ventilated discs at the front, drums at the back, and for the handbrake that sailed past the hill parking test and didn't complain at assisting with rapid hairpin negotiation. I just couldn't fade the brakes at all.

To put it mildly, your 'visuals' are certainly taken care of. The Turbo is hardly the sort of car people are likely to fail to notice. Apart from the stick-on works sidewinders (in fashion, it seems, with both quick BMW and Porsche fans) advertising 'Turbo', there are also spoilers for both ends. The frontal one, behind which lurks the oil cooler, is huge and is again bedaubed with stripery, while the boot-top one, consisting of a much more subtle speed lip, is curiously made out of pliable rubber, no doubt to satisfy the increasingly severe German vehicle legislators. To cover the wide tread area of the 5½ J rims, in place of the Tii's 5 Js, 'afterthought'-looking wheel spats are surprisingly, and rather crudely, screwed on each wheel-arch eyebrow. One must hope that all this paraphernalia is actually functional; the spoilers, it's said, help to keep both ends glued to the deck—they must contribute to making the Turbo some two hundredweight heavier than its less sophisticated 2002 stablemates.

Although this is all rather new for the showrooms, BMW have in fact been involved with turbocharging for some years. As long ago as 1969, a turbocharged 2002 won the European Touring Car Championship; the works cars produced as much as 275 bhp for long events such as the saloon classic of them all, the Spa 24 hours, and up to 400 bhp (briefly!) on test rigs. When looking back even over ordinary production 2002s, one realises the extent of their progress in the area of power outputs. The early model, with its single twin-choke carburettor, produced 100 bhp at 5500 rpm: at the time, impressive. With a pair of twin-chokes, the figures rose to 120 bhp at 5800 rpm, and recently, with the Kugelfischer fuel injection, the Tii owner has been purchasing 130 bhp at 5800 rpm. The Turbo gives its supporters another 40 bhp on top of that!

In passing, very sadly, I must report the following with my test example: the exhaust smoked profusely at tick-over, water consumption became progressively more alarming throughout the week's loan, and the glassfibre front air-dam was starred with fatigue cracks—all with less than 20,000 on the clock. Perhaps, though, one should make some allowance for the fact that a dozen or more other wild young scribes had no doubt abused this particular motor car along the way; with your very own Turbo, qualifying as a most definite 'one careful owner' example, you'd most probably evade such fault features.

The final question with any car such as this that one's asked to review is 'Would you buy one?' Well, if I had a family (which I don't) who couldn't fit into the Dino or 911 (neither of which I happen to own at the moment) and I still wanted performance of this sort of level, then maybe I just might be tempted. But if the boy-racer trim, however up-market, offended—as I suspect it would in the eyes of people with the kind of money required—then I'd plump for extremes of specification such as an XJ12 for the family and a superbike for me. After all, why compromise at all? I'd have thought that anybody who could afford to blow £4,299 on such a sports saloon could afford not to.

The turbo looks basically similar to other 2002 models apart from the dress-up kit. Inside, there is one dial in the centre of the dash that tells you it's a turbo, while outside there are the pop-riveted wheel arch extensions and air dam. From behind, the small boot spoiler and badge give it away. Under the bonnet there is very little to see that excites you.

BMW 2002

FROM BAVARIA

WHEN the four door BMW 2000 series came out in 1966, it was greeted with great enthusiasm. Here was a luxurious car whose performance can shame even some of the sports cars, yet docile enough to carry the family comfortably around town.

But there is a section of the motorists who would like a sporty two-door car that pleases the missus as well as their sports car egos. The BMW people had this in mind when they put the 2002 on the road on 1968.

To differentiate the two door version from the four door two litre car, the figure '2' was added.

Here is a two door car with all the virtues of BMW but costs much less. The running gear is practically the same as the four door version and the BMW 2002 we tested is the most basic of the series. It has a single choke Solex downdraught carburettor enough to allow the 1990 cc engine to churn out a modest 100bhp (DIN) at 5500rpm.

In its basic form, the BMW 2002 is uninspiring in its performance figures but it handled a lot better than most of the cars available here today, thanks to its all independent suspension with front and rear anti-roll bars.

STYLING AND ENGINEERING

The current BMW 2002 looks like the early models except for minor face lifts in and around the car. The boat-shape body now has a matt black grille in the nose and oblong tail lamps. Previous models had all chrome grille and round tail lamps.

Straight waist line running from nose to tail distinguishes the "Bee-Emm" from its many curvaceious competitors. It's bonnet and boot lid are large and quite flat pieces stretching from side to side and from the base of the front and rear screens to the nose and tail respectively.

From outside, it is obvious that a greater area of the 2002 is occupied by the passenger compartment. The base of the front and rear screens straddle a great proportion of the car's length.

A new set of 5J x 13in. steel rims comes with the latest models. The ventilated naves, exposed chrome nuts and small chrome wheel centre increased the sporty image of the 2002.

Beneath the forward opening bonnet is the familiar two litre four-cylinder in-line engine. In our test car, the power output is a modest 100bhp (DIN) at 5500 rpm due mainly to the single fixed choke Solex downdraught carburettor.

Unseen from the outside are the minor renovations of the single overhead camshaft engine. Larger valves are used then in the previous models. Eight counter balance weights on the crankshaft allow a smooth revving engine.

On the transmission side, the established BMW front engine rear wheel drive system is retained. Excellent gear synchronising and slick gear changing are still the delightful features of the German gearbox.

Independent suspension all round enables the 2002 to behave well on the road. The front suspension is by MacPherson struts and coil springs while the rear end is by semi-trailing arms and coil spring.s

Anti-roll bars fitted to the front and rear keeps the BMW more upright through corners. Only the two litre version of the two door BMWs has the rear anti-roll bar.

PERFORMANCE AND ECONOMY

Those looking for wheel spinning performance would be better off with the Tii (fuel injected) or the Tik (turbo-charged) versions than the bsaic version we tested. The 2002 we had would be ideal to those looking for a good handling two litre car with a reasobable performance.

Top speed is 162km/h (about 100mph) which is not the fastest around town and neither is it embarasing for a two litre car.

A reward for the modest performance is the good fuel consumption. We recorded a figure of 11.2 litres per 100 km (25mpg, to you) overall and if gentle on the accelerator pedal, 10 litres per km (28mpg).

HANDLING, STEERING AND BRAKES

As is expected of BMWs the handling is exceptional. Well designed suspension system with the right spring ratings, damper and anti-roll bar settings easily put the Bee-Emm in the sports car class.

Discerning drivers would smile approvingly at the gradual breakaway of the tail when pushing the car to the limit.

When we got behind the steering wheel, we thought that the rear anti-roll bar would contribute to a more sudden and big breakaway when the 2002 reached its limits of adhesion. On the contrary, the anti-roll bar enabled the rear end to absorb greater cornering forces and yet allow a gradual breakaway.

The original tyres fitted were the Phoenix P110 Ti radial ply tyres, a German brand that is hardly heard of in Malaysia. Its block tread pattern squealed a lot on hard cornering but held the Bee-Emm well.

Twirling the four spoke safety padded steering wheel was a fairly light process. It took just 3½ turns from lock to lock and parking manoeuvres were not difficult.

At high speeds, the steering wheel could be left alone even when belting over some uneven surfaces. This demonstrated the incorporation of steering geometry and the good stability of the German car.

Twin servo units and dual circuit brake system makes sure that the BMW stops in time. The pedal pressure was light but not over servo-ed. When we stood on the brake pedal at high speeds, the rear wheels did not lock up and the car stopped in a straight line.

DRIVING POSITION, CONTROLS AND INSTRUMENTS

The BMW seats are quite high from the floor and when the driver is properly seated, he would easily has a commanding view out of the car.

A stalk sprouts out from either side of the steering column. That on the right, controls the wipers and wipe/wash facilities while that on the left, the directional indicators, parking lights, headlamp flasher and dip switch.

Except for the cigar lighter, blower cum backlight heater and lights switches which were mounted on the nacelle, the rest of the controls are within easy reach.On the nacelle are three circular dails. They are the tachnometer on the right, the speedometer (in kilometres per hour) in the centre and a segmented dail for fuel and water temperature gauges and warning lights for oil pressure, charge, main beam and directional indicators on the left. They were legible even at a galnce when driving.

Due to the broad and high transmission tunnel, it was impossible for the driver to stretch the left foot on long journeys. This would be tiresome to many drivers.

Sporty drivers will find it quite difficult to heel and toe due to the brake and accelerator pedals lying in different planes.

RIDE, COMFORT AND NOISE

The ride of the BMW 2002 can easily be described as smooth. Bumps and rough surfaces were easily soaked up by the independent suspension. It was just like riding in the bigger Bee-Emms.

We were not too happy with the seats. They were too firm even for our young bones and there was no lateral support. We would have hoped for a softer and figure enveloping seat even though those original seats were of robust construction.

On each side of the passenger compartment, for the rear passengers, there were provisions for resting the elbows. Grab handles were part of the front door arm rests while the rear ones were roof mounted.

Three adults can easily be seated at the back with adequate legroom. Thanks to the almost level roof, the rear passengers would not be easily bumped on the heads when the car goes over heads when the car goes over humps or the like.

A weak point with the BMW 2002 is the ventilation system. Though the front quarter lights are retained to help in the supply of fresh air to the passenger compartment, they are useless on rainy days. The doors do not have window frames and hence, it was not possible to fit rain visors.

The three-speed blower was not powerful enough to bring in fresh air and for South-east Asian countries, perhaps an air-conditioner would ease matters.

Mechanically. tje BMW 2002 was quiet even when stretched. Road noise was in tolerable amount but not wind noise. Slight opening of the door windows would be sufficient to distract the occupants and further openings can result in deafening roars when the car is at high speed. However, the thick rubber door surrounds provided adequate seals between the glass and the metal body.

EQUIPMENT AND IFTTINGS

As is typical of German and particularly BMW car manufacturers, each of their cars is well equipped. The 2002, though in its most basic form, does not disappoint.

Current models have built-in rear Screen heaters, hazard warning flashers, matt black wipers to eliminate reflection, among the other previous standard equipment like centre consoles and coat hooks,

Though the lap and diagonal safety belts for the front were fitted, it was awkward to fasten. Protrusions from the door side of the seats (like the safety catch for tipping the squab and the hinge for the reclining squab) hindered the belt from freely pulled across the user and fastened. It took some time to clear the belts from the tight space between the door and the protrusions.

Thick carpeting on the floor from edge to edge were well fitted and invited the plastic drop-opening glove box was unlockable and was small for a two litre car. In order to operate the bonnet release lever, the glovebox had to be opened.

LIGHTING

Though only two round headlamps up front, the light was very bright due to the use of halogen bulbs. We had no trouble cruising at about 140kph (approx. 86 mph) during a night drive.

While the spread of the headlamps were good, there were definite 'shadows' in the pattern, probably due to the design of the headlamp glass covers.

Top: The simple fascia has padding at the right places.
Inset: Owners of the 2002 should handle the doors with care because of their rimless windows.
Above: Access to the rear is adequate in spite of the thickly padded front seats.
Right: The cockpit is built with the driver's safety in mind.

Bee Emm's potent lungs give it the go.

The light of the instrument panel can be adjusted by turning the light switch on the nacelle.

It is a pity that there was no provision for lighting the engine compartment.

The boot can be illuminated by switching on the sidelights and the transparent tail lamp covers allowed light to be shone into the boot.

BOOT

The BMW 2002 may look like a shrunk four door version but on opening the boot, the spacious luggage capacity made one think twice. A 15 cu. ft. boot would satisfy most potential 2002 owners.

To retain the rear and rigidity, the rear panel is as high as the rear mudguards and a fairly high sill results. Bags and any thing for the boot will have to be lifted over the sill.

On the nearside floor is the well for the spare wheel while the other half is occupied by the ten imperial gallon fuel tank.

MAINTENANCE AND SPARES

Mechanically, the BMW is fairly straight-forward and most competant mechanics would have no problem in tackling even major work on the car.

What matters most, particularly in Malaysia, is the availability of spares. In Singapore, the 2002 is selling like hot cakes and most of the stocks of spares, according to some BMW mechanics, are for the two litre variety.

PRICE

Price at just over $ 23,000, the BMW 2002 is quite an expensive car. Due to the differences in import duties, BMW 2002 bought in Malaysia is slightly cheaper han bought in Singapore.

Most of the money you would pay for the Bee-Emm goes to the government and for the money you would be paying for a car, the Bee-Emm has to compete with some of the more luxurious locally assembled cars if you happen to be in Malaysia.

On top of that, the onus of getting tha rare import permit lies with the Malaysian buyer and can be a difficult process.

SUMMARY

Fro some $ 23,000, the BMW 2002 does not have the performance that is expected for that amount of money. Perhaps those looking for a good handling quality luxurious two-door car, the BMW 2002 would be the answer.

The Tii (fuel injected) version costing some $ 4,000 more and has some 30bhp more than the 2002 now being tested, could perhaps be a more interesting Bee-Emm to own.

2nd	:	100km/h (approx. 62mph)
3rd	:	148km/h (approx. 91mph)
4th	:	162km/h (approx. 100mph)

ENGINE:
No. of cylinders	: Four
Material (cylinder head)	: Aluminium alloy
Material (block)	: Cast iron
Main bearings (number)	: Five
Valve gear layout	: SOHC
Capacity (cc)	: 1990
Bore (mm)	: 80
Stroke (mm)	: 89
Compression ratio	: 8.5:1
Carburettors	: One fixed choke downdraught
Max. Power	: 100bhp (DIN) at 5500rpm
Max. Torque	: 115.7 lb ft (DIN) at 3500rpm
Cooling system	: Water

TRANSMISSION:
Gearbox	: Four speed, fully synchromeshed, floor shift
Ratios 1st	: 3.764
2nd	: 2.02
3rd	: 1.32
4th	: 1
Reverse	: 4.096
Final drive ratio	: 3.64
Clutch	: Single dry plate, hydraulically operated.

CHASSIS & BODY:
Type and method of construction	: Steel, unitary

DIMENSIONS:
Wheelbase	: 98.5 (2500mm)
Front track	: 52.7 (1348mm)
Rear track	: 52.7 (1348mm)
Overall width	: 62.5 (1590mm)
Overall length	: 166.5 (4230mm)
Overall height	: 56.5 (1410mm)
Ground clearance	: 6.5 (160mm)
Front legroom (seat forward/back)	: 19/27 (483/686mm)
Rear legroom (seat forward/back)	: 17¾/10¾ (456/273mm)
Seats	
Front squab width	: 22½ (571mm)
Front seat width	: 22¼ (565mm)
Front seat length	: 18¼ (464mm)
Rear squab width	: 51 (1295mm)
Rear seat width	: 50½ (1282mm)
Rear seat length	: 18 (457mm)
Rear wheel arch intrusion	: None
Luggage capacity (cu.ft)	: 15 cu.ft.

REPLENISHMENT & LUBRICATION:
Engine sump capacity	: 7 imp. pts. (4 l.)
Engine oil change interval	: every 4000mls. (6000km)
Gearbox capacity	: 1.8 imp. pts. (1 l.)
Final drive capacity	: 1.7 imp. pts. (0.95 l.)
Grease points	: None
Lubrication interval	: —

LIGHTING:
Headlamps (Type and power)	: Quartz iodine H4 bulbs
Battery	: 12V 44AH
Charging system	: 630W alternator

WHEELS & TYRES:
Wheels	: 5J x 13
Tyres	: 165 SR 13 Phoenix P110 Ti
Pressure	: 26psi (1.8 atm.) all round

BRAKES:
Type	: Disc/drum
Servo	: Twin
Circuit	: Dual
Rear valve	: Yes
Adjustment	: Self-adjusting

SUSPENSION:
Front	: Independent, MacPherson type struts, coil springs, lower wishbones, anti-roll bar.
Rear	: Independent, semi-trailing arms, coil springs, telescopic dampers, anti-roll bar.

STEERING:
Type	: ZF Gemmer hourglass worm and roller
Ratio	: 17.57:1
Lock to lock	: 3½ turns
Steering wheel diameter	: 15½ in.
Turning circle	: 31½ ft. (9.6m)

FUEL CONSUMPTION:
Overall (mpg)	: 25 (11.2/100km)
Driven carefully	: 28 mpg (10/100km)
Fuel grade	: Premium
Range (miles)	: approx. 250
Tank capacity	: 10 imp. gals. (46 l)

PRICE STRUCTURE:
List price	: $ 23,000 plus
Free service & when	

EQUIPMENT:
Safety belts	: Yes
Tool kit	: Yes
Heater	: Yes
Blower	: Yes
Rear windscreen heater	: Yes
Cigar lighter	: Yes
Map light	: No
Clock	: No
Fresh air ventilation	: Yes
Hazard warning	: Yes
Sun visors	: Yes
Tachometer	: Yes
Vanity mirror	: Yes
Reversing lights	: Yes
Coat hooks	: Yes
Grab handles	: Yes
Reclining seats	: Yes
Wide/wash facility	: Yes
Map pocket	: Yes
Boot light	: Yes
Engine compartment light	: No
Adjustable steering wheel	: No
Oil pressure gauge	: No
Oil temperature gauge	: No
Ammeter	: No
Electric winding window	: No
Petrol filler lock	: Yes
Fuel low level warning	: Yes
Underseal	: Yes
Glove box lock	: No
Parcel shelf front	: Yes
rear	: Yes
Headrests	: No
Steering lock	: Yes
Parking lights	: Yes
Door armrests	: Yes
Rear centre armrest	: No
Front centre armrest	: No
Dipping mirror	: Yes
Laminated windscreen	: Yes

PERFORMANCE
Max. speed in gears	
1st	: 59km/h (approx. 36mph) ∎

In 1972 the BMW sold for $4097. The following year this jumped to $4963. By '74 the price was a staggering $5845 and many people were shaking their heads. "It's a great car," they said, "but there's no way they can sell many at that price." Well the suggested retail price of the base 2002 is now a heady $5995 and they're still selling like hot cakes. With everyone else's prices shooting up, it appears that the discriminating buyer is still willing to pay a little extra for quality. And make no mistake about it, the BMW 2002 is a quality automobile.

It is one of those vehicles that appears small on the outside but has bags of room inside. Though it may not be as wide as some cars on the market, legroom in the back seat for instance is just as good as an intermediate size automobile. Headroom is excellent both front and rear while the trunk has a capacity of 15.9 cubic feet — again comparable to a domestically produced intermediate.

Styling for the most part counts for very little in Germany, witness the Beetle, Mercedes-Benz products and the BMW 2002. A car must be functional first of all, then you can worry about the shape of the exterior sheet metal. And the function of an automobile is to transport people and their belongings from one place to another with a reasonable degree of comfort and speed. The BMW 2002 conforms to this philosophy.

The outer shape of the car is boxy, but this provides for a comfortable seating position for the car's passengers and space for all their luggage. With many cars today, the rear passengers must sit with their knees tucked under their chins and their heads rubbing the roof. All

DRIVING ROAD TEST

A genuine two-door sports sedan

Photos by John Plow

BMW 2002

Space, taste, function and quality are all fitting terms for the BMW interior

because the car has been designed with a sexy fastback roofline. In the BMW 2002 you sit upright, armchair fashion whether you're in one of the front bucket seats or the rear.

Though there haven't been too many changes to this car that meet the eye, the steering wheel is one thing they've altered. In the past we constantly complained about the size of the truck-like, steering wheel. The latest 2002's come with a sporty, four-spoke wheel of smaller diameter. The turn indicator stalk has been shifted from the right side of the steering column to the left. — the position that most of us are accustomed to. But other than that, the dash remains much the same as it has for years. Again functionality plays a key role in the design of the dash. All instruments are ledgible without having to search for them. A large speedometer and tach dominate the section directly behind the steering wheel. A convenient shelf runs from the outside edge of the instrument cluster to the right hand door providing a handy place to put loose change or cigarettes. Presumably the reason for not changing the dash layout is that it would be pointless to tamper with something that is doing the job it was intended to do. But the design fails in one area.

Flow-through ventilation has been markedly improved in many cars the last five years or so, but the BMW's has not. There are no face level vents in the dash face for instance. Well there's a slight gap at each end of the dash, but it doesn't do much good. On warm days it is necessary to have the rear quarter windows popped open (they flip out rather than roll down) and the fan on one of the three settings. Unfortunately the fan is quite noisy on its two highest settings.

The remainder of the interior appointments are excellent. The front bucket seats are wide and well padded with the rear bench, also spacious and comfortable. Between the two front seats there is a neat console arrangement with a box-like section for storing extra loose articles. The shift lever is a short handle protruding from the top of a rectangular shaped rubber boot. All in all a practical if not flashy passenger compartment.

From a mechanical point of view the 2002 is a sophisticated automobile — and so it should be for the price. A single overhead camshaft, four cylinder engine provides the power. The block is cast iron while the cylinder head is an aluminum alloy

The BMW's unmistakable chunky shape is not affected by the bigger bumpers

with hemispherically shaped combustion chambers. Compression ratio is 8.5:1 and the manual calls for premium fuel. At 5500 rpm the 121 cubic inch engine puts out 98 horsepower, which is sufficient to propel the car as fast as 105 mph. Gasoline consumption is in the neighborhood of 25 mpg. Power assisted disc brakes are used up front and the suspension is fully independent.

The transmission shift linkage is the smoothest there is on the market. It's a joy to use. Throws are short and crisp with no vagueness at all.

In the morning, the BMW 2002 starts without any problems, though it sometimes stumbles when cold. Once up to temperature the engine runs smoothly except for a slight hesitation below 3000 rpm when the throttle is depressed quickly. Occasionally we noticed that the engine was reluctant to start when hot. All of these problems of course are a result of the latest emission control devices. Even so, the 2002 bears up rather well. Acceleration is startling with 0-60 mph achieved in 10.6 seconds.

But for all its practical considerations and boxy, sedan-like body, the BMW 2002 is actually a pseudo-sports car. Thanks to the independent suspension it handles better than many so-called sports cars. Everything about the car feels right. Within minutes of sliding behind the wheel you feel comfortable and at home. Spirited driving holds no surprises for the driver. The car is completely predictable. Over rough roads or railroad tracks it feels exceptionally solid. There are no rattles or complaints from the body-work or suspension.

Today there are many small cars on the market. Each has redeeming qualities but none combine the attributes of practicality and quality as well as the BMW 2002. If the car has a fault, it is its price, but as we men-

Unfortunately the 'spare' is under the floor

In spite of emission controls, engine still provides lively performance

tioned at the start, there are those who do not mind paying a little extra if they're getting good value for money. There is no question that the 2002 is a sophisticated automobile employing high quality materials and assembled to exacting standards. It has an excellent reputation and enviable track record. Little wonder then, that the sales continue to be brisk.

DRIVING ROAD TEST

technical data

BMW 2002

SPECIFICATIONS

ENGINE
Location	front
No. of Cylinders	four, in-line
Valve operation	single overhead camshaft
Carburetion	one, two-barrel
Compression ratio	8.5:1
Bore	3.50 ins.
Stroke	3.15 ins.
Displacement	121.4 cu. ins. (1990 cc)
Power	98 hp.@ 5500 rpm
Torque	116 lb. ft. @ 3500 rpm

TRANSMISSION
No. of forward speeds	four
Gear ratios	1st-3.76; 2nd-2.02; 3rd-1.32; 4th-1.00
Final drive	3.64

BRAKES
Front	disc; power assisted
Rear	drum

DIMENSIONS
Wheelbase	98.4 ins.
Track	53.07 ins. front and rear
Width	62.6 ins.
Length	166.5 ins.
Height	55.5 ins.
Weights as tested	2430 lbs. (52.5% front; 47.5% rear)
Fuel capacity	11.4 gals.
Tires	165SR x 13, radial

STEERING
Type	worm and roller
Turns lock to lock	3.5
Turning circle	34.2 ft.

SUSPENSION
Front	independent, McPherson strut, coil springs, tubular shock absorbers
Rear	independent, coil springs, semitrailing arms, tubular shock absorbers

CALCULATED DATA

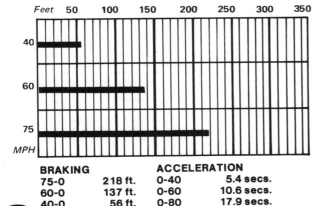

BRAKING
- 75-0: 218 ft.
- 60-0: 137 ft.
- 40-0: 56 ft.

ACCELERATION
- 0-40: 5.4 secs.
- 0-60: 10.6 secs.
- 0-80: 17.9 secs.

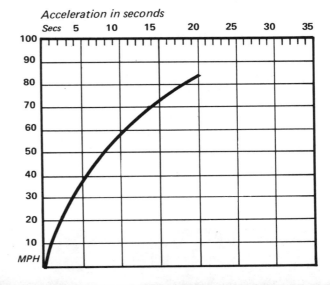

Shell Fuel consumption 20-26 mpg (Shell Premium)

SUGGESTED RETAIL PRICE $5995

Road test
by John Bolster

Hand built in limited numbers the BMW 2002 is only available in left-hand-drive. Progress is unbelievably smooth on the road.

BMW Turbo—flexible racer

For many years, engineers have been trying to design a car which would combine the traffic manners of a low-compression hack with the open-road performance of a competition machine. High-density induction is the obvious answer and Mercedes-Benz built cars in the nineteen-twenties and thirties with a supercharger which could be put in action by merely engaging a clutch. Unfortunately, a blower driven from the crankshaft consumes a lot of horsepower, so a relatively small gain in performance was accompanied by a large increase in fuel consumption.

Yet the idea is immensely attractive and the problem has now been solved by harnessing the normally wasted energy of the exhaust gases to drive the supercharger. The extremely high revolutions of the exhaust-driven turbine suit the centrifugal type of supercharger, which is far more efficient than positive-displacement compressors and needs no wasteful step-up gears in this instance. Fuel injection overcomes carburetter pressurising problems and the turbocharger can work as effectively with petrol engines as it has for a long time with diesels.

The BMW Turbo was originally a competition car, but its racing achievements are too well known to require discussion. The subject of the present test is a production car, though at this stage hand-built in limited numbers and only available with left-hand steering. I tested it simply as a fast touring vehicle and found that it has an enormous potential as the ideal high-performance car of the future. It has the same sort of speed and acceleration as the hottest versions of the 3-litre BMW, for instance, but its small size is a great advantage on British roads. Above all, its fuel economy is incomparably better, which is what modern motoring is all about.

For commuting, the engine is generally kept below 4000 rpm, when the static compression ratio of 6.9 to 1 is representative of the actual figure, ensuring extreme flexibility on the high gears. If the revs are allowed to enter the interesting part of the spectrum, there is a positive boost of 7.8 lb, which puts the effective compression ratio up to about 9.5, the same as the Tii, which is the HUCR (highest useful compression ratio). Now power is required from the engine and the higher volumetric efficiency increases the output from 130 to 170 bhp at the same revs.

On the road, the results are outstanding, the engine being unbelievably smooth, quiet and flexible for a four-cylinder unit. There is a suggestion of whistling wind occasionally before any real boost is developed, but most of the time the Turbo is quieter than a normal 2002, for the turbine acts as a most effective exhaust silencer.

The engine is deliberately de-tuned to give a long life with complete reliability and an ignition cut-out avoids any risk of over-revving. This is an essential piece of equipment because the engine accelerates so fast under full boost that it would be all too easy to go right into the red.

On the test car, the cut-out began to function above an indicated 136 mph, which is about 130 mph against the watch. For me, the gearbox was rather too far away and the synchromesh baulked on fast changes, but I did a 0-60 mph in 6.7 s with quite a leisurely change from first to second. Similarly, I did the standstill to 100 mph acceleration in well under 20 s, with a minimum of wheelspin and unhurried gearchanges. The driving position is excellent, apart from the too remote gearlever, and the seats give good lateral location.

The behaviour of the Michelin tyres was outstanding on wet roads and one can really use all the power all the time, almost irrespective of surface conditions. On a previous occasion, I did experience some brake fade with a Turbo, but I failed to worry the brakes during this test. The car is well balanced on corners, the big Michelins giving some extra cornering power compared with the standard 2002, but the ride has not been stiffened up and the Turbo is perfectly comfortable for shopping trips.

It would be possible to criticise the Turbo because full boost is not instantly available at virtually zero revs. In practice, this need not apply, for it is normally possible to foresee when full acceleration will be required and get things turning over nicely in preparation. There is a knack which is soon acquired but during hard driving the revs can be kept up

The Turbo suffers from too much "go faster" decoration making it dangerously conspicuous.

Road test

Below: The BMW's compact size is a great advantage on congested British roads.

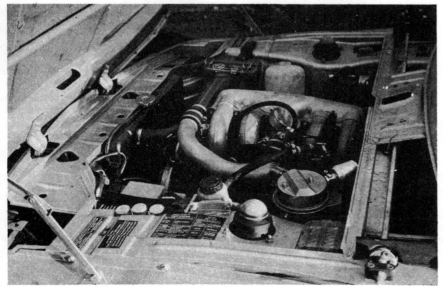

Under the bonnet turbocharger installation boosts power to 170 bhp DIN.

sufficiently so that there is no appreciable delay in response.

In my opinion—and there will be those who disagree—the Turbo has too much "go-faster" decoration. It could be driven faster on the road if it were less conspicuous, the coloured stripes performing no useful function while the flared guards over the wide tyres could be blended into the body sides more neatly. Similarly, the front and rear spoilers evidently do their job efficiently, judging by the high-speed stability of the car, but I am sure that they could be just as effective if they were less noticeable. Some owners may wish to emphasise the car's racing background, but the police may find it rather too interesting.

The BMW 2002 has been with us for a long time but the addition of the turbo-charger has made it one of the most effective and economical ultra-fast touring cars that has yet been produced. I had an overall average of 23.5 mpg, including taking the performance figures, and most owners will do far better than this, probably approaching 30 mpg on occasion.

Personally, I don't mind a "left-hooker," but I'm sure there is sufficient demand in England to justify the production of some right-hand-drive examples. Surely it's time that some eyeball ventilators were added, too, for we have all grown out of swivelling quarter lights. Perhaps if the production were increased, the car could be sold at a more realistic price, for it's a pretty expensive 2-litre, even though it is a 130 mph car.

The BMW 2002 Turbo has six-cylinder flexibility with only a four-cylinder thirst. It can be driven all day in London without ever wetting a sparking plug and on the open road it can see off almost anything. It handles remarkably well and gives a comfortable ride. I like it!

SPECIFICATION AND PERFORMANCE DATA

Car tested: BMW 2002 Turbo 2-door saloon, price £4220.84 including car tax and VAT.
Engine: Four-cylinders 89 x 80 mm (1990 cc). Compression ratio 6.9 to 1, 170 bhp (DIN) at 5800 rpm. Inclined valves operated by chain-driven overhead-camshaft and rockers. Schafer mechanical fuel injection with exhaust-driven turbo-supercharger.
Transmission: Single dry plate clutch. 4-speed all-synchromesh gearbox with central change, ratios 1.0, 1.279, 1.861 and 3.351 to 1. Hypoid final drive, ratio 3.36 to 1. 40 per cent self-locking differential.
Chassis: Combined steel body and chassis. Independent suspension of all four wheels by coil springs and telescopic dampers, with MacPherson front and semi-trailing arm rear geometry. Anti-roll bars both ends. ZF-Gemmer worm and roller steering. Servo-assisted dual-circuit disc/drum brakes. Bolt-on steel wheels fitted 185/70 VR 13 tyres (light-alloy wheels extra).
Equipment: 12-volt lighting and starting. Speedometer. Rev-counter. Fuel, temperature and blower-pressure gauges. Clock. Heating and demisting system. Flashing direction indicators. 2-speed windscreen wipers and washers. Reversing lights.
Dimensions: Wheelbase 8 ft 2 in. Track 4 ft 5.5 in. Overall length 13 ft 10 in. Width 5 ft 6 in. Weight 2281 lb.
Performance: Maximum speed 130 mph. Speeds in gears: Third 95 mph, second 62 mph, first 33 mph. Standing quarter-mile 15.2 s. Acceleration: 0-30 mph 2.2 s, 0-50 mph 5.4 s, 0-60 mph 6.7 s, 0-80 mph 12.2 s, 0-100 mph 19.5 s.
Fuel consumption: 23 to 29 mpg.

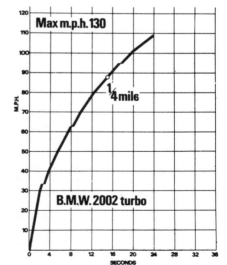

BAVARIAN MOTOR WONDER

Engineering and evolution combine in the BMW 2002 to make it the most refined super coupe on the market, a virtual Wundercar

ROAD TEST

If Clark Kent drove a car to suit his dual nature as mild mannered reporter for the *Daily Planet* and Superman, Champion Of Peace, chances are it would be a BMW 2002, a veritable Q-ship of the highways.

Possessed of a mild, ten-year old body design grown familiar to even the most unobservant of American drivers and an unrivalled combination of controls, power, and handling, this smallest BMW can turn any twisty bit of road into an exciting experience, even if it can't leap over tall buildings with a single bound and isn't faster than a speeding bullet.

That kind of performance hasn't come overnight to BMW. Looking back through ten years of *RT* reports on the car's predecessors—first the 1600, then the 1800, and finally the 2002—you can trace the growing confidence and ability of the car as the talented engineers turned what was originally an economy design into the whispering bomb you can buy today. It is, given BMW's unique heritage, a transformation to be expected, for ever since the formation of the current company in the '20s, racers have been winning in and on machines carrying the famous spinning propeller badge, itself a carryover from the parent company that built WWI aircraft engines.

Indeed, much of what goes into a BMW of any stripe today is still aircraft-quality, not only because of BMW's industrial connections, but because that dedication to engineering so typical of German cars finds many of its greatest proponents among BMW designers and technicians. It is with them a tradition, and a very strong one at that.

Much of that *engineered* feeling comes across immediately through the shape of the 2002. You can almost *see* the sober designer who laid out the basic lines, as he chopped a little here, added a little there and generally worked over what everyone else thought a small sedan should look like, until he had what seemed to be a reasonable compromise in looks, aerodynamics, and interior room.

The body has a smooth, timeless feel about it that gives the lie to its real age, much like a Volkswagen Beetle or Rolls Royce. The rounded corners, excellent visibility of the tall greenhouse and lack of protruding bits and pieces make it hard to call the shape *styled,* so clear is the impression of computer-like design.

Body detailing, in keeping with what we have learned to expect from German cars, is of the highest order. Access to the engine is gained through a hood that seems less like a panel than the whole front end of the car, but the joint is cleverly hidden by the chrome bumper strips on the sides of the body, a theme carried out for the trunk as well. Some years ago BMW introduced flow-through ventilation and opted for a still-unique design in interior air evacuation by ducting it out under the trunk lid. Like we said, it's *engineered.*

The same no-nonsense concept is used for the interior itself, in everything from the (orthopedically-designed, naturally) seats to the (matte-black non-reflective vinyl) dashboard. The only fripperies apparent are the wood inlay of the dash panel and the wood gearshift knob, which you get the feeling were forced on the reluctant engineers by adamant marketing people who felt that nobody would buy the car without any wood. They were wrong. The BMW has such a well-integrated interior that no feeling of austerity is likely to be felt by anyone driving it, which is a remarkable achievement considering the number of cars laden with dials and black surfaces that only make you think you're in a stripped submarine.

Seating positions are also typically German; high and rather upright, providing unhindered visibility through the generous glass area on all sides. The seats themselves are a little firm, but with excellent valences to promote good lateral support for hard cornering, a wise move considering how most BMW drivers behave when presented with a curve or two. From the driver's seat everything is in easy reach with the important and annoying exception of the gear lever, which seems about an inch too short; going Jackie Stewart-wise from second to third can be a shoulder-hurting experience if you're not careful and if you're very short. Otherwise there is no control out of reach, and most are grouped so that your hands need only be away from the wheel a moment.

Foot controls, the brake, clutch, and accelerator are not suspended, but pivoted on the floor. This makes heel-and-toeing an absolute cinch and leaves plenty of room to the side for your left foot to rest although there is no dead pedal.

An option that our test car didn't have was the padded steering wheel, and it's one we sorely missed. During a fast run down a winding road on a hot day, you can work up a lot of sweat on your hands, and with the slippery, small-rimmed plastic wheel, good control takes conscious effort for a good grip to avoid slipping. In town it's not a problem but out where the car changes from Clark Kent to Superman it would be appreciated.

With all that glass area, you'd expect some heat buildup, and you'd be right. Our test car had dark upholstery that breathed very well, but still couldn't overcome the blistering rays of the Southern California sun. The air conditioner, a Frigiking dealer-installed option, provided great gusts of cold air, but only through the vents on the vertical driveline console, which would freeze your elbows solid very nicely, but wouldn't help your outside (sunside) limbs. BMW reports that help is on the way, though. Aside from the hot-weather problem, ventilation and defrosting were very good, the flow-through system effectively getting rid of unwanted dead air.

Mechanically, the 2002 is an evolutionary rather than revolutionary car, just as in the bodywork. The engine pumps out a respectable 98 BHP at 5500 rpm, and

is responsive enough to put down most of it at nearly any rev range. Working in conjunction with a slick ZF four-speed manual gearbox also helps performance, of course, but even the automatic 2002s we've driven seemed like healthy movers. Here, too, the car exhibits the Kent/Superman syndrome, because if you drove the car around a city all day long, you'd never suspect that it was capable of laying down such good times at the strip; zero-to-sixty in our test car averaged out to about 10.3, with quarter-mile times in the mid-seventeens. Not blazing-tires stuff, to be sure, but pretty good considering the seemingly-endless tractibility of the engine.

Fuel consumption, a matter on everyone's mind, should please any prospective BMW owner. We managed around 25 mpg for the whole test period, including the track sessions, which gives you a fair idea of how you'd do without lead-footing it. Discretion, as always, is the key to good gas mileage.

The all-independent suspension has also reached its finely-tuned level by constant refinement rather than instant fixes. The 1600 tested by *RT* a decade ago exhibited everything from wheel-hop to lifting inside wheels under hard cornering, but the 2002 that is sitting in your dealer's showroom has none of its forbears faults. Like the engine, the suspension shows none of its true ability until pressed to the limit, which in this car is not reflected by its skid pad performance, something it shares with another car tested in this issue, the Lotus Elite. And like the Lotus, it shines in how it can thread its way through a high-speed needle. The harder and faster you drive, the better it gets. Steering is by worm and roller, but feels more like rack-and-pinion, so precise is the directional control.

Any prospective BMW 2002 owner has a choice between Michelin XAS and Uniroyal Rallye tires, and our car was shod with the Uniroyals. They proved to be excellent, giving a smooth ride around town with little noise, while working very well at speed in the bends. To be fair, though, they worked best when they got really hot, and paradoxically, when it was raining. In between, the tires were doing their duty, but without any special ability.

The performance and safety of the little coupe was demonstrated to us in a most unusual way while we were testing at Orange County raceway. While circulating around the handling course, one wheel dropped into a trough next to the road at about 75-80 mph. The trough proved to be very deep and worse yet, ended in a vertical face laced with jagged rock; the result was that the tire burst and the rim was severely dented on impact. To us in the car, (driver and two passengers) however, it only seemed like a blowout. The car was stopped without any wandering or lurching around, and we got out to look at what we expected to be a torn tire casing. Needless to say, the extent of the damage was unexpected—and impressive. And what that kind of stability has to say about the car is equally as impressive . . . imagine the same thing happening to you on a freeway; the ability to hold the car in a straight line without fighting the wheel could mean the difference between life and death.

That unplanned testing accident proved a clincher for what we were beginning to suspect about the 2002 . . . that it was the best, the very best, of that class of small cars known as super coupes, in every way. Which is as it should be: the BMW 2002 was the *original* super coupe, the granddaddy of the whole bunch. And it's still the best.

BMW 2002
PERFORMANCE DATA

Acceleration, sec:
- 0–30 mph 3.5
- 0–40 mph 5.5
- 0–50 mph 7.8
- 0–60 mph 10.3
- 0–70 mph 14.8
- 0–80 mph 20.2
- Standing start, ¼ mile 17.65
- Speed at end ¼ mile, mph 76.2
- Avg accel over ¼ mile, g 0.197

Speeds in gears, mph:
- 1st (6400 rpm) 32
- 2nd (6400 rpm) 60
- 3rd (6400 rpm) 91
- 4th (5700 rpm) 104
- Engine revs at 70 mph 3800

Speedometer error:
Electric speedometer	Car speedometer
40 mph	40 mph
50 mph	50 mph
60 mph	60 mph
70 mph	70 mph
80 mph	80 mph

Brakes:
- Min stopping distance from 60 mph, ft 139
- Avg deceleration rate, g 0.865

Fuel economy:
- Overall avg .... 24.5 mpg
- Range on 13 gal tank .... 318 miles
- Fuel required .... regular

Skid pad:
- Max speed on 100-ft rad, mph 32.05
- Lateral acceleration, g 0.686

Interior noise, decibels (dBA):
- Idle 54
- Max 1st gear 84
- Steady 40 mph 68
- 50 mph 73
- 60 mph 73
- 70 mph 76

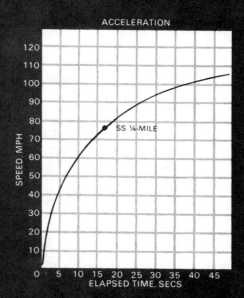
ACCELERATION

Graph Of Recorded Data Expressed In Percentage of 100 (100 = best possible rating)*
- Acceleration
- Brakes
- Skid Pad
- Interior Noise
- Tire Reserve
- Fuel Economy
- Overall Rating
- Highest Rated Domestic To Date — Chevrolet Corvette
- Highest Rated Import To Date — Lotus Elite

SPECIFICATIONS

Engine:
- Type .... SOHC L-4
- Displacement, cu in 121.3
- Displacement, cc 1990
- Bore x stroke, in .... 3.50 x 3.15
- Bore x stroke, mm .... 89.0 x 80.0
- Compression ratio 8.3:1
- Hp at rpm, net .... 98 @ 5500
- Torque at rpm, lb/ft, net .... 106 @ 3500
- Carburetion 1 2-V

Emissions, gm/mile:
- Hydrocarbons 2.4
- Carbon monoxide 35.0
- Nitrogen oxides 1.6

Prices:
Factory list as tested
West Coast $6703.75
Accessories included in price: $110—dlr prep; $69—tach; $39—anti-sway bars; $86—tinted glass; $550—air cond (dlr installed); $9.75—tire excise tax; $207—am/fm radio

Manufacturer's guarantee and service:
- Warranty (mos/miles) 12/12,000
- Oil change (mos/miles) .... -/5000
- Lubrication (mos/miles) 10,000
- Tune-up (miles)
  - Minor .... as req'd
  - Major .... as req'd

Driveline:
- Transmission .... 4-spd man
- Gear ratios:
  - 1st 3.764:1
  - 2nd 2.020:1
  - 3rd 1.320:1
  - 4th 1.000:1
- Final drive ratio 3.64:1
- Driving wheels .... rear

Wheels and tires:
- Wheels .... 5J x 13
- Tires 165 HR13
- Reserve load, front/rear, lb 622/618

General:
- Wheelbase, ins 98.5
- Overall length, ins 176.0
- Width, ins 62.6
- Height, ins 55.5
- Front track, ins 52.8
- Rear track, ins 52.8
- Trunk capacity, cu ft 15.9
- Curb weight, lbs 2400
  - Distribution, % front/rear 54/46
- Power-to-weight ratio, lbs/hp 24.5

Body and chassis:
- Body/frame construction .... unit
- Brakes, front/rear .... disc/drum
  - Swept area, sq in 249
  - Swept area, sq in/1000 lb 103.75
- Steering .... worm and roller
  - Ratio 17.57:1
- Turns, lock-to-lock 3.7
- Turning circle, ft 34.2
- Front suspension: Independent; MacPherson struts, coil springs, tubular shocks, anti-sway bar
- Rear suspension: independent; semi-trailing (diagonal) arms, coil springs, tubular shocks, anti-sway bar

Test Equipment Used: Testron Fifth Wheel, Esterline-Angus recorder, Ammco decelerometer, General Radio Sound Level Meter

* Acceleration (0-60 mph): 0% = 34.0 secs., 100% = 4.0 secs.; Brakes (60-0 mph): 0% = 220.0 ft., 100% = 140.0 ft.; Skid pad lateral accel.: 0% = 0.3 g, 100% = 0.9 g; Interior noise (70 mph): 0% = 90.0 dBA, 100% = 65.0 dBA; Tire reserve (with passengers): 0% = 0.0 lbs., 100% = 1500 lbs. or more; Fuel economy: 0% = 5 mpg, 100% = 45 mpg or more.

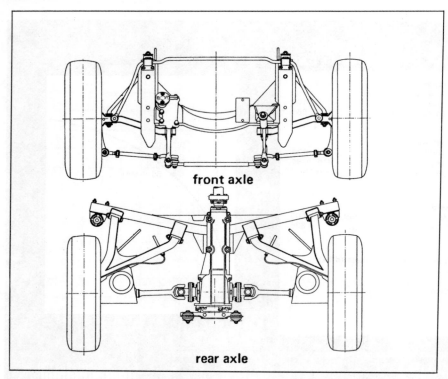

front axle

rear axle

TECH SCAN

That many fine, expensive automobiles currently available bear more than a passing resemblance to this BMW in basic design is no accident. Now nearly ten years old, it originated with the 1600 model, and its influence on other cars from Mercedes on down is unmistakable. It was among the first, if not the first, to use a sub-frame (or stub frame, as it has lately been called) for the rear as well as the front suspension. This allows the use of unit body construction with its benefits of light weight, stiffness and freedom from squeaks and rattles but without the noise, vibration and harshness that had been the major objection to unit construction. The sub-frames attach with resilient mounts where required to isolate "road noise" from the main body structure.

The rear suspension uses control arms to govern wheel movement in a somewhat analogous manner to what has been used on front suspensions for years. This is far superior to swing-axle designs formerly widely used on independent rear suspensions because it eliminates built-in camber changes that are the bane of swing-axles. With the control arm design, roll steer, camber change etc. can be tailored to just what suspension engineers deem desirable for optimal handling and tracking. The semi-trailing or diagonal control arms BMW uses accomplish much the same things as the uprights with upper and lower A-arms used on modern race cars without intruding into trunk space.

The now almost-commonplace MacPherson strut front suspension was fairly novel when first used on the 1600. Being a design compaction in that several of its components perform multiple functions thus eliminating the total number of parts required, it is very light and uses little space. BMW uses the struts in conjunction with a stub frame which bolts to the unit body. The engine also mounts to this member which isolates vibration from the main chassis. The engine is located quite far forward for a rear drive car which permits generous interior room within a compact overall length.

The 2002 engine, like the rest of the car, is also an outgrowth of the 1600. It is a sturdy 5-main bearing design using a cast-iron block and light alloy cylinder head. Somewhat unusual is the distributor drive which is taken off the single overhead camshaft. It leaves the distributor high and dry and easy to get at. The cam is driven by a conventional duplex roller chain. The oil pump is driven with a short chain from the nose of the crankshaft. The pump is driven with a short chain from the nose of the crankshaft. The pump itself is of a rotary (no gears) design and delivers at a pressure far in excess of most contemporary cars. (The car has no pressure gauge, just a warning light.) The rods are forged as is the crankshaft which is also nitrided all over for additional strength. This engine, like all other BMW engines, is pretty much bullet proof and in racing applications is capable of absorbing prodigious power increases without the need of additional beefing up.

But the really remarkable thing about the 2002 engine is the way it performs despite the fact that it meets emissions requirements same as any other. The key to this is the cylinder head and combustion chamber which were redesigned for the 1973 model. The valves are staggered to one side of the combustion chamber, and the spark plug placed opposite them. There are 3 spherical surfaces, 2 for the valves and 1 for the spark plug. They work in conjunction with a hump on the top of the piston to induce turbulence and clean, complete burning. The word BMW uses to describe the chamber, *Dreikugelwirbelwannenbrennraum*, itself creates quite a bit of turbulence and means literally tri-spherical turbulence-inducing combustion chamber. The chamber gives the rare combination of low emissions, good power output and good fuel economy, a veritable engineering triumph in this day and age. Originally developed for the 6-cylinder BMW engine, the new chamber permitted doing away with the air pump and thermal reactor which created both a drag on power and ill-tempered running.

There is a 2-stage exhaust gas recirculation system to control oxides of nitrogen. The first stage starts to operate at around 2500 rpm and 3.5 inches of vacuum. The second stage is both vacuum and temperature controlled, and below 63°F. ambient or below 113°F. cooling system temperature it is inoperative. The system recirculates a small amount of gas at low loads, increasing during acceleration and higher loads. At wide-open throttle the complete EGR system is inoperative to provide maximum performance.

Carburetion is by a single 2-stage downdraft unit with automatic choke. (A 2002tii version with Kugelfischer mechanical fuel injection and 27 more horsepower is also available, but only in 1974 models.)

The manual transmission with aluminum housing is of conventional design and uses Borg-Warner syncromesh units instead of the Porsche patent servo-mesh split rings previously used. There is no official explanation for the change. A 3-speed ZF torque converter automatic transmission is also available. The clutch for the manual transmission is of conventional diaphragm design. The final drive unit uses a hypoid gearset.

With the 2002's suspension and chassis design anything from super luxurious riding to race car handling can be had depending on what spring rates etc. are selected. As brought into this country, the 2002 is something in between. Front and rear anti-sway stabilizer bars are used, giving the car more roll stiffness than is usually found in sedans this size. The springs are also somewhat firmer than usual. The ZF–Gemmer worm and roller steering is very crisp and light. The suspension calibration and the steering (and the lively, responsive engine) make the car very nimble and agile and a delight to drive for anyone with a sporting corpuscle or two in his blood. It is this character that makes the car a sports sedan, and in truth there have been lots of so-called "pure" sports cars built that couldn't match this car's handling, not to mention ride. ●

TEST

BMW 2002 Turbo

1,990 c.c.

Exhilarating performance without any temperament and little loss of economy over standard 2002Tii. Taut ride and excellent handling and brakes give startling journey time potential. Poor ventilation and fairly high noise level familiar shortcomings of 2002 design. No conventional performance competitors at the price. Regrettably, available only in left hand drive form.

FEW WAYS of increasing the power of an engine look simpler in principle than turbo-charging. Harnessing the exhaust gases with a small impeller that is used to blow air into the engine to increase the density of the charge seems so logical, that one wonders why it is not employed more often. There are however, some snags. The turbochargers themselves cost a lot of money, and there are few companies producing them. There are also abundant problems in the installation of such a device, not all of which are overcome even in the BMW's case. The biggest problem concerns the induction system's ability to meter fuel appropriately, regardless of whether it is being drawn into the engine as the result of manifold depression, or whether the charge is being forced into the engine as a result of pressure from the turbocharger. In this, the BMW turbo is conspicuously successful, but in respect of the accommodation of the necessary large-bore exhaust system, the design is not as good, as there are a number of annoying vibration periods that result from poor location of the system. However, problems of this nature are readily soluble and we would hope that only early production examples suffer these annoyances.

BMW first gave notice of their interest in turbo-charging when they campaigned cars so equipped in the Touring Car Challenge way back in 1969. In the 2002 saloon, they knew that they had a chassis with a proven success record in saloon car racing; all that was needed was more power.

The success of the project culminated in winning the European Touring Car Challenge, and in those days, Porsche 911s raced in the same class, so BMW's win was no mean achievement. Having proved their point by winning the Championship, the emphasis in BMW's racing programme moved on to developing the bigger saloons and coupés into race-winning cars, and turbocharging seemed to have died a death. Then, in September 1973, what should be the star of the Frankfurt Show but the 2002 turbo—now developed as a road car with plans for the building of at least 2,000 examples.

Compared to the car as seen at the time of the original announcement, the Road Test example is mechanically identical; only the external decorations have been toned down since the first car appeared with "turbo" in large letters in mirror image on the front air dam. As sold in the UK, the 2002 turbo has just a restrained side flash in BMW's racing colours with "turbo" set in it and a matching colourful flash across the front air dam that helps to disguise the extreme depth of the front of the car. On the tail of the car there is restrained identity lettering, but the most obvious external features are the add-on glass fibre wing extensions that hide wheels of greater offset than the normal 2002 wheels. At the rear of the car there is a full-width soft rubber lip on the trailing edge of the boot and of course at the front, the deep front air dam and the absence of a bumper completely change the appearance of the nose.

In the class of genuine 4-seater sports coupés/saloons such as the Triumph Dolomite Sprint, the Fiat 124 Coupé, the Ford RS2000 and the Alfa Romeo 2000GTV, the normal 2002Tii is most competitive already. By increasing the performance so dramatically as in the turbo's case, BMW have moved the car into a very different echelon of rivals. In fact, the first comparable performer to the 2002 turbo is the Porsche 911 and that is nearly £2,000 more to buy, and one must go up as far as the the Porsche 911S before finding one that is appreciably quicker than the 2002 turbo.

Power output from the 1,990 c.c. engine is no less than 170 bhp (DIN) at 5,800 rpm and the maximum torque figure is 117 lb. ft. at a not unconservative 4,000 rpm. The explanation for the increased torque is that the effective cubic capacity of the engine is increased by the effect of the improved charge density that the turbocharger confers, and, of course, torque is directly related to capacity. Thus there is no lack of flexibility in the turbo-charged car and it only feels less tractable than

Some roll is noticeable when cornering hard but general ride and handling inspire the greatest confidence

BMW 2002 Turbo

the normal 2002tii at very low revs because it is higher-geared. Thus, in town or during slow running in traffic, the 2002 turbo shows no special shortcomings and indeed, the excellent gearchange and ready response of the engine enable smart progress to be made in broken traffic.

The maximum pressure that the turbocharger is allowed to create in the inlet manifold is 1.55 atm (approx 7 psi above ambient pressure) and a valve opens automatically when this pressure is reached. Although adjustable, this valve is set by the factory and if it is touched other than by a dealer, the warranty is invalidated.

From 4,000 rpm onwards, pressure in the inlet manifold is at 1.55 atm, the torque is at its maximum, and the power is building up quickly to its peak. On the road, a steadily increasing surge is felt as 4,000 rpm is exceeded and this continues to build up as the revs rise. Thus it can be judged that for optimum performance it is best to keep the engine turning in the range 4,000 to 6,000 rpm, and the well-chosen close-ratio gears enable this to be done successfully.

Performance

Nose-heavy weight distribution and relatively narrow tyres for the power output mean that it is easy to get appreciable wheelspin when attempting to get optimum acceleration times through the gears. After much experiment it was found that dropping the clutch suddenly at 4,800 rpm produced just the right amount of wheelspin to avoid dropping the engine speed below 4,000 rpm while at the same time avoiding an excess of wheelspin that would have wasted time. With this treatment, the car leapt from rest to 30 mph in 3.0 sec, and there was just a brief squeak of wheelspin when taking 2nd gear in which the turbo raced on to 60 mph in just 7.3 sec. The rev limiter allows a maximum in 2nd gear of 63 mph when 3rd gear can be taken and held to 6,400 rpm, giving 98 mph. The time to 100 mph would have been slightly quicker had the cutout not operated just before reaching this speed, but even so, a time of 20.7 sec from rest to 100 mph is fast by any car's standards.

Not surprisingly, acceleration tails off from 100 mph onwards, but the turbo is just among the elite of high performance cars capable of reaching 120 mph in under half-a-minute. Acceleration in each gear shows more obviously how the effect of the increasing assistance of the turbocharger can be felt. In third gear, for instance, the standard 2002Tii is faster for each 20 mph increment up to 50 mph, giving a time of 5.6 sec from 30-50 mph against 6.2 sec for the turbo. The times from 40-60 mph are similar at 5.7 sec for the Tii against 5.5 sec for the turbo, but then the latter just marches away as its times gets less and the Tii's get longer. Both 50-70 mph and 60-80 mph are covered by the turbo in 5.0 sec and 70-90 mph is covered in 6.0 sec. Proof that it is worth hanging onto third gear even until the rev-limiter operates is given by reference to the acceleration times in top gear. They show the equivalent 70-90 mph time in top gear to be 7.7 sec and in practice this means, that on the road one can sprint past and away from most cars and such acceleration in either third or top gear nullifies the disadvantage of the left-hand-drive. About the only speed range at which the turbo is at a disadvantage is when following traffic at around 55-60 mph when little useful acceleration is left in second gear and third gear acceleration is still building towards its optimum. This can be overcome by hanging back slightly on the approach to a likely overtaking place and then rushing up to overtake in anticipation of a clear road. The use of such dodges and the turbo's reserves of acceleration in any gear, give the potential of remarkably short journey times.

Roadholding and ride

Such considerable qualities of acceleration are limited if they are not matched by good roadholding, and fortunately in the turbo's case, this is of a very high standard. Nose-heavy weight distribution and fairly stiff front end roll control mean that the turbo, like the basic 2002Tii is set up to give initial understeer. However, with so much available power, a transi-

DICK ELLIS
M.S.I.A.

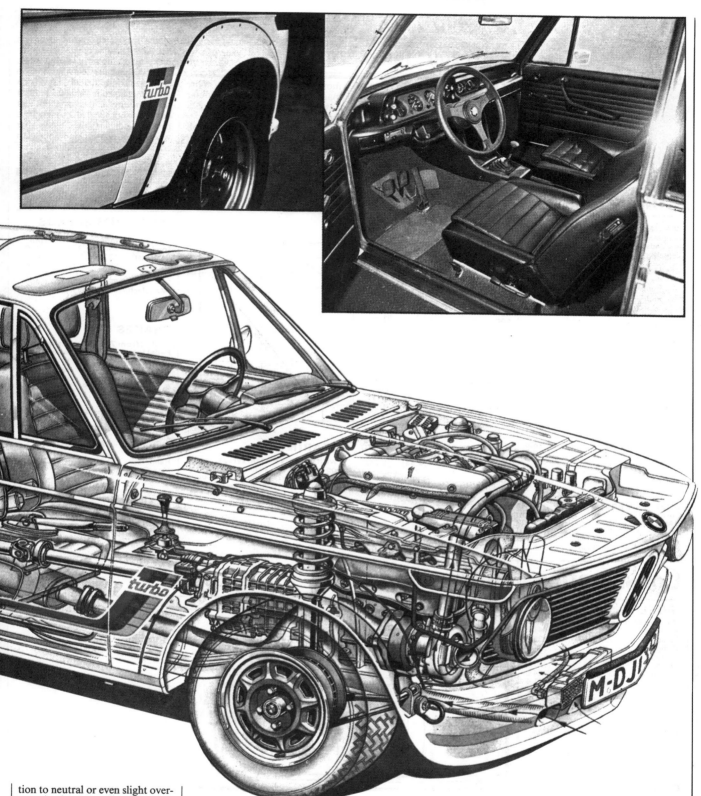

tion to neutral or even slight oversteer is always available, and indeed at slow speeds especially on wet roads, some caution must be taken to avoid too much oversteer. The accurate and lightweight ZF-Gemmer worm-and-roller steering allows plenty of feel of what is happening at the front of the car, and in addition, the steering is free from kickback on rougher surfaces. The steering ratio is the same as that of the normal 2002Tii at 3.7 turns lock-to-lock for a compact turning circle a little over a mean 32 feet. For serious road or competition use, higher ratio steering would be a help and as the steering effort required is low, even at parking speeds, a higher ratio would probably be a significant all-round advantage. The only further point that might be made in the context

Dick Ellis' cutaway of the BMW 2002 turbo shows the installation of the turbocharger immediately below the heat shield that surrounds the exhaust manifold. The arrows show how the air is taken into the compressor and sent along thin wall piping to the plenum chamber atop the engine. The air taken in through the hole in the glass fibre front air dam is used to cool the turbocharger. As suspension differences compared with the standard 2002Tii are confined to different settings only, the only visible sign of the 2002 turbo's improved road performance is the ventilated front disc brakes. In the other colour pictures on this page can be seen the distinguishing details of the turbo, such as the red instrument surround, detachable glass fibre wing spats, the soft rubber spoiler on the bootlid and the effective bucket seats that have head restraints as standard

BMW 2002 Turbo

of handling concerns the effect of the peakiness of the power output. Effectively, this means that as one is accelerating through a long corner, one is getting a steadily increasing amount of power, tending to tighten the line more and more and this effect is more marked than in a car of similar power output but possessing a flatter power curve.

The springing, damping and roll resistance of the turbo are all firmer than the normal 2002Tii which moves the point at which the semi-trailing arms begin to jack up to higher speeds. When allied to the turbo's limited slip differential, the firmer suspension gives more predictable control of the handling at the car's limit of cornering power as, unlike the normal 2002Tii, the inside rear wheel does not lift and spin with consequent loss of speed. Even if

Comparisons

MAXIMUM SPEED MPH
Jaguar V12 Series III
 Roadster . . . . (£3,743) 143
Broadspeed Ford Capri
 turbo. . . . . (£3,419) 140
BMW 2002 turbo . . . . (£4,221) 130*
Porsche 911 . . . . (£6,124) 130*
Alfa Romeo 2000GTV . . . (£2,945) 120
Manufacturer's claim

0-60 MPH, SEC
Jaguar V12 Series III Roadster 6.3
Broadspeed Ford Capri turbo 7.0
BMW 2002 turbo 7.3
Porsche 911 . . . . 7.8
Alfa Romeo 2000GTV 9.2

STANDING ¼-MILE, SEC
Jaguar V12 Series III Roadster 14.9
Broadspeed Ford Capri turbo 15.4
Porsche 911 . . . . 15.8
BMW 2002 turbo 16.0
Alfa Romeo 2000GTV 16.4

OVERALL MPG
Porsche 911 . . . . 23.2
BMW 2002 turbo 21.7
Alfa Romeo 2000GTV 21.1
Broadspeed Ford Capri turbo 17.1
Jaguar V12 Series III Roadster 15.0

Performance

ACCELERATION

True speed mph	Time in Secs	Car Speedo mph
30	3.0	33
40	4.3	43
50	5.7	53
60	7.3	64
70	9.8	74
80	12.5	84
90	15.7	94
100	20.7	104
110	27.2	114
120	39.8	125

Standing ¼-mile
16.0 sec 91 mph

Standing Kilometre
28.9 sec 112 mph

Mileage recorder: accurate

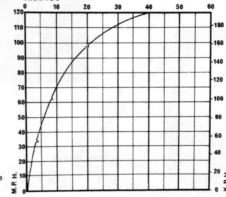

GEAR RATIOS AND TIME IN SEC

mph	Top (3.36)	3rd (4.44)	2nd (6.79)
10–30	—	7.7	4.2
20–40	10.0	7.0	3.8
30–50	9.1	6.2	3.3
40–60	8.6	5.5	3.1
50–70	8.4	5.0	—
60–80	7.7	5.0	—
70–90	7.7	6.0	—
80–100	8.5	—	—
90–110	11.3	—	—
100–120	18.6	—	—

GEARING
(with 185/70VR 13 in. tyres)
Top . . . . . . 20.36 mph per 1,000rpm
3rd . . . . 15.41 mph per 1,000rpm
2nd . . . . . . 10.08 mph per 1,000rpm
1st . . . . 5.41 mph per 1,000rpm

MAXIMUM SPEEDS

Gear	mph	kph	rpm
Top (mean)	130*	209	6,400
(best)	130*	209	6,400
3rd	98	158	6,400
2nd	63	101	6,400
1st	35	56	6,400

*Manufacturer's claimed maximum

BRAKES

FADE (from 70 mph in neutral)
Pedal load for 0.5 g stops in lb

1	30–25	6	35
2	35–25	7	30–35
3	35–30	8	35
4	35	9	35
5	35	10	40–30

RESPONSE from 30mph in neutral

Load	g	Distance
20lb	0.20	150ft
40lb	0.51	59ft
60lb	0.86	35ft
80lb	0.96	31.4ft
Handbrake	0.30	100ft

Max. Gradient 1 in 3

CLUTCH
Pedal 42lb and 5in

Consumption

***FUEL**
(At constant speed—mpg)*
30 mph 41.6
40 mph 38.2
50 mph 34.0
60 mph 29.7
70 mph 27.0
80 mph 24.0
90 mph 21.4
100 mph 18.9
*N.B. Manufacturer's claimed figures—fuel injection incompatible with *Autocar* equipment

Typical mpg 23.0 (12.3 litres/100km)
Calculated (DIN) mpg 24.6
 (11.5 litres/100km)
Overall mpg 21.7 (13.0 litres/100km)
Grade of fuel Premium, 4-star
 (min. 97 RM)

OIL
Consumption (SAE 20W/50) 1,500mpp

TEST CONDITIONS
Weather: Dry, clear and fine
Wind: 5-10 mph
Temperature: 21 deg C (70 deg F)
Barometer: 29.7 in. hg.
Humidity: 62 per cent.
Surface: Dry concrete and asphalt.
Test distance: 1,700 miles

Figures taken by our own staff at the Motor Industry Research Association proving ground at Nuneaton.

Dimensions

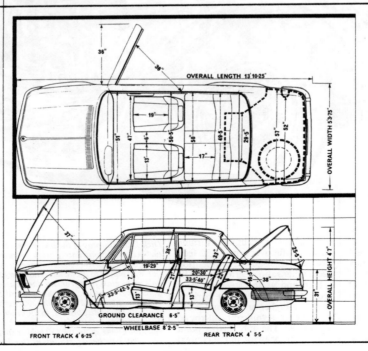

STANDARD GARAGE
16ft × 8ft 6in.

TURNING CIRCLES:
Between kerbs
L, 33ft 11in.; R, 31ft 0in.
Between walls
L, 35ft 2in.; R, 32ft 3in.
Steering wheel turns, lock to lock 3.7

WEIGHT:
Kerb Weight 21.7 cwt
(2,429lb-1,103kg).
(with oil, water and half full fuel tank).
Distribution, per cent
F, 54.7; R, 45.3
Laden as tested:
24.6cwt (2,754lb-1,251kg)

the driver lifts off; the feeling that the normal 2002Tii gave of being "on tiptoe" is absent in the turbo version and one can give only unreserved praise to the car for its handling. A certain contributor to the high cornering potential is the aerodynamic aid that the front dam and rear spoiler give, for in addition to contributing to good straightline stability, they ensure no loss of precision as cornering speeds increase.

In the search for good handling, the ride may suffer but in the turbo's case, this is only marginally worse than the high standard set by the 2002Tii at low speeds, and praise is due to BMW for finding that elusive compromise between ride and handling. Praise is due too for the Michelin XWX tyres with which the test car was fitted, for there are no conditions of dry or wet road under which these tyres give any cause for complaint.

Brakes

We have criticised previous 2002 BMWs for the speed sensitivity of their brakes and also a tendency to go "soggy" when used frequently and hard. The adoption of ventilated front discs and the introduction of a pressure-limiting valve into the rear circuit has removed these criticisms, and the system is now well up to the standard of the rest of the car. An added bonus is that the softer pad material that the ventilated front disc brakes use improves the initial response of the brakes compared to the normal 2002Tii and less than 30 lb

---

Specification

BMW 2002 turbo

FRONT ENGINE, REAR-WHEEL DRIVE

ENGINE
- Cylinders: 4, in-line
- Main bearings: 5
- Cooling system: Water; pump, fan, thermostat
- Bore: 89mm (3.504in.)
- Stroke: 80mm (3.150in.)
- Displacement: 1,990 c.c. (121.4 cu. in.)
- Valve gear: Single overhead camshaft and rockers
- Compression ratio: 6.9 to 1. Min. octane rating: 98RM
- Induction: Schafer mechanical injection with exhaust gas-driven KKK/BLO turbocharger
- Fuel pump: Kugelfischer high pressure type PL04
- Oil filter: Full-flow, replaceable cartridge
- Max. power: 170 bhp (DIN) at 5,800 rpm
- Max. torque: 117 lb. ft. (DIN) at 4,000 rpm

TRANSMISSION
- Clutch: Fichtel and Sachs, diaphragm spring, 7.9 in. dia.
- Gearbox: *4-speed, all-synchromesh, optional 5-speed
- Gear ratios:
  - Top 1.00 / *Fourth 1.00
  - Fourth 1.24 / Third 1.28
  - Third 1.58 / Second 1.86
  - Second 2.16 / First 3.35
  - First 3.37 / Reverse 4.00
  - Reverse 3.65
- Final drive: Hypoid bevel, 3.36:1
- Mph at 1,000rpm in top gear: 20.36

CHASSIS and BODY
- Construction: Integral with steel body

SUSPENSION
- Front: Independent; MacPherson struts, lower wishbones, coil springs, telescopic dampers, anti-roll bar
- Rear: Independent; semi-trailing arms, coil springs, telescopic dampers, anti-roll bar

STEERING
- Type: ZF-Gemmer worm and roller
- Wheel dia.: 15 in.

BRAKES
- Make and type: ATE ventilated disc front, drum rear, pressure limiting valve in rear circuit
- Servo: ATE vacuum
- Dimensions: F, 10.08 in. dia. R, 9.8 in. dia. 1.57 in. wide shoes
- Swept area: F, 210.4 sq. in., R, 88.6 sq. in. Total 299 sq. in. (244 sq. in./ton laden)

WHEELS
- Type: Pressed steel disc 5½J in. wide rim
- Tyres—make: *Michelin XWX/Pirelli CN36SM
- —type: radial ply tubed
- —size: 185/70VR 13

EQUIPMENT
- Battery: 12 volt 44 Ah.
- Alternator: 53 amp a.c.
- Headlamps: 110/120 watt (total)
- Reversing lamp: Standard
- Electric fuses: 12
- Screen wipers: Two-speed, wash/wipe facility
- Screen washer: Standard, electric
- Interior heater: Standard, water valve type
- Heated backlight: Standard
- Safety belts: Standard, inertia reel, static in rear
- Interior trim: Cloth seats, pvc headlining
- Floor covering: Carpet
- Jack: Screw-pillar type
- Jacking points: 2 each side beneath sills
- Windscreen: Laminated
- Underbody protection: Tectyl underbody sealing to floor pan and wheel arches

MAINTENANCE
- Fuel tank: 15.4 Imp. gallons (70 litres)
- Cooling system: 11 pints (inc. heater)
- Engine sump: 7½ pints (4.4 litres) SAE 20W/50 Change oil every 2,000 miles. Change filter every 2,000 miles.
- Gearbox: 1.8 pints. SAE 80EP. Change every 16,000 miles.
- Final drive: 1.7 pints. SAE 90 (Hypoid) Change every 8,000 miles.
- Grease: No points
- Valve clearance: Inlet 0.006–0.008 in. (hot) Exhaust 0.006–0.008 in. (hot)
- Contact breaker: 0.016 in. gap; 62 ± 3 deg. dwell
- Ignition timing: 25 deg. BTDC (stroboscopic at 2,500 rpm)
- Spark plug: Type: Bosch W200 T 30. Gap 0.24 in.
- Tyre pressures: F, 30; R, 30 psi (normal driving) F, 30; R, 33 psi (high speed and full load)
- Max. payload: 892 lb (405 kg)

*Fitted to test car

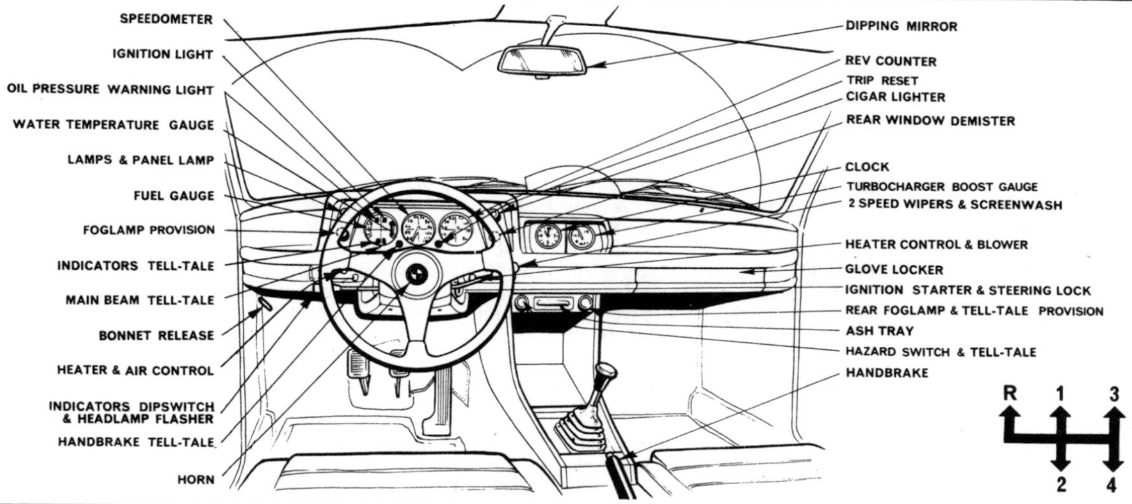

Dashboard diagram labels:
- SPEEDOMETER
- IGNITION LIGHT
- OIL PRESSURE WARNING LIGHT
- WATER TEMPERATURE GAUGE
- LAMPS & PANEL LAMP
- FUEL GAUGE
- FOGLAMP PROVISION
- INDICATORS TELL-TALE
- MAIN BEAM TELL-TALE
- BONNET RELEASE
- HEATER & AIR CONTROL
- INDICATORS DIPSWITCH & HEADLAMP FLASHER
- HANDBRAKE TELL-TALE
- HORN
- DIPPING MIRROR
- REV COUNTER
- TRIP RESET
- CIGAR LIGHTER
- REAR WINDOW DEMISTER
- CLOCK
- TURBOCHARGER BOOST GAUGE
- 2 SPEED WIPERS & SCREENWASH
- HEATER CONTROL & BLOWER
- GLOVE LOCKER
- IGNITION STARTER & STEERING LOCK
- REAR FOGLAMP & TELL-TALE PROVISION
- ASH TRAY
- HAZARD SWITCH & TELL-TALE
- HANDBRAKE

Gear pattern: R 1 3 / 2 4

Servicing

	2,000 miles	4,000 miles	8,000 miles
Time Allowed (hours)	1.17	1.33	3.40
Cost at £3.30 per hour	£3.86	£4.39	£11.22
Engine oil	£2.00	£2.00	£2.00
Gearbox oil	—	—	£1.00
Oil filter	£1.79	£1.79	£1.79
Air filter	—	—	£15.24
Contact Breaker Points	—	—	£0.77
Sparking Plugs	—	—	£1.59
Total Cost:	**£7.65**	**£8.18**	**£33.61**

Routine Replacements:	Time hours	Labour	Spares	TOTAL
Brake Pads—Front (2 wheels)	0.50	£1.65	£5.59	£7.24
Brake Shoes—Rear (2 wheels)	1.00	£3.30	£19.44	£22.74
Exhaust System	1.00	£3.30	£71.55	£74.85
Clutch (centre+driven plate)	2.50	£8.25	£87.80	£96.05
Dampers—Front (pair)	2.50	£8.25	£54.75	£63.00
Dampers—Rear (pair)	1.00	£3.30	£26.20	£29.50
Replace Drive Shaft	0.50	£1.65	£54.77	£56.42
Replace Alternator	0.50	£1.65	£50.16	£51.81
Replace Starter	1.00	£3.30	£106.13	£109.43

AUTO TEST

BMW 2002 Turbo

The fuel injection plenum chamber, a multiplicity of pipework and various controls for varying atmospheric pressures and changing mixture strength with throttle opening mean a confusing under bonnet scene. However, items requiring routine maintenance are all within easy reach. To the engine's right are the air filter and generous water reservoir and the coil is tucked away on the bulkhead, well away from spray and dirt

pedal pressure is required for normal check braking. The accelerated brake fade test did not reveal any tendency for the braking efficiency to fall off, and indeed, the optimum braking figure of 0.96g given by the brakes when cold was repeatable at the same pedal pressure immediately after the last fade test application had been done. The handbrake gave a useful 0.30g which would be sufficient to stop the car on its own from 30 mph in 100 ft. As would be expected, the turbo had no difficulty in restarting on a 1-in-3 hill, rushing away with wheelspin after holding securely on the convenient and efficient handbrake.

Fuel consumption

With an overall consumption of 21.7 mpg, the turbo is only beaten among the world's outstanding performance cars by the remarkable Porsche 911 which managed 23.2 mpg overall. Inevitably, this figure includes a great deal of very hard driving, for this is what the car is all about. The best figure obtained was 24 mpg on a long journey including much motorway and this is the sort of consumption that most owners are likely to get with some regularity. The turbo's fuel tank holds 15.4 gal so there is a really useful range of over 350 miles available. Although the Kügelfischer high pressure injection pump is not compatible with our Petrometa test equipment, the BMW claimed figures have been used in the data panel since the calculated DIN consumption figure is acceptable enough to indicate that these figures are accurate. The fuel tank fills well to the neck without blowback.

Comfort and controls

Immediate differences inside a turbo compared with other 2002 models are the bucket seats, a red surround to the instruments and the turbocharger boost gauge; otherwise, the interior is the same as the current 2002Tii.

The bucket seats are a deliberately tight fit for people of average build upwards (or outwards). Thus they give excellent lateral support which the high cornering forces demand. The backrests are finely adjustable for rake, and there is more than adequate rearward movement. However, with the front seats right back, little kneeroom remains for rear seat passengers and some compromise is necessary. For the driver, the seating position has to be quite high as, on this left-hand-drive model, the pedals are floor-pivoted (as opposed to top-hung on the right-hand-drive BMWs that are imported). The pedals themselves are comfortable to operate, and there is space to rest the left foot beside the clutch and adequate spacing of brake and accelerator to allow heel-and-toe operation. The smart leather-rimmed steering wheel has a thick, hard rim and is undished. Ahead through the wheel is a full display of round instruments.

All the controls are within easy reach of the driver and as the car is fitted with inertia reel seat belts as standard, the controls are also accessible to a front seat passenger. Heating and ventilation controls are sited on the facia rail either side of the steering column. A poor feature is that the heat control is on the driver's left, making it difficult for the front seat passenger to operate, although the volume and heater fan control are accessible to both in the centre of the car.

The turbo was given all the improvements for 1974 cars that were announced at the 1973 Earls Court Motor Show which include slight revision to the rear seat cushion to provide more legroom than hitherto and the door trims are colour-keyed to the remainder of the interior trim.

Living with the 2002 turbo

Usually in the case of the highest performance version of a particular model, there is some clue either from the exhaust noise or the more exotic appearance to give a clue as to the car's status. In the turbo's case, the exhaust is in fact quieter than the standard car as a result of the very big bore system that must be used to avoid back pressure on the turbocharger. The exhaust is made in special heat-resistant material to combat the very high temperatures that result from the turbocharger installation. So the only clues are the external bodywork changes and of course, the blistering performance.

The increased performance is gained at a quite reasonable increase in fuel consumption and the only penalty that one has to pay is in the more frequent service attention that the car requires. The very high temperatures in the turbocharger require that the engine oil be changed more frequently for fear that the oil's qualities will be destroyed more rapidly. Thus, the oil and filter must be changed every 2,000 miles and as the filter requires a special tool for removal, this has to be done by an authorised dealer. Apart from this requirement, only a change of rear axle oil at 8,000-mile intervals is an exception to normal servicing procedure for similar cars. BMW insist that any attention to the turbocharger and the mechanical fuel injection be carried out by authorised dealers only but emphasise that once set up accurately, the Schafer (Kugelfischer) fuel injection should be free of problem. Certainly in the case of the Road Test car, it always started first turn, and ran without temperament thereafter, providing exemplary drive-away characteristics. All the controls are light and all work well with a feeling of solidity. The gearchange especially has short definite movements back and forth and across the gate.

Some of the test staff felt that a longer gearlever would be an improvement in providing greater leverage to make the change even faster and to help to overcome synchromesh resistance when changing down into first gear on the move—an operation that occurs often on very winding roads when seeking optimum performance. The clutch is not too heavy in its operation, and the driving position allows a good leverage on the pedal. Its take-up is progressive and as the transmission is free of snatch or wind up, smooth progress is easily possible.

Access to items requiring routine maintenance is good beneath the big forward hinged bonnet, and even the dipstick can be easily unravelled from the mass of pipework and trunking that surrounds it. The combined reservoir for the brakes and clutch is brought to the nearside front wing top for easy visual checking and for replenishment if necessary.

The boot size is less than that of the normal 2002 as the greater capacity of the tank means that the floor is higher. However, the shape is still regular and a big suitcase can still be fitted in.

Conclusions

One unwelcome feature of most high performance cars is their size, as in general, the big engines used demand plenty of space. In the BMW turbo's case, like Porsche, tremendous performance is available in a handy package, and it is surprising how much this helps in making good progress in broken traffic.

In this country, a disadvantage that will weigh heavily against the car is the fact that it is only available in left-hand-drive form, which is heartily disliked by insurance companies if they do not already dislike the performance potential. Where the car is most likely to fail in marketing terms is in being the top of a range that starts with quite mundane examples in a performance sense and already possesses a similar model that appeals to the discriminating sporting motorist. However, the 2002 turbo is most civilised and sacrifices nothing in the provision of so much performance at a price unmatched by any rivals. It is a pity that something as simple as excessive heat in the region of the steering box should be the reason why an acceptable right-hand-drive version is not available. □

MANUFACTURER: Bayerische Motorenwerke AG, Munchen 13, West Germany.

UK CONCESSIONAIRES: BMW Concessionaires (GB) Ltd., BMW House, 361-365 Chiswick High Road, London W4.

PRICES	
Basic	£3,607.55
Special Car Tax	£300.63
VAT	£312.66
Total (in GB)	**£4,220.84**
Seat Belts	£32.25
Licence	£25.00
Delivery charge (London)	£17.22
Number plates	£5.00
Total on the Road (exc. insurance)	**£4,300.31**

Insurance	Group 7
EXTRAS (inc. VAT)	
Five speed gearbox	£255.04
Steel sliding sunroof	£125.25
Fitted to test car	
TOTAL AS TESTED ON THE ROAD	**£4,300.31**

What Car? rarity/BMW 2002 Turbo
Boost for a new generation?

One quick glance at the picture shows that this is a very special 2002 — the Turbo, a guaranteed rarity in Britain since only 180 of these cars, the first series produced turbocharged car in the world, are to be imported.

Any resemblance to a normal BMW 2002 stops at the body silhouette, for by the theoretically straightforward expedient of bolting on a turbocharger BMW have transformed what was already a very lively machine into a four wheeled bullet. Turbocharging is an excellent way to boost massively the power output of a car without having to resort to large and heavy engines or high degrees of tuning on smaller units. In essence it is a form of supercharging; the exhaust gases drive a turbine which forces air under pressure into the engine. (In conventional supercharging the turbine is driven directly by engine power through belts, absorbing some power. Turbocharging is more efficient since only the waste energy of the exhaust gases is used to drive the turbine.)

The result is a power output of 170 bhp, compared with the 130 bhp of the fuel injected 2002 Tii. And in contrast with what one would normally expect to find in a highly tuned, small engine the power comes in through a wide band and the engine is completely tractable and docile at all revs.

From the outside, the Turbo will attract the attention of friends, girls, and the local police force thanks to its flared wheel arches, covering wider than standard wheels, large front air dam, boot mounted spoiler and immaculate but eye catching paintwork — white with candy stripes down the sides and on the air dam. It is not a car for those of a retiring disposition!

Inside the Turbo is fitted out as one would expect of a comfortable saloon, the black, very well shaped bucket seats and small, thick sports steering wheel being further clues to its nature. Instrumentation is as on the standard 2002, except that the instruments are set in a red surround and the speedometer reads to 150 mph. The one additional instrument is a pressure gauge for the turbocharger and this is the one that you watch as the needle swings smoothly through the white section into the narrow green band in-

The excellent roadholding and braking capabilities of the BMW 2002 Turbo inspire great confidence when driving at high speeds

The neat installation of the turbocharger unit does create accessibility problems

dicating maximum boost.

As soon as the needle hits the green the Turbo is transformed from a tame tiger into a ferocious one. The driver can feel the car surge forward at the sort of speed one normally expects from a Porsche 911 or V12 Jaguar. Peak boost starts at 4000 rpm, which corresponds with maximum torque. Maximum power is at 5800 rpm and the rev counter red-lined at 6500 rpm, so for optimum performance it is best to keep the engine turning between 4000 and 6000 rpm and the perfectly spaced gears allow this to be done successfully — changing up through the gears at the immaculate gearbox at the red line drops the revs back to just 4000, keeping the turbocharger boost at maximum.

The straight line performance figures are outstanding, a real demonstration of turbocharging's virtues: 60 mph comes up from rest in 6.9 secs and the car accelerates on to 100 mph in just 18.2 secs. Remarkable figures that are beaten only when one reaches the class of Porsche 911S and E-types, and giving the Turbo almost unbeatable value for performance at £4221. In £s for mph terms, only the Jaguar can beat it.

This sort of performance would not be so healthy without roadholding and braking to match and we can commend the Turbo on both counts.

Although looking rather garish, the Turbo is in fact a very civilised car. There is comfortable seating for rear passengers, an excellent ride and a boot large enough for a big suitcase. Nor are noise levels at all objectionable. Again, fuel consumption is good for a car of this performance. During the hard driving of performance testing we achieved 18.7 mpg, but a more realistic figure for the owner/driver is our touring figure of 23.5 mpg.

With so much praise, there just has to be a problem — and there is. The Turbo is only available in left-hand drive form, which obviously limits its appeal. The reason is that turbochargers produce a great deal of heat — the Turbo's exhaust manifold glows red hot when being driven hard — and in right hand drive form there would be excessive heat round the steering box.

But for those who are willing to put up with the inconvenience of left-hand drive, the Turbo is a fine car, and there are only 180 in Britain, so it has marked its place in motoring history.

BMW's LITTLE SEDAN WITH THE BIG SQUIRT...

2002 Tii —

THE TEN TENTHS TOURING CAR

IT'S TEN-TENTHS because it needs to be driven with complete concentration and care. Sure, it's a rapid little fun car, but it is *not* good in the BMW tradition. Matt Whelan took a close look at the often-praised small Bee-Em, and came back to report that a combination of the Australian Design Rules and age had almost ruined it.

THE BMW 2002 HAS a long-established reputation as a desirable, rapid little sporty car — and after hearing and reading the constant praise of its abilities I was looking forward eagerly to my first 'real' drive of the big-power small Bee-Emm, in fuel injected form.

It disappointed me. Maybe it was because I'd not long before spent 2000km in the 2002Tii's big brother, the brilliant 3.0Si, or maybe it was because modifications to comply with Australian Design Rules had taken the edge off the small car — but it certainly wasn't the hotshoe I had expected.

Previous brief drives in 2002s had left me believing in the legend, but an extended run in the more powerful version left me disappointed. Perhaps I was expecting too much — it *was* fun to drive, but just didn't match up to my pre-programmed expectations.

Sydney BMW expert David Bones threw some light on the problem for us. When I complained about the relatively poor (and *relatively* is the qualifying word) handling, and the performance, he explained much of the edge had been taken off local versions of the 2002Tii by ADR compliance modifications.

"One major factor is that the five-speed gearbox, commonly used on this model in Europe, isn't available here — it transforms the car," he said. "But the ADR's have really done the rest. A spacer has been inserted in the front suspension struts to make the car comply with ride-height regulations, which imbalances the handling to a noticeable degree.

"Pollution control equipment also required, combined with the unavailability of the five-speed box, has taken the sting out of the motor," he said.

The test car had only covered 1800km when we picked it up, and may have been too new to be giving of its best, performance-wise. And though we couldn't detect a difference with it turned off, the Hitachi air-conditioner (you could almost call this one a refrigerator!) could also have had an effect on engine performance.

The 02-series body is a little outdated in concept — whether the 3-series replacement is substantially better remains to be seen. I'm sure the 5-series body, though heavier, is much more practical for local conditions. The 2002 gains performance from its light weight, but it's simply too small for all but the sports-car afficianado who wants a roof over his head.

It could well be called a two-plus-two — certainly if driver and passenger are tall. The rear seat is difficult to get into, and cramped when you get there. There *is* a large boot (huge, in fact, for this size car) and front seatroom is excellent — but it isn't a family car.

The 2002Tii *is* a strong little performer, but in its as-tested condition it just didn't live up to expectations. It accelerated from 0-100km/h in 11.3seconds — a very snappy time, but in fact four-tenths of a second slower than the factory-quoted figure for the standard 2002.

The factory lists the Tii's time to this speed as 9.4seconds — just under two seconds faster than our car! You can usually bet on the accuracy of figures quoted by the West German manufacturers — but, even allowing for a little optimism on BMW's part, the Tii we tested obviously wasn't performing as it should.

Our first brief acceleration checks, using just the speedo and a stopwatch, achieved a time better than the factory's — but when we compared these times with our figures, recorded by the MM Fifth Wheel, we realised the fault. The little Bee-Em had a big, big speedo which wanted us to believe we were doing 100 when actual speed was 90km/h.

Don't be misled by my disappointment, though — the Tii is still a goer. It whips through the gears like a racer, the engine spinning quickly to its 6200rpm redline. Highway power is excellent for high-speed cruising, overtaking, or cornering.

The test car achieved almost identical speeds-in-gears figures to those quoted by the factory — 54km/h in first, 90 in second, 140 in third, and 180 in top.

Interior noise is subdued at most speeds, but as the willing four gets close to peak revs the normally-pleasant howl becomes intrusive.

The Tii's motor is the same basic unit which powers all the 02-series cars. In this case, it is in two-litre form (actually 1990cc), with Kugelfischer fuel injection adding to the OHC unit's already-proven sting.

The unit produces 96kW (DIN) at 5800rpm, and 181Nm at 4400rpm. It is rather peaky — power jumps from

BMW 2002 Tii

35kW at 2000rpm to over 70 at 4000 and 90 at 5000 — but is still flexible, thanks to a relatively broad spread of torque. It can get a little "cranky" when asked to lug lower than 40-50km/h in top, but otherwise behaves well at city speeds and in slow traffic situations.

Fuel consumption is excellent for a car of the Tii's performance potential — the DIN standard measured consumption claimed by the factory is 8.8 1/100km (31.6mpg), and we find that quite believable in the motor's European form. I'm not convinced it will achieve the same as we see it here, but under full-bore test conditions we achieved a low of 20mpg and a high of over 26, creditable figures indeed. In normal day-to-day driving situations, we would expect it to return between 24 and 30mpg fairly easily.

It's a real pity we can't get the five-speed version of the Tii here — from all reports it does wonders for performance. The five-speed unit is a close-ratio version, retaining the direct drive top gear with the four indirects more closely spaced to achieve better acceleration and allow better use of the power at speed.

The four-speed box is fine for normal use, but doesn't offer the same scope for control in hard driving. The gearbox itself is smooth, ratios are well-chosen, and the shift direct and accurate. It's worth levelling a complaint here against the BMW gear-lever knob — it seems a standard feature of the manual cars, and I would probably replace it with an accessory shop unit. Though the circular disc-topped knob suits interior styling, it is uncomfortable to use.

The four-speed gearbox uses Borg-Warner syncromesh as opposed to the Porsche syncro in the five-speed unit — what difference it makes to this model we don't know, but the unit in the test car was quite satisfactory.

Handling is another area where the 2002, according to "legend", shines — and another area where the test car was disappointing. Possibly this particular unit was badly set up, but I feel the spacers inserted in the McPherson struts (to increase ride height for ADRs) *have* had an effect. The car just doesn't have the same balanced feel of other BMWs — the traditional 'feel' which inspired driver confidence and the capabilities which have saved many a desperate cornering situation for BMW owners. It is fussy on corners — I suppose you could call it responsive — but to the degree where it is almost hairy.

Corners have to be judged and lined up perfectly or it tries to get itself out of shape. The Tii is extremely responsive to throttle application — it could be called twitchy. It is great fun for someone who *knows* what he is up to and simply wants to *play* the car through corners on the throttle, but for the less experienced, relying on BMW's fabled handling, it could be dangerous.

With power on coming into a corner the little car tends to understeer, but the instant you back off it turns immediately into a strong oversteer characteristic. It is fun to set it up through a corner like this — but it needs perfect judgment.

Obviously, these instant response reactions aren't evident at all times — you only notice it when going hard, particularly in tighter corners. At

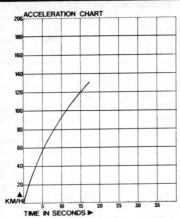

ABOVE: Instruments appear confusing, but in practice they are easy to read at a glance. RIGHT: Access is excellent everywhere but to the rear seat. The design is practical for two people, less so as the numbers increase.

ROAD TEST DATA & SPECIFICATIONS

Manufacturer: Bayerische Motoren Werke A.G. Munich, Germany
Make/Model: 2002 Tii
Body type: 2-door sedan
Test car supplied by: Capitol Motors, Sydney

ENGINE
Location: Front
Cylinders: Four in-line
Bore & Stroke: 89mm x 80mm
Capacity: 1990cc
Compression: 9.5 to 1
Aspiration: Kugelfischer PL104 fuel injection
Fuel pump: Mechanical
Valve gear: SOHC
Maximum power: 95.6kW @ 5800rpm
Maximum torque: 181Nm @ 4500rpm

TRANSMISSION
Type/locations: Four-speed ZF manual
Driving wheels: Rear
Clutch type: SDP

Gear	Direct Ratio
1st	3.764
2nd	2.020
3rd	1.320
4th	1.000
Final drive	3.640 to 1

SUSPENSION
Front suspension: Independent by McPherson struts with anti-roll bar
Rear suspension: Independent by semi-trailing arm
Shock absorbers: Direct-acting telescopi
Wheels: 5J x 1
Tyres: 165SR x 1

STEERING
Type: ZF worm and rolle
Turns lock to lock: 3.
Wheel diameter: 39cr
Turning circle: 10.4n

BRAKES
Front: Disc. Diameter: 24cr
Rear: Drum. Diameter: 25.6cr
Servo assistance: Vacuum standar

DIMENSIONS AND WEIGHT
Wheelbase: 250cr
Overall length: 423cr
width: 159cr
height: 141cr
Track, front: 135cr
front: 135cr
Ground clearance: 16cr
Kerb weight: 1010k

CAPACITIES AND EQUIPMENT
Fuel tank: 50 litre
Cooling system: 7.1 litre
Engine sump: 4.2 litre
Battery: 12V 44 A/
Alternator: 630

speed it is a delight — but whatever the speed, if you push it, you need to take care.

One of the most annoying aspects of the car's handling is the tendency to life the inside rear wheel on tight corners. I found I couldn't negotiate one of my favourite corners (an uphill super-tight lefthander) as fast in the Tii as I could in a 1900 Torana. It entered the corner at enormous speed, but at the apex picked up that wheel and spun it until it slowed considerably. A limited slip diff would certainly be a help in this sort of situation.

I can't remember a Torana being as quick in any other situation — although I can't remember it being as twitchy either. Again, I must point out that most of my criticisms are based on the fact that the car isn't as good as it *should* be (and probably is, overseas) — with the name it has and at its price — but it *is* still superior to any of our bread-and-butter sporties.

The Tii's suspension is independent all-round, with McPherson struts, lower wishbones, rubber auxiliary springs and a torsion bar stabiliser at the front, and a box-section beam

THE 02-series body design is neat, and still attractive, but it isn't really practical. Best part of the design is the incredibly-good vision afforded, and the size of the boot.

LEFT: Dash, wheel, and control layout is spot-on, but it's hard to get a good driving position. ABOVE: The slanted OHC four is almost covered by the Kugelfischer fuel injection plumbing. ABOVE RIGHT: Typical of BMW's attention to detail is the 'odds-and-sods' shelf, with rubber strips at regular intervals to prevent moving around on corners. That gearshift knob looks good, but is uncomfortable.

CALCULATED DATA
Power to weight: 10.5kg/kW
Piston speed at max. rpm: .... .992m/min
Specific power output: 48kW/litre

ODOMETER
Start: 1506km
Finish: 2497km

PERFORMANCE

FUEL CONSUMPTION

	Litres/100km	(MPG)
Average for test:	12.84	(22)
Best recorded	10.66	(26.5)
Possible best (under optimum conditions):	11.3-8.8	(25-32)

ACCELERATION
0-40km/h 2.8s
0-60km/h 4.6s
0-80km/h 7.8s
0-100km/h 11.3s
0-110km/h 13.6s
0-120km/h 15.4s
0-130km/h 17.3s

OVERTAKING TIMES (holding gears)

Km/h	2nd	3rd	4th
40-70	3.4s	5.3s	—
50-80	3.4s	4.8s	7.0s
60-100	4.5s	6.5s	9.5s
80-110	—	5.3s	8.1s
100-130	—	6.0s	8.3s

STANDING 400M
Average: 17.5s
Best Run: 17.3s

SPEEDS IN GEARS

	Max. km/h	rpm
1st	54	6400
2nd	90	6400
4th	140	6400
5th	180	6400

BRAKING

Three maximum stops from 50km/h:

Stop	G-force	Pedal pressure (kg)	Distance (M)
1	1.12	20	8.8
2	.95	22	10.3
3	1.26	20	7.8

Five maximum stops from 100km/h:

Stop	G-force	Pedal pressure (kg)	Distance (M)
1	.94	20	41.7
2	.96	19	40.8
3	.99	24	39.5
4	.96	22	40.8
5	.94	22	41.7

SPEEDO CORRECTIONS

Indicated:	50	60	80	100	110
Actual:	45	55	73	90	99.3

BMW 2002 Tii

supporting semi-trailing arms, coil springs, rubber auxiliary springs, and torsion-bar stabiliser at the rear.

Braking is one aspect of the Tii I can't fault — and it provides ample illustration, to local manufacturers, that disc/drum systems can be made a whole lot better than they are giving us.

Yes, that's right — it's a disc/drum braking system. You'd expect four-wheel discs from BMW on a high-performance car such as this, wouldn't you? But it makes little difference — the car stopped brilliantly time after time, achieving an incredible best from 50km/h of under eight metres. Remember, our Fifth Wheel figures give only the distance it takes *the brakes* to stop the car, and don't include reaction time — even if you add a *slow* reaction time to that you have an excellent stop. The best stop from 100km/h was a shade under 1.0g deceleration, taking 39.5m with 24kg pedal pressure.

The car showed little tendency to lock wheels — sure, you could lock them by stomping on the pedal, but in emergency-stop situations they react better than most.

The servo-assisted system applies pressure through a dual master-cylinder to front discs, increased in size (to 256mm diameter) over the standard 2002, and 200mm rear drums with self-centering shoes. The handbrake operates mechanically on the rear drums, and has individual adjusting nuts for each cable — these are reached by lifting the rubber sleeve on the handbrake lever.

The interiors of all BMWs, except for the special luxury versions, have always been pretty stark — and the Tii is no exception in basic form. Our test car had been optioned up, with wool seatcovers, air-conditioning, and stereo cassette-radio, but it was still pretty basic. The Germans are, after all, a practical people . . .

The seats are wrong, dead wrong, for tall people. They give absolutely no support to the top half of the back, the shoulders, nor the neck — and it becomes tiring very quickly. We removed the seat-covers almost immediately — simply because they filled in the recessed centre of the seat back, which was designed to give lateral location to driver and passenger. The result, when combined with inertia-reel seatbelts, was almost funny — both front seat occupants had to struggle desperately to stay in place on corners.

Of course, the Germans have given typical attention to detail — the controls are where they should be, the finish is excellent, and the quality undoubted. Even the boot was an example, with strong covered-board lining everywhere to keep the whole thing neat and in shape. But attention to small details doesn't help when a few major points are missed — like those seats, and the fact that there's not enough rear seat room for anyone over a metre and a half tall.

One of the major strong points of the now-dated design, however, still exists. You notice it immediately — the all-round vision is almost impossible to believe. Very slim pillars and a huge glass area allow excellent vision from every angle — and you can see every corner of the car clearly. It is probably the easiest car I've ever had to park, and likewise probably has the best vision of any car I've been in.

I enjoyed my time in the BMW 2002Tii — it was *fun*. I only complain because I don't believe its merits justify the near ten-grand asking price. When I tested it the basic Tii was selling for just over $8000 — and even at that price would only have been worth the money to a real BMW enthusiast.

It's unfortunate that expensive imports suffer the most from duty increases, but that's the way it is — and I suppose those who want a 2002Tii badly enough will still be glad to shell out all those bikkies — and be happy they did.

COST SCHEDULE

Make/Model: . . . . BMW 2002 Tii
Pricing (basic): $9862
Pricing (as tested): $11380
Options (prices):
Metallic paint $273.00
Air conditioning Hitachi $700.00
Radio/cassette "Clarion" . . . $365.00
Heat insulated glass $90.00
(now standard on new models)
Stripe $90.00
Registration: $121.55
Insurance Category: 4
Rates quoted below are for drivers over 25 with 60 percent no-claim bonus and where the car is under hire-purchase. This is the minimum premium level — decreasing rates of experience and lower age groups may have varying excesses and possible premium loadings.
Non-tariff: $342.87
Tariff: $329.00
N.R.M.A. $368.25
Warranty: 6 months/10,000km
Service:
Initial service is free. This covers the first 1000km and includes lubrication and maintenance.
Other Services:
Lubrication and maintenance services every 8,000km. Parts and labour is chargeable.
Spare Parts:
(recommended cost breakdown)
Disc Pads (set of four) $17.80
Muffler (front) $34.50
(rear) $37.00
Windscreen $145.00
Shock Absorbers: Front $29.50
Rear $20.85
Headlamp Assembly $31.75
Taillamp Assembly $46.00
Bumpers: front $79.75
rear $97.50
Front Guard $85.00

WE LIKED the boot, we did not like the front seats! The rearward curve of the top half of the seat means it gives no support to the upper half of the back.

USED CAR R&T CLASSIC

BMW 1600 & 2002, 1967-1976

They are great fun and appreciating quickly

BY THOS L. BRYANT

PHOTO BY SCOTT MALCOLM

IN THE 10-year span covered in this Used Car Classic, we probably devoted as many or more pages to the BMW 1600/2002 series as to any other model. And from the very beginning it was obvious that these are exceptional sport sedans. In the first road test of the 1600 (May 1967), we talked about the car's "excellent handling and stability," and concluded, "At the risk of becoming tiresome, let us say just once more that the BMW 1600 is a great automobile at the price" (then just a little more than $2600 on the west coast).

The 1600 was a pacesetter car for BMW, reviving the firm's flagging financial position and introducing significant numbers of Americans to the joys of the marque. The single-overhead-cam 4-cylinder engine displaced 1573 cc and was rated at 96 bhp at 5800 rpm, with 91 lb-ft of torque at 3000. Our original road test report described it as "one of the smoothest 4-cylinder engines we've encountered and there is just enough roar on acceleration to make it sound meaningful. It winds smoothly right through the rev range and is still pulling eagerly at the 6000-rpm redline." We were also impressed with the appearance of the engine, appreciating its aluminum cam cover, clean castings of the head and block and the obvious care taken in installation. Fuel delivery was by a single Solex downdraft carburetor.

Just a year later (May 1968), we published a road test of the 2002, which was essentially the same car but with a 2-liter (1990-cc) engine. The 2002 was meant as a replacement for the 1600 TI, the dual-carburetor, highly tuned version that boasted 118 bhp at 6200 rpm. It seemed clear that BMW foresaw that the TI model was going to have difficulty conforming to the coming U.S. emission standards, but that the 2-liter engine from the 2000 TI sedan would fill the bill. With its exterior dimensions being virtually the same as those of the 1600 engine, there were no installation problems.

The 2002 was somewhat disappointing to our road testers of the day who found it "Noisier and not much quicker than 1600 . . . usual BMW traits of quality, handling, ride . . . modern sports car performance for four passengers at a reasonable price." The 2002's performance was obviously affected by the air injection and carburetor/distributor changes needed for the fledgling emission standards, and 0-60 mph acceleration was accomplished in 11.3 seconds, compared to 11.6 sec for the previous year's 1600. We did point out, however, that the 1600 with similar emission controls probably would not be that quick. Nevertheless, the initial 2002 did not gain the editors' nod over the 1600, and the road test report ended: "Our conclusion, then, is that the 1600 remains the best value in the BMW line." This reflected the roughly $300 difference in price between the two at the time.

In October 1969 we revealed the results of an Owner Survey on the 1600 and 2002 (more about that report later). We didn't road test either model in 1970 (unbeknownst to us at the time 1970 was the last year in the U.S. for the 1600), but in 1971 there was much to report. In our August issue the BMW 2002 was named the Best Sedan, $2200-4000, in our list of the Best Cars in the World, 1971. The description of the car that appeared in that article is worth repeating: "The BMW 2002 is not just a sedan for transportation but a car for the enthusiast who needs space for four people. A family sports car, you might say. And while the 2002 is not big on carrying capacity, it will carry a family if need be and when it's not busy with such mundane chores, it provides the driver with great motoring. Its 2-liter engine is torquey and smooth if not quiet, its gearbox is a delight to operate, its suspension supple, yet competent, and its brakes reassuring. For a long time we've said that BMW has the best set of mechanicals in the world for a medium-priced sports car, sitting right here in this upright sedan."

By October of that year, BMW had introduced the 2002tii version into the U.S. and that called for an examination. The tii was the high-performance 2002, with Kugelfischer mechanical fuel injection instead of the single carburetor. The compression ratio was down to 9.0:1 compared to the 10.0:1 of the earlier 2002 models because of the toughening emission-control standards, but the tii did offer impressive horsepower and torque gains over the carbureted version: 140 bhp at 5800 versus 113 at 5800, and 145 lb-ft torque at 4500 compared to 116 at 3000. It was obvious, however, that the fuel-injected engine had to be wound rather tightly to get the most out of its capabilities.

The fuel delivery system was not the only change, as the tii also was fitted with different gearbox ratios for the standard 4-speed as well as having a 5-speed transmission available as an option. The 4-speed's ratios were slightly taller than those of the normal 2002, as was the final drive (3.45:1 versus 3.64), making the tii a car more readily suited to high-speed cruising. The 5-speed model came with a 4.11:1 final drive.

The other differences also played a part in giving the tii a more sporting character: larger brakes, 1.4-in. greater track width with wider wheels (5.0 in. instead of 4½ in.) and Michelin XAS radial tires with an H speed rating in place of the S-rated tires; and slightly beefier chassis components.

In our performance testing of the fuel-injected 2002, we found that it was not significantly quicker to 60 mph or through the standing-start quarter mile than the normal model. The 0-60 time for the tii was 9.8 sec and the quarter mile was covered in 17.3 sec at 78.5 mph—these times were 0.6 and 0.3 sec quicker, respectively, than for the carbureted 2002 of the day. The road testers reported that the tii used for that acceleration testing was not as thoroughly broken in as they would have liked, and mentioned too that there was "particular difficulty getting the 2002tii 'off the line' smoothly (BMW rear suspension always lets the wheels patter badly anyway) and when the wheels finally settled down the engine bogged."

The tii's handling garnered its share of praise in that road test, complementing high marks for the car's responsive engine, nice gearbox and good controls: "Then comes handling, and for a rather tall sedan the 2002tii is quite good. Its steering seems somewhat lighter than previous 2002s we've driven, despite the new car's greater weight; it's still not the lightest thing going but is very precise and quite quick enough. The tii comes with anti-roll bars front and rear, which keep body roll down to a moderate level, and the wider track, wider wheels and slightly stiffer tires all conspire to make the tii significantly better-handling than a 2002."

Our next encounter with the 2002 occurred in May 1972 when we compared the tii version with the Alfa Romeo 1750 Berlina and the newly introduced Mazda RX-2; a test designed to discover if the new Japanese sedan could compete with two of Europe's most successful sport sedans (we concluded it could). We were dismayed at the price increases that had shoved the small BMW sedan from a $2500 car when introduced as the 1600 in 1967 to $3803 for the 2002 and $4360 for the tii version. Nonetheless, the BMW continued as a favorite of ours, and it

Teutonic simplicity marks the interior design of the 1600/2002.

BRIEF SPECIFICATIONS

	1967 1600	1968 2002	1971 2002tii	1974 2002tii
Curb weight, lb	2050	2210	2310	2420
Wheelbase, in.	98.4	98.4	98.4	98.4
Track, f/r	52.4/52.4	52.4/52.4	53.8/53.8	53.8/53.8
Length	164.5	166.5	166.5	176.0
Width	62.6	62.6	62.6	62.6
Height	54.0	54.0	55.5	55.5
Engine type	sohc 4	sohc 4	sohc 4	sohc 4
Bore x stroke, in.	3.30 x 2.79	3.50 x 3.15	3.50 x 3.15	3.50 x 3.15
Displacement, cc	1573	1990	1990	1990
Horsepower @ rpm	96 @ 5800	113 @ 5800	140 @ 5800	125 @ 5500
Torque @ rpm	91 @ 3000	116 @ 3000	145 @ 4500	127 @ 4000

PERFORMANCE DATA
From Contemporary Tests

	1967 1600	1968 2002	1971 2002tii	1974 2002tii
0-60 mph, sec	11.6	11.3	9.8	9.5
Standing ¼ mi, sec	18.2	17.9	17.3	17.7
Avg fuel consumption, mpg	25	22-27	22.7	23.5
Road test date	5-67	5-68	10-71	6-74

TYPICAL ASKING PRICES*

1967-1970 1600	$1500-2500
1968-1970 2002	$2000-4000
1971-1973 2002	$2750-4500
1971-1973 2002tii	$3000-5000
1974-1976 2002	$4000-6000
1974-1976 2002tii	$4500-8000

*Prices are estimates based on cars that are in reasonably good condition but not restored to like-new. Cars in excellent condition will, of course, command higher prices, while poorly maintained cars will be cheaper than the range given here.

earned first-place votes in the comparison for handling, ride, gearbox, outward vision, body structure, interior and exterior styling, and overall finish, as well as being rated best overall "by a comfortable margin."

The 1972 emission controls were tough on imported cars because unleaded fuel became the order of the day: "The familiar and popular BMW was such a victim," we reported in a brief test in January 1973 when we examined the changes for the new year that would help alleviate the unpleasant side effects in driveability that had begun to plague the carburetor-equipped model. The 1972 car's compression ratio had fallen to 8.3:1 and the air pump for emission control was engendering backfiring on deceleration, exhaust back pressure, and extreme underhood temperatures.

To help correct these symptoms, BMW adapted its tri-spherical turbulence-inducing combustion chamber (used in the 6-cylinder BMW engines) to the 2002's engine—"The chamber apparently gives good emission characteristics, power output and fuel economy," we noted in the update. "Our test car ran with practically no symptoms of lean carburetion and went as well as the 2002 we tested for our June 1968 issue with a 10.5-sec 0–60 mph time and a 17.8-sec quarter mile. The 1968 car had done 10.4 and 17.6 respectively, and differences that small aren't significant," we concluded.

In June 1974 the 2002tii again figured in a comparison test—sports cars versus sports sedans—as it went up against the Jensen-Healey. The report noted that the 2002, in carburetor and fuel injection models, had received an extensive facelift that year. The grille was redesigned and new larger taillights were the immediately obvious alterations along with some improvements to the interior trim. Perhaps the most significant alteration to the 1974 2002, however, was the newly required bumpers to meet U.S. safety standards. Although they were cleanly designed and made of aluminum, they still added 9.5 in. to the car's overall length and more than 100 lb of weight—and BMW purists, of course, thought them unsightly, to say the least.

Our assessment of the 2002tii, however, was that its "character remains essentially the same: a practical, fairly roomy car for four people and their luggage with very sporting engine, suspension and brakes, exceptional solidity of construction and an unmistakably German precision about its controls. This is a car which tells you in no uncertain terms what it's doing but stops just short of making you uncomfortable in the process—a combination that pleases a lot of enthusiastic, fast drivers."

In its confrontation with the Jensen-Healey, the 2002tii got the nod in 0–60 mph acceleration (9.5 sec compared to 9.7) but the British sports car won the quarter mile with a time of 17.4 sec at 80.5 mph compared to the German sedan's 17.7 sec at 81.0. On the winding road portion of the test, the Jensen-Healey "does have the greater capabilities," we concluded, although "the BMW is still highly satisfying and sporty to drive..."

It's interesting to me that perhaps the best summary of the 2002 came in the form of a eulogy when we presented the initial road test of its successor, the 320i, in the December 1976 R&T: "Despite a base price that had risen to more than $6500, the 2002 was a brisk seller even in its final days. It was out of date in some important areas, particularly ventilation, but its reputation for reliability combined with ride, sporty performance and handling were hard to beat. It was the ideal car for the practical enthusiast who refused to give up the joy of driving simply because his automotive requirements dictated room for more than two people and generous trunk."

Buying a Used BMW 1600/2002

ONE OF our earliest Owner Survey reports featured the 1600 and 2002 (October 1969), and we received nearly 1000 responses to our Used Car Classic Questionnaire (September 1979) from BMW owners, so we have a considerable amount of background information on points to examine when choosing one of these cars.

First, from the Owner Survey report, there were two problem areas mentioned by more than 10 percent of the owners: clutch throwout bearing and emission-control-system ills. The clutch troubles were cited by 13 percent and usually required one or more replacements. The emission control complaints were voiced by 17 percent of the 1968–1969 model owners (the 1967 1600s had none) and generally dealt with poor driveability rather than actual failures, although the air pump's anti-backfire "gulp" valve did fail on some cars—this item was mentioned by a number of people who completed the Used Car Classic questionnaire too for models from 1968 through 1971.

Other problems listed in our Owner Survey included easily broken interior fittings—specifically window winders—mentioned by 10 percent, while 8 percent reported failed door latches, and 7 percent listed problems with three items: speedometers, wiper motors or gearbox (usually synchronizers). Mufflers were mentioned by 25 percent of the 1967 1600 owners but didn't make the 5-percent minimum figure for the 1968–1969 models; however, our Used Car Classic questionnaire results showed mufflers and exhaust system problems were mentioned for every year from 1967 through 1975. Overheating was mentioned by 22 percent of the 2002 owners in the Owner Survey but didn't seem to afflict the 1600 significantly, although in our 1979 questionnaire, overheating and radiator troubles were commented on by owners of 1600s and 2002s for each model year except 1973 and 1975–1976.

Other problems that surfaced in our review of the Used Car Classic questionnaires included front-end shimmy related to tie-rods (1967 through 1971), gearbox output shaft and flange wear (1969, 1972 through 1975, with 1973 models seeming to suffer most in this area), and cylinder head leaks resulting from overheating and warpage of the aluminum head (1970 through 1972 models). Carburetion problems and/or dissatisfaction surfaced most commonly on 1971 through 1973 models, and again on 1975–1976 2002s, probably reflecting driveability complaints related to emission standards.

In pointing out things to look for when buying a used 1600 or 2002, many owners also cited rust around the front turn signal housings, wheel wells, rocker panels, the rear shock towers and along the bottoms of the doors. Owners of early 1600s found the 6-volt electrical system a trial, and almost everyone was resigned to the gearbox synchros going bad, especially in 1st and 2nd.

It may seem that I've listed more problem areas for BMWs than is common with previous Used Car Classic articles, but I suspect this is actually a case of simply having more input from more owners than ever before. Most BMW drivers wrote glowing comments about the joys of their 1600 or 2002. Even though the 2002 has not been out of production for very long, it has already begun to assume cult-car status, as many BMW fans find the newer models are not perhaps as sporting, and because many

The BMW 2-liter engine is impressive in looks as well as performance.

enthusiasts have simply been priced out of the market for a new model. As one New York owner wrote on his questionnaire, "Please don't buy these cars! I need a lifetime supply and can't afford the new BMW junk." Or, in the words of an Orlando, Florida owner: "Being an avid car enthusiast, I have owned and driven many diverse specimens and still consider the 2002 to be the greatest all-around car *ever*, and worthy of any necessary expense."

That brings up a final point about buying a used BMW 1600 or 2002. The price of parts is high and getting higher the longer the cars are out of production. Also, BMW specialists generally charge relatively high labor prices for their expertise. So the potential buyer should keep in mind that service, parts and repairs will generally not come cheaply, and factor that into any budget plans before purchase.

Driving Impressions

DAVID ANDERSON is a Los Angeles attorney who is a BMW *aficionado* of the first order. He drives a 1972 2002tii that he bought new and on which he has rolled up 128,000 miles. He also owned a 1968 1600 that he bought used in 1969 and kept and enjoyed for 10 years—clearly not a man given to changing cars willy-nilly. His 2002tii is what I think a Used Car Classic example should be: clean, well cared for, mechanically excellent and, most importantly, used daily for driving pleasure.

Anderson admits to driving his BMW quite hard; it's his belief that's what cars of this sort are designed for, and the best way to enjoy them. His engine developed worn rocker shafts at 87,000 miles, so David had it rebuilt, blueprinted and balanced by Hyde Park BMW in Los Angeles just over 40,000 miles ago. Along with Koni shocks and adjustable anti-roll bars, at 105,000 miles on the odometer he invested in a new set of springs and was impressed with the restoration of the supple ride and handling characteristics for which the car is famous. He has also been running different distributor advance springs (either BMW 1600 or Corvette distributor springs, he says) and finds this modification improves performance and fuel mileage both. One of the first things he did was replace the radiator core with an American-made one to give greater cooling capacity and prevent chronic overheating for which 1600s and 2002s are famous. He has also put in a Recaro seat at the driver's position, and runs Phoenix Stahlflex 205/60-13 tires on 6.0-in. alloy rims. Anderson feels the wider wheels and tires give his car that extra bite in cornering and braking, and after driving the car I heartily agree. David also tipped me off that there is an aluminum disc in the front shock towers of 1974–1976 2002s to raise the front end in compliance with U.S. bumper height rules. Removing those discs, Anderson says, results in better handling.

Anderson's 2002tii is an exciting car to drive. It has the taut feel of a well made, well maintained car, and the engine and chassis are both first-class in response. The engine revs freely right up to (and through) the 6500-rpm redline without missing a beat. The acceleration is not blindingly quick but the 2002tii is no slouch getting up to speed. David's car, like many if not most 1600s and 2002s, suffers from bad gearbox synchros, especially noticeable in making the 1-2 shift. But, as he says, there's really no known cure for this that will last any length of time. You can redo them, but it doesn't take very long for them to go bad again, so you simply learn to live with this condition. Other than the synchros, however, the gearbox is a delight to use, and has a strong, positive feel.

The handling is stimulating and made me want to press on with more and more daring. The 2002tii is a car that feels comfortable right from the start and I didn't have any sensation of needing time to get used to any idiosyncrasies beyond the characteristic oversteer when the car nears its limits. David's wheel/tire combination seems a perfect choice to me, as driving at speed down a winding lane with a touch of water and sand here and there was a lark, with the speed limited only by body roll (not inordinate) and common sense.

David Anderson's philosophy about his 2002tii is that it's a car built to offer maximum driving pleasure. He believes that it's a car that requires the finest maintenance and he's quite fastidious about it, both mechanically and cosmetically. He admits that it isn't an inexpensive car to keep in tip-top shape, but it gives him so much driving fun that it's worth it. Also he can't think of a new car that offers near the pleasure he derives from his 2002tii. That's understandable—it is an exceptional automobile.

PHOTO BY DOROTHY CLENDENIN

Ventilated vinyl was all the rage in the early seventies

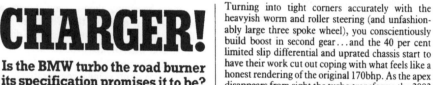
A few of the early turbos had '2002' and 'turbo' in mirror writing on the front spoiler, later deleted

The special Mahle wheels are to BMW specification

CHARGER!

Is the BMW turbo the road burner its specification promises it to be? Jeremy Walton puts his courage to the test

Typically, the photographer had selected the nastiest, hilliest and iciest bend to be found in the December Kent countryside. Trustingly, L&C Managing Director Tim Hignett had handed over the polished white 2002 turbo with an encouraging "Don't worry about the smoke on the over-run. Drive it as hard as you like, it's done over 50,000 but we have yet to replace a major mechanical component within the engine, gearbox, or rear axle. I'll take full responsibility if it blows up."

Back on that Kentish hillside there was plenty of smoke, but 50 per cent of it came from the hard-pressed front disc brakes. The carefully renovated 2002 went on to romp through a 100 mile assessment with more speed and better handling than many much-vaunted modern sports saloons.

For the 2002 turbo is still a fast car, even though it was announced ten years ago and rests within a body and suspension system from the sixties. Without the benefit of current electronic ignition systems and a large intercooler, you have to develop a sensitive 'turbo toe' to get the best out of that remarkable 2-litre motor. Careless full throttle applications will set the boost needle firmly in the red section and the engine pinking, especially if the car's being asked to fly uphill sideways in second gear! Ease the throttle slightly, the pinking disappears and the 2002 gobbles up the ensuing straight just as quickly as would an 1800 Golf GTi, but with a lot more spectacle when it comes to slowing down...

The brakes, a mixed disc/drum system rather than a thoroughbred four wheel disc layout, actually work superbly and repeatedly. Yet the driver will be very busy, even with the four speed gearbox, given some serious low gear corners. The chances are that top will have to be exchanged for second (which provided an indicated 65mph at the 6400rpm limit) on the way into a corner that would normally call for third in a conventional hot saloon. You *need* second because the turbo doesn't provide real boost until 4000rpm, and because you don't drive a car like this without seeking some excitement in life! The turbo provides it all.

Turning into tight corners accurately with the heavyish worm and roller steering (and unfashionably large three spoke wheel), you conscientiously build boost in second gear... and the 40 per cent limited slip differential and uprated chassis start to have their work cut out coping with what feels like a honest rendering of the original 170bhp. As the apex disappears from sight the turbo transforms the 2002 into a sliding simile of a rally car, slithering through a turn on opposite lock with exhilarating balance.

Stir in the occasional patch of ice and you can imagine that the tail-happy 2002 provided some amazing motoring during that 100 miles. They were all memorable miles, for the turbo also sat through traffic jams with 1100rpm sewing machine grace and covered some indecent motorway miles, when I discovered I had been so wrapped up in the pleasure of swooping through country roads that a good 30 miles and five minutes stood between me and the appointed return time!

The turbo is particularly suited to motorway cruising of course, the aerodynamic aids apparently squashing the car to the road, such is its stability. This is true whether you are travelling at an amiable 4000rpm and 85mph, a lustier 5000rpm and slightly over 100mph, or a full blooded 6000rpm and an indicated 120mph. It is noisy and harsh by modern standards – my XR3i sounded *quiet* on the way home and appeared to have a smooth ride by comparison!

150mph speedo

Contributing to the fun is the clean cockpit. LHD only was the ex-factory offering and the layout is largely 2002, with the frivolity of a red fascia and a central instrument pod for the boost gauge and a clock. In front of the driver are the ergonomically efficient pull switches, a 150mph speedo (240km/h in German market models) and an 8000rpm tachometer with that redline before 6500. This provided an indicated 35mph in first, 65mph in second and a solid 100mph in third at only 6000rpm. The white, green and red sector boost gauge of the test turbo lacked the original German market markings in atmospheres.

The Rentrop sports seats were exceptionally comfortable and efficient, while the German-labelled heater controls, and lack of ventilation eyeballs, reminded one instantly of the substantial progress BMW made in cabin comfort in the step from 2002 to 3-series. Even on a December day you can work up a lather in a turbo 2002! And enjoy every minute of it...

Tim Hignett has been a BMW dealer for 14 years and has restricted his choice of collectable examples of the marque to models of that period. Thus the S-registered turbo shares accommodation with the most original British-based CSL coupé I have seen (with plenty of odd features, such as electric rear windows only, confirming that BMW fell into much the same homologation-run bodges as their rivals), a superb 36,000 mile 2500 saloon and an M1 that has spent some time at Beaulieu.

This turbo was bought in secondhand about three years ago for £2500. Mechanically it needed little work – a Garrett turbocharger had been fitted, so that was abandoned in favour of the original KKK – but the body needed to be taken down to basics and was "sprayed from the floorpan upward." Most people are fooled by the rare Mahle 6J × 13 alloy wheels (steel 5½J were a more common sight) but they are to an original BMW specification. They wear the correct Michelin 185/70 XWX rubber. All but one of the wheels came via BMW Italia.

The body wears some new glass-fibre parts, such as the original shape front spoiler. The screw-on spats add a real air of seventies aggression. However, the controversial 2002 and turbo badges (that were reversed out in mirror writing for the front spoiler and caused so much concern in fuel-crisis-stricken Germany that they were deleted) were not present on this car either.

At the rear is a black soft rubber spoiler and the front item wears one cooling duct. The first Munich-registered examples did without front ducts completely but one duct was worn for the later press handout pictures. These turbos had the same spoiler painted white with both BMW central badge and turbo lettering to the right, whereas this British-registered machine carries the turbo badge on the boot panel with the usual turbo side-stripes. Since the spoiler striping was different from shot to shot in the original brochure, concours judges could find that any of three front spoiler layouts are actually as per original factory intentions!

Tim Hignett has restored a mechanically sound machine to eminently usable everyday condition, within a body that was notable for its clean lines before BMW went a little over the top on the turbo decoration. When they get the time L&C will probably take the engine apart to attend to the valve seals to cure the over-run smoke, but that work, plus the acquisiton of a five speed gearbox (originally offered optionally) may well have to wait behind the purchase of another favourite BMW. "I'd really like an M535i," says this reformed agriculture graduate, who set out for a life far away from the motor trade, "but I'll wait until they become unfashionable and unwanted to get one at the right price," concluded Tim Hignett shrewdly.

CLASSIC CAR

BMW 2002

In the era of Warwick Yellow Monaros and Candy Apple Red Falcon GTs there was a small two-door sedan that took a different road to high performance. John Wright tells the tale . . .

You would look in vain for a speed stripe. There were no fancy hubcaps. The basic styling was functional, unostentatious and certainly unadorned by any badge that read GT. The interior was well trimmed in a spartan style but with reasonable room for four adults.

About the only clues to the nature of the machine were the tachometer calibrated to 8000rpm with no redline and a faintly sports-inspired three-spoke steering wheel.

What a delightful piece of teutonic understatement! What a machine to give substance to the cliche, "wolf in sheep's clothing." This Cortina-sized two-door sedan-coupe was too rakish a concept to encapsulate those no-nonsense lines — was more of a Q-car than the Cortina GT or even the giantkilling Mini Cooper S had ever been.

In the early '60s music was changing. New rhythms, new beats, a whole new style emerged as Mersey moved into Beatlemania. And in the second half of the '60s the automotive world experienced a similar revolution. In almost the way one explosive Beatles hit succeeded another, hot new cars seized our attention. Some were brightly coloured and awash with stripes and badges, while others took the Q-car route.

With the '70s a mere flick of the calendar away, the very notion of high performance was changing. Only five years earlier we had thought the EH 179 was the Real Thing. Even the Cooper S was suddenly desperately outclassed, and one of the key players in the new performance was a small two-door German sedan, the BMW 2002.

Introduced in Europe in 1968, the 2002 appeared on Australian new-car lists within a few months. Now let's just imagine for a moment that Ford Australia had not got around to building its Falcon GT and that the General hadn't brought his Monaro to market: in early 1969 that diminutive Bavarian machine would then have been just about the hottest property on the downside of an E-Type Jaguar. To find a 17.3 second standing quarter mile and zero-to-60 miles per hour in 9.5 seconds you wouldn't even have contemplated a 2.0 litre four-cylinder car, no matter what its badge read. And how about 93 mph (149km/h) in third gear?

All of which raises another matter; in early 1969 BMW was hardly the prestigious presence in this country (or anywhere else in the known automotive world) that it is two decades later. There was always the chance you could boast that you'd bought a BMW and people would say, "A what?". And who would have guessed that this slab-sided plain looking number would provide the basis for the world's first production turbocharged car?

Naturally, the 2002 couldn't live with the high-performance V8 heavy metal when it came to the clicking of stopwatches, but it was nevertheless a most evocative way to travel. And where the Monaros and Falcon GTs were becoming almost commonplace as the '60s moved over for the '70s there was still no other car in the world that bore direct comparison with the BMW 2002.

While BMW cannot be credited with inventing the idea of dropping a comparatively big engine into a lightweight body — the Americans and Jaguar had been doing it since the '50s — the 2002 was something different. What was new was the remarkable combination of characteristics in so outwardly bland an offering.

This car performed near as dammit as well as a Lotus-Cortina, but was far more sophisticated and much quieter. It could be used as a comfortable touring car in the same kind of manner as, say, a Rover 2000 but had a reserve of power beyond the Rover's sweetest dreams and probably more luggage room as well.

Despite putting its modest engine capacity to impressively good use, the 2002 remained a quiet and refined machine. It could be used as a docile town car with never a hint of temperament or a blurt of exhaust to distract or irritate occupants who didn't realise they were inhabiting a real sports sedan.

BMW was arguably the first manufacturer to combine so many motoring purposes in the one machine: sports sedan, grand tourer, economy runabout and family hack. In the last of these roles, a couple of extra doors would have counted in its favour but this was about the only area of compromise.

Consider, too, that this brilliant result was achieved by mixing and matching from within its own model range. Outside experts need not apply, danke.

CLASSIC CAR

By placing the engine from the middleweight 2000 sedan (BMW's rival to the Mercedes 250, the Rover 2000 and the Triumph 2000) into the lightweight two-door 1600 body, the company found a quick and easy way into the fast lane.

For the most part the mechanical specification was straightforward: front engine, rear drive, four-speed gearbox, front discs/rear drums. The front end was a well arranged MacPherson strut system with lots of travel built into the coil springs and an anti-roll bar. At the rear, BMW employed its by then already familiar semi-trailing arm/coil spring independent configuration, while fitting an anti-roll bar which wasn't common practice in the late '60s.

The suspension was on the firm side, which made for somewhat bouncy progress over corrugations but did temper the familiar BMW tendency to oversteer at the limit. Yes, the 2002 could be provoked to oversteer, but the transition from mild understeer through neutrality to final oversteer was reasonably progressive; far more so than in the softer

BMW 2002 got its performance from a large motor in a lightweight body.

Three Series cars which eventually superseded the 2002, at least in the marketplace if not the hearts of dedicated drivers. The bigger 2500 and 2800 models were also more wayward in their behaviour at the limit, requiring quick responses when the outside rear wheel began to tuck under.

Looking back 20 years we might find it odd that neither four-wheel-disc brakes nor a five-speed gearbox were included in the specification. After all, the Alfa GTV and even the upright little Giulia Super provided both. But to drive a 2002 is to remember how effective a four-speed gearbox could be when the ratios were appropriately spaced and the final drive was also right.

Such was the nature of BMW's now-legendary chain-driven single-overhead-cam four that four speeds were entirely sufficient, sacriligious as that claim sounds. (Later a five-speed box was offered as an option on the injected 2002 Tii but most testers preferred the four-speed.)

Off the line, 3000rpm on the tacho and a dropped clutch took you almost immediately to 55km/h. Into second and you'd passed the 80km/h mark in 6.8 seconds. We were non-metric in the Australia of 1969 and zero to 60 miles per hour (97km/h) was still a favoured yardstick. The 2002 was one of the few cars that could achieve this feat with only one gearchange and one of even fewer to accomplish it in less than 10 seconds.

Among the front-running four-cylinder machines of the era were the Fiat 124 Sport and the 125. Despite twin-overhead camshafts, neither could come close to the 2002. An even second after the 125 hit 50mph, the BMW was doing 60, and the 186S-powered Holden Monaro GTS was a full two seconds slower through the standing quarter!

The secret of the performance was, of course, the power-to-weight ratio. On the scales the 2002 stopped the needle at a trim 940kg, which was some 100kg lighter than the Alfa Giulia Super which also gave away 400cc. That's why BMW didn't need to fit dual carburettors or an elaborate exhaust manifold to its 2-litre engine to endow the 2002 with breathtakling performance, by the standards of the late '60s. That's also why the 2002 has always responded so brilliantly to conventional tuning techniques; dual Webers, a reworked cylinder head and a hotter cam turn it into a serious hotshot.

Later, the factory produced a twin Solex-equipped version known as the Ti, which in turn gave way to the injected Tii. Both are rare in this country with the former virtually non-existent.

> "Like all good engines, the 2002's was notable not just for what it could do but the manner in which it performed."

The standard 2002's maximum speed of about 172km/h was less impressive than its sprint to get there. BMW was hardly at the forefront of the race for superior aerodynamics when this body was designed. But that strong acceleration only began to taper off at 160km/h, meaning that all the performace was readily usable. An all-day jaunt down the autobahnen was easy work at 160, at which speed the engine was turning over at just over 5300rpm.

Like all good engines, the 2002's was notable not just for what it could do but the manner in which it performed. Back in the late '60s 2 litres was considered large for a

four, but this single overhead cam unit was superbly smooth and fuss-free.

Generally, the dynamics were excellent. Over broken surfaces, the 2002 displayed poise and its long travel suspension made it ideal for goat tracks such as our Highway One, particularly as it was back then. On flat, twisty bitumen, its limits were high indeed and through the tighter corners there was enough power in second gear to induce a tail-out attitude at will.

Driving a 2002 was always fun. Despite its sedate demeanour it was a car that begged to be thrown around. It felt solid, as if carved from a rock, and the engine thrived on having its neck wrung. And yet, used gently, it was delightful in its torquey effortlessness — the most satisfying spinoff derived from building a high-performance car by dialling in extra cubes.

Because the engine seldom had to work to propel the 2002 at respectable velocities, outstanding fuel economy was another welcome attribute, and the figures were seldom worse than Uncle Arthur was recording in his Hillman Hunter.

The weakest link in the 2002's dynamic makeup was the recirculating-ball steering. Turn-in was quite good, but there was less feel here than in anything from Alfa Romeo or even — dare I say it? — that old British battleship, the MGB. At high cruising speeds, a certain sensitivity to crosswinds combined with lost motion in the steering to produce wandering. We look back from the lofty heights of 1989 and wonder why BMW's engineers hadn't simply specified a rack and pinion system.

As for the brakes, they were well up to the car's performance and delightful to use, thanks to a most progressive pedal. The only flaw here — despite the rear drums — was one that is still common to many high-performance cars; thanks to a tight suspension and the consequent reduction in weight transfer, it was rather too easy to lock the front wheels during serious braking.

In 1989 those BMW initials carry a formidable status, but even now we don't automatically associate the badge with luxury. This is one area where the 2002 had no pretensions, fitting squarely into the German idiom of combining high-quality fittings with a lack of warmth. No, you fools, I don't mean that there wasn't a heater — in fact, it was a unit that could roast your toes in short order. I mean that the prevailing tone was austerity.

The seats were, well, seats. They certainly didn't err on the side of softness and the vinyl in which they were trimmed looked (and felt) as if it would last forever. There was no armrest built into the rear bench, which was a surprising omission in a car that competed on price with the 2 litre English luxury models and the local heavy weight Fairlane 500. The carpet, too, was of good quality but would probably have served almost equally well as fabric for outdoor use.

Access to the rear seat was reasonably good for a two-door car. Ventilation was primitive with no face vents on the dash, while the quarter vent windows produced a predictable increase in wind noise at higher speeds and were fiddly to operate.

In summary, the 2002 would not have impressed buyers looking for the most features and flash for their money; it required driving to reveal its virtues, preferably hard driving.

It's perhaps not stretching plausibility too far to compare the Datsun 1600 with the BMW 2002. Of course the Bavarian machine went harder, handled better, used better quality materials and carried more status, but it was also hugely more expensive. Nissan got its plot surprisingly right with the 1600 — as if its engineers had scored a sneak preview of what their counterparts at BMW were doing.

Now let's extend the analogy. Having got it so nearly right with the 1600, the Nissan designers lost their way with its successor, the 180B, which was fatter, glitzier, less nimble and far less charasmatic. The judgement might seem harsh, but the Three Series BMWs in their early guises (320i and 320 automatic for Aussie consumption) bore the same kind of relationship to the 2002 they superseded as did the 180B to the old 1600.

What the BMW designers had achieved with their now classic 2002 was a masterpiece of understatment, a car that did nearly everything excellently and absolutely everything more effectively than its looks implied it could. It was a Q car; a machine whose very blandness of appearance became source of delight and inspiration to those who knew what fine engineering lay beneath.

Unlike its successors, the 2002 never pretended to be a luxury car. It was a sports sedan, end of story.

Of course, the Three Series cars got better with the passing years, but the emphasis gradually shifted towards the six-cylinder

323i while the 318i languished as an under-powered price-leader to the increasingly comprehensive BMW lineup.

This story does have a happy ending. After disappearing from the Australian market for a couple of years, the 318i was reintroduced late in 1988. Suddenly it looked cheap at $36,000, the handling had been sorted out some years before, and finally the performance too, was right.

Driving the latest 318i enthusiastically, I am put in mind of the old 2002. The return to the original philosophy of a lusty four-cylinder engine in a compact body has been a long time coming!

CAR

PRACTICAL CLASSICS BUYING FEATURE

BARGAIN BMW

Mention BMW to most people and it conjures up images of fast, expensive and luxurious sports saloons for which spare parts and maintenance are very costly. However, while this is generally true there is one model group which provides a notable exception to the rule – the 2002s. These cars were the forerunners of the 'modern' 3-series vehicles and, more importantly, they played a crucial role in establishing BMW on the lucrative American market.

The 2002 series was given a face-lift in 1973 and, as part of this, a black plastic grille became standard as seen on this 2002 Touring.

The first of the BMW 2002 models appeared on sale in 1971 and was powered by an ohc 1990cc four-cylinder engine with single carburettor which produced 100bhp at 5500rpm. It followed in the footsteps of the 1600cc version and, like this earlier model, featured MacPherson strut and coil suspension at the front with a semi-trailing arm and coil independent set-up at the rear. In addition they boasted front and rear anti-roll bars to tighten up the handling. Stopping power was provided by discs at the front and drums at the rear.

An alternative at this time to the ordinary 2002 was the 2002TI but this was officially only available in Europe. Its engine enjoyed the benefit of twin Solex carburettors and a higher compression ratio. In 1971 the more generally available 2002Tii (Chassis No. 2750001 onwards) was born which was pow-

Chris Graham discovers that the BMW 2002 can provide a surprisingly cheap, distinctive and practical entry into the classic car scene

ered by the 1990cc 'E121' engine with the added advantage of a Kugelfischer mechanical fuel injection system. This boosted its power output to an impressive 130bhp at 5800rpm. Added to the four-speed manual and three-speed automatic gearbox alternatives was a five-speed manual option. However, this is very rare today and worth its weight in gold.

Later that same year the Touring style body was released with its distinctive hatchback design but, despite being available with three engine options (1600, 1800 or 2000), it did not prove to be a great success with the buying public. The model range then remained the same until 1973 when several important changes took place. First the '2002 Touring'-badged car (Chassis No. 3441102 onwards) was introduced (previously it was known as the 2000 Touring) and this was closely followed by a Targa variant. Then, later in that year, all models got a face-lift and the engine was changed to the 'E12' version (Chassis No. 2770006). This was a very important turning point because components from the E121 and E12 engines are not interchangeable. Therefore it is vital that you know which you are dealing with.

The visible changes to the face-lift cars included black plastic grilles, a rear light cluster change from round to square, simulated wood facias, improved instrumentation, front head rests and a four-spoke steering wheel. Also at this stage BMW introduced the Lux option which was basically a trim package. It was available on all models apart from the Touring versions (because of their folding rear seats) and consisted of improved seating, wooden door cappings, glove pockets in the doors and a rear arm rest.

The final two options were the cabriolet (converted for BMW by Baur) and the very sporty turbo (only available in left-hand-drive) which became available in 1973. For this model BMW added a turbocharger to the fuel-injected 2-litre engine and managed to produce 170bhp at 5800rpm!

Rust problems in the boot are common. Mud traps behind the rear wheels lead to the floor rotting out in the rear corners.

Another problem area can be the section of rear inner wing inside the boot which supports the suspension top mounting. A common trick is to cover up any corrosion with carpet so beware!

Production of the 2002 continued until 1975 when the fuel crisis started to bite. Unfortunately it nipped the turbo and Tii models in the bud and sent a shock wave through BMW. They reacted by replacing the complete 2002 series with the bottom-of-the-range 1502 that was powered by a 'cooking' 1573cc unit.

What you'll find

To get the inside story on the 2002 range I visited independent BMW specialists ETA Engineering (Shorne Service Station, Shorne Crossroads, Gravesend Road, Shorne, Kent, Tel: 0474 822810). Geoff Weeks and Eddie Wynne, who run ETA, are genuine 2002 enthusiasts who both run examples on the road and track. Their first point to me was that the 2002 range of cars is structurally very good. They all last well and rarely fail MoTs on corrosion related problems. In most cases any rust found will be largely cosmetic – front wings, door skins, rear wheelarches etc.

However, if you are looking at a car with a view to purchase it is certainly worth making a thorough inspection nevertheless. You should pay particular attention to the following points.

To begin with you can find problems with the front panel. If you lift the bonnet and inspect the area between the radiator mounting panel and the front outer panel (behind the headlamps) you may well find trouble. Moving down you may also discover that the front valance is rusted. However, putting such problems right is not terribly difficult. While repair skins are available for the front panel, Geoff and Eddie recommend the replacement of the whole thing which, with the cost of a factory replacement being about £130, really is quite a practical proposition.

The front wings are bolted into place and so any corrosion in this region is relatively simple to deal with. Although glassfibre replacements are available for about £30 I am sure that most enthusiasts would prefer to opt for the genuine steel replacement at only about £46 a piece! Geoff and Eddie tell me that you will rarely find any corrosion affecting the inner wings.

Another point to watch for at this end of the car is a rusty bulkhead. Although this is a fairly rare condition the cause is water lying in the heater collector box just ahead of the windscreen. This box, of course, does have drain holes but they inevitably become blocked and so the water becomes trapped. It is definitely something to check for if you are looking at a car which has been off the road for a long time and stored outside.

Bonnets rust around their edges and are worth inspecting carefully because genuine replacements cost about £140. The doors can suffer from leaky weather strips and blocked drain holes which leads to rotten skins. Putting this damage right is not too expensive though, with pattern skins costing about £25 each and repair sections starting at £7.50.

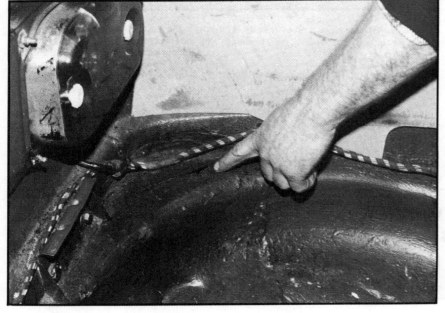

Interiors in the 2002s are plain but functional. PVC seat covers were the standard fitting but cloth was an option. Leather, as in this case, was made an option on the late cars and is very rare. The Tii models feature a 140mph speedometer instead of 120mph one used on the basic 2002.

Fortunately, however, in all but the worst cases, any corrosion is likely to be confined to the skin and the door frame itself should be intact. This is good news because replacements cost about £130. Inside there should be few corrosion problems and Geoff and Eddie told me that the floorpans hardly ever rust.

Turning to the sills, the main sections do not usually pose problems. However, at the back the sill disappears behind the front of the rear wing and there are three sections of panelwork which come together here. This is a recipe for disaster and corrosion will often be evident. Putting this to rights is relatively cheap under normal circumstances but, if you are dealing with a cabriolet model, the job will be a good deal more complicated and expensive. These cars featured special strengthening membranes within the sills which must be replaced correctly and cost more than £100 each.

At the back of the car it is the boot area which is likely to cause the most trouble. Condensation within the boot and mud traps on either side of the floor lead to corrosion. On the offside there is a mud trap between the wing and the fuel tank and the situation is made worse by a stiffener bracket for the boot floor also located here. On the nearside it is the spare wheel well which provides the problem and, in both cases, rust develops and eats out through the rear wing and up through the boot floor in the corners. Fortunately new boot floors are available from the factory for about £60.

The inner rear arches, where the shock absorber top mounts are located, can cause problems and need to be checked carefully. These are visible from inside the boot but be particularly suspicious if you find them covered in carpet. They were never treated like this at the factory and so any such addition is likely to be hiding important corrosion. A repair section is available for this area and costs about £15.

Mechanics and interior

Assuming that the car under inspection is capable of being driven then you should go through all the usual checks. Make sure that the oil light extinguishes quickly (there is no pressure gauge) and listen for any rattles when the engine is cold. Any knocking is likely to be piston slap which will often disappear once the engine has warmed. Also be on your guard for timing chain rattle and clattering valve gear which is a symptom of poor maintenance. If you get a chance to drive the car at motorway speeds check the rear view mirror as you release the accelerator pedal and watch for tell-tail blue smoke from the exhaust. This can point towards valve stem seal wear which is quite serious. Putting it right involves removing and stripping the head and this is time-consuming work.

When driving make a point of listening for gearbox noise. The rear mainshaft flange has a tendency to become loosened on the earlier models and this leads to the car jumping out of second gear. The prop shaft centre bearing can also wear and cause vibration as the drive is taken up.

If you are looking at a Tii and notice that the engine is running roughly and not idling properly, then this may well indicate that the complicated linkage between the throttle valve and the mechanical injection pump is worn. There are lots of joints involved which all become loose with time and make the correct idling and mixture settings very hard to achieve. A replacement set of linkages costs about £100 but Geoff and Eddie say that it's well worth it to save the endless fiddling.

Generally speaking though, the 2002's engine is pretty 'DIY-friendly' and it should not pose too many problems to the experienced enthusiast. The turbos, of course, require rather more specialised treatment for which you would be wise to consult the experts. Engine parts, on the whole, are reasonably priced but the key to engine longevity is regular maintenance with par-

Pre-1973 cars had round rear light clusters like this. These models are desirable today. If you have them take care because replacements from BMW cost just over £80 each! Most badges and brightwork are available from specialist sources so don't be too concerned if you find bits missing. If you find damp in the footwells it is likely that the rear window sealing rubber is perished and leaking.

ticular emphasis on oil and filter changes. The engines are strong and trustworthy if they are looked after properly. To illustrate this Geoff told me of a 2002 Touring he owned which covered an impressive 190,000 miles with just one head overhaul during that time!

You should encounter few problems with the brakes or steering systems. However, be prepared for cars which have been sitting around for ages to be suffering from seized calipers. These are expensive to replace but service kits are available and represent a much cheaper alternative. Discs are about £14 each and so having to replace these is not a problem. The biggest trouble with the rear brakes concerns their adjustment. The adjusters seize up and the front one, in particular, is awkward to reach because it is tucked away behind the radius arm. You may also notice that the handbrake is not all that it might be. The lever sometimes pulls away from its mountings and cracks the floor. In most cases you will probably find the whole assembly pretty sloppy (it pivots on a single pin) so welding in a new bracket and repairing the floor is usually the simplest answer. The cables have a habit of seizing and replacements cost about £10.

BMW certainly did not go mad with the 2002 interior but, although it's all pretty basic, everything usually works and lasts well. The instruments are generally good and usually only suffer from niggling little problems like flickering illumination caused by bad earthing.

As far as the trim is concerned the only parts that wear noticeably are the cloth seats. The vinyl and leather versions are much tougher and more durable. Replacing cloth covers is not a problem inasmuch as the original material is still available from the specialists but, if you are unable to tackle the job yourself, the cost can be high if the services of a professional trimmer are required.

What to pay

It's still true to say that the BMW 2002 models, apart from the cabriolet and very rare turbo versions, are still amazingly cheap to buy. At the bottom end of the market restoration projects and general 'basket cases' can be found for as little as a few hundred pounds. Moving up the scale a little to cars which are in a reasonable condition but require some work increases the asking prices to anything between £500 and £1,000. Basically sound examples which are original and have a history seem, at the moment, to be changing hands at about £2,500, while ordinary saloons in class 1 condition are fetching prices in the region of £4,000.

The rather more specialised models are a different kettle of fish. A cabriolet, if you can find one, will cost about £3,500 in unrestored condition and you will need to part with £6,000+ for something smart. Most cabriolets seem to have been looked after well so it is unlikely that you will find one that has been allowed to rot away. The turbo models appear to be somewhat of a law unto themselves. Even though they were all made to left-hand-drive specifications they still command at least £4,500 in fair condition and up to £15,000 for top quality examples.

Specifications

	0-60mph	0-100mph	Max mph	Mpg	No. Prod.
2002	13secs	41.5secs	108	25	348,988
2002TI	9.5secs	33secs	115	25	16,488
2002Tii	9.3secs	31secs	116	30	38,703
2002 turbo	7.9secs	22secs	128	23	1,672

Rust like this at the front is common. All panels are available.

Conclusions

There can be no question that the BMW 2002 range of cars currently represent good value for money. They are very affordable to buy and restoration, with genuine factory parts, can be a surprisingly cheap operation. In terms of pure desirability the general concensus of opinion is that the Tii and Touring models are the ones currently to go for but, in truth, all offer a great deal of car for the money.

Another benefit of the present low prices is that it scares off the speculative restorer whose only intention is to 'tart-up' the car and sell it on for a quick profit. This is good news because it means that most of the vehicles now available are fairly original and should provide excellent restoration projects for those genuinely interested. □

The 1990cc ohc engine used for the 2002 should be capable of well over 100,000 miles if it is properly maintained.

The writer wishes to thank Geoff Weeks and Eddie Wynne of ETA Engineering for their help with this feature.

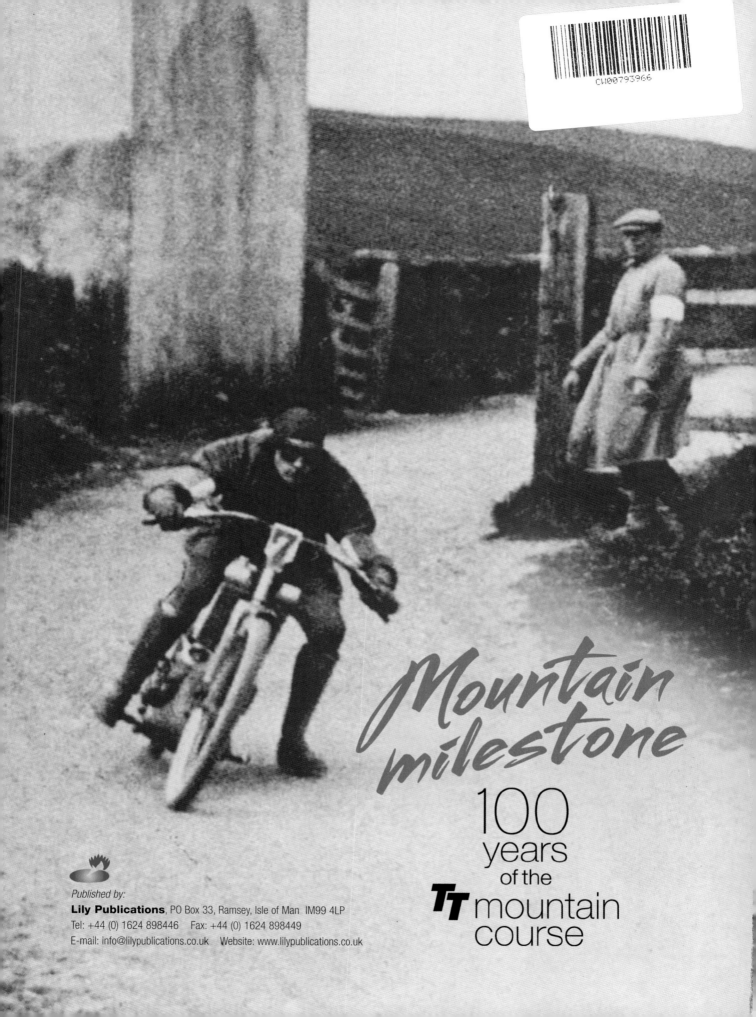

Mountain milestone

100 years of the TT mountain course

Published by:
Lily Publications, PO Box 33, Ramsey, Isle of Man IM99 4LP
Tel: +44 (0) 1624 898446   Fax: +44 (0) 1624 898449
E-mail: info@lilypublications.co.uk   Website: www.lilypublications.co.uk

Acknowledgements

The author's knowledge and memories of the Mountain Course have been augmented by those of Stan Basnett, Ian Baxter Ross, Malcolm Black, BP Archive, Paul Bradford, Ralph Crellin, Jim Davidson, Alan and Mike Kelly, Exxon Mobil Archive, John Foster, the Hanks Family, Ray Knight and his 'TT Riders Guide', the Manx Museum Library, Mortons Media Ltd, Leslie Quilliam's 'Gazetteer of the Isle of Man', Walter Radcliffe, Bill Snelling, VMCC Library, Paul Wright, Miles Cowsill, Dave Collister and the many riders from the past century who recorded their thoughts on this most demanding of race circuits. To all I proffer thanks.

Much effort has gone into supplementing the text with relevant photographs and these have come from Stan Basnett, Geoff Cannell, Eddie Cawley, Ron Clarke, Dave Collister, Miles Cowsill, Juan Cregeen, John Dalton, Pat & Tony East, FoTTofinders of Laxey, Wolfgang Gruber, Alan & Mike Kelly of Mannin Collections, Peter Kelly, Ian Kerr, Ted Kneen, Ray Knight, Yuki Kobayashi, Adam Lowe, Mike McDonald, Mortons Media Ltd, Terry Nester, Richard Radcliffe, Walter Radcliffe, Ken Sprayson, Rob Temple, John Turton, the VMCC, John Watterson and from the author's collection.

Special thanks go to Vic Bates who took many photographs to suit the text and also drew the course maps and those appearing at the start of each chapter.

The author and publisher have made every effort to locate and credit the source of the almost five hundred illustrations used in this book. If any have not been acknowledged in a correct fashion, please accept our apologies.

Contents

Foreword
  By Nick Jefferies 4

Introduction 5

Place names 6

Early roads and riding 8

The Start to Quarter Bridge... 22

Quarter Bridge to Ballacraine... 46

Ballacraine to Cronk y Voddy 66

Cronk y Voddy to Ballaugh 82

Ballaugh to Sulby Bridge.. 102

Sulby Bridge to Ramsey Hairpin 116

Ramsey Hairpin to the Bungalow................................ 134

The Bungalow to Creg ny Baa..................................... 152

Creg ny Baa to the Finish 170

Index 192

Produced and designed by Lily Publications Ltd

PO Box 33, Ramsey, Isle of Man, British Isles, IM99 4LP

Tel: +44 (0) 1624 898446   Fax: +44 (0) 1624 898449

www.lilypubs.co.uk   EMail: info@lilypublications.co.uk

© Lily Publications 2011/Second Edition 2014

All rights reserved. No part of this book or any of the photographs may be reproduced or transmitted in any form or by any means, electronic or mechanical, including photocopying, recording or by any means of information storage and retrieval system without written permission from the publisher, save only for brief extracts which may be used solely for the purpose of review.

Foreword
by Nick Jefferies

In a long and successful competition career, all-rounder Nick Jefferies holds the unique record of having won the Manx Two-Day Trial, the Senior Manx Grand Prix and a Formula 1 TT.

I have a confession to make. Those many dark hours in the school classroom were not so dark for me. If I wasn't actually secretly reading the "Shell History of the TT", I was probably memorizing the salient facts, dates, winners, and record laps of the previous years. At one time in my life, I could recite every winner, every fastest lap, every record lap, probably even the winning number. Such was my enthusiasm for this unique event.

Hardly surprising then, that School became the past with little academic accolade and my young adult life became a strong petrol filled mixture of trials, enduros, and road racing. The highlight for me in the sixties and early seventies was late August when I sailed off to the Manx, to compete in the Manx Two Day Trial, and watch the Manx Grand Prix, an event I have always supported. Eventually, I plucked up enough courage to enter the Production TT, and I still remember the amazing feeling, the mixture of fear and elation as I first travelled down Bray Hill, to follow the heroes of my youth, and become a TT rider.

Thus, for me, it really is a privilege to pen the Foreword to David Wright's new book 'Mountain Milestone'. I feel as though I have lived and breathed the event since my first visit in 1953, and I have always been interested in discovering new facts about racing on this hallowed track. I am sure that racer or spectator, marshal or official, there is something in this book for everyone. This book is literally a minefield of interesting facts from 1911 to the present day. I love the little titbits, the minutiae, the surprising revelations. David Wright has properly researched this book, and you will find it hard to put down once started.

So many people have been involved in the TT and the Manx Grand Prix since 1911 and I pay tribute to the dedication of the marshals, the wise counsel of the officials, the enthusiasm of the spectators, and the courage of the intrepid competitors. I sometimes ask myself the question: is the TT *the* most amazing sporting event in the whole wide world? Ladies and gentlemen, I think you will agree. It probably is.

The Jefferies family have such a strong connection with the TT. My father Allan was runner up to the incomparable Geoff Duke in the 1949 Senior Clubmans TT. Brother Tony would undoubtedly have won more than the three TTs that came his way, which included his famous victory on "Slippery Sam" in 1973. Who knows how many more victories my Nephew, David would have achieved?

I must close by paying tribute to Ian Hutchinson, winner of FIVE TTs in 2010. I am proud to know him, to have once employed him at the Allan Jefferies Motor Cycle Shop in Shipley, and wish him every success in his future career.

I trust that you will enjoy reading this book as much as I have

Here's to the next 100 years!

Nick Jefferies leaves Ramsey on his Castrol Honda to tackle the Mountain climb.

Introduction

The Isle of Man has a mystical attraction to fans of motorcycle racing, for the action and excitement witnessed at places like Bray Hill, Barregarrow, Ballaugh, Gooseneck, Windy Corner and Creg ny Baa, has filled the memory banks of successive generations during a century of competition over the world-renowned Snaefell Mountain Course. Home to the famed Tourist Trophy and Manx Grand Prix races, no other race-track compares in distance, nature of going and longevity – it is truly unique.

To celebrate one hundred years of racing over the Mountain Course, these pages tell how and why it came into use, what conditions were like a century ago, the changes that have taken place in roads, bikes, riders and race organisation, the way it has spurred machine development, the lap speed barriers that have been broken 50, 60, 70 .. 100 .. 130 mph, how it has written itself into the Manx landscape and culture, how its world-wide reputation for triumphant and at times tragic racing has been earned and, most of all, it tells the countless stories, old and new, that have been generated by its use.

This is an unhurried lap of the course spanning the past 100 years and sharing the thoughts of great riders like Charlie Collier, Graham Walker, Stanley Woods, Harold Daniell, Geoff Duke, Mike Hailwood, Peter Williams, Charlie Williams, Joey Dunlop, Steve Hislop, David Jefferies, John McGuinness, Ian Hutchinson, plus sidecar drivers like Dave Saville, Roy Hanks and Dave Molyneux. These courageous men weighed the high risks associated with racing over the most demanding circuit in the world, they accepted its challenge and all proved victorious. Not only do they tell here how it is ridden, but they also recall the incidents that have befallen them and many others as they raced over the 37¾ miles that make up a lap of the Mountain Course. It is a racing lap like no other, that tests riders and machines to their limits as it twists and turns through towns, villages and countryside, before climbing over bleak mountain and moorland, then plunging back down almost to sea-level.

Whether or not you are one of the world-wide 'brotherhood' of TT & MGP fans who comfortably fill their conversations with talk of Doran's, Rhencullen, Glentramman, Brandywell and Governor's, (and maybe even with Cronk Urleigh, Brough Jairg and Gob ny Geay!), you will discover new information here from all around the Mountain Course, for after 100 years of racing there is barely a yard of its going that does not have a story to tell.

David Wright

Mountain milestone

Place Names on the current TT Mountain Course that are used in this book.

A name in italics is an alternative to the one above it, or is no longer in common TT use.

Start
Glencrutchery Road
Grandstand & Pits
Nobles Park
St Ninians cross-roads
Parkfield
Ballanard Road
Bray Hill
Quarter Bridge Road
Ago's Leap
1st mile-marker
Woodlands
Alexander Drive
Eyreton
Quarter Bridge & Hotel
River Glas
Port y Chee Meadow
River Dhoo
Jubilee Oak
Joey Dunlop Foundation Building
Braddan Bridge
Kirk Braddan
2nd mile-marker
Snugborough
Cronk Doo
Railway Inn
Union Mills
Mullen Dhoo
Mwyllin Doo Aah
Ballahutchin
Elm Bank Hill
3rd mile-marker
Ballafreer
Glen Lough
Ballagarey
Elm Bank
Glen Vine
Glen Darragh
Ballabeg
4th mile-marker
Crosby Left
Vicarage Corner
Cedar Lodge
Crosby Village
Crosby cross-roads
The Crosby (Hotel)
Crosby Hill
Ballaglonney Hill
Halfway Hill
5th mile-marker

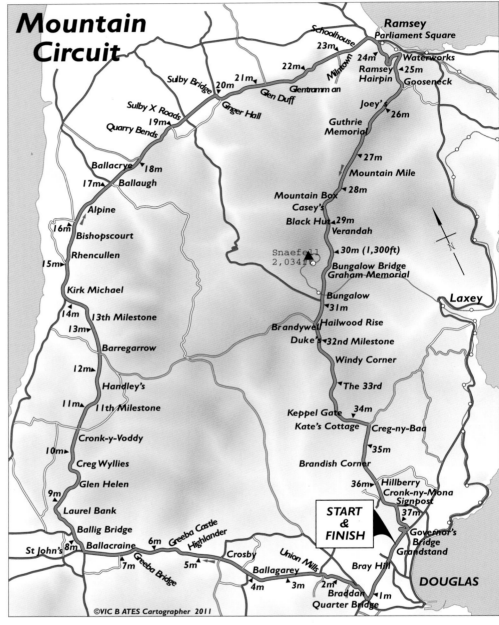

Halfway House
The Waggon & Horses
St Trinians
The Highlander
Greeba Verandah
Greeba Castle
Hall Caine's Castle Bends
Pear Tree Cottage
6th mile-marker
Appledene
Shimmins Corner
Central Valley

Greeba Gap
Greeba Young Mens Hall
Greeba Bridge
The Hawthorn
Knock Breck
Cronk Breck
7th mile-marker
Gorselea
Ballagarraghyn (Bridge)
Ballacraine
Ballacraine Hotel
Ballacraine Farm

Ballaspur
Ballig Farm
Tynwald Mills
8th mile-marker
Ballig Bridge
River Neb
Doran's Bend
The Rock Face
Laurel Bank
Cronk ny Killey
Creg ny Killey
9th mile-marker

Glen Moar Mill
Black Dub
Quarter-distance marker
The Vaish
Glen Helen
Sarah's Cottage
Creg Wyllies (Hill)
10th mile-marker
Lambfell Moar
Lambfell Beg
Cronk y Voddy Straight
Cronk y Voddy cross-roads
Ballakaighen Road
Mens Institute
Burnside
Stockade
Cronk Bane
11th mile-marker
Eleventh Milestone
Drinkwater's (Bridge)
Handley's Corner/Cottage
Cronk y Fessage
Cronk ny Fedjag
Ballamenagh Corner
12th mile-marker
Shoughlaigue Bridge
Ballaskyr
Cronk Aashen
Barregarrow cross-roads
Little Bray
Barregarrow bottom
13th mile-marker
Cammal Farm
Thirteenth Milestone
Cronk Urleigh
Ballallona Bridge
Westwood
Ballakinnag
14th mile-marker
Douglas Road Corner
Glen Wyllin Corner
Kirk Michael
The Mitre
The White House
Motorcycle Museum
15th mile-marker
Birkin's
Rhencullen
Holly Ridge
Bishopscourt
Bishopscourt Straight
Bishopscourt Glen
Bishopscourt Farm Bends
Orrisdale North

16th mile-marker
Bishop's Dub
Alpine
Alpine Cottage/House
Brough Jairg
Balla Cobb
Ballaugh
17th mile-marker
Ballaugh Bridge
The Raven
The Railway Hotel
Gwen's
Ballacrye
18th mile-marker
Quarry Bends
Ballavolley
Wildlife Park
Gob Y Volley Quarry
Sulby Straight
Half-distance marker
19th mile-marker
Sulby Glen Hotel
Sulby cross-roads
Glen Kella Distillery
Sulby Bridge
20th mile-marker
Ginger Hall (Hotel)
Grangey
Kerromoar
Abbey Ville
21st mile-marker
Glen Duff
Glentramman
Temple's Corner
Water Trough Corner
Glentramman Loop Road
Churchtown
Lezayre Church
22nd mile-marker
Ballakillingan
Conker Fields/Trees
'K' Tree
Sky Hill
Milntown Cottage
Pinfold Cottage
Milntown Bridge/jump
Glen Auldyn Bridge
Milntown
Gardeners Lane
23rd mile-marker
Lezayre Road
Schoolhouse Corner
Crossags
Parliament Square

Taubmann Street
Albert Road
Ramsey Depot
Queens Pier Road
Cruickshanks
May Hill
Whitegates
24th mile-marker
Stella Maris
Ramsey Hairpin
Mountain Road
Waterworks
Ballure Reservoir
Tower Bends
Albert Memorial
25th mile-marker
Gooseneck
North Barrule
Joey's
26th mile-marker
Guthrie's (Memorial)
The Cutting
27th mile-marker
Mountain Mile
28th mile-marker
Three-quarter distance marker
Mountain Box
East Snaefell Gate
Snaefell
29th mile-marker
Casey's
Rice's Corner
George's Folly
Black Hut
Stonebreaker's Hut
Shepherd's Hut
John Smyth Shelter
Verandah
30th mile-marker
Laxey Valley
Bungalow Bridge
Stonebridge(s)
Graham Memorial
Laxey River
Bungalow Bends
McIntyre Box
The Bungalow (Hotel)
Murray's Museum
Joey Dunlop Memorial
31st mile-marker
Hailwood Rise
Hailwood's Height
Mullagh Ouyr
Brandywell

West Mountain Gate
Second Mountain Box
Beinn y Phott
Duke's
Thirty-second Milestone
32nd mile-marker
Baldwin Valley
Windy Corner
Carn Gerjoil
Nobles Corner
Nobles Park Road
Slieu Lhost Quarry
Thirty-third
33rd mile-marker
Keppel Gate
Clark's Corner
Kate's Cottage
Shepherd's House
34th mile-marker
Creg ny Baa
Keppel Hotel
Gob ny Geay
Sunny Orchard
The Cutting
35th mile-marker
Brandish
O'Donnell's
Upper Hillberry
Hillberry
36th mile-marker
Glendhoo
Hillberry Road
Cronk ny Mona
Johnny Wattersons Lane
Scollag Road
Signpost Corner
Bedstead
Rectory Corner
Bennett's Corner
The Nook
37th mile-marker
Governor's House
Bemahague
Governor's Bridge
Governor's Dip
Glencrutchery Road
Greenfield Road
The Finish

Mountain milestone

TT mountain course

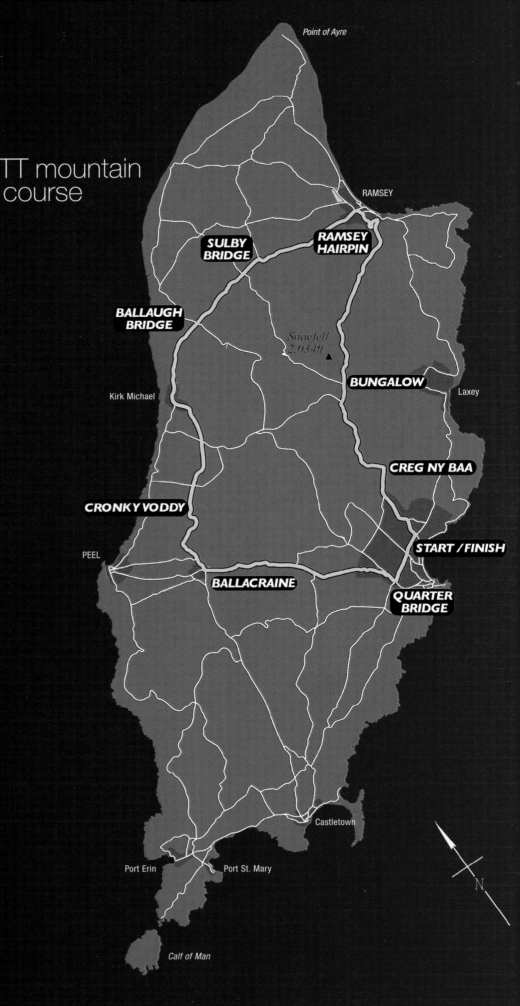

CHAPTER ONE

Early roads and riding

The Isle of Man Tourist Trophy races have existed for almost as long as motorcycles have been ridden on British roads, but just as the motorcycles of today are very different from those of over 100 years ago, so the pioneering TT riders would find the Mountain Course – though still comprised of the same roads – very much changed. Before starting out on a lap to experience old and new, it is worth setting the scene by looking at how and why the Mountain Course and the TT races came into being.

In the last few years of the 1890s when the overall speed limit on Britain's roads was lifted from 4 mph to 14 mph, many were experimenting with powered two-wheelers to create a new type of personal transport in the form of a motorcycle. Most efforts were based around the attachment of primitive engines to modified pedal-cycles; using steam, paraffin and petrol as fuel, with the latter becoming the most popular and resulting in almost all early machines being fitted with small four-stroke engines that came with direct, single-speed transmission that lacked any form of clutch.

Pioneer designers had problems to overcome in adapting the early internal combustion units for motorcycles, for as well needing to improve the rudimentary cast-iron engines of the day, they had to devise new forms of ignition, carburation and lubrication to withstand the rigours of roads built for the horse and cart.

As the 1890s passed into history and a new century dawned, the early 1900s saw British manufacturers like Ariel, Riley, Raleigh, Enfield, Triumph and Matchless striving to devise motorcycles that could be produced for commercial gain. There was much experimentation with engine location, but eventually accepted practice saw the power unit located vertically behind the front down-tube, just above or in front of the cycle-style pedals, the latter being retained to provide help in starting and to allow the rider to contribute extra propulsion on gradients.

Individualists

Motorcyclists of today will feel a link with riders of early powered two-wheelers, for they were individualists who sought the freedom to travel when and where they wanted without having to comply with the timetables of railway and omnibus companies, on machines that offered a challenge and a dash of excitement. If the opportunity came to pit their skills and the performance of their machines against other riders, many took it, notwithstanding the overall 14 mph speed limit and the risk of being charged with "furious riding" by horsey-minded magistrates sitting on the local Bench.

The competitive urge generated by motorcycles extended to manufacturers, and they sought competitions that allowed them to enter their products against those of other makers. Successes gained them publicity, whilst failures provided information on where they needed to improve and, often by study of the opposition, how to achieve such improvements.

Road Racing

In most of Europe manufacturers could help develop their motorcycles by contesting races over public roads, but such activity was prohibited in Britain. The organised competition available here was confined to long-distance reliability trials, occasional hill-climbs and sprints, plus racing on a few cycle-tracks. But though opportunities for motorcycle competition were limited, sufficient competitive spirit existed for the Automobile Club of Great Britain and Ireland (later the RAC) to establish the Auto Cycle Club (later the ACU) in 1903, to issue licences for competition, adjudicate on record attempts and encourage motorcycling in general.

It was the Secretary of the Automobile Club of Great Britain and Ireland, Julian Orde, who while regretting

Mountain milestone

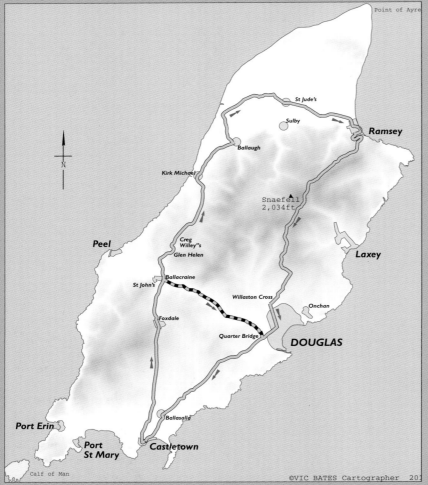

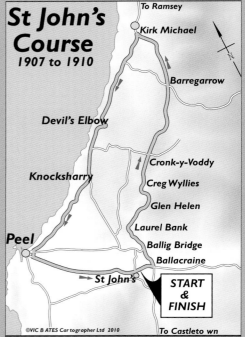

Above: The 52 mile circuit used by cars in 1904 and the shorter version used by motorcycles in 1905.

Top right: A typical motorcycle of the early 1900s

Middle right: The St John's course used for the first running of the TT in 1907.

Below: Dot Tilbury of the Manx Post Office responds to Geoff Cannell's invitation to say a few words at the TT Re-enactment ceremony in 2007.

how Britain's motor manufacturers were disadvantaged by legislation which prevented them competing over ordinary roads, remembered that his cousin Lord Raglan was Lieutenant Governor of the Isle of Man, a self-governing Crown Dependency in the middle of the Irish Sea.

Lord Raglan was an enthusiastic motorist and when Julian Orde visited him on the Isle of Man in February 1904, he was sufficiently interested in Orde's suggestion that the Island might close its roads to allow cars to race over them that he lent him his Daimler, introduced him to several motoring-minded Manxmen and invited them to seek out a possible course.

The map on page 10 shows the 52 mile circuit that Orde chose; it being one that he felt would test machine reliability and driver skill, as much as out and out speed. With Britain due to field a team of cars in the prestigious Gordon Bennett Cup meeting in Germany later in the summer of 1904, Orde persuaded Lord Raglan to use his influence and have Manx highway legislation amended to permit the use of the Island's roads for car racing. By cutting legislative corners Raglan achieved this in an exceptionally short time, and in May 1904 eleven cars took part in eliminating trials over 6 laps of Orde's 52 mile course, plus a hill-climb and sprint, to choose a team to represent Britain in the Gordon Bennett Cup.

The International Cup

Motorcyclists gained their equivalent of the Gordon Bennett Cup when a competition called the International Cup was run for them in France in 1904. This led to motorcycles joining cars on the Isle of Man in May 1905 to take advantage of the four-wheelers' organisation and to run eliminating trials for two-wheelers, to choose a team for the 1905 International Cup. Although called eliminating trials, the event was a race, indeed it was a historic one, for 'The Motor Cycle' described it as "the first legalised road race for motorcycles ever held in these (British) islands" and it ran over a shortened version of the car route, turning right at Ballacraine and returning to Douglas, as shown by the dotted line on the map on page 10.

The Tourist Trophy

While British motorcycle manufacturers were keen enough to pit their products against continental opposition in the International Cup, they were unhappy with the format of the event, for its regulations stipulated a maximum machine weight of just 110 lb/50 kg, but put no limit on engine size.

The Collier brothers, Charlie and Harry, were keen competitors on Matchless motorcycles built by their family company, but felt that races such as the International Cup were taking motorcycle development in the wrong direction. They were not alone and in 1907 the Auto Cycle Club decided to run a new international competition which was to be "a race for the development of the ideal touring motorcycle", the winner of which would receive a Tourist Trophy. It was a momentous decision, though they could hardly have dreamt that they were setting the 'International Auto-Cycle Tourist Trophy' meeting – soon abbreviated to TT – on a high-speed journey that would last for over 100 years. It was a journey which would test the skill and courage of many riders and capture the imagination of generations of race fans, whilst the silver Tourist Trophy showing Mercury poised on a winged wheel has become one of the most widely recognised images in the world of motorcycle racing.

The first motorcyle TT

The first Tourist Trophy meeting for motorcycles, took place on the Isle of Man on the 28th May 1907 over a course in the west of the Island known as the St John's Course - nowadays often called the Short Course. The organisers were aware of the limited performance available from the single-speed machines of the time, but still chose a testing 15½ mile lap of undulating loose-surfaced roads to be covered ten times. There was a compulsory ten minute stop for riders at half-distance when they refuelled and took refreshment, and to emphasise that this was an event for touring machines rather than out-and-out racers, each rider received a fixed amount of petrol based upon 1 gallon per 75 miles of race distance for multi-cylinder machines and 1 gallon per 90 miles for single-cylinder models. Although multis and singles ran together, they were in two distinct races and the Tourist Trophy was to go to the winner of the single-cylinder race.

On a dry but cold day, 25 competitors left Tynwald Green in pairs at one minute intervals to contest the first running of the race for the Tourist Trophy. After eventful rides that included strenuous use of their bikes' pedals on hills, the mending of punctures, repairs to belt drives, extinguishing of fires and recovery from spills, it was Charlie Collier who brought his single-cylinder Matchless to victory and took the Tourist Trophy in a time of 4 hours 8 minutes and 8 seconds at an average speed of 38.22 mph, whilst Rem Fowler (Norton) was first home in the multi-cylinder race, taking 4 hours 21 minutes and 53 seconds, averaging 36.22 mph.

Progress

The TT returned to the St John's course for the years 1908 to 1910, when amendments to the rules saw pedals banned, fuel consumption restrictions initially tightened then dropped, the need for silencers discarded, the two races for singles and multis merged into one in 1909, and the winning average race speed rise to 50.63 mph.

Whilst there had clearly been progress in terms of speeds achieved in the period 1907-1910, some of the urge had gone out of motorcycle development. It was true that several manufacturers had moved to all chain drive, there were experiments with dry-sump lubrication and steel was being used for some components instead of cast iron, but the TT organisers were keen to promote more technical advancement. As an example of the problems demanding attention, when operated at large throttle openings, early iron engines tended to overheat causing loss of power and component failure. The Rex marque was so concerned about this at the TT that it came to an arrangement with a lady who lived at Laurel Bank (just prior to Glen Helen and Creg Wyllies hill), for her to leave buckets of cold water outside her cottage, so allowing its riders to stop and pour cooling water over their engines in the hope of regaining power before tackling the steepest hill on the course. That was not a realistic option for an ordinary road rider and Rex eventually went on to produce water-cooled engines.

The hill called Creg Wyllies was a stern challenge for TT competitors and there were occasions when riders of the early single-speed racing bikes would fail to get up as they ran out of speed and their engines stalled against the single gear being pulled. Some would try and avoid stalling by jumping off and running alongside, but those that did stall would either dismount and push to the top, or coast back down the hill and try again. As the ordinary motorcyclist would encounter steeper hills, what both racer and tourer needed to help hill-climbing were improved transmissions, in particular multiple gears and a clutch. In keeping with its original TT philosophy of using racing to help develop the motorcycle, the ACU forced manufacturers to progress the matter of gears by deciding that for 1911 it would adopt a much tougher course for those seeking to win what had already acquired prestige in the world of motorcycling, the Tourist Trophy.

A Mountain to Climb

In a bold move, organisers shifted the Tourist Trophy meeting for 1911 from the St John's Course to one with more than twice the lap distance and with the mountain of Snaefell to climb for good measure. The move had been

Mountain milestone

These two Scott riders took part in the Snaefell hill-climb in 1910.

forecast in a report soon after the 1910 races which said "There is great speculation that next year's event will be held on a new course over Snaefell Mountain and if this proves to be correct then much thought will have to be given to variable speed gears and the more exacting demands that such a course will put upon both man and machine". Looking ahead, the report continued "this year's race was won at over 50 mph. Even on a new course and with the anticipated progress of machine development there must be limitations to the endurance of the riders. Is a 60 mph or even 70 mph TT possible in the future"?

Some manufacturers contesting the TT were not keen on the new Mountain Course, for they had already received an indication of Snaefell's demands when the ACU ran a 6 mile hill climb up its flanks from the outskirts of Ramsey to the Bungalow Hotel (almost the highest point on the Mountain Road) after the 1909 and 1910 TTs.

Brushing aside concerns, the organising ACU announced the running of an additional TT race for singles up to 300cc and multis up to 340cc at the 1911 event, and then ignored pleas that pedals should be permitted in the new race!

The reason for the differentiation in capacity between singles and multis, was to copy the rules of the existing 'big' class where multi-cylinder bikes (mostly twins) had been allowed a capacity advantage since the multi and single-cylinder races on the St John's Course were merged in 1909. That differential was reduced in 1910 and again in 1911 when capacity limits in the big class were set at 585cc (multis) and 500cc (singles).

The two races in 1911 were given the titles of Junior and Senior, with the smaller class attracting 37 entries and running on Friday 30th June over four laps, and the big bikes, of which there were 67 entries, going on Monday 3rd July over five laps. The event was called the Coronation Jubilee TT in celebration of the coronation of King George V and the golden jubilee of the Borough of Douglas. It formed part of a week of celebratory events on the Island.

Welcome

The Borough of Douglas made no secret of the fact that it welcomed the move of the TT from the St Johns to the Mountain Course, for it expected to gain trade as a result. The 'Isle of Man Weekly Times' was equally open when it wrote in 1911: "It cannot be denied that the races cause some inconvenience on the Island; but one would cheerfully put up with that because of the results received in the shape of the money which is spent, and the advertisement the Island receives". Indeed, when the race organisers announced in 1911 that practice would take place over just one week, it was the newspaper that led a successful campaign for it to be extended to two weeks, so that more trade would result.

Riders started and finished their practice laps from the Quarter Bridge on the outskirts of Douglas, and although official practice was initially limited to the hours of 3.00 - 7.00 am, the roads were not closed for riders' use and were subject to normal traffic. One report of a practice hazard described "a trio of herring carts that made a special feature of their capacity for monopolising the centre of the road". At the time of the 1911 TT the Island was debating its speed limit which was still 14 mph, unlike the United Kingdom where it had been 20 mph for several years. Competitors paid little heed to the 14 mph limit during their practice laps over open roads and, believing that a 20 mph limit would be difficult to enforce, the Island authorities abandoned any form of overall limit, a situation that prevails to this day.

The 1911 TT Course

With a distance of $37\frac{1}{2}$ miles to a lap in the form used in 1911, the new TT Mountain Course offered widely varied going as it passed through the towns of Douglas and Ramsey, plus the villages of Crosby, Kirk Michael, Ballaugh and Sulby, whilst also crossing undulating open countryside and climbing over bleak mountain and moorland, before plunging back down almost to sea level. Many early machines took an hour to complete a lap, even under race conditions.

The roads of a century ago may have followed almost the same route as those of today, but they offered totally different going. They were the earlier creation of a mainly agricultural community and had been brought into use to get beasts and goods to local markets, and allow passage by foot, horse and cart between the major towns. Many were not formally constructed, they just followed the firmest route and were much narrower than today. When the occasional bog or soft spot eventually caused too much inconvenience and could not be by-passed, it might have received a cartful of stones gathered from the adjoining fields or sea-shore by a local farmer, but that was the limit of maintenance. Initially there were few bridges and river crossings were by fords that would be impassable at times, whilst un-culverted streams would wash away stretches of road in winter.

By the 18th century local authorities were charged with maintaining roads, but they probably operated in the same ad hoc manner of filling holes and perhaps digging and clearing roadside ditches to allow water to drain away. The appointment of the first Supervisor General of Highways shortly after the Island's Highways Act of 1753 brought a greater degree of central control and prescribed standards were eventually adopted for the construction of new roads. However, in respect of many of the roads adopted for the TT course it was said in a report to Tynwald in 1912, "they require considerable expenditure to restore ruined foundations or to provide proper foundations where none previously existed". It was a slow process and only the occasional stretch that became virtually unusable was reconstructed to proper standards, the remainder being dealt with by patching. That was certainly the case at the time of the 1911 TT, when of roadside kerbing there was little and of tarmac (first used in Britain in 1904) there was none.

Macadam

While the Manx roads of a century ago were improved from earlier times, the all-important upper surface of the road was only rolled macadam. There was no tar-binding and macadam was a composition of graded sharp-edged stone and finer material that was dampened and compacted by roller, the theory being that if laid to form a crown in the centre of the road, rainwater was shed to the edges where the camber was steepened to help it drain away. The highway authority endeavoured to select the best road making material, but the over-riding principal was to use what was reasonably close at hand, because until well into the early years of the 20th century, almost everything had to be shifted by horse, cart, wheelbarrow and occasional steam-engine. The Mountain Course is dotted with small quarries, but use of local material meant that the type and quality of the road surface varied from mile to mile, and then the passage of horses and iron rimmed cart wheels quickly destroyed any newly laid macadam top surface, causing holes to form. They filled with water which accelerated the break-up, so calling for another round of stone patching.

Wet weather presented riders with roads that acquired a surface layer of mud which covered man and machine, whilst also finding weaknesses in the waterproofing of a bike's electrics. Things were little better in the dry, for the passage of a race machine created hedge-high clouds of dust that lingered for minutes. This added to the difficulties facing riders and made overtaking hazardous, for they were forced to ride blind into the dust cloud of the one in front. In truth, motorised vehicles needed a different surface from the one used in 1911.

At the time of the 1911 TT there were approximately 200 motorised vehicles registered on the Island to serve its residents and half a million summer visitors. That was more than double the number registered at the running of the first motorcycle TT in 1907, but motoring was still for the few and despite such expansion in numbers, the principal form of personal conveyance was still by horse-drawn vehicles or pedal cycles. The Island did have a comprehensive public transport system of trains, trams, etc, that met the needs of many residents in both town and country, so when road improvements received consideration, the factors uppermost in the decision-makers minds were whether fishing, agriculture and what it called the 'Visiting Industry' (tourism) would benefit from any expenditure – private motoring was low on the list.

Difficult Conditions

On early motorcycles the rider sat high, the puncture-prone tyres were of narrow section and whilst many machines adopted primitive forms of gears or variable transmission for the 1911 event, the state of the roads, the gradients faced and the distance involved in a TT race was truly daunting. Whilst the fastest bikes managed only 70 mph downhill, they required as much courage and skill to handle as today's 200 mph missiles. Tyres bursting and leaving the rim were a recognised hazard and with a large assortment of controls to deal with, many of which were mounted on the petrol tank, taking a hand off the handlebars was a risky business.

Most of the Mountain Road was un-fenced, allowing sheep to roam freely and on much of the lowland section chickens were accustomed to scratch in the road outside farms and cottages, while cats and dogs ran loose, and even cattle were not always confined. After practising in 1911, Jimmy Alexander complained about "a lot of cattle on the road between Ballaugh and Ramsey" and an Island newspaper criticised local farmers for not controlling their beasts. As narrow as today's country lanes, the 'main' roads were frequently overhung by hedges and trees that no-one felt a responsibility to cut, and the grass banks were left untrimmed.

Given its 37½ mile length, there was little that the organisers could do to prepare the TT course for the racing, although prior to the 1911 event they inspected it with Highway Board officials, noting repairs to the surface over the Mountain. However, 'The Isle of Man Weekly Times' wrote at the time "The Mountain Road is dreadful – the repairs which the Surveyor had done on the road lately has made them worse than ever". Before the start of practice riders were told "All dangerous parts of the course will be marked by banners some distance from the point of danger". Those banners spanned the road

This rider approaching May Hill on the outskirts of Ramsey in 1912 has used the footpath on the approach to the corner to avoid the many water-filled potholes in the road.

before difficult corners and were supported by a post each side. Initially made of canvas, they were vulnerable to strong winds and so were superceded by fishing nets or wire netting, with letters proclaiming 'DANGER' affixed. By 1920 there were 12 locations for the banners that were erected and taken down on a daily basis by Banner Marshals.

A Tough Course

With its formidable length and extremely varied going, the Mountain Course presented a far greater challenge to man and machine than the original St John's Course. It used part of the old course (from Ballacraine to Kirk Michael) but most of its lap was new, so riders were understandably keen to get as much early morning practice as possible to learn its twists and turns, with a report of the time claiming "between the hours of three and seven in the morning, when most people are wasting these good hours in bed, scenes of excitement will take place, and the roars of the engines will be heard". Of course, not everyone on the Island agreed with that view and a contrary one expressed in a letter to a local newspaper said: "I think there is a great deal too much time spent by grown-up people playing with adult mechanical toys . . . spending their time riding to nowhere and back simply for the fun of riding when they could be better employed".

Encouraging Development

As the organisers planned, and as riders and manufacturers feared, the first TT over the Snaefell Mountain Course presented an extreme challenge; but later generations had much to thank them for because noted historian 'Ixion' wrote in 'The Motor Cycle' some years after: "The 1911 race had more influence on the motorcycle's development than any other single event in its history, for the entire industry was thereby driven to develop engines of greater flexibility and to equip them with reliable and efficient variable gears". Indeed, it was the American Indian concern who showed the way forward, by taking the first three places in the Senior TT of 1911 using two-speed countershaft gearboxes and all-chain drive, to finish ahead of makes like Ariel, Bradbury, NSU, Rex, etc, whilst a twin-cylinder Humber was first home in the Junior race.

Before the 1911 TT it was reported that "The ACU are taking steps to arrange for a dust-laying material to be laid on part of the track, so as to minimise the dangers of the course". Hopefully it did not have the side-effects of one used previously on the St John's Course, which attacked riders' clothing, causing holes to appear.

Mountain milestone

Charles Franklin looks relaxed prior to the start of the first Mountain Course TT in 1911. Note the spare inner-tube tied around his waist, for most riders expected to collect a puncture during the race

The 1912 Junior TT was the first to be run in really wet conditions. Here winner Harry Bashall (Douglas) tackles the experimental tar-sprayed surface on the exit from Quarter Bridge.

Tar spraying

Just after the first Mountain Course TT the Isle of Man Highway Board made its annual report to Tynwald the Manx Parliament and said: "Motor car traffic is increasing and in order to keep our roads in satisfactory repair for such traffic, the Board is of the opinion that something must be done for the prevention of dust during the summer season. Tar spraying and tarmacadam is now generally adopted as a means for dust laying by road authorities throughout the United Kingdom, it is said with satisfactory results. In next year's estimates it is proposed to set aside a sum for experimental tar spraying of the road surfaces through villages".

So keen was the Highway Board to try this new system that soon after the 1911 TT it authorised its Surveyor General "to tar paint that portion of the road from Quarter Bridge to Braddan Bridge as an experiment towards dust laying". It was a job estimated to cost £13 and was seemingly a success, for in September a tar-spreader and boiler were purchased.

Place names

Before setting out on a lap of the Mountain Course, it is as well to point out that there is no standard list of locations that identifies every twist and turn to riders and spectators. Also, the names found in use range from traditional ones applied on a day to day basis by the Manx people which derive from their Celtic and Viking ancestry, those that have developed during a century of racing and are known to and used by the average TT enthusiast world-wide, the 'official' list used by the race organisers that today recognises over 100 primary locations, and more detailed lists used by the 12 Chief Sector Marshals to identify the many points within the Sectors into which the course is sub-divided for the purpose of marshalling, that total over 200 locations. That figure is somewhat more than the 22 marshalling points which were manned at the first running of the Mountain Course TT in 1911.

Wide variation in spelling is found, even of common place names, while some spots carry more than one name and some names that were adopted for a few years have dropped out of use. Occasionally, a rider or spectator will get the wrong name for a place stuck in his mind and use it for years, before discovering the mistake, whilst others christen their favourite locations with names that mean something specific to them.

Much of the Island's history is contained in the place names that occur on the route of the Mountain Course, for it passes through 9 of its 17 ancient parishes. The derivation of some of the important ones is given within the text as they are met with on a racing lap, the primary sources for which have been Leslie Quilliam's 'Gazetteer of the Isle of Man' and the writings of a great enthusiast for Island racing, Canon Stenning.

Occasional mention is made of spectating facilities, but no attempt is made to make this a complete spectator's guide, for viewing points are subject to frequent change.

Learning the way

The task of learning the almost forty miles of twists, turns and gradients of the Mountain Course in sufficient detail to risk their lives, has always been a huge challenge to riders. It is not just the direction of bends that have to be learnt, for to make the swiftest and safest progress, riders have to be

Hutchinson tyres were a popular make in the early days of the TT. At the top of the advertisement can be seen a cross-section of the butt-ended inner tubes that riders carried and which allowed them to replace a punctured tube without removing the wheel.

confident that they know all the braking and peel-off points, what gear to be in, the nature of the cambers, white lines, awkwardly positioned manhole covers, solid and splayed roadside curbs, changes of road surfaces, shiny bits, permanent wet patches, lingering dampness, major bumps, passing points, slip roads, marshals posts, etc.

Occasionally a rider appears and shows race winning potential from the start of a TT career, but most need several years riding experience over the course before becoming really confident and the general advice to newcomers has always been to "make haste slowly". Not everyone takes such advice, for by nature racers are confident individuals who believe that they are as good, or better, than the next rider and this has often led to their undoing. TT and MGP winners like Jimmy Simpson, Manliff Barrington, Harold Daniell, Ken Bills, Don Crossley, Bob McIntyre, 'Milky' Quayle and others, have all explained how, imbued with the dangerous combination of over-enthusiasm and lack of course knowledge under racing conditions, they each suffered a crash in their Island debuts. All lived to race (and win) another day, but some dashing young racers with more enthusiasm than course knowledge have been less fortunate.

Which Way?

In the past, many riders got their first sight of the Mountain Course when they started their opening lap of official practice. Indeed, when multi-TT winner Wal Handley arrived for his first one in 1922, not only had he not been around the course, but with race officials half asleep in the just breaking dawn he was allowed to set off in the wrong direction – towards Governor's Bridge not Quarter Bridge.

The underlying principles of learning the Mountain course are to seek the best tuition, complete as many laps as possible, and not go too fast too soon. The need for riders to get in many familiarisation laps was brought home to Travelling Marshal Albert Moule when the fallen rider he attended at Creg ny Baa during the 1964 MGP told him: "I forgot the corner". Clearly a case of too fast too soon, for he forgot what is a right-angled bend approached downhill at near maximum speed!

Crossley's Tours

Since the 1965 TT it has been compulsory for all newcomers at the TT and MGP to take an official instructional lap on a coach, where they receive advice from experienced former racers on how to ride the Course. Sidecar newcomers go on a different coach to solo competitors.

These coach trips are still nicknamed 'Crossley's Tours' after the man who started them in the 1950s, former MGP winner Don Crossley. But, as riders will be told, that compulsory lap must be supplemented by many voluntary ones to learn the direction of the multitude of corners. Inexperienced Newcomers can be disillusioned by the way an experienced rider just disappears away from them through a twisty section, but increased course knowledge will see them doing much better, for as Ray Knight tells in his 'TT Riders Guide': "There are some amazingly fast places . . ., depending upon your machine, course knowledge and sheer nerve".

Qualifying

Official practice sessions initially extended over two weeks, all of which were early morning ones. The time available has been reduced down the years, and although the amount of practice time still seems generous, if a rider strikes machine trouble in a few sessions and spends much of them sitting at the side of the road, not only does it make course learning harder, but he runs the risk of not meeting the qualifying times set by the organisers.

Qualifying standards were introduced soon after the TT moved to the Mountain course. In 1914 each rider had to complete a minimum of 6 laps in

Father and son Ossie and John Wade rode HRDs in the 1926 Senior TT. They arrived a week before the two weeks of official practice and so spent four weeks on the Island for their rides.

practice, one of which had to be in less than 70 minutes (Junior) or 60 minutes (Senior). Qualifying times have gradually reduced down the years and in 1925 they were 50 minutes (Junior) and 45 minutes (Senior). As additional classes were created they were also given qualifying times that, by the Diamond Jubilee in 1967 were 30 minutes (Junior), 29 minutes (Senior), 37 minutes (Sidecar), 35 minutes (Lightweight 125), 33 minutes (Lightweight 250) and 45 minutes (50cc). Today, solo riders must achieve lap times within 115% of the third fastest practice lap set in their class, and sidecars must be within 120% of their class figure.

By 1930 practice was down to ten early morning sessions starting on a Thursday, and soon after it was said of NSU newcomer Otto Steinbach: "he does six laps in the morning and six laps in the afternoon on a road bike to learn the Course, plus official early morning practice on his race machine". When the BMW 'works' team decided to contest the 1938 Senior TT, it sent riders, road machines, a mechanic and a physical fitness trainer to the Island for a fortnight in April and in 1953 the NSU factory provided road machines for the experienced Bill Lomas to give course instruction to its riders a couple of months before TT.

German competitors have a reputation for thorough preparation and 24 year old rising star Walter Zeller was on the Island in March 1953 and aiming to cover about 150 learning laps at the rate of eight a day on a road-legal BMW, before making his TT debut. Prior to the 1956 Clubman's TT races, double winner Bernard Codd took five weeks off work from the family farm and did at least seven learning laps of the course a day on a Matchless G9 roadster.

The first-time riders on the Mountain Course described above were the lucky ones, for in the immediate post-war period the average working man in Britain was allowed only two weeks holiday a year, plus a few Bank Holidays. A man short of time for course-learning before his first Island ride was Mick Grant. Although ultimately to win several TT races, before his debut in the 1969 MGP Mick borrowed a 'tired' Vincent Rapide a few weeks before the race and, with wife-to-be Carol on the pillion, set off for a weekend on the Island. The Rapide was not in the best of health, having a suspect big-end and being generally out of sorts, while Mick was low on funds. In his words "all we had was this rattly old Vinnie, two sleeping bags and about 3/6d in change". Sleeping rough for their short stay, they nevertheless got the Vincent around the course a few times and Mick went away with a reasonable idea of how the corners were strung together. Come the MGP and did his reconnaissance trip pay dividends? Well, yes and no. Velocette mounted and with a top speed of about 110 mph, Mick suffered a lengthy pit-stop with points trouble and finished 48th out of 48 starters. As he said "After that things could only get better".

What Method?

Many riders seek to learn the course in stages, one being Geoff Duke. He described the methods he used for his debut at the 1948 MGP with: "Determined to leave nothing to chance, I arrived on the Isle of Man one week before official practice began, having ridden up to Liverpool on my scrambles bike, suitably fitted with road tyres and higher gearing". His first trip around the course made him realise what a challenge he had taken on, so he decided to learn it in three sections "from the Grandstand to Kirk Michael, Kirk Michael to Ramsey, and from there back to the start. Allowing myself two days for each section, I set off on my scrambles machine and stopped at every significant bend to study the general surroundings, walking back and forth along an imaginary racing line. When a meal time came or darkness fell, I would complete the lap and then continue from the same place at the next session. Endeavouring to remember every bend from the start line to the finish, by the end of that first week I knew in my mind exactly where I was on the course at any given time".

Many years after Geoff, the outstanding Steve Plater told how he made five or six trips to the Island before his 2007 TT debut "and did lap after lap". Then Mick Grant "suggested that I try and break the course down into sections rather than concentrating on the lap as a whole" and Steve found this helped his course-learning. Many former riders serve as mentors to Newcomers, and whilst learning the course cannot be rushed, the process can be accelerated with the aid of a good coach.

To be really worthwhile, learning laps must involve many stops, retracing of steps and consideration of alternative lines, for to get up to race pace riders must acquire the knowledge and confidence to fit a series of bends together so that they can be ridden in one totally committed move. Newcomers to short circuit racing often give the impression that they learn their limits by going over them and in the resultant falls get safely absorbed

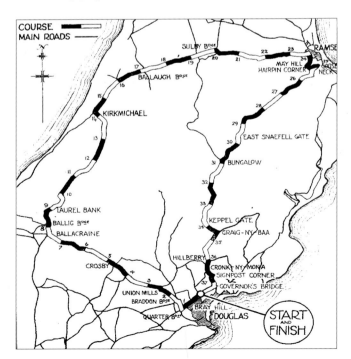

Bill Lomas (left) and Werner Haas (second left) with their road-going NSUs in 1953.

by huge gravel traps, but that option is clearly not available to the rider who wants a long career on the roads – a different mind-set is needed.

A top racer of the 1950s and 1960s was Derek Minter and, even after he had been racing the TT for several years, he would go round in his van as often as possible, frequently stopping and walking the corners. Derek was convinced that the reason so many riders came to grief on the Island was that they did not learn the course thoroughly before trying to race over it. He was the first man to lap the TT course at 100 mph on a British single, and also won the 1962 Lightweight TT on a Honda, so he was a quick rider. In his view: "Really and truly your mind was one step ahead of your body. You're in this corner but your mind is actually in the next corner so you know what part of the road you need to be on when you get there". With the massive increase in speeds that have taken place since the 1960s, that situation of mind being well ahead of bike and body must be even more relevant today, particularly as the view of the course ahead is often quite short and so riders must be confident of what is around the next blind bend and have the mental strength to ride in fully committed fashion, or as American newcomer Chris Crew so graphically put it after his debut in 1989 "You have to be willing to bet your life that you know where the next corner is going over the crest of the next hill".

Current Honda race-team supremo Neil Tuxworth has over 40 years association with the TT that included riding in 78 Island races. He had never seen the Mountain Course when he arrived for his first TT in 1971, but in his words "I was quite fortunate to make friends with Ray Pickrell (TT winner) and he showed me round the course quite a bit, although he was lapping the course quicker in his van than I was on my race bike!".

Alex George had a long and successful MGP and TT career, yet he wrote in the early 1980s that he never stopped learning when riding the Mountain course, telling "On every single lap I am trying to sort out ways to increase the flow, ways to leave the braking for an extra five yards or so for corners, in fact anything that will knock the slightest fraction off the time".

Yet another top-man who felt that there was always something to be learnt from an extra lap of the course was Joey Dunlop. By 1986 he had been competing and winning at the TT for 10 years. Nevertheless, one day during practice week he went out on the Formula 1 bike during early morning practice and on two of his other bikes during evening practice. Not content, he and brother Robert then completed two laps in a car in the dark. His Manager at the time, Davey Wood, said: "Joey likes to do this very late at night, as he still feels that he can learn something every time". Sometimes

This map of the Mountain Course comes from an official race handbook of 1922, but identifies relatively few places by name.

Early roads and riding

Well-meaning types have added markers around the course to help inexperienced riders recognise peel off points.

Joey would go alone, but at other times he would take passengers. One friend tells of a late night session when there were six in the car and Joey pushed the pace. On the completion of the lap Joey got out of the well-loaded car, extracted a cigarette, bent down and casually lit it from the glowing brake disc.

Bends and Straights

Top riders are often asked for advice on how to ride the Course and that of eleven-time TT winner Steve Hislop was: "work at the bends to get the best out of the straights so that you can carry the speed with rhythm and flow, that's the key to TT success". His advice was almost the same as that given by six-times winner John Surtees some 35 years earlier, and John was a master of choosing the line through a series of bends that would give him the fastest exit onto the next straight, thereby achieving a time-saving couple of hundred more rpm than someone with a lesser line. In respect of learning the way Steve Hislop added: "there's no real secret, it's just about putting in the time and having an aptitude for it", while Guy Martin makes the point that in contrast to short-circuits, "riding the TT isn't about how hard you can brake or get on the throttle, it's about carrying momentum".

With the interval starting system employed at the TT and MGP, competitors get a better chance to ride their own race, unlike on short circuits where as part of a jostling pack they spend as much time trying to spoil their opponents' lines, as they do on riding their own. Indeed, the front runners in a Mountain Course race often ride in comparative isolation and are riding against the clock – aided by the signals they receive – as much as against other competitors. It is so different to circuit racing and is much enjoyed by exponents of the art of road-racing, for as Peter Williams said in his autobiography 'Designed to Race' ". . . I seldom had anyone in my way, I had the road to myself, I could get on with the job . . ."

There are riders who favour a certain part of the Course, and both Joey Dunlop and Steve Hislop liked the twisty 'between the hedges' stretch from Ballacraine to Ramsey better than the featureless miles over the Mountain, while nine-times TT winner David Jefferies really enjoyed the Mountain, saying: "As I come out of Ramsey, it's tail up and head down. I just go for it and blast over the Mountain as one section". World sidecar champion of 1977 George O'Dell was another man who liked the Mountain section of the course, as did Carl Fogarty who thought that the open bends he could see through suited his short-circuit style. But Phillip McCallen was a rider who did not believe in favouring any particular section, saying: "start to like or dislike sections and you start slowing down here, going faster there and you're soon in trouble at a place like the TT".

'Isle of Man TT Riders Guide'

Plenty has been written down the years on how to ride the course, although some efforts were more casual guides than serious instruction. Now a bit dated, the 'Isle of Man TT Rider's Guide' of former TT winner and long-time competitor on the Mountain Course, Ray Knight, really presented a from-

TT and MGP winner Richard 'Milky' Quayle makes a serious point to Steve Plater during a course-learning lap before Steve's TT debut in 2007.

Mountain milestone

the-saddle account of how to ride the course, for the benefit of those who were about to tackle it from a similar position.

Today many riders watch video recordings and play video games of the races, and whilst that has not replaced the need to put in real learning laps, their virtual rides give them a head-start. But sidecar racers need to be careful with their viewing because three-wheelers take a different line to solos in some places, and they also have to learn the racing line as a team.

A successful sidecar driver from the pre-video era was Colin Seeley and one of his course-learning methods prior to his debut in 1961 was to drive round it in the dark in his van at speed, using only dipped headlights. In Colin's words "that really made you concentrate". A year before Colin made his debut, the first Suzukis entered the TT. Team manager Jimmy Matsumiya visited the Island for 10 days in chilly February to reconnoitre the course and plan accommodation, garaging etc. During one of his laps, Jimmy delegated to an assistant the job of lying on the bonnet of a borrowed Morris Minor and filming the line through each corner.

Wet and Windy

There is much more to learn about the Mountain Course than the direction of the corners. Experienced riders get to know what to look for in respect of the Manx weather. They are aware that conditions can change from sunshine to cloud to rain during a race, and that all such conditions may be encountered on just one lap at different parts on the course. They also get to know the effect of the often strong wind – what impact it will have on gearing (and thus fuel consumption), where it will hit them when passing gaps in hedges, gateways, etc, and where it will result in the road becoming littered with leaves from overhanging trees, for all can affect their preparation and race performance. Whilst they must be prepared to cope with such obvious adverse conditions, bright sunshine can also bring problems, for during practice there are places where their view of the road ahead can be seriously reduced by the low sun shining straight into their eyes.

Joey Dunlop prepares to set out on yet another lap of the Mountain Course.

First Lap on Closed Roads

New for the 2004 MGP was that the opening lap of practice was reserved for Newcomers and to help their course learning they rode it under the control of Travelling Marshals, who headed batches of them in a swift but steady ride around the Course. The same procedure was extended to TT Newcomers in 2005 and it was a rather more controlled version of what Travelling Marshals would do informally in the early post-war years, gathering half a dozen first-timers and conducting them round after other riders had started practice. Newcomer at the 2006 TT was young Manxman Conor Cummins. He really did not need an introductory sighting lap behind a Travelling Marshal, for living in Ramsey and working in Douglas, he went to work via the Mountain section of the course and came home via Ballacraine, Kirk Michael and Sulby, thus putting in a full lap each day.

Surprise

When a Newcomer has put in long and expensive hours of course learning involving visits to the Island prior to the event, official laps of practice, many hours of laps outside official practice, and has reached the stage where the appearance of the course has been committed to memory along with all the course-side features to be used as braking points, etc; race-day can bring a nasty surprise. This is caused by a large influx of spectators. They crowd the roadside banks, obscuring previously learnt braking points and their presence creates a completely different backdrop to corners than those fixed in the rider's mind. It is just another example of what makes Island racing so different.

The length and complexity of the TT lap is part of its attraction, but when it comes to its learning, it must be a sobering thought for a Newcomer that one lap of the Mountain Course is nearly equivalent to the combined lap

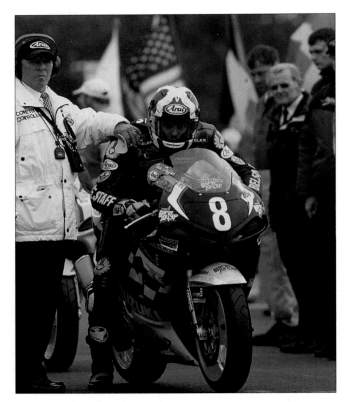

Share the rider's tension at the start of a race.

Early roads and riding

Rob Fisher knew where he was going when he rode to victory with Rick Long in the 2002 Sidecar TT.

distance of all the circuits used in a year of Moto GP meetings. Various attempts have been made to count the number of bends with Geoff Davison claiming 219 in the late 1940s, while others seem to accept a figure of about 260 today and the 'Guinness Book of Records' saying 264.

Even when a rider has gained the necessary course knowledge to ride to the maximum over this demanding of circuits, that knowledge must be regularly polished, both to maintain the pace previously achieved and to shave fractions of a second off subsequent laps, for quoting Steve Hislop: "you never stop learning at the Isle of Man" and Steve's dedication was such that he tells how he "even used to lie in bed at night and visualise my way round it – every corner, every bump, every straight", lap after lap until he fell asleep. Perhaps that is just part of the reason that Steve, tragically killed in a helicopter crash in 2003, took eleven TT wins in his racing career.

TT Landmarks

The dedication of great riders to master the TT Mountain Course during a century of racing has seen lap speeds rise enormously:

First	50 mph lap during a race	Frank Philipp	1911
	60 mph	Jimmy Simpson	1924
	70 mph	Jimmy Simpson	1926
	80 mph	Jimmy Simpson	1931
	90 mph	Freddie Frith	1937
	100 mph	Bob McIntyre	1957
	110 mph	John Williams	1976
	120 mph	Steve Hislop	1989
	130 mph	John McGuinness	2007

Course preparation

Growth in speeds has been achieved against an ever-changing background involving successive generations of riders, on vastly different machines and with constantly changing course conditions.

With the exception of the forced stoppages created for several years by two world wars and some shorter interruptions, the wheels of Island racing have really never stopped turning during the past century. It is virtually a 365 day-a-year job for the ACU and Manx authorities to run the actual races, allied to much forward planning in maintaining and preparing the many miles of the Mountain Course to make it a stage fit for the greatest road-race in the world.

For the Manx People

In 1911 the population of the Isle of Man was less than 50,000 and there were only some 200 private cars and motorcycles registered for use. Today it is home to some 82,000 people who enjoy an exceptionally high degree of personal mobility.

The primary job of the Manx Highway Authority (that has gone by several names down the years) is to keep the roads of the Mountain Course fit for use by the Manx public, because for 48 weeks of the year those roads are not a race circuit. However, discussion with engineers currently and formerly responsible for highway works, show that the TT and MGP are never far from their minds when they consider works that affect the Mountain Course. Indeed, to keep the roads fit and available for racing involves a rolling procedure, year on year, that requires advance planning, the implementation of legislation, budgetary provision, regular maintenance, alterations and improvements; all of which calls upon the century of experience held by local people and organisations in the running of the massive enterprise that is Island racing. It does not just happen!

Getting Ready for Racing

Legislation exists that permits the Island's roads to be closed for racing, but application to Tynwald is required each year, for confirmation that they can be held.

Over each winter, representatives of the ACU (organisers of the TT) and the Manx MCC (organisers of the MGP) carry out course inspections and identify any small items of work they would like the Highway authority to carry out. Most will be done in the Spring, although major repairs and alterations may involve long periods of works outside race times and the regular programme of surface-dressing with tar and chippings that sees several stretches of the course treated each year, is normally done between TT and

Mountain milestone

MGP. The days of the horse and cart to haul materials and an army of men with wheelbarrows to move and spread them is long gone, and virtually every process is now mechanised. Even in 1922 a report at the end of April told "the Island is literally buzzing with steam-rollers".

That mechanisation extends to the cutting of countless miles of grass on the roadside banks in the two weeks before racing, plus constant sweeping of the course prior to and during race fortnight, for heavy wind or rain can bring down leaves and branches and create standing water from blocked drains. That is not acceptable for racing, so out go the sweepers. Such activity seriously affects the daily travels of Manx road users, who become very aware that racing is on its way again.

A pre-race process that has yet to be mechanised is the re-painting in black and white of roadside kerbs on bends. Another important job is the erection of nameboards (some are left in place all year), warning arrows, notices, air-fencing and straw bales; work that commences in April each year.

Ready to Race

The old procedure of closing the roads for practice and racing by sending an official car around the course is no longer used, because today the roads close at times prescribed by the Road Closing Orders that are published in the local press. For many years practice has opened on a Saturday evening (but not at the 2011 TT), and as 18.00 hours arrives on the appointed date, the roads of the Mountain Course go strangely quiet. If all is well, with the correct number of marshals in place and no errant cars parked on the circuit, the first bikes blast into life some twenty minutes later, heralding a fortnight of high-octane activity over a course that has no equal.

TT opinions

Most riders who have raced over the Mountain Course are positive about the experience, whilst also recognising the risk involved. Here are just a few opinions from down the years:

- Jake de Rosier was a top American rider who rode his only TT in 1911. When a local reporter asked his opinion he said "I think the course is an ideal one for testing a machine, and the man who wins deserves all he gets". After a short pause he added "the course ain't no birthday party".

- TT star of the early 1930s Tim Hunt, also successful on Continental circuits, rated Island racing the best, describing it as "the only course in the world".

- Talking about the establishment of the World Championships in 1949 (of which he won the 350cc class), TT winner Freddie Frith said "The TT was the Blue Riband of racing then, worth more than a world championship to people in the sport".

- Walter Kaaden was the respected race engineer of the East German MZ concern, and in the 1960s he described the Mountain Course as "the ultimate test in the world for a two-stroke".

- After a spate of machine problems with the fast but temperamental Yamaha two-strokes of the late 1960s, Phil Read said "That Mountain Course was designed to produce great bikes as well as great riders".

- Peter Williams "… there is no doubt that it provides the greatest experience of fast motorcycle riding in the world."

- Multiple TT winner Mick Grant claimed: "It's the finest race in the world – unconditionally, the name of the game is road-racing, not circuit".

  Mick had much short circuit racing success up to GP level, but it was at the TT that he got most out of plying his craft as an exceptional racer, describing the course as "the most hallowed stretch of tarmac in motorcycling history".

- Writing in the 1980 TT Programme, star of the time Alex George said "The Isle of Man to me represents the ultimate challenge in racing that is unequalled at any Grand Prix or any other race circuit in the world.

  "It really is the greatest test for man and machine against the clock and fellow riders, and it could not be organised anywhere else but the Isle of Man".

- Sidecar world champion and TT winner of the 1980s Jock Taylor said simply "I still regard the mountain circuit as the ultimate challenge".

- "The normal risks of racing are multiplied ten-fold at the TT" said eleven-times TT winner Steve Hislop, who added "It is undoubtedly the most dangerous racing event in the world . . . it's an endurance test as much as anything else and you can't afford to lose concentration for a split second or you are quite literally taking your own life in your hands. It is an event like no other on earth".

- One of Steve Hislop's biggest rivals was Carl Fogarty who, through the activities of his TT-racing father George, had been brought up with Island racing.

  Carl went through a range of feelings towards the TT, starting as a young racer and claiming "The TT, an event that I had wanted to compete in ever since I was a young kid", through "riding a 37½ mile circuit over mountains and through tiny villages, at touching distance from the spectators and occasionally touching the kerbs, is a fantastic feeling".

- "You've got to ride your own race, respect the course and ride as smoothly as possible to be fast. You can't force it" says Jason Griffiths who achieved multiple podium places and now runs a motorcycle business in Castletown on the Island.

- "It's the best track in the world – the TT is the best thing in the world – ever", said Guy Martin in 2008, as he strove to meet his life's ambition of a TT win.

- After over 80 years of close involvement, Murray Walker wrote in 2008 "In terms of stature, history, spectacle and human endeavour, the TT is still right up there amongst the greats of motor sporting events – a fantastic and charismatic phenomenon that I believe is without equal in the entire history of two and four-wheel racing".

Mountain milestone

MOUNTAIN COURSE

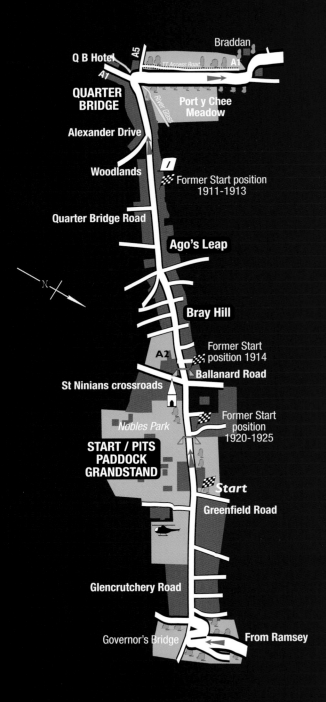

CHAPTER TWO

The Start to Quarter Bridge

The Start, Pits, Paddock and Grandstand are at the hub of race activity, being located where everyone gathers to deal with practice and pre-race preliminaries like signing-on and scrutineering, before riders are set off over the Mountain Course, either to hone their skills in practice sessions, or to challenge for a Tourist Trophy on race days. Also serving as a centre for press and broadcasting activities, for some spectators the Grandstand is the only place to watch the racing, but as the following 37¾ mile journey will show, the Mountain Course offers a wide selection of spectating opportunities and the TT experience is not complete until several have been sampled.

The ACU's intention that use of the new Snaefell Mountain Course would push the development of variable transmission met with success and at the 1911 TT it was possible to see examples of epicyclic hubs, epicyclic engine-shaft gears, sliding countershaft gears, and expanding drive pulleys. They were modest developments, but a great improvement over the previous single fixed gear arrangement, which compromised both hill climbing and speed on the level.

The course used in 1911 was almost the same as today's, except that at Cronk ny Mona riders turned right and rode through Willaston to St Ninian's crossroads where they turned right down Bray Hill. The section of course that now includes Signpost Corner, Governor's Bridge and Glencrutchery Road was not brought into use until 1920.

1911 Start

Four distinct starting positions have been used for motorcycle TT races run over the Mountain Course, all in Douglas. The first was situated on the flat section of Quarter Bridge Road after the bottom of Bray Hill, before it started its descent to the Quarter Bridge and was used between 1911 and 1913. Scoreboards were positioned at the side of the road to record riders' progress, the timekeepers had a small wooden shed to work in, the press occupied a room in an unfinished house close to the line, spectators gathered in a temporary grandstand and Clerk of the Course Mr J.R. Nisbet controlled proceedings with the aid of a megaphone. As there was no room for pits to be provided at the start for the 1911-1913 races, refuelling was allowed at two other locations: one in front of Braddan Church and the other shortly after leaving Parliament Square in Ramsey.

Top of the Hill

In 1914 the Start was moved to the top of Bray Hill, just south of St Ninian's crossroads, and a big entry of 159, (48 Junior and 111 Senior) showed that the TT had really caught the imagination of riders and manufacturers, with almost 30 different makes entered.

The new location was used for pre-race machine inspection, where the 'Clerk of the Scales and Official Measurer' ensured compliance with regulations that included minimum weight limits of 132½ lbs for 500cc machines and 110¾ lbs for 350s. Scoreboards were erected and Pits were provided alongside the start for refuelling - the only place where the taking on of petrol was permitted - whilst behind the Pits a temporary Grandstand was provided to hold 1,000 people at 5 shillings a head. After experimenting in 1913 with running each race over two days, the Senior and Junior events of 1914 reverted to single races run on separate days, the former covering 6 laps and 225 miles, the latter 5 laps and 187½ miles. All riders (except those on Scotts which had kick-starters) had to bump-start or paddle their machines with cold engines, so the slight down-grade away from the Start was welcomed.

Although the start line was on Bray Hill, the timekeepers' box was placed 80 yards down Ballanard Road, which was just before riders who were finishing the race took the right-angled bend onto Bray Hill. It was this

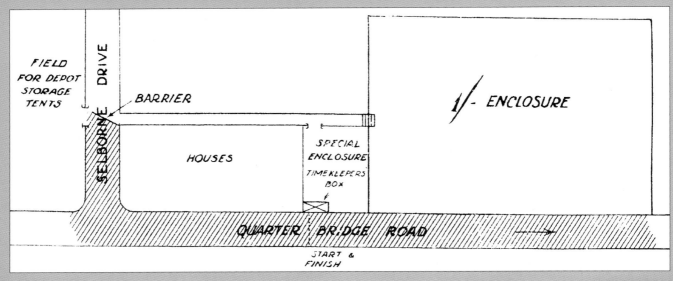

Top: 1911-1913 Starting area.

Middle: : Riders getting away at the start of the 1921 Junior TT.

Left: The current Grandstand, Control Tower and Pits.

about a ruling that engines had to be cut at a pit-stop.

♦ In the 1920s, spectators in the Grandstand were entertained by a band that played before the race, at slack intervals during the race, and at the finish. In 1922 they listened to "a small orchestra" and as Manxman Tom Sheard crossed the finishing line to win the Junior race it struck up with 'See the Conquering Hero Comes'.

♦ After several podium finishes, Graham Walker finally achieved a TT win in 1931 when he brought his Rudge home first in the Lightweight race. A report of the time said "Mrs Walker and her boy watched the race from the Grandstand today". The boy was Murray Walker whose TT and MGP connections started in 1925 and extend to this day.

♦ In the early post-war Clubman's TT races, riders were required to kick-start their machines at the start. For many this was the toughest part of the race and, if unsuccessful, they had to push to their pit and change a plug. It was the time of lowly 72 octane Pool petrol and some riders sought to increase their chances of getting a good start by squirting neat lighter fuel into their carbs just before the off to get more volatility into the mixture, whilst in his Senior Clubman's TT winning ride of 1949, Geoff Duke used a kickstart lever into which, contrary to the rules, he welded an additional 2" of length for increased leverage.

♦ When sidecars were still bump-started, passengers had to be quick to jump aboard after the engine fired. Several have been left behind.

♦ The 1970 Production TT featured a Le Mans type start where riders lined up on the opposite side of the Glencrutchery Road to their machines. Upon receiving the starting signal they sprinted across to where helpers held their bikes, then started their engines and disappeared in a weaving and jostling massed getaway. Previous year's winner on a Triumph Bonneville, Malcolm Uphill, had an injured left foot, so was given dispensation and allowed to have someone else do the running for him. Whilst Malcolm could not run when off his bike, he proved a speedy 'runner' when on it, bringing his Triumph Trident home in first place in the 750cc class ahead of Peter Williams and Ray Pickrell on Norton Commandos.

♦ Never far from controversy, Suzuki-mounted Graeme Crosby described what happened on the start line for the 1981 Formula 1 race: "My bike got a puncture as we were joining the field on the road at the start. The mechanics had to take the bike out of the line-up and fit a new rear wheel, changing the sprocket at the same time. I stood at the start-line waiting for the bike to return and we were ready to roll just 10 seconds after I should have started at number 16".

Officials decided that Crosby had to start at the back of the field, without any credit for the 4 minutes or so that he lost. With Joey Dunlop taking an early lead and Ron Haslam second, Honda were quickly in control of the race while Crosby was left with the task of fighting his way past the 60 riders who started ahead of him. This he did to good effect, setting a new lap record on his final lap and seemingly finishing third. Haslam had taken over the lead from Dunlop at mid-distance after Joey had tyre troubles and changed a wheel at his pit-stop, so 'Rocket Ron' came home for what he hoped was his first TT win. Indeed, he was garlanded and feted as the winner, much to Honda's delight.

After a read of the rule-book, Suzuki submitted a protest over the time-penalty imposed on Crosby at the start of the race. The Race Stewards decided that Suzuki were right, that Crosby's race-time should be calculated on the period that he was actually racing (excluding the time he was waiting to start), and that as he was the rider who completed the race in the shortest time, he was the true race winner by 2 minutes and not the luckless Ron Haslam. Honda did not take kindly to the decision and for the prestigious Classic race at the end of TT week, their riders came to the line in black leathers and their bikes were painted black to register a protest against the

The descent of Bray Hill when the TT first used the Mountain Course.

organisers earlier decision. It did them little good, for Suzuki took first and second places with Graeme Crosby and Mick Grant

Whilst the Grandstand offers the best viewing of the startline and refuelling action - at a cost - free viewing is possible from behind the wall of Nobles Park as the riders leave the line, pass the Grandstand and head for St Ninian's crossroads, although commercial activities have restricted the space available of late.

BRAY HILL

Given the signal to start, competitors leave the line at the present Grandstand under maximum power, changing up through the gears as they swiftly gather speed on the slight downgrade of the Glencrutchery Road.

Riders are barely settled on their machines before they are airborne over the crossroads by St Ninian's Church, streaking under the pedestrian footbridge and then facing a slight right and left curves as the road plunges out of sight down the notorious 1 in 8 gradient of Bray Hill.

With a name derived from the Celtic word Brae meaning a rough hillside above a river and formerly called Brae Road, today's Bray Hill is talked of with awe by spectators and with bravado that thinly disguises a background of apprehension by riders, for it is a fierce introduction to the demanding lap that lies ahead, having been likened to dropping off the edge of a cliff.

The course now runs between kerbs, pavements, garden walls and houses on either side and has a relatively smooth and cambered tarmac finish, but in the TT's early years Bray Hill was still part of the open countryside, being an unsurfaced road that ran between hedges and banks until it reached the first houses located at the bottom of the hill, at what was then the outskirts of Douglas. A local newspaper reported at the opening of practice for the first TT over the Mountain Course in 1911, "Bray Hill is in a shocking condition and unless it is seen to an accident will occur. The Borough Surveyor promised yesterday to have it brushed and watered but it has not been done".

Although required to be taken at speed, for many years a rider's progress was governed by the hill's bumps, with those created where side roads joined being particularly bad. Former TT winner Graham Walker wrote in 1935 of there being "three humps, each one steeper than its predecessor", whilst another report of the time said "the fast machines appear to hit the high spots only and to be in the air for the greater part of the time". Former winner Wal Handley was spectating at the Junior race in 1937 and wrote of Jimmy Guthrie's passage "he appeared over the brow of the hill at such a speed . . . he swept right across the road and came down the hill in a series of terrific

Maximum concentration from John McGuinness, as he hits 180 mph between the walls and kerbs of Bray Hill.

The Start to Quarter Bridge

Giacomo Agostini in full flight on his MV Agusta.

between large Edwardian houses, passing the one mile completed marker and the spot used as the start for the first TT races over the Mountain Course from 1911-1913. It was here that 'Mavro' Mavrogordato experienced the component failure that every rider dreads on the Mountain course. Entered for the Senior class of the 1927 Amateur race (predecessor to the MGP) on a 500cc Scott that delivered 26 bhp, he described how "On almost the last day of practice, I had just breasted the rise after Bray Hill, flat out and 'howling' well, when it happened. There was no warning whatever and the first inkling I had that everything was not going to plan was when I found myself somersaulting through the air. Then I rolled - how I rolled! - along the road, which is slightly downhill at this point, with the bike and the accompanying fireworks display careering along behind". With no one around, Mavro had to rely on marshals from Quarter Bridge to come and gather up him and his

Despite the easing of Ago's leap, today's riders still get the front wheel up, as John Burrows shows.

machine, although neither was badly damaged. The reason for parting company with his Scott was that the bicycle-style bracket securing the front mudguard to the underside of the fork crown fractured and allowed the mudguard to drop onto the wheel.

The boundary wall to the large house 'Woodlands' now looms high over riders as the cambered road curves left and seems to narrow, just before it begins to drop at the point where Alexander Drive meets the course. This area is steeped in motoring memories, for very early car races, which slightly pre-dated the motorcycle TTs, started from outside the entrance to 'Woodlands' which is on Alexander Drive. There drivers enjoyed a useful downhill run on what some called Brown's Hill, before joining the present course on the Quarter Bridge Road.

At the junction of Alexander Drive with Quarter Bridge Road (some people know the spot as Eyreton and until recently that was the name of a house there) can be found the Wal Handley Memorial Seat, which was placed there at the time of the 1948 TT.

While Wal is best remembered for the spot near the 12th milestone called Handley's Corner, his name is also attached to a TT award called the Walter Handley Trophy. Given variously to the rider setting the fastest lap in the Senior and then the fastest lap by a sidecar, it now seems to have fallen out of use. Winner of four TT races, Wal was killed while flying for the Air Transport Auxiliary force in 1941.

Approaching Quarter Bridge

A report from 1936 told "the approach work from 'Woodlands' was amazingly fast, and machines were flying with the greatest of ease at the rapid top bend". Still travelling immensely fast, today's riders are faced with reducing speeds by well over 100 mph as they plunge down the sharp drop to the Quarter Bridge. Partially overhung by trees creating changing light conditions, and well laced with oil and diesel droppings from traffic queues

Mountain milestone

Tony Wright has to stand and watch as his BSA Gold Star is consumed by fire from a full tank of petrol at Quarter Bridge, soon after the start of the 1960 Senior MGP.

which form here, it is a stretch that has always required hard but skilled use of the brakes, as the rear wheel hops over the bumpy and, at times, slippery surface.

QUARTER BRIDGE

No longer looking like a conventional bridge but nevertheless spanning the River Glas, Quarter Bridge originally marked the boundary between the two Douglas quarterland farms of Ballabrooie and Ballaquayle, thus acquiring its name. It was described in 1911 as "dangerous, because it was banked the wrong way, there being a drop of one foot from the centre of the road to the gutter on the left".

From 1911 to 1913 Quarter Bridge was the official point for the despatch and timing of riders on practice laps and in an unusual show of officialdom, in 1911 an early-rising policeman checked riders' licences and machine registrations here. Seemingly, the despatch of riders in those early years was not quite the orderly one-at-a-time business of today, for a report from 1913 said that at the official practice start time of 4.30 am, 15 riders started off together "with quite a hearty crackle of exhausts". What fun they must have had over the ensuing 37 miles!

A Busy Spot

With primary routes into Douglas converging at the Quarter Bridge it has always been subject to much traffic. Reports tell of it being widened by fifteen feet on the south side in 1912 at a cost of £400, and in 1922 it was said in respect of the racing at this point: "Quarter Bridge may be taken at speeds up to 30 to 32 mph by a good rider".

A roundabout was formed here in 1959 that was revamped in 1963, a second roundabout was added in 1987 and it has been subject to several more alterations as Highway Engineers struggled to cope with increasing traffic flows, including works prior to the 2011 TT costing some £130,000. That work was in lieu of a more grandiose scheme estimated at £4 million pounds involving the creation of one big roundabout and demolition of the Quarterbridge Hotel. While that project has been abandoned, as with many roadworks on the TT Course it is good to know that the engineers started their design of the new roundabout from the riders' apex of the Quarter Bridge. So that point was to remain a constant, everything else was to flow from it and the riding experience would have remained largely unchanged. Similarly, when the £130K project was announced, the official statement explained that ". . . the carriageway was getting rutted, particularly on the racing line and would need to be renewed". How many Highway Authorities give consideration to the racing line in their works?

Careful!

The first lap of a race can prove particularly testing at Quarter Bridge, for it is a moment when pent-up aggression released by the start is still looking for an outlet, bikes are carrying a full load of fuel, while tyres and brakes are not up to working temperature. All this has to be handled over a surface that despite being smoothed and re-laid at regular intervals, quickly develops a new set of bumps. Experienced riders know that they are not going to win the race here on the first lap but that with slight misjudgement they could easily lose it. Relative caution is called for, but spills and other incidents are fairly frequent.

Back in 1914 T. Greene arrived here with an overfull tank after a mid-race stop for fuel. Surplus petrol splashed out over the hot cylinder and man and machine were engulfed in flames. Spectators managed to 'extinguish' Greene but his Rudge burned out.

Even riders with many laps under their belts find this a tricky spot, and several well-known names have come to grief, including TT winners like Tim Hunt, Geoff Duke and Ian Hutchinson.

Having changed from the Norton machinery that he had been riding so successfully for the previous four years, Geoff Duke was still getting to know his 'works' Gilera at the 1953 TT. Being pressed hard by Ray Amm in the Senior race, Geoff held the lead on the fourth lap as he entered Quarter Bridge, but accelerating hard on the exit he drifted a little wide and his rear wheel hit a patch of molten tar. Geoff tells in his autobiography 'In Pursuit of Perfection' how "The Gilera went into a full-lock slide and, unaccustomed to a light flywheel, I snapped shut the throttle and the engine stopped. The rear tyre bit, the machine jerked upright and went into a lock to lock wobble before we hit the deck". Geoff split the petrol tank of the Gilera in the fall, so was out of the race.

Barry Sheene made his first Isle of Man appearance to contest the TT in 1971 but fell from his Suzuki on the exit from Quarter Bridge in wet conditions, whilst contesting the lead of the Lightweight 125 race. Although 'The TT Special' reported that "he received no injuries and was as cheerful as ever afterwards", Barry never returned to race at the TT.

Not all falls at this spot are due to rider error. A practice report from 1925 tells how "C W Johnston (348 Cotton) came down the hill practically all out, relying on his brakes to pull him up for the corner, but with both brakes hard on he found himself sliding in a straight line at almost undiminished speed". The machine hit a refreshment stall at the bottom and 'Paddy' Johnston was thrown over the counter, bruising his ribs and side, besides grazing his knees. It turned out that his failure to stop was not down to the Cotton's brakes but to the fact that since the previous day's practice the Highways Board had sprinkled loose stones and sand to cover molten tar".

Probably the most expensive fall at Quarter Bridge was that of Ian Hutchinson when lying third on the last lap of the 2009 Senior TT. With two wins to his name from earlier in the week, 'Hutchy' was on course to take the Joey Dunlop TT Championship Trophy for the best performance of the week, worked out on a points basis in relation to finishes. Perhaps more important than the prestige to be earned from winning the trophy was the fact that it came with a cash prize of £10,000. When added to the £9,000 he would have won for finishing third in the Senior and unknown amounts of 'bonus' money that would have come his way, his simple spill on someone else's oil -

No problems for a smooth Jimmy Guthrie as he takes Quarter Bridge on a Norton in 1936.

Mountain milestone

marshals were showing warning flags - cost him well over £20,000. Undeterred, 'Hutchy' returned in 2010 and rode into the history books with five TT wins from five starts.

A Popular Position

Quarter Bridge has always been a popular spot for spectators and at the first TT on the Mountain Course in 1911, the landlord of the Quarter Bridge Hotel sold tea and cakes at early morning practice sessions. For many years there was a temporary grandstand in front of the hotel, but in later years spectators stood behind bales and then behind metal barriers. Quarter Bridge still has the desirable ingredients of easy access, free viewing and refreshment facilities (though the former The Nook Café is now closed), whilst also allowing the option to move during or between races.

Some spectators choose to sit on the grass bank on the left of the entrance to the bend, while others used to crowd behind the protective bales in front of the Quarter Bridge Hotel, so getting a head-on view of riders coming down the hill, before they turned right and headed towards Braddan Bridge. For many years intrepid local youngsters crossed under the bridge during early morning practice to get to the other side.

Norman Norris on his Dunkeley Precision in 1914. His crash-helmet has obligatory cut-outs around the ears to allow him to hear anyone overtaking. Note the leather toolbox supplemented by smaller containers strapped around his waist, the spare inner-tube around his shoulders, spare drive belt on rear carrier, hooked ends to the handlebars and the simple stirrup front brake.

Mountain Course Moments

After the 1911 TT the organising ACU sent a letter of thanks to the Island's Highway Board and made a donation of £42 to its funds, a figure that dropped to £25 the following year then increased to £50 in 1913.

◆ In 1911 the magazine 'The Motor Cycle' was already referring to the Isle of Man as "The Motorcyclist's Mecca" and by 1914 'Motor Cycling' magazine organised day trips to the TT by train and boat, something it did for 50 years.

◆ Given the primitive state of the roads, competing at the TT in the early days was as much a challenge to a rider's courage and stamina as it was to his machine's speed and reliability. Competitor 'Pa' Applebee said of the time "a rider was considered amazingly fresh if he could stand at the end of a race". His point was particularly relevant in 1912, for when practice started, local rider Duggie Brown announced after one lap "the roads are shocking, almost unrideable". Things got worse when that year's Junior race was held in very wet conditions which proved particularly challenging to those still riding belt-driven machines, for the combination of rain, mud and, usually, a bit of oil from the engine, was more than enough to cause belt-slip on inclines, resulting in loss of power and making the rider wish for chain drive; something that many concerns were moving over to.

This is Joey Dunlop tackling Quarter Bridge early in his TT career in 1977. He was to ride the pronounced cambers here many more times.

The Start to Quarter Bridge

The passenger could be urging the driver to wind it on as they exit Quarter Bridge and have a nice straight run to Braddan.

◆ Details of Marshals, Medical and Telephone Stations were given in a race handbook produced by the ACU for the 1913 event, and it said "At the following points of the course, regarded as more or less unsafe, Marshals will be stationed". It listed a mere 22 Marshals Stations, with 6 Medical Stations and 7 Telephone Stations, the latter being augmented by the use of the Railway Company's telephones at Kirk Michael, Ballaugh and Sulby. That was the expanded coverage for race-days, during practice it was less.

◆ Frank Bateman was killed in the 1913 Senior TT when a tyre burst on the Mountain descent. For 1914 the race regulations required the fitting of security bolts to rims, however tyres were still very basic and punctures and bursts were quite frequent. There were other accidents at the 1913 TT that raised concerns about safety and some asked for a slower race, possibly to be achieved by restoration of the fuel ration that applied to the 1907 and 1908 events. The organisers declined to take such a step, but whereas previously there was no standard form of rider headgear, in 1914 the ACU insisted that all competitors wear crash helmets to a specified standard. The recommended design of helmet cost 18/9d and came with an ACU stamp of approval.

◆ For many years the 'weighing-in' procedures were carried out the day before a race and riders' bikes were impounded overnight in a marquee in the Paddock, under the control of an ACU official and a patrolling policeman. However, in 1914, rumours of possible disruptive overnight action by suffragettes, saw a squad of soldiers with fixed bayonets take up guard duties around the marquee at nine o'clock on the evening before the Junior race.

Ruth Sheard quotes in her book 'The Modest Manxman' about her TT-winning grandfather Tom Sheard, that on the night before the 1913 race "employees of the Highway Board found considerable quantities of broken plate glass scattered on the road at Kirk Braddan, Pear Tree Cottage and Bray Hill. On Creg Willys some dangerous nails had been placed in the road. The indications pointed to the glass and nails having been planted of deliberation with a view to wrecking machines. Whether the militant suffragettes are responsible for the outrage there is no evidence to show". One of the most militant of them, Mrs Pankhurst, was born on the Island.

Mountain milestone

TT MOUNTAIN COURSE

LEAVING QUARTER BRIDGE... BRADDAN BRIDGE... SNUGBOROUGH... UNION MILLS... BALLAHUTCHIN... BALLAGAREY... GLEN VINE... CROSBY... THE HIGHLANDER... GREEBA CASTLE... APPLEDENE... GREEBA BRIDGE... THE HAWTHORN... GORSELEA... BALLAGARRAGHYN... BALLACRAINE

CHAPTER THREE

Quarter Bridge to Ballacraine

Having plunged through the built-up area of Douglas, the course now turns west, taking the A1 Peel Road through the mixed going of the Island's central valley. As riders complete the turn at Quarter Bridge they are keen to open up for the half mile straight and mostly level stretch running alongside a former railway line (now converted to the TT Access Road) that leads to Braddan Bridge. However, they must not get too far out to the left on the exit at Quarter Bridge, for the course is bounded by a substantial stone wall that continues for the whole length of the straight on both sides, giving a tunnel-like impression at speed and being bounded by a pavement on the right-hand side. The wall on the left has several nasty stone projections that receive a fresh coat of white warning paint before each TT, while former noticeable undulations in the road surface that saw riders airborne have been eased down the years.

Over the far side of the left-hand wall and former railway line is the River Dhoo, which runs to join the River Glas before they enter the sea together at the place formed from their combined names - Douglas. Alongside the right-hand wall is a fine line of trees planted in 1935 to commemorate the Silver Jubilee of King George V. These are on the edge of Port y Chee Meadow, a flat open grassy area of which Lesley Quilliam says in 'A Gazetteer of the Isle of Man' that "though usually translated from the Manx Gaelic as harbour of peace, it is possible that it is fort of the fairies" and Canon Stenning wrote of Manx legend telling "that on a bright moonlight night the little people who are by description like Scandinavian trolls, may be seen gambolling all along the ancient fortress-ridge, covered with trees, towards Braddan Bridge". The hundreds of TT visitors who camp in this field will know little of such history or that it was once a tidal arm of the sea, but they do know that they are close to the race action as riders pass within a few feet of their tents and that they can enjoy the facilities of the Douglas Rugby Club, which today occupies the meadow with a pitch and clubhouse.

BRADDAN BRIDGE

Whilst it is a straight stretch of road leading away from Quarter Bridge, the full-throttle action which it encourages can create problems for riders when it veers slightly right as they reduce speed, drop one or two gears, before heeling left and right through the swooping bends over the former railway bridge. In 1950 the motorcycle magazines were critical of the riding standards of many Clubman's TT riders and of an evening practice session 'Motor Cycling' reported: "Of several phenomenal misjudgements of Braddan Bridge, easily the most outstanding was that of F.C.J. Collinge (Triumph), who, after a 70 yard skid, failed to take either the course or the Strang slip road, but hit the parapet". Riders who do take the slip road have to miss the substantial Jubilee Oak in the centre of the road, planted to commemorate Queen Victoria's 50 years on the throne in 1887.

Gordon Pantall slid off his Yamaha at Braddan in the 1971 Junior TT, and after taking a moment to pick himself up and dust himself down, he found a policeman wheeling his bike away. Prising it from his grasp, Gordon rejoined the race without great delay, but the spill may have robbed him of victory, for he finished in second place, just 36 seconds behind winner Tony Jefferies.

The approach to Braddan Bridge was increased in width in 1923, then the entire stretch of road from Quarter Bridge was reconstructed in 1924 and the bridge itself, that was built in 1793, was considerably widened on the right in 1927. The latter works, according to top man of the day Wal Handley, saving riders several seconds on their lap times.

A few years after these major works the spot was described as really fast with a magnificent surface that allowed the quick men to take it at a full fifty. In later years the stone parapets of the bridge were replaced by metal railings.

Braddan Bridge appears straightforward, but as former TT winner Ray Knight points out in his 'TT Riders

Mountain milestone

In this early morning practice photograph from 1935, Jimmy Guthrie has cranked left into Braddan Bridge and then leans his Norton to the right on the exit.

Guide': "the entry point is not immediately apparent" and "braking, too, is somewhat deceptive". It has caused many riders difficulties down the years and even Joey Dunlop admitted "I've nearly been off there a couple of times". But one of the saddest incidents involved French Newcomer Serge Le Moal at the 2004 TT. Having achieved his dream of riding the Mountain Course, Serge was just 1½ miles into his first practice lap when he lost control approaching the Bridge, crashed and was killed. It was an event which underlines the fact that racing motorcycles is such a potentially dangerous past-time, with perils waiting in even the most innocuous looking locations.

Braddan Bridge is visually dominated by the two stone Braddan Churches (old and new) that have both gazed down upon a century of racing. The location is sometimes referred to as Kirk Braddan, the name being derived from Saint Brendan or Brennan. In the Island's hey-day as a tourist venue, thousands of people would come from Douglas on summer Sundays to attend

A modern race-day scene from a much changed Braddan Bridge.

Quarter Bridge to Ballacraine.

Braddan Bridge is popular with spectators at race times, just as it was in this photograph from 25 years ago.

open-air church services here, amongst them TT fans during race fortnights. Such services are no longer held, but race spectators continue to come in great numbers, for it is a good place to watch riding styles varying from the smooth, swift and confident, to ragged and hesitant. The most popular viewing area is in front of the old church which provides raised viewing facilities for hundreds who, at small cost, can enjoy the timber benches and associated refreshment facilities provided to enhance their day's racing. Should racing be delayed, there may be opportunity to fill the time by viewing the collection of Celtic and Scandinavian stone crosses within the old church and the gravestones of some interesting Manx characters outside.

After its erection in about 1920, a stone war memorial served as a traffic island at the mouth of Vicarage Road that goes off to the left between the churches. A report told that riders "just scrape by the Memorial Stone", which has now been relocated to the churchyard.

Pit crews in attendance at Braddan when it was used for refuelling

Mountain milestone

Geoff Duke's line through Union Mills was illustrated in a Castrol booklet of 1950..

Re-fuelling

It was the area immediately in front and to the side of where spectators now sit that saw 'Braddan Depot' used as an official refuelling point at the 1911-1913 TTs, for there was no space at the start area on the Quarter Bridge Road. For some years after this refuelling point was dropped, spectators would crowd the outside of the bend, standing in the road in front of the present wall, confident that every rider would get round without incident.

Flip-flop

For riders, all attention must be focussed on the road at Braddan's flip-flop 'S' and although aware of the banks of spectators, they must concentrate on getting its left and right-hand bends to perfection, perhaps sacrificing a little speed with a close, almost shoulder rubbing the railings left-hand entry, to achieve a line permitting the fastest exit from the right-hander, thus gaining better drive and higher speed on the up slope of the Snugborough straight that follows; with the power of today's bikes seeing them lift the front wheel as they depart.

The bends here are the sort that encourage riders to push a little harder each time they ride them. One who pushed too hard, too early, was Reay MacKay on his 500cc Vincent back at the 1967 TT. Blaming his resultant spill on a new front tyre that was inadequately scrubbed, Reay collected two damaged ankles in the fall, though the bike was relatively undamaged. A substitute rider was found for the ageing Vincent - perhaps the last girder-forked bike to contest the TT and certainly the last Vincent to do so - and Melvin Rice brought it to the chequered flag for a Finisher's Award.

Despite the many works that have altered the shape of the road here, the bends of Braddan Bridge would still be recognisable to the riders of a century ago, even though they have been widened, better surfaced, walls have been replaced by railings and there are many white lines on the road. In addition, two mini-roundabouts were created in 2008 for ordinary traffic, without affecting the racing line.

SNUGBOROUGH

So far the ordinary road rider has been subject to a 30 mph speed limit from the start of the lap, but leaving Braddan Bridge this increases to 40 mph, for a short distance. Racers ignore such day-to-day constraints and feeling that engines, tyres, etc are warming up, they have the throttle wide open as they hit the stretch that leads to Union Mills and pass the second milestone. Having a continuous low stone wall on the left and footpath with hedge on the right as the road curves left, this stretch offers virtually no spectating facilities. It was once called Cronk Doo but is today known as Snugborough, and though now relatively straight with a slight uphill gradient, it was described 50 years ago as "a twisty stretch" having two speed-restricting bends in its latter part that were removed and the area widened in 1976, with further slight widening in the late 1990s.

The larger-than-life Liverpudlian Bill Fulton was a MGP and TT regular during the 1960s and 1970s and he had reason to remember the

Steve Hislop approaching the end of the Snugborough Straight in 1990.

Snugborough bends, for during Thursday afternoon practice for the 1970 TT he fell here and broke a leg. It was on the occasion that the organisers experimented with sending riders off in groups of 9 at 90 second intervals. Bill did not blame the new system for his downfall, but rider opposition meant that it was not adopted for racing.

UNION MILLS

Snugborough ends with a sweeping right-handed approach to Union Mills, then riders first curve slightly left, before dropping into the village with its elongated S bend that requires experience, skill and confidence to tackle 'on the limit'. On these demanding bends it is easy to pick out those who know where they are going from those who are doubtful that they have the right line. Indeed, a Newcomer who is passed by a top rider just before Union Mills can become thoroughly disillusioned, because although he may have been on the star rider's tail entering this testing location, the top man is likely to be just a fast disappearing speck in the distance when the novice finally exits the bends.

The first warning boards for Union Mills were erected in 1936 but with speeds on the approach governed by the Snugborough bends of the time, riders could focus fairly clearly on the tunnel-like entrance here and be confident of where they were going. Today the straighter road and faster bikes brings higher approach speeds, making it harder to read the entrance to the bends. Riders know that the sweeping right-hand approach past the entrance to the Snugborough Trading Estate is followed by a slightly curving left, before the main Union Mills bends of a falling right-hander followed by what is almost a double left. However, at high speed the entry detail is lost in overhanging trees, enclosing walls and houses, so there are now substantial white direction arrows on orange backgrounds to serve as reminders of the basic directions.

It was on the approach here that seasoned TT runner Ian Lougher came off at high-speed on the first lap of the 2004 Junior race. Following riders feared for Ian's well-being as they passed the scattered wreckage of his Honda, but he escaped injury and even went out and rode a practice lap after the finish of the race. Just over 40 years earlier Honda were into an all-conquering period at the TT, but the dangers of the Mountain Course were brought home to them in 1962 when Kunimitsu Takahashi crashed his 125 on the right-hander in Union Mills, wrecking the bike and seriously injuring himself. Going back many more years, when Tommy Hatch rode through on his Scott on the first lap of the 1928 Senior he got into a spectacularly long weave and skid, so decided to stop and pump some more air into his tyre!

Prior to the 1960s riders used a drainage pipe in the left-hand wall on the

Dropping into Union Mills.

entrance to the Union Mills section as a peel-off point for the falling right-hander that took them close to the right-hand kerb, past 'The Railway Inn' and over the old railway bridge and River Dhoo. The upper surface of this bridge received major remedial works in early 1998, resulting in the closure of the road to traffic for several months before the TT, but the bridge still causes race bikes to leave the ground here while cranked over.

Duke's Way

Sixty years ago Geoff Duke was the top TT racer of his era and he described riding through the main Union Mills bends from the peel-off point, saying, "Come across from the left of road, accelerating at about 70 mph in second gear. Important not to enter too fast to avoid drifting towards the wall beyond the cottage. Keep to centre of road till just before crown of bridge, then swing left, taking first left sweep wide to cut in close to the mill, thus avoiding outward drift on the exit, where change to third gear at 85/90 mph". Other riders of his and later eras stressed the need to get entry speed right, Mike Hailwood saying "go into them too fast in the wrong gear and you're lucky to come out".

The faster a rider's progress through Union Mills, the more physical effort needed to haul the bike upright after the entry right-hander to be able to crank it over to the left past the local post-office and filling station. Not everyone gets it right and the post-office has had unwelcome two-wheeled visitors at its door, but as Ray Knight put it "the fact that incidents seldom occur here is testimony to the restraint that its frightening aspect imposes on

How the buildings of the Union Mill used to crowd the edge of the course.

Union Mills in about 1910.

Mountain milestone

The sweep through Union Mills allows a fast passage for experienced riders.

the right-hand". Back in the 1930s some of the area in front of the post-office was heavily sandbagged to a height of nearly four feet (reducing the available width of road) and people were allowed to stand behind the wall thus created. In later years it was home to the young Gibb brothers, who went on to form the 'Bee Gees' pop group.

Steve Hislop stressed the need for the correct line right through this series of bends with "I've seen loads of lads go hurtling into the top of Union Mills, they're all on the wrong side of the road, they struggle to change direction and by the time they get to the post-office, they're actually shutting the throttle when they should have been driving from back at the Railway Inn".

After completing countless laps of the TT course, Joey Dunlop had the way to tackle these bends firmly in his head. He did not bother with detailed explanations, for he spoke of visualising and taking them as "one, two, three" - no problem!

Why Union Mills?

There was a mill here for centuries and as there was little else the spot was known as Mullen Dhoo or Mwyllin Doo Aah (the mill of the black water). When the mill was developed for the manufacture of clothing it became known as Union Mills, the name coming from the mill's products that were sold under the trade name 'Union'. Its buildings butted up to the left edge of the course before the present filling station, later being used as a laundry and eventually demolished.

BALLAHUTCHIN

If there is one point that the riders of yesterday and today are agreed on, it is that the rate at which they exit Union Mills determines their speed for the next mile and more. This is because after a couple of hundred yards of flattish going under an arch of massive beech trees, they are faced with the long uphill drag of Ballahutchin - known to some as Elm Bank Hill. One hundred years ago this was a demanding climb, with riders adopting a crouched position to reduce wind resistance while juggling their fuel mixture and spark controls in search of the best possible settings, as they plodded their way up past the third-milestone marker with leather drive-belts clacking and exposed valve-gear clicking. The rules of the time required them to start with cold engines, which meant there was always the worry of the bike seizing when subject to the heavy loading imposed by this upgrade early in a race, so those pioneer riders would give a couple of extra pumps of the manual oil-pump on the climb, adding a smoke haze to the dust cloud their passage had already created.

Engines could still seize on the Ballahutchin in later days, and an example came during practice for the 1985 TT when Pete Bateson went up flat-out on his 250 Rotax. Just before the crest his engine seized and although quick to grab the clutch-lever he had the double mis-fortune to have the clutch cable break. With engine and back-wheel locked and no way of freeing them, Pete fought the bike before parting company with it, sliding two hundred yards up the road on his back, and having both Joey Dunlop and Mick Grant come speeding by far too close for comfort, as he gathered his wits before scrambling off the road.

Today's bikes storm up the hill with front ends going light over its 'stepped' climb with an ease and speed that would have astonished TT pioneers; but it is still vitally important to come out of Union Mills as fast as possible, because a rider who does so can easily be travelling 15-20 mph faster up the hill than one on an equivalent machine who did not get the earlier bends right.

The prefix Balla - as in Ballahutchin and Ballafreer - can be found in many

Quarter Bridge to Ballacraine.

Wal Handley (8) overtaking Leo Davenport (5) on Ballahutchin in 1932.

for although such engines have a race life of some 620 miles on short circuits, the Mountain Course makes greater demands and they cannot be guaranteed to last the approximately 500 miles involved in practice and racing at the TT. As it is a lengthy and specialised workshop task to rebuild the most highly tuned engines, it presents a problem for those who are hard on them. Of course, lesser lights have to make their engines last rather longer and have no option but to 'make-do' with Paddock repairs.

Just as of old, riders pass nothing more than a few isolated properties as they climb between the hedges up the Ballahutchin and curve slightly right at Ballafreer without easing the throttle, before they reach the top of the hill.

As the course then starts a short but steep downhill run, it passes the entrance to Glen Lough Farm, where hundreds of visiting race fans pitch their tents in the field abutting the course. No longer do they get the wake-up call of a race bike passing just a few yards away during early morning practice as they did of old, but perhaps many are secretly glad of that, having spent convivial nights on site with other campers, or at nearby hostelries.

Manx place names and means homestead or farm. On Ballafreer's land are the remains of an old chapel (known as a Keeil) and legend has it that when heading to this Keeil, St Patrick stood on a thorn. He responded by laying a curse on the field where this occurred, saying it would never carry crops of grain.

Hard on Engines

Sidecars put big demands on their engines on climbs such as Ballahutchin and during the second Sidecar race at the 2007 TT, previous winner Nick Crowe set a new outright lap record of 116.67 mph before the engine of his LCR Honda blew here on the third lap. Discussion with his engine tuner concluded that Nick had run the engine for too many miles, so from thereon he vowed that he would not do more than four laps of the Mountain Course on an engine without a rebuild. A similar problem afflicts solo users of the current generation of the most highly-tuned 600cc engines,

BALLAGAREY

The straight downhill run past Glen Lough is followed by a short and very fast upgrade, after which riders must tackle one of the most fearsome corners on the course. Known to some as Glen Vine and earlier as Elm Bank, in official TT terms this is called Ballagarey, although it has been known to generations of racers as 'Ballascarey', as it sweeps right-handed between roadside walls that have nothing more than a few bales between them and a rider who gets it wrong at close to maximum speed.

Unforgiving is the word that best describes this point, although riders have been heard to use far stronger ones, particularly when first trying to come to terms with it. Although so very demanding, for it is approached "mega flat-out in sixth" (as John McGuinness puts it) over a brow where the front goes light, this is just an ordinary bend on an ordinary road to Manx motorists, and thus it receives very little special attention. The right-hand curb has been splayed and painted black and white on the apex, but that and the bales on the exit are the only concessions to racing.

Approaching the top of Ballahutchin is a right-handed sweep past Ballafreer.

Mountain milestone

RIDERS VIEWS . . .

Ballagarey may be a nasty wall-bound, right-hander on an ordinary Manx road, but it did not prevent Harold Daniell writing of how he took it "full blast in top" on his record-breaking 91 mph lap when winning the 1938 Senior TT.

◆ Twenty years later and John Surtees called this spot Glen Vine and wrote of it as "a corner I consider to be one of the most difficult to get right on the TT course. I always took very strict control of myself and the bike there, especially as the corner always seemed to catch whatever sun was going, and in those days the road surface would often break-up on a hot day. Then the bike would get very light going in at the peel-off point, but if you left it too late you'd be scratching off that wall on the exit in the most almighty fashion. There was also a very worrying telegraph pole bang in the line of fire on the way out. But if you did it all right, and went through absolutely on the limit, not only was it one of the most satisfying things on earth, but you also got a good run at the hill up into Crosby and down past The Highlander". It is interesting to note that John expected a good passage through Ballagarey to benefit his speed on the almost two mile stretch to The Highlander that followed.

◆ In Ray Knight's view back in the early 1990's, what he called Glen Vine was the only spot between Union Mills and Greeba Castle where a rider "should allow any slack in the throttle wire", and he advised first-timers to allow a fair amount until they were quite sure how fast it was. Speaking soon after Ray, Steve Hislop reckoned to go through in top on a 600 but to hook back a gear and take it through in fifth on the big bike, a view confirmed by Guy Martin ten years later with "I keep thinking I'm going to get Glen Vine flat but . . . no", whilst Ian Lougher explained "it's back a gear on the Superbike . . . there's a huge difference from the 600". Ian Hutchinson shared those

Riders lean into the start of Ballagarey in this photograph from 1991.

views, feeling that fifth allowed him to drive through the corner and thus hold a tighter exit line.

It was Guy Martin who brought the name of Ballagarey to everyone's notice when he crashed here while challenging for the lead in the 2010 Senior TT, miraculously escaping with minor injuries. Guy rated it a 170 mph get-off!

The day before Guy's accident, Paul Dobbs fell at the same spot during the postponed second Supersport race and was killed. Thus is the lottery of a high speed racing accident that occurs in a confined location bounded by kerbs, lamp-posts, garden walls, trees, etc.

◆ The last view of Ballagarey is from sidecar ace Dave Molyneux who, on his ultra low-slung outfit explained "I'm turning in before I can see

John McGuinness (3) and Ryan Farquhar (1) ride the apex of the long Ballagarey bend in the 2007 Junior TT.

Road works by Cedar Lodge in Crosby in 1926.

the corner, and the bike's in a three-wheel drift, flat out in top - I love this bit, so bloody fast".

The apex at Ballagarey is a pronounced point just before the traffic lights, but it is a 'long' corner and riders are cranked over well before and after the apex.

TT Racer?

There is little provision for spectating, but this is a spot where an ordinary road rider who perhaps fancies his chances of being a TT racer after having had a fast blast over the sweeping and open bends on the Mountain section of the course, may well find himself revising his views after seeing the degree of commitment required to effect the all-important fast passage through wall-bound 'Ballascarey'. For racers, this corner encapsulates everything that is good and bad about the course. Getting it right gives another adrenaline rush and the best possible speed over the fast section that follows, but getting it wrong just does not bear thinking about.

GLEN VINE

So, having come flat out through Ballagarey - or not, depending on whether it is a 600 or 1000, on the rider's degree of course knowledge, skill, courage, road conditions, etc, the full throttle drop that follows into Glen Vine (that was known to some as Glen Darragh) takes them past the fourth-mile marker, located close to the Glen Darragh Road junction, known earlier as Ballabeg.

Mile Markers

All round the course the organisers provide markers at each elapsed mile from the starting point at the Grandstand. On this stretch of the A1 from Quarter Bridge to Ballacraine they almost coincide with the permanent milestones of the Highway Authority, but the latter do not measure the distance from the start of the TT Course, for their starting point is from a milestone which used to be on the corner of Parade Street and the Red Quay in Douglas.

Surprise!

The excitement to be had from riding fast on the closed roads of the TT course is not confined to racers, for motorcycle mounted Travelling Marshals and even drivers of the 'Roads Open' cars have had their share. One former motorcycle Travelling Marshal who moved on to cars, still recalls an incident here with an almost unbelieving shake of the head. At the end of one practice session he left St. Ninian's crossroads in the 'Roads Open' car to wind up the racing and hand the roads back to the patient Manx public. At the top of the Ballahutchin he took a firm grip on the steering-wheel and set the car up at 100 mph for the demanding right-hander at Ballagarey. It was just after the corner as the road began to fall away downhill that, with gathering speed, he spotted the little grey saloon car tootling along at 30 mph - on closed roads! Fortunately it was going in the same direction as the 'Roads Open' car that had already started to pull towards the right-hand side of the road for the stretch through Glen Vine, but it did not prevent a flutter of the heart and the acquisition of a few more grey hairs for our driver.

CROSBY

Riders rush down into Glen Vine and keep to the right into Crosby Village which is in the Parish of Marown. The road here was widened in the early 1930s and it has one bend that is known to riders as the 'Crosby Left', but what a bend it is! This long, sweeping, wall-bound left-hander probably troubled the first riders to use the Mountain Course as they hit 60-65 mph over loose macadam, but today's heroes tackle it at almost three times that

Mountain milestone

Antonio Maeso gets crossed up when landing at the top of Crosby Hill in 2010.

speed over a very ordinary road surface by racing standards.

The surface has improved down the years, but as speeds have risen so it has been a continuous challenge for riders to get through flat out in top gear; something that ten-times TT winner Stanley Woods claimed to do in the 1930s, when he called the spot Vicarage Corner, although in the late 1970s Tom Herron seemed undecided about which gear, saying "sometimes top, sometimes not, I never seem to get it right". The bend requires a classic line that starts from the right (where there are a lot of manhole covers), before sweeping across to the apex and drifting out to give riders another 'high' and feeling of achievement, particularly if they have been able to overcome the little inner demon that tells them to roll it off.

It seems incredible that as maximum speeds climbed from 65 mph, past 100 mph, then over the 150 mph mark, riders were still able to take the bend in Crosby flat out. However, today's big bikes are so fast that riders do now have to ease the throttle. In John McGuinness words: "Some places that used to be flat out just aren't any more. Crosby used to be - simple. But the big bike now - Christ, it's a handful. Back off, then on it again".

Black Thursday

It was an incident here during Thursday afternoon's practice in 2003 that made this forever a TT black-spot in the minds of followers of racing over the Mountain Course. David Jefferies had established himself as the top rider of the early 2000s, with nine TT victories made up of hat-tricks in 1999, 2000 and 2002, but during early practice sessions in 2003, set-up problems meant he was slow to build up the pace on his TAS Suzuki, whilst team-mate Adrian Archibald was fast from the outset. With Thursday afternoon's practice being broadcast on Radio TT, spectators out on the course waited in anticipation as the commentator described Archibald and Jefferies passing the Grandstand nose to tail at 170 mph. Those watching at Quarter Bridge, Braddan Bridge and Union Mills were thrilled to see the two aces pushing each other hard, but as Archibald streaked across Crosby cross-roads and up the hill out of the village he had lost his shadow, for David Jefferies (known to many as just 'DJ') had crashed at the ultra-fast left-hander, hit a roadside wall and the finest TT rider of his generation was killed.

The incident brought following riders to a halt. Many were close friends of 'DJ' and were deeply shocked as they witnessed the aftermath of his accident, as were the many thousands of TT fans who eventually got to hear what had happened. David had always taken a pragmatic approach to the dangers associated with riding the Mountain Course to win. Particularly poignant are the words he gave to 'Motor Cycle News' in a pre-TT interview in 2003 where he said: "At the TT I ride well within my limits. If you get it wrong you either hit a stone wall . . .".

There will be few riders not aware of the sad history of this spot, but it is something that they must put out of their minds as they flash through with concentration and riding skills tested to the limit. Bringing the bike vertical after the bend, they straightline the following slight curves as they pass over Crosby cross-roads and storm up the hill out of the village, passing the popular 'Crosby' public house on the right and knowing that there are some nine miles of rural going ahead, before they hit the next village at Kirk Michael.

Before going over the summit of Crosby Hill, still known to some as Ballaglonney Hill and to others as Halfway Hill, riders will receive more warnings during evening practice that the setting sun will be in their eyes as they go over the crest and there will be further warnings at several points on the way to Ballacraine, where the course eventually turns from west to north.

Keep it Straight!

Soon after riders pass The Crosby public house, they reach the fifth milestone and then burst over the blind summit of Crosby Hill, where it is a fight to keep the front down at today's speeds. Some 40 years ago a top rider said of his front wheel "it doesn't go very high, it just waves in the wind a bit", but with riders now coming over the top almost 40 mph faster, they need to be very careful, for the bike seeks to stand on its back-wheel and sometimes requires a touch of rear brake to prevent the front coming too high.

Whilst up on one wheel, there must be total concentration on keeping in line for a controlled touch-down because, as Steve Hislop found when he got crossed up landing his Honda one year, the resulting feet off the footrests tank-slapper can be very un-nerving!

THE HIGHLANDER

Crosby Hill may be all in a day's work for experienced riders but it can be daunting for Newcomers to the Mountain Course. Experienced GP runner Jamie Robinson made his TT debut in 2008 and described the bit that followed as "like falling off the face of the earth", for the road goes sharply downhill, passing the former Halfway House hostelry (also known as The Waggon and Horses), curving left and eventually flattening before riders pass The Highlander at full throttle, after which it rises slightly and bears right and left on the run to Greeba Castle.

The Buggane

With heads under their screens on the downhill stretch, today's riders will be too busy to spare a glance to the right as they pass the ruined St Trinian's chapel in a nearby field, whilst in earlier days they would have been too busy hanging on over the bumps to notice it was there. St Trinian's is a place of Manx legend which, in brief, tells how monks tried to build a chapel here, but how an evil buggane (a Manx ogre) from nearby Greeba Mountain, repeatedly ripped off the roof; so the chapel was never finished.

A few ogres, mostly mechanical, have beset riders on this stretch down the years, for it comes after the testing full throttle section from Union Mills, through Glen Vine and Crosby village, meaning that engines have been under continuous heavy load and making this downhill approach to The Highlander one of the fastest parts of the course. A rider of the 1920s who was better known as an accomplished sprinter, record-breaker and tuner, was Maurice

Davenport. He entered the 1928 Senior TT but had the bottom spindle of his Webb girder forks shear when passing The Highlander at over 90 mph. His subsequent slide along the road gave him severe friction burns and marked the end of a short TT career, although he did tune the Excelsior with which Syd Crabtree won the 1929 Lightweight race. This probably provided some compensation for his spill, because Crabtree was desperate for a TT win and promised Davenport £1,000 if he produced a bike that would win the race, a huge sum for the day. Maurice Davenport's brother Leo rode the TT with success between 1930 and 1934, taking victory in the Lightweight 250 race of 1932 on a New Imperial.

Although fast, the downhill section approaching The Highlander used to be bumpy, narrow and was not so straight. That curbed speeds for some, but not all, for in the late 1930s, Harold Daniell told of "taking The Highlander full-bore in the middle of the road and hitting the bumps head on". Even before Daniell's exploits 'Motor Cycling' told of the effects of the undulations in 1929 with: "Generally speaking, of the faster machines - and many were doing about 95 to 100 mph - the Nortons seemed to be the steadiest. Stanley Woods and Jimmy Simpson executed prodigal aerial feats without a flicker of unsteadiness, and were travelling at indecent speeds". Stanley himself spoke of enjoying the same spot with "to feel the machine airborne and then do a nice landing a few dozen yards further up the road".

Roadworks saw some of the bumps removed in 1954, but at the Golden Jubilee TT of 1957 there were still reports of the "tremendous leaps taken by the fast machines as they come down the hill and hit the bumps near the hotel". More work was done a few years later and nowadays, the whole stretch is relatively straight and smooth.

For those who like to see riders at speed there are spectating opportunities behind the walls of the nearby fields, but the days when they could sit or stand on the small forecourt of The Highlander to watch the racing are long gone. Legendary Australian motorcycle designer Phil Irving (Vincents, Velocettes, etc) first visited the Island in the early 1930s and wrote "at the popular Highlander, there was nothing between the riders and the customers but the glasses of Manx ale held in their hands". Manx ale remained a big attraction for spectators and a report of a practice session of 1952 said "almost race-day crowds were gathered at this popular vantage point to witness the winding-up of the evening practising, the car park opposite the pub and the new park in the adjacent field being jammed solid with coaches, cars and motor bikes".

Excitement

Spectators expected excitement here and a couple of years earlier 'Motor Cycling' wrote of an evening practice session for the Clubman's TT: "At The Highlander, two of the greatest thrills were provided by Vincent riders. First, C. Howkins went through such a sickeningly frightful repertoire of wobbles that spectators Allan Jefferies, George Rowley and Bert Perrigo (all former top motorcycle competitors) had, in their own words, to switch instantly from beer to whisky, whilst A. Phillip almost caused them to give up drinking altogether when he brought the house to its feet with a simply staggering performance. He took the whole section - down the hill, across the bump and out to Hall Caine's Castle Bends - with the throttle hard against the stop. The sight of the thousand twin roaring through the air like some giant flying saucer compelled even the oldest inhabitant to confess that he'd never known anything like it".

For many years average speeds were recorded over the virtually flat out two-mile stretch from Glen Lough to The Highlander, but it now seems to have fallen out of favour as a place to take speed readings.

Some of the top speeds recorded in earlier years were:
1951   Geoff Duke, 500cc Norton, 122.48 mph.
1964   Mike Hailwood, 500cc MV Agusta, 150 mph

Geoff Duke flies his Norton over the bumps near The Highlander

Mike Duff had good bikes in four classes for 1964 and recorded times of 125 Yamaha 125 mph, 250 Yamaha 143.4 mph, 350 AJS 122.9 mph, 500 Matchless 137.1 mph.

1969 Helmut Fath, 500 cc URS was fastest sidecar at 135 mph, some 10 mph faster than his nearest BMW-powered rivals.

As late as 1997 Phillip McCallen considered that this stretch was the fastest on the course, which means it is rather too fast to take eyes off the road to read a signalling board, although some of those on Classic bikes at the MGP find it a convenient spot to receive information.

No longer a public-house, The Highlander is now a pleasant country restaurant that fills its tables during race-weeks at the TT and MGP, but finds that the road closures during the evenings of practice week are not so good for trade as they were reported to be 50 years ago.

GREEBA CASTLE

After passing The Highlander, riders head for the twists and turns of Greeba Castle, home in the TT's first years on the Mountain Course of author Hall Caine, many of whose best-selling novels were set on the Island. He was not a fan of the TT, which is a pity, for he might have woven the races into one of his plots. Just prior to the 1912 TT it was reported that "the corner just before Mr Hall Caine's residence has been widened, and now is an 'all out' curve".

On the tree-lined approach to Greeba Castle, the brave and knowledgeable move to the right and keep it flat out for longest, passing what for a few years was a spectating spot known as Greeba Verandah. Nowadays there is little scope to spectate on this stretch, so only marshals are able to admire the skills and courage on show.

This undulating S bend is generally known as Greeba Castle from the gothic-style residence that towers above, but in earlier days it was often called Hall Caine's or Hall Caine's Castle, and today some know the spot as

Mountain milestone

Pear Tree Cottage from the more modest residence that sits at road level. Anyone riding a road bike through here can experience a thrill if they are 'trying', whilst also realising how much more exciting it would be if they had the entire road available to them like the racers. The slight rise on the approach makes it a blind bend and Ray Knight wrote of the problems facing competitors down the years with "No matter what you know about where it really goes, and how much road there is, it's a constant war with the little man inside not to sit up early and brake", for riders know that they must go down through the gears and engage in some serious angles of lean through the 'S' that follows.

Tommy Spann rode the TT from 1923-1934 and was a decent runner, though he had more retirements than finishes. In a wet race of the late 1920s he started off at a good pace, but was surprised to come up behind top TT runner Alec Bennett on the approach to Greeba Castle. Lining up his AJS to pass Bennett, something told him to hold back, for he should not have been travelling faster than the man who had won more TT races than anyone else at the time. Tommy was pleased that he did back off, for as they went through the bends they came across several fallers on what he called the "Greeba grease".

In 1931 this S bend was reconstructed, the camber was reduced and the surface was said to have been super-elevated. A few years later and Graham Walker described how taking it in third on a fast lap "you didn't drop below ninety" and 'Motor Cycling' reported prior to the 1938 TT that again "the tricky Greeba corner has been super-elevated", but riders were quick to describe the so-called improvements as deceptive, for although the spot looked faster, they were unable to ride it any quicker than before.

John Surtees described these bends as "a real teaser that you had to handle in clinical fashion, choosing your braking and gear-changing points very precisely". At the summit of the slight rise on the approach, riders change down, peel off and drop down to the left past the mushroom farm, then haul the bike vertical, start to accelerate on the climb out and heel over to the right, being very aware of the unyielding stone wall on the outside of the bend, behind which marshals stand on an elevated platform. There is no footpath on that left-hand side and in days gone by the wall was unprotected, although later it did receive a row of straw-bales and today it is covered with air-fencing.

Riders still get an 'adverse' feel from the camber as they exit Greeba Castle, and the spot requires care with the throttle - wet or dry.

Bill's Mistake

Top Lightweight racer of the 1960s, Bill Ivy, was required to make an appearance in front of the Island's High Bailiff one year to explain how, in a non-racing incident, he managed to scrape the side of his expensive sports car for quite a distance along this wall. The incident provided the accompanying Mike Hailwood and two young lady passengers with quite a fright and cost Bill a fine of £12 payable to the Manx Treasury, plus a larger sum for bodywork repairs.

Smoother

The bends at Greeba Castle have been improved on occasions by the highway authority but their testing shape remains essentially the same. Works in 1964 were described with "The roadmaking machine which lays such a fabulously smooth surface had been to work here; though the road looks very nice, it is doubtful if there really has been an improvement, especially from the safety point of view. The road has been narrowed slightly on the exit to the right-hand sweep because the pavement is now wider on the right. Consequently the fast men finish up nearer the left-hand wall on their way to Appledene, and this wall - unprotected by sandbags or straw bales - is a

In the early 1950s a rider skims a roadside cottage at Appledene.

nasty, jagged, stone affair. Furthermore this new surface does not give as much grip as one might imagine, especially if at all wet. The marshals there, including a very experienced Manxman who knows the course better than most, thought the new arrangement positively dangerous".

So, although the 'Blaw-Knox' machine-laid tar and bitumen macadam surfaces were smoother than before, riders had concerns and back in 1960 the Highways Board presented a five-page report to the Manx MCC with a detailed analysis of experiments carried out to determine the skid resistant qualities of the road surfacing used on the course, and such tests are carried out to this day. But all they do is say that the surface meets the criteria laid down on a piece of paper, and they were certainly not set by a rider cranking a racing motorcycle over to its limits on two small pieces of rubber.

As smoother surfaces were introduced the organisers advised riders of one of the consequences with this note from the 1961 MGP: "The new road surface runs from the top of Bray Hill to Crosby Village. You will find yourself approaching Braddan Bridge, Union Mills and the right-hander at the top of the hill out of Union Mills faster than before. Be prepared for this, that extra 10 mph could be quite a handful".

Today's sidecars find the ups, downs and twists of Greeba Castle quite demanding and tend to go through in fourth gear with "the whole bike moving about . . . the backend comes up and wheelspins on the exit" says Dave Molyneux. Sidecar star of an earlier era, Mick Boddice, was challenging for the lead in the second Sidecar race of 1986 when passenger Chas Birks fell from the chair here and broke his thigh.

APPLEDENE

Riders will have received earlier warning of the dangers from the setting sun as they head west during evening practice, though the demands of the Greeba Castle bends may have driven it from their minds, but as they power out of the final right-hander they receive a dazzling reminder that hits them straight in the eyes near the six-mile marker.

They are then faced with a sinuous stretch which cannot decide whether to go left, right, or straight on, but which when taken at speed makes considerable demands upon physical fitness as the bike is hauled first one way, then the other.

It is Appledene's bends which have to be tackled first here. Called Shimmin's Corner in the 1920s, in the 1930s it was a place for second gear which Harold Daniell described as "the corner where the Course dodges round a cottage" and Stanley Woods called "a right-hander with its cottage entrance almost on the road itself". It was just before the 1954 TT that the

Quarter Bridge to Ballacraine.

Michael Dunlop attacks Greeba Bridge.

cottage - named Appledene - was demolished, thus allowing the road to be realigned and riders' passage speeded past the point, to the stage where Nick Jefferies said in the early 1990s "Trying to get Appledene right on a 750 is a bit awesome. It's so fast, and you're trying to get the front wheel down to change direction, having to really heave the bike over", whilst John McGuinness confirms that "Appledene is hard work, as it's all off camber. You steer through with the balls of your feet".

Wettest?

Demanding enough in the dry, tragedy struck at Appledene during the closing stages of the 1961 Junior MGP, which was one of the wettest and coldest races recorded. Leader Fred Neville lost control of his AJS here in a fatal crash, and Frank Reynolds went on to win a race in which over half the entry retired, whilst those that reached the finish were almost paralysed by the cold. It was a time when the Manx MCC thought it was acting in the best interests of the MGP's amateur competitors by refusing to permit the use of fairings at its event on the grounds of cost, even though it was long after they were in established use elsewhere. Indeed, many MGP riders actually had to remove their fairings for the event. The weather-protection they offered would certainly have made a contribution to rider safety at the 1961 Junior MGP, and the Manx MCC permitted their use the following year. It was all too late for talented Fred Neville, and legend has it that his grieving friends buried his plum-coloured helmet at the edge of the Mountain Course.

Albert's Nasty Moment

The road twists and turns between walls and high hedges as it continues its run through the Central Valley (sometimes referred to here as Greeba Gap),

and riders build speed on the way from Appledene to Greeba Bridge. It was on this stretch that Albert Moule riding an Earles-forked Mondial in the 1953 TT, had a major moment that he remembered to the end of his long MGP and TT riding career, which stretched from 1936 to 1967 and was then followed by many years as a Travelling Marshal. The Mondial was quick for its time and Albert was pushing it to the limit when it suddenly became uncontrollable. An eye-witness said "by dint of will-power he managed to clear the bordering stone wall until he finished up in a heap with the complete front section of the bike disappearing backwards over his head to clout him in the back as it landed". Albert had suffered a major mechanical failure when the twin down-tubes of the frame fractured, jamming the throttle open and making the bike uncontrollable. It was a serious incident, but one he would recount with humour in later years as he spoke of the broken shoulder-blade and ribs that he collected, and with being so badly knocked about and winded that when a Travelling Marshal stood over him and pronounced him dead, he was unable to disagree.

With his distinctive black and blue helmet painted in roundels, Albert was a Travelling Marshal through the late 1960s and 1970s, riding mostly Triumph Tridents before moving on to Hondas in 1977. Doing this job meant that he rode through Appledene at speed on many more occasions and he told how when he passed the scene of his accident, he always gave the place a nod of respect.

John Surtees spoke of doing the whole twisting stretch from Greeba Castle to Greeba Bridge in third gear on his MV in the late 1950s whilst, riding in the same era, Bob McIntyre reckoned to exit Greeba Castle at close to 110 mph and notch fourth for the ride to Greeba Bridge on his Gilera. Both bikes had four completely open exhausts which ended in sound-enhancing megaphones that sent the fabulous note of their engines bellowing up and down the scale when ridden on opening and closing throttles on this twisty stretch; no doubt rattling the corrugated iron building known as Greeba Young Men's Hall (or just Greeba Clubroom), that sits at the edge of the road.

GREEBA BRIDGE

Having dealt with the twists and turns of Appledene and beyond, today's riders are travelling very quickly as they approach the left sweep of Greeba Bridge. With a low stone wall and then the bridge parapet on the left and a conventional kerb and footpath on the right, it looks fairly straightforward, yet the history of the spot is one of accidents, including fatalities. Top riders do not rate it as difficult, although Ray Knight makes the point that "If there is a problem here, it is simply that of judging where to brake and where to peel off" and the solution he offers is to use the black and white painted kerbs to peel off to the left. Steve Hislop did the same, saying "You'll brake probably just about 100 yards short of them, jab back two gears to fourth and then just as you get to the black and whites you start to peel in". Some riders say that the spot is faster than it looks, so they work on achieving a fast passage to benefit speed on the run to Ballacraine that follows.

It was probably when speeds were climbing due to the better road surfaces of the late 1920s that Greeba Bridge began to catch riders out, for the road was narrower than it is today. Jimmy Simpson was one of the fastest men of that era and he spilled here during the 1929 TT, sustaining injuries that kept him out of racing for the rest of the year. There was no medical care available on the spot when Jimmy and three other riders came off in the race and he had to wait for over four laps until the race finished, before receiving proper attention. Sadly, one of the other riders later died from his injuries. Speeds continued to rise here in the mid 1930s and another experienced man to get things wrong was Jack (CJ) Williams, when he and his Velocette went through the hedge on the exit during the 1938 Junior. Winner of the Senior later the same week, Harold Daniell described it as a second gear bend "where an awkward camber merits due care and attention". A year later and the Highways Board must have decided to do something about the awkward camber and were using the term super elevation to describe their works of improvement at Greeba Bridge.

Robbed?

The helicopter rescue service was introduced at the TT in 1963 and Tony Godfrey has gone down in history as its first customer, after he was air-lifted from Milntown during Monday's 250 race. The service has subsequently proved to be of tremendous value, particularly at spots which are remote from side roads that can be used to evacuate a rider. However, when Pete Munday fell from his Bultaco here in Wednesday's 125 race in 1963, he was robbed of the honour of being the helicopter's second passenger, because low cloud meant that it could not land.

Mistake!

Before widening took place here in the 1960s, Geoff Duke told of entering at about 50 mph in second gear and leaving at about 60-65 mph, being careful not to let the slight hump of the bridge cause the bike to drift out to the right and being prepared for it to unload the rear suspension. Whilst the road widening helped make a rider's passage much faster, the extra speed reduced the chances of recovering from a mistake. Later to be a Sidecar TT winner and World Champion, George O'Dell was in his second year at the TT in 1971 when he made a mistake that cost him dear at Greeba Bridge. The mistake George made was in failing to stop at the end of an early morning practice lap to check a mystery squeak from the front wheel of his outfit. He explained the consequences with: "We had just done a good, quick lap . . . I was on top of the world . . . approaching Greeba Bridge I had been motoring just that little bit harder. I knocked it off, dropped down a gear, just touched the brakes". That was the moment the front wheel collapsed, the outfit ploughed straight into the bridge, through a fence, into someone's garden and finally finished upside down back on the road. Passenger Pete Stockdale broke both ankles, and George broke two fingers on his right hand, plus his right foot and nose. With the outfit a complete write-off, that was clearly the end of his Sidecar TT for 1971, however hard man George also had an entry and had qualified to race in the Production TT on a Triumph solo and, despite his injuries, was bitterly disappointed when the doctor refused him permission to race.

It was at the 1984 TT that solo rider Malcolm Lucas fell here after a front wheel puncture. Picking himself up, Malcolm's first words were "Thank God it went there and not somewhere really fast". Most people would have considered a fall at Greeba Bridge to have been fast enough, for it is one that solos take hard in third (like John McGuinness) or fourth, with sidecars using fourth.

Previously offering a small amount of viewing space, Greeba Bridge is now closed to spectators.

THE HAWTHORN

The straight that faces riders after Greeba Bridge has always attracted the full throttle treatment. Half way along on the left is The Hawthorn public house (occasionally spelt Hawthorne and sometimes referred to as The Hawthornes and the Hawthorn Inn in the past). It is a popular spot for fans at race times, for they can sit with pint in hand separated from the flat-out racers by just a low stone wall.

Sidecars leave Greeba Bridge during evening practice.

Quarter Bridge to Ballacraine.

The AJS team at their Hawthorn base in 1914.

The Way it Was

The Hawthorn has existed for over a century and back in 1914 the AJS team based itself here for the TT, working on its bikes in the open garden and lighting the shed that served as a garage with acetylene lamps when it got dark.

It is interesting to compare the requirements of a 'works' team contesting the TT in 2011 with that of AJS in 1914. Today's team comes in an emblazoned pantechnicon incorporating a fully-fitted workshop, generators supplying power and its working space extended with awnings complete with their own floors. Then there are spare bikes, engines, wheels, abundant tyres for wet and dry conditions, impressive tool-chests, tyre-warmers, hydraulic benches, wheel-stands, starting rollers, computers and every other conceivable racing aid, all 'manned' by liveried support staff with access to specialist mechanics available to deal with engines, chassis, suspension, electrics; backed by even more staff chasing around in hire cars and staying at the best hotels.

AJS were a medium-sized firm that had previous TT experience but made a late decision to compete in 1914. Four of the Stevens brothers who ran the firm came to the Island to stay at this country inn, brother Jack bringing four 'works' bikes by rail and sea (no van, he would have used porters) and the others coming in two sidecar outfits that contained all the spares and tools to support five bikes (for they also assisted private owner Billy Jones) and their personal belongings. They had the services of just one recognised mechanic.

Harry Stevens wrote a fascinating account of AJS efforts in 1914, telling how carefully the bikes were built with everything "polished like silver inside". The first practice session revealed that the compression ratio of 6:1 was too high, so it was reduced to about 5:1. There then followed a series of problems with big-ends, pistons, lubrication and tappets closing up.

Make Your Own Crankpins

As if that were not trouble enough for a small team operating from a shed at the side of the course, late in practice they experienced failure of several crankpins. Inspection of their spare crankpins revealed minute cracking and there was an air of despondency in the camp. Harry tells: "after a good deal of argument and discussion, it was decided that the engines should be removed from the frames, and that Joe should visit Douglas, and see if he could either get some firm or repairer to make us half a dozen crankpins, or make them himself, if he could not find a man capable. These he must insist upon being made from ordinary common mild steel. It was arranged also that if he was successful in getting a firm to make them, he should get a supply of prussiate of potash, which we should use for flash-hardening of them".

It took until mid-day on Saturday of the final practice week to get the replacement crankpins, and with all the dismantled bikes required to be up and running, tested and prepared for weighing-in on Monday prior to racing on Tuesday, "superhuman efforts" were needed to get the bikes ready. Those tasks were achieved and the race turned into a triumph for the AJS concern which took the first two places, with their three other supported bikes also finishing. Harry concluded his account by admitting that celebrations were on a scale justified by a TT win, and after doing the rounds of various establishments in Douglas they returned to Greeba with half a dozen boxes of champagne that, eventually, was being drunk out of pint glasses.

Joey Out

The Hawthorn has seen its share of rider retirements, including that of Joey Dunlop in the 1995 Lightweight 125 race. It was a cool morning and although Joey had put tape over part of the radiator of his Honda to compensate, he was soon aware that it was not running hot enough and the

Mountain milestone

Three riders power away from Greeba Bridge towards The Hawthorn, adding their black tyre marks to the road surface.

bike seized. Would an extra piece of tape have allowed him to keep running? It will never be known, but what is known is that in the days of water-cooled 125 two-strokes on the Mountain Course, the black art of applying tape to the radiator could mean the difference between success and failure. It may seem a crude method of controlling radiator temperature, but it was actually quite precise - if you knew exactly where and how much to apply.

GORSELEA

Soon after passing The Hawthorn, riders are faced with two similar challenging right-hand bends. Top riders do not make a fuss about these, but they do ride them very much quicker than the average competitor. These two right-handers come into the same category as a number of other bends on the course, in so much as they are faster than they look, providing you have a good handling bike, plus the experience, skill and courage to tackle them at your personal limit. Both are unforgiving of errors, with absolutely no run off space. Nick Jefferies is not given to overstatement, but in his words "on a 750 this is dramatically fast", whilst John McGuinness says they are "mega-quick right-handers - the first one's easy but for the second you need to be fully committed to making a late apex". Sounds technical, but you can see what he means if you look at the line of the inside kerb, though John does not bother to mention that there is a slight rise into the first one that can have the front end going light as the rider is cranked over and that overhanging trees can create unknown degrees of dampness on the exit from the second.

The section containing these two bends is known as Gorselea, with the first sometimes referred to as Knock Breck, although marshals use the radio call-sign of Harold's in recognition of local farmer Harold Leece, who has welcomed generations of spectators to his farm Cronk Breck. It is the second bend that is formally recognised as Gorselea from nearby Gorselea House. Because they are so unforgiving, they really are the sort of bends where riders need to practice what they often preach, about riding with a bit in hand on the Mountain Course, and this was just as much the case over 60 years ago when speeds were slower but the road was notoriously bumpy here.

The Worst Point

The Highways Board has always carried out alterations and repairs on the 37¾ miles of the Mountain Course to improve things for local traffic and to counter the natural deterioration that occurs, but because of the wear and tear factor, the 'worst' point on the course in terms of road conditions for racers varies from year to year. Consensus at the first post-war TT in 1947 was that the bends after The Hawthorn were the most difficult due to their poor condition. Jock West explained that "the road at this point is comparatively narrow and extremely bumpy; in addition, the surface is ridged. In order to take these bends at the maximum possible speed these ridges have to be crossed at an acute angle". Aware of the conditions, Jock was into his second lap of an evening practice session and tells of approaching Gorselea on his AJS "thinking that perhaps, after all, this section wasn't so bad as we had supposed, when the machine lurched badly and almost unseated me. My first impression was that the frame had broken . . . almost immediately I was aware that it was not the frame that had broken but the right handlebar. Unfortunately the throttle control was attached to the broken-off portion and it was impossible to reduce speed". Jock managed to negotiate the corner, but vowed to replace the light alloy handlebars with steel.

Although there have been accidents here down the years, thankfully, they have been relatively few. At the 1994 MGP local man David Black had his best bike blow in Thursday afternoon's practice, and so he put together another one for the Senior race. It was before the days when the organisers ran practice sessions after the races as they now do, so the bike was wheeled out for the Senior with virtually untried engine, new tyres, etc. Starting carefully, David began to push on a bit after Crosby and was getting into his stride by Gorselea. Tipping the big Kawasaki in at something like 140 mph, the front suddenly went away, he was low sided and sent sliding up the middle of the road, whilst the bike hit a telegraph pole on the left, bounced across the road, caught fire and was burnt out. Dave counted himself as lucky to get away with just a broken wrist and dislocated shoulder and, with the humour that riders are able to bring to such serious situations, claims that the rescue helicopter probably took longer than usual to arrive because they did not hurry; the reason being that it is not a spot where riders are expected to survive an off! James McBride was another man to have a high-speed fall here. His came in Monday evening's practice for the 2008 TT and John McGuinness, who was on his tail, feared the worst as James disappeared at speed into roadside trees. Although hospitalised, James amazingly escaped serious injury and was hobbling about at the Grandstand before the end of the week.

In an interesting experiment at the 1991 TT, Shaun Harris was fitted with a heart-beat monitor that was said to have "shot up to astronomic levels the moment he saw Bray Hill" and thereafter showed blips at other tricky parts of the course. Shaun asked if the monitor showed a big blip at Gorselea, because his most frightening moment of the lap occurred when Steve Hislop came past him there and got into such a wobble that it looked as if he was off.

Ted Mellors aviates his Velocette over Ballagarraghyn in 1939.

BALLAGARRAGHYN

There is sometimes a vagueness amongst race fans as to the location of Gorselea and occasionally it is referred to as Ballagarraghyn, but that comes a few hundred yards later after the road has straightened and at a point where the road bridges a watercourse. After Gorselea's right-hander, riders approach Ballagarraghyn flat-out on the level, but immediately after the bridge the road has a short but steep drop that has always caused bikes to become airborne. In earlier days this spot was called Ballagarraghyn Bridge, and although the hump was eased considerably in 1954 and has been since, ever faster race bikes means that they still fly here.

Some Like It, Some Don't

The downhill 'step' in the road at Ballagarraghyn is not the sort of thing that all riders like to encounter at high-speed, for whilst some have the technique to cope with it, others do not. This means that many of the latter have suffered frightening moments, with front wheel landings and major wobbles. In the mid 1930s Graham Walker wrote of tackling it on a rigid racer "full bore in top you have that lovely leap into the air which is such a fine barometer as to the state of tune of the engine", the latter presumably gauged from the distance jumped, which Stanley Woods said was some 20 yards. It was Jock West, this time riding a 'works' BMW in 1937, who discovered that the 'step' did not suit his new telescopic front forks, for it caused them to bottom out. This meant that he had to ease the throttle just before the jump, a fact that John 'Crasher' White almost found to his cost when he was slipstreaming Jock at this point. 'Crasher' was on the new-to-the-Island double ohc 'works' Norton and no doubt wanted to measure its speed against the ultra-fast BMW. Despite BMW's front suspension problems here in 1937, Norton replaced the girder forks on its racers in 1938 with

Alex George rounds Knock Breck with Gorselea to follow.

Mountain milestone

When Ballacraine was approached from St John's - 1907 to 1910.

primitive telescopic ones that relied solely on springs to control movement (no hydraulics), and that gave just two inches of travel.

Approach to Ballacraine

It is a tree-lined straight for a few hundred yards after Ballagarraghyn, before riders resort to heavy braking for the right-hand turn of Ballacraine, having completed the base of the roughly triangular shaped TT Course.

Today this stretch is virtually a spectator-free zone, but it has not always been. Even after the roads were closed to ordinary traffic for practice sessions from 1928, it was still the custom for spectators to use the course-side footpaths to move from one vantage point to another during practice. By 1930 the organisers felt that something had to be done about this potentially dangerous activity and the 'Official Car', that usually carried a police sergeant, engaged in a campaign of stopping and warning people off the paths. One morning in 1930 the car stopped at Ballacraine corner and the policeman in the car spoke to a fellow officer stationed there. The outcome was that six spectators walking towards Ballacraine were booked. The action was condemned in the local press, where the 'Isle of Man Weekly Times' wrote: "When the Manx Legislature closed the roads for practices, it was not thought it would affect pedestrians walking along a footpath on a straight, but only vehicular traffic on the roadway itself. The police seem to be unnecessarily officious in this paltry matter and deserve to be reprimanded for their conduct". They were strong words, but the organisers felt justified in taking action to protect spectators from their lack of awareness of danger. It was something that they had previously done by way of polite safety notices in the race programme, one example of the time reading: "spectators would help the riders very considerably by refraining from crossing the road during the races".

Although spectators are now much better behaved, notices and police action are still not sufficient to curb the activities of mother ducks that occasionally seek to cross the course from the fields of Ballacraine Farm with a line of ducklings in tow. Marshals do their best to discourage such activity at TT time.

Mountain Course Moments

The behaviour of a few of the motorcyclists who came to spectate at the TT in its early days was regarded with disfavour by some Manx people, particularly when young ladies were invited to ride side-saddle or in over-crowded sidecars on Douglas Promenade. To curb such activity, a bye-law was introduced for 1914 that said "No person is allowed to ride on the back of any motorcycle, or not more than one passenger to be carried in a sidecar, for the 14 days prior to the races, on the Promenades of Douglas or Ramsey".

◆ The First World War (1914-1918) put a stop to motorcycle racing and the TT did not return until 1920. Expenditure on the Island's roads was

Spectators crowd the edge of the course on the approach to Ballacraine.

John Turton's photograph captures the atmosphere of the TT at Ballacraine, as Carl Rennie rides the corner in 2010.

restricted during the war and there were many complaints of the poor condition of the Mountain Road. The Island lost some of its roadmen to the Army during the war, but because it housed several camps for interned aliens, the Highways Board was able to make use of alien labour for road repair and quarrying of stone, at a cost of one shilling a man a day.

◆ In 1921 a rider who went over for a preliminary look at the course in early May reported: "the Mountain Road is rather bad, but will not be repaired too early in case of rain rendering further repairs necessary".

◆ Like any other event, Island racing can be affected by outside influences. In 1921 a coal strike at TT time restricted boat sailings to 3 per week, while a general strike in 1935 saw early arrivals for practice carrying their own baggage to their hotels and dining by candlelight. In 1955 a rail strike prevented many travelling to the event, in 1966 a seamen's strike caused the TT to be run at the end of August and early September just before the MGP, and in 2001 racing was cancelled on account of foot and mouth disease in the United Kingdom.

◆ In the early years of the TT, riders finishing within half an hour of the winner were awarded gold medals. From 1922 those were replaced by Silver Replicas of the main Tourist Trophy that were awarded to those who finished within a specified fraction (later a percentage - currently 105%) over the winner's time. Bronze Replicas were added to the awards in 1934 and those who complete a race outside Replica time receive a Finisher's Award. The rider who sets the fastest lap of a race also gains a Silver Replica.

◆ The banners that used to span the road as a warning of dangerous corners were largely replaced in 1922 with eight feet high warning boards mounted at the edge of the road.

◆ In 1922 members of the Island's Highways Board went on a week's tour of England to assess road construction, quarries, working practices, etc, but the topic that would have pleased TT racers was the interest they showed in tarmacadam surfacing as an improvement to the many miles of waterbound macadam still used, and the basic tar-sprayed finish used in towns and villages, on which more than 1,000 barrels of tar were spread in 1922. The need for upgrading was revealed in a specialist's report which told that 80% of the Island's main roads (which included the TT course) required reconstruction and then resurfacing.

◆ There have been many men of small stature who starred at the TT, among them being Charlie Dodson, Luigi Taveri, Bill Ivy and Robert Dunlop. In 1923 the organisers introduced a minimum weight limit for riders of 130 lbs. The new rule was not expected to trouble many, but in 1924 N Houlding (Matador) was found to be underweight and was given a yard of lead piping to carry as ballast. However, after donning full riding kit and being re-weighed, it was found possible to reduce the ballast to two 8" lengths of pipe. At the 1958 TT at least seven riders had to add weight, with the diminutive Luigi Taveri putting on 21 lbs, and during a spell of ill-feeling between Phil Read and little Bill Ivy in the mid 1960s, Read threatened to inform race organisers that Bill was under the minimum weight; something that Bill had forgotten to mention.

◆ In 1923 the ACU asked for permission to run evening practice sessions at the TT, but they were refused and it was to be 14 years before they were introduced. The same year the Manx MCC ran the first Amateur race which was the predecessor of the Manx Grand Prix.

MOUNTAIN COURSE

BALLACRAINE ... BALLASPUR ... BALLIG BRIDGE ... DORAN'S BEND ... LAUREL BANK ... BLACK DUB ... GLEN HELEN ... SARAH'S COTTAGE ... CREG WYLLIES ... CRONK Y VODDY

CHAPTER FOUR

Ballacraine to Cronk y Voddy

One of the TT's best known corners, Ballacraine has not only been part of the renowned Mountain Course for 100 years but it also featured in the very earliest TT races that ran on the St John's (or Short Course) from 1907-1910. In those days riders started from Tynwald Green some half a mile away from Ballacraine to the west, and so this was the first corner they faced. On their single-speed and clutchless machines, pioneer riders like Rem Fowler, Jack Marshall and the Collier brothers strove to maintain momentum on the ninety-degree left-hand turn at the Ballacraine Hotel to get the best run at the ascent of Ballaspur that followed. Unfortunately, a ditch and wall on the outside of the bend acted like a magnet to those who took it too fast and spills were frequent.

In an attempt to help riders, the organisers constructed a timber banking on the outside of the corner for the 1910 race to cover the ditch and part of the face of the wall, so allowing them to maintain speed in relative safety. But just before the banking was finished a report said "opinions differ as to the safety of it", followed on completion by the news that "Several riders tried it out this morning. Sproston on his Rex took it at speed and paid the penalty of coming off". In the race itself, H.H. Bowen (BAT) moved into the lead after the refuelling stop, raised the lap record to 53.15 mph, but crashed on the Ballacraine banking and put himself out.

When the TT moved from the St John's to the Mountain Course in 1911, Ballacraine changed from being a left-hand corner to a right-hand one approached at full throttle from the direction of Douglas. Being at the cross-roads of several of the Island's major routes, Ballacraine is an accessible spot for spectators, although with limited viewing facilities it is perhaps better suited to those who want to watch a couple of laps there and then move by back-roads to another position. It has always been popular and a report in 'The TT Special' of 1936 said "The real TT fever . . . was in evidence at Ballacraine for today's big race. Hundreds of spectators crowded to the corner from an early hour, and other vantage points close by were lined with people". Those gathered showed complete confidence in the ability of riders to reduce speed for Ballacraine, crowding against barriers on the outside of the corner in areas that are now kept clear to allow a run-off area.

Telephone

When the Mountain Course was first used, Ballacraine was one of few points provided with a telephone during racing. It was a vital communication link, for in the words of Canon Stenning (later Archdeacon) who first marshalled there in 1912 and later became President of the Manx Motor Cycle Club: "There were no organised posts between Douglas and Ballacraine, so anything that happened between these points had to pass unnoticed" - that is 7.8 miles of the course! Instructions to marshals said: "the actual telephoning should be left to the Post Office Operators stationed at each point". Even by 1920 there were only three telephones available during practice and seven on race days.

In Canon Stenning's words "Ballacraine corner was a major risk. The road was steeply cambered; on both sides was a grass verge; in this verge ran a considerable drainage ditch; the road surfaces under the trees had patches of vegetation growing in countless places. The actual corner right in front of the farmyard had a very unpleasant wall. Riders dared not swing out to enable a view round it because of the steep slope down to the Foxdale road at that point, so it was a matter of hugging the farmyard wall in faith and hope and fear. The side-roads were not roped or barred, and the crowds were always thick there . . .

Race Reports

There were no radios in the early years of the TT and the interval starting system meant that spectators had only a hazy idea of how the races were going, but at Ballacraine Canon Stenning would announce any

Mountain milestone

Trusting crowds line the outside of the corner as riders bank over at Ballacraine in the early 1930s.

telephoned information he received to the crowd. Then came the single-source public address commentary from the Grandstand through loudspeakers located here and when that was supplemented with commentary from several points around the course, Ballacraine was one of the places chosen for commentators like Geoff Cannell and ex-racer Tommy Robb to provide information-hungry spectators with times, positions and other race information.

For race fans who could not get to the TT, the BBC broadcast the meeting world-wide from the 1930s, because the TT was a high-profile sporting event; it being said in the 1950s that the man in the street was as interested in hearing the results of the Isle of Man races as he was the Cup Final. Initially the BBC gave full race commentaries and listeners would include workers at the factories of successful British racing firms like Norton and Velocette. Mick Woollett recounts in his book 'Norton' how former employee Charlie Edwards told him that on race days "They used to put a radio on the shop floor and we all gathered round. The firm supplied lemonade and if we won they gave us the rest of the day off".

The BBC eventually scaled down its broadcasts, dropping them completely in the 1970s. Then, to give Manx Radio's commentators more time to assess times and race positions, the Ballacraine commentary point was moved further along the course to Glen Helen in 1991.

It was mentioned above that at early TTs Ballacraine was attended by a Doctor from Peel and reports of 1913 also tell of the presence of an ambulance sidecar "which has performed useful work". The very basic medical care available was supplemented on race-days by a train at nearby St John's station that maintained steam in case it was needed to convey an injured rider to hospital in Douglas. It was all very different from today's on-the-spot ambulance and medical staff, backed by speedy rescue helicopters that are just a radio call away and can deliver an injured rider to hospital within minutes.

Much the Same

Whilst the race organisation has moved up a few gears in the past 100 years and road surfaces are greatly improved, much has remained the same at Ballacraine. The buildings that today stand on three of its four corners: Ballacraine Hotel, Ballacraine Farm and the private dwelling to the south, have witnessed a century of racing, although the Hotel is now a private dwelling, meaning that customers no longer watch the races from the window of the bar.

Mind The Wall

The forbidding wall-bound exit from Ballacraine's right-hand corner has always been there, although it was initially lower. With an adverse camber in front, the stonework serves as a solid restraint to riders. Photographs from the mid 1930s show there were no protective bales in this area, but today both the former Ballacraine Hotel and the exit wall are well padded. The 1959 TT Programme warned of "a fast corner with very little favourable camber", and Geoff Duke tells how he "slid right across the road and almost cannoned into the wall" after his oil-tank split and coated his tyre on his debut Island ride in the 1948 MGP.

In modern everyday use Ballacraine is a busy crossroads where the apex has been eased and where traffic lights were added in 1976, so as riders complete their braking and start to lean over for the corner they have to remember the oil and diesel droppings left by the many vehicles that have been stopped at the lights. This requires particular caution in the wet, when there can also be water running across the bend from the uphill section they are about to turn into.

RIDER VIEWS . . .

Wal Handley told in the early 1930s how he slowed to about 40 mph for the bend, and in the mid 1930s Graham Walker "used the gutter at 45 and just missed the pub wall on the left", whilst in the early 1950s Geoff Duke advised going down into bottom gear on a Manx Norton and warned of there being "no favourable camber". Nearly 20 years later Peter Williams recommended second gear on a Matchless G50 and "brushing your shoulder against the (inside) wall", then using all the road on the exit but "don't stick your neck out".

Mick Grant and Joey Dunlop were in agreement that braking on the approach was hard to judge, whilst Nick Jefferies offered the reminder that it was a "frequently dirty" corner where it was not worth trying too hard.

All stressed the need for hard braking on the approach - although each had different marker points - followed by sensible riding through what they considered a fairly straightforward corner.

Out of interest, driver of the Roads Open car in the late 1950s, Lieut-Colonel Harry Kissack, reckoned it was taken in a four-wheeler "at a good 45", no doubt with accompanying tyre-squeal.

By the time they reach Ballacraine, riders know if their bikes are running well. Many a rider who has spent hours rebuilding an engine or attempting to sort a mysterious electrical fault, has discovered in the 7.8 mile run from the Start that the problem still exists and so has left the course here and returned to Douglas by the back roads. However, not all riders stay in Douglas and certainly in the early days of the TT, teams like Triumph and Levis stayed at Peel (3 miles to the west).

From about 1920 riders staying in Peel could officially start their practice laps at Ballacraine. Reporting their intentions to a marshal, the information would be passed by telephone to Race Control in the Grandstand, but the rider would only be timed from Douglas to Douglas. The concession was withdrawn after the 1964 MGP.

The Ballacraine Hotel was used as a TT base by the occasional private entrant, including the A.E. Reynolds equipe from Liverpool that ran Scotts in the 1930 Senior event. Team rider, Ernie Mainwaring, pushed his Scott as hard as it would go in the race, until on the fifth lap he pushed too hard and crashed - at Ballacraine!

Today Ballacraine is still an important location on the Mountain Course and is a 'station' for one of the team of Travelling Marshals who assist at incidents if called to do so by Race Control.

IT HAPPENED HERE . . .

The excitement generated by early racing on the St John's Course was too much for one Manxman, with the 'Peel City Guardian' reporting "Most probably as an outcome of the motorcycle race practising, the death occurred on Friday evening of Mr John Crellin, dyer of Ballacraine, aged 73. He was lying against a hedge, where he had a full view of the riders in their difficult task of rounding Ballacraine corner. Suddenly he collapsed and fell. Quickly lifted up, in a few seconds he was found to be dead". Official practice was limited to early mornings when the St John's Course was used from 1907-1910, but riders actually practised at all hours.

◆ Already a two-times winner over the St Johns Course, Charlie Collier was the crowd's favourite to win the first TT over the Mountain Course in 1911. After holding the lead on his Matchless in the early stages, spectators were surprised to see him stop at Ballacraine and take on additional petrol. As the race regulations prescribed that refuelling could only take place at official Replenishment Depots located at Braddan and Ramsey, Charlie was disqualified from the race and stripped of his second place finishing position.

◆ Mounted on a Sheffield Henderson for his TT debut in 1922, Ernie Searle described how he got things wrong at Ballacraine, "attempted to climb the wall" of the Hotel and left marks five feet up it.

◆ In the spoof TT film 'No Limit' made on the Island by George Formby during the 1935 TT period, George, the gormless hero of the film, was shown crashing his 'Shuttleworth Snap' racer through the door of the Ballacraine Hotel and finishing at the bar.

A nasty moment for this rider on the banking at Ballacraine in 1910. The car of Dr Macdonald of Peel - parked on the course - is ready for any fallers. Note the handles of a stretcher sticking out of the window.

◆ In a far more serious incident in the 1971 Senior TT, Brian Finch crashed his Suzuki into the Hotel and was killed. It was thought that his rear brake pedal snapped and also that the slip-road was blocked. Brian's name is remembered by a plaque on the front of the building.

◆ With its high-speed approach, Ballacraine puts much stress on braking systems. Already a pre-war TT winner and destined to later become world champion, Freddie Frith experienced a problem in practice for the first post-war TT in 1947. Approaching the corner on his twin-cylinder 'works' Moto Guzzi, Freddie had the front brake lock, he was thrown through the air, broke his shoulder on the kerb and thus ruined most of his racing season.

◆ Joey Dunlop was late arriving for his first TT, having ridden at Cookstown on the Saturday before practice started and crossed to the Island on a friend's fishing boat. In 'Joey Dunlop - A Tribute', author Ray Knight quotes Joey's own words with: "I went out in Monday morning practice and didn't know which way the track went. I got to Ballacraine crossroads on my first lap and didn't know where to go so I stopped and waited for a rider who knew his way. I then followed an experienced man until he was out of sight, then I slowed and waited for another one and so on. That was how I completed my first ever TT lap".

◆ Veteran sidecar racer Roy Hanks was passengered for many years by Dave Wells. On one occasion when Dave leant over the bike's rear-wheel for Ballacraine's right-hander, he grabbed a handful of the wild Valerian flower that grows there on the roadside bank. Proffering them to Roy on the climb to Ballaspur he received a shake of the head, so tossed his bouquet out of the back of the sidecar, much to the surprise of the following crew.

◆ For many competitors, mechanical problems identified at Ballacraine mean retirement. Riding his Greeves in Thursday afternoon practice for the 2004 Classic MGP, Ian Rycroft had his primary drive belt snap as he approached the corner, but as he lived just 3 miles away in Peel, he found someone to drive him to his home, collected a new belt, returned to his bike, fitted the belt, got away again and completed the lap in 63 minutes.

As riders leave Ballacraine and head north towards Kirk Michael on the A3, they ride a six and a half mile stretch of road that formed part of the

Mountain milestone

Under the trees at the top of Ballaspur. Spectators are no longer allowed here.

original St John's course used from 1907 - 1910. The non-stop swerves and undulations of the first few miles of this stretch make it universally recognised as one of the hardest sections of the Mountain Course to learn and ride, with Carl Fogarty describing it as "technically difficult, but very satisfying when you string it all together" and Joey Dunlop saying he "really liked this bit". It is where experienced riders make time, but where the inexperienced are advised to forget about time and concentrate on getting the sequence of bends firmly in their minds, for it is unforgiving of mistakes.

BALLASPUR

Having rounded Ballacraine's corner, riders are faced with a short climb that in 1911 would have seen them push the throttle lever open to its maximum and plod slowly upwards, whilst 50 years ago they would have achieved maximum revs in second gear and hit nearly 90 mph. Today's riders just point the bike, twist the grip and hold on tight as, with power to spare, they storm the rise to Ballaspur and its rock-faced left-hand curve, followed after its summit by a downhill right-hander. There are no footpaths on this stretch and the road is comparatively narrow as it runs under trees, through a roadside mixture of bluebell and garlic-covered banks, farms, fields and woodland.

Sidecar passengers can feel vulnerable at Ballaspur, because as they climb the hill and prepare to shift their weight to the left for the first rock-faced bend, they know that their drivers will be tempted to cut the corner as close as possible. George O'Dell was to take a Sidecar TT victory in 1977, but the 1976 event with Alan Gosling in the chair had not gone well after an incident at Ballaspur. In George's words "I am afraid I took a big chunk out of him. I did get close to the wall but it wasn't so necessary for him to lean out that far. I believe that it was an error by both of us. He could have ducked in more while I perhaps should have given him an extra 12 inches. In training we had been fastest and had gone through the tight left-hander at Ballaspur so many times without incident". A few years later and Dave Hallam inflicted similar punishment on passenger Mark Day.

Milky's Mistake

George O'Dell spoke of having gone through Ballaspur on many occasions without trouble and this is a point often made by riders who are puzzled when they crash at a spot with which they are familiar. One such was MGP and TT winner Richard 'Milky' Quayle, for the Manxman was caught out by Ballaspur on the second lap of the Formula 1 race at the 2003 TT. Bikes accelerate so hard and fast that Ballaspur has inevitably become far more demanding than it was at the time when Bob McIntyre set the first 100 mph lap on his Gilera in 1957 and spoke of riding it "with discretion in third". Today's top riders try to take it with just an easing of the throttle but 'Milky' got it wrong, hit his shoulder against the left-hand rock face and was propelled diagonally across the road to strike the bales protecting the wall on the other side. Losing control, he and the bike cartwheeled down the road under the horrified gaze of spectators watching from the edge of the road near Ballig Farm. Sustaining serious injuries that prevented him working for several months and, being a family man, 'Milky' decided to call a halt to his racing career. However, the crash did little to diminish his interest in the TT and given that it was captured on video and quickly got onto the web, it spread his name worldwide. Today he acts as a rider liaison and recruitment officer for the event, while the race organisers now seek to give riders some protection by cladding the exposed rock on the left at Ballaspur with sheets of plywood, while the bales on the right have been replaced with air-fencing.

Watching the racing from this area is now very restricted, much to the regret of regulars who liked to stand on the roadside bank and experience the ultra-close sight and sounds of raw road racing, before walking the short distance to a completely different world in the form of the civilised refreshment, parking and shopping facilities at nearby Tynwald Mills. The buildings there originated as a dyeing and woollen mill in 1836. Expanding to employ 100 people, the Manx tweed it produced was exported worldwide.

BALLIG BRIDGE

Having successfully negotiated the curves of Ballaspur, where bikes go light over the top, riders then apply full throttle for an exhilarating downhill rush of a few hundred yards through what seems like a tree-lined tunnel, to where Ballig Bridge crosses the River Neb which is making its way towards the sea at Peel.

An early report told how the hump that existed at Ballig Bridge "adds such terror to the race", although its severity was eased slightly in 1920 when the dip on the far side was partially filled in. However, it remained a place of spills, and when Alec Jackson went over the bars of his Sunbeam here in 1922, landing on his head but getting up unhurt, a report said "the crash helmets have now many lives to their credit, and do not tear off the men's ears, as prophesied". The bridge parapets were popular locations for the advertising banners of concerns like Dunlop tyres and BP fuel in the early 1930s, but the bridge was reconstructed in 1935 to remove most of the hump and its alignment changed to straighten the road; although a rider of the time claimed that the spot was "still full of danger".

Before the reconstruction work, Noel Pope wrote of his MGP debut in 1933, "Ballig Bridge in those days was a real bridge and the biggest jump on the course. Most of the star men slipped over it with no fuss at all, but the lads amused themselves by seeing who could do the biggest jump". It was one of the star men, Stanley Woods, who told that "one's right-hand handlebar would have been over the parapet of the bridge if you were travelling at the maximum speed and on the best line". He omitted to mention that you needed to be well and truly airborne at something like 50 mph, for your handlebar to be over the parapet, that there was a distinct curve to the right after the tyre-flattening landing and that in the 1925 Junior race his handlebar fractured upon touch-down. In the days of rigid frames, machines took a heavy pounding when landing from this jump and there are other recorded instances of frames breaking, chains snapping and wheels collapsing, whilst Alf Brewin who rode an OEC in the 1933 TT recalled "I made a record jump of 60 feet at Ballig and when I came down everything happened at once . . . the wheels came up, hit the mudguards and ripped them off. The front brake cable went flying".

A man from 'The TT Special' was at Ballig for the opening practice session of the 1936 TT. It was the first year of use of the widened bridge and he reported that all riders changed down on their first lap but thereafter took it "full bore". In the opinion of leading riders, the levelling and straightening of Ballig Bridge saved them 3-5 seconds a lap, and other changes to the course contributed to an overall reduction in lap times of between 10-15 seconds for 1936.

Watching

In the 1930s spectators could sit on the banks that sloped away from the bridge on the right. One writer described it as a "bluebell covered grandstand", but today it is forested. They could also stand in the field on the right, as well as crowding the mouth of the road on the left that leads off to

Charlie Collier and his Matchless raise the dust at Ballig Bridge in an early TT.

Peel. If the water level in the river was low the intrepid could cross the road under the bridge whilst racing was in progress, and thus move locations. Whilst that is still theoretically possible, restrictions placed on spectating on the right-hand side of the course means that there are no longer any viewing points there, although there is still a restricted view available from the mouth of the road leading off to Peel.

Ballig has had its share of rider retirements, amongst the more notable being Fergus Anderson stopping with a broken con-rod on his Moto Guzzi, whilst leading the 1951 Lightweight. Then in 1999, Jim Moodie rode his 'works' Honda all out from the start of the Senior race, probably in a tactical move by Honda to break the Yamaha of David Jefferies that they saw as a major threat to them winning. So fast was Moodie's pace that he smashed the lap record from a standing start at 124.45 mph. However, so furious was his riding that just over 8 miles into his second lap, Jim was forced to pull in at Ballig with a shredded rear tyre, his race over. As Honda feared, David Jefferies took his Yamaha to victory.

Riders hardly feel the bridge today, but a group rushing down from Ballaspur creates an exciting spectacle and a wall of noise. They then ease throttles to line up the following Doran's Bend, some of their exhausts breathing flames on the over-run, as they take a line that brings them tight to the right-hand gutter. Watching free of charge from behind the roadside wall here used to provide the sort of close-up race thrills that just could not be experienced by spectators who paid to sit in the Grandstand at the start and finish area. It was regular TT leaderboard man Alan Bennallick who, forced to watch rather than race in 2000, said of the spot: "I couldn't believe how scary it looked from a spectator's point of view".

In the early years of the TT the stretch from Ballig, through to Glen Helen and up onto the Cronk y Voddy straight presented difficulties for the organisers, for there were no side roads to allow access in case of incidents. To help with communications, Boy Scouts (or Scouts as they are known today) were enlisted, and armed with flags and whistles they were stationed at intervals to announce the approach of a rider or pass information. They also served as message runners, something they did elsewhere on the course before the widespread use of telephones, and then radios.

DORAN'S BEND

This long sweeping left-hander immediately after Ballig Bridge was probably included in the name Ballig until AJS 'works' rider Bill Doran crashed here in 1950 after a full-bore, two-wheeled slide and fall that saw him break a leg and leave his name at the spot. Although now known to most as Doran's, it took some time to become so, for the 1957 BP publication '50 Years of TT History' talks of "the very fast left-hand sweep . . . sometimes referred to as Doran's Bend". There is still no nameboard identifying it and few remember Bill's more noteworthy TT achievements of second place in the Senior events of 1948 and 1951 on Norton and AJS machines.

A 'Moment' for Stanley - and others

Stanley Woods was a much better known rider who achieved 10 TT wins before the Second World War, and it was some 25 years before Mike Hailwood bettered that figure in the 1960s. Stanley and his Velocette almost came to grief here when pressing hard in the 1938 Senior race in pursuit of the Nortons of Harold Daniell and Freddie Frith. In his words "I was in third gear with the engine screaming at almost peak revolutions, when suddenly, without the slightest warning, I was in the middle of a tremendous broadside skid! I was travelling at well over a hundred miles an hour and you will have some idea how quickly things were happening". Going on to tell how he steered into the skid, then "I was almost in control again when, unexpectedly, the machine hit a bump and immediately I found myself fighting a terrific wobble. I was standing right up on the footrests, for the sudden jerk as she came out of the broadside had very nearly thrown me off . . . and in due course I got her straight again". Stanley estimated "I was completely out of control for one hundred and fifty yards - perhaps more".

With a pavement, low stone wall and river on the right-hand side, Doran's is a blind bend of which riders must be absolutely confident if they are to ride it in the totally committed manner of top men like Stanley Woods. It takes courage, for speeds are high and back in 1935 Graham Walker described it as having a permanent appearance of melting tar, while 'The Motor Cycle' said the road surface here was "somewhat wavy" in 1954 and the following year Ellis Boyce fell whilst challenging for a win in the Senior MGP. He told how "In a split-second I lost it, mounted the kerb and fell at 120 mph. By some miracle I wasn't hurt but the Norton was absolutely destroyed". A few years later Bob McIntyre spoke of detesting the nasty long curve of Doran's "almost as much as Bill (Doran) did", while Peter Williams wrote of this spot in 1971, "Doran's is a very deceptive bend with adverse camber going out and it keeps going round. This is where Alan Barnett fell off when I was following him in 1970. I was quite amazed how fast he went round the first time, and I couldn't believe when he went round even faster the lap after. Then he fell off". Peter also described it as very, very bumpy, and as a corner that has to be ridden in a manner that sets a rider up for the best line to tackle the right and left-handers that follow. Thirty-five years later, Ian Lougher described it as a good bend to get right, but a really dangerous one if you get off line.

Surprise!

At least two men who thought they had the line right at Doran's found unexpected danger when taking the classic sweep from the right-hand gutter to clip the left-hand apex, for what they thought was an ivy-covered bank on the inside was actually an ivy-covered stone wall. Local rider John Crellin found this out the hard way during practice in 1988 when he clipped it with his shoulder, but it was an even more frightening experience for Steve Hislop who, employing his usual hanging off the bike style, actually scraped his helmet on the wall as he rode it close, too close, at 120 mph during the 1988 Formula 1 race. Admitting afterwards that he always thought it was a hedge, Steve added that it had given him "a real shaking and it affected me for a couple of laps".

There is still much stone in the area of Doran's from old mine workings and huge sloping banks of it are located above the road on the left-hand side. Some of that probably came in handy over the winter of 1930, for after damage to the road edge caused by autumn flooding, the road was widened and strengthened on the outside where it abuts the river Neb. Whilst Doran's is acknowledged as being a demanding corner, history records two top riders who actually fell on its high-speed approach. Harold Daniell totally wrecked his Norton there during practice in 1938, and in the early 1980s Chris Guy described how the bike "just got out of my hands" before they parted company and slid a long way down the road. Though both men were badly knocked about, they were back racing after a few days.

It is worth dwelling on Harold Daniell's spill and the aftermath, for things are rarely as straightforward in TT racing as are shown by bare race results. He was down to ride 'works' Nortons in 1938, but as they were late arriving he started practice on the bikes of his brother in law, ace tuner Steve Lancefield, and it was his 500cc bike that he wrecked. With injuries that prevented him practising for several days, Daniell got minimum time on the new 'works' Norton and after the company's run of seven Junior TT victories was brought to an end by Velocette earlier in race-week, the pressure to do well in the Senior was considerable. There were always unseen tensions in the Norton camp between individual riders, between riders and manager Joe

some riders even do a little 'show-boating' for the crowd by pulling the front wheel higher than necessary.

The small hall that stands adjacent to the spectating area here is the home of the St John the Evangelist Men's Institute. Still in use, it is believed to have been erected in 1909 when the TT was in its formative years, but buildings on the other side of the road that formerly housed a Chapel of Ease and a school, are now used as private dwellings.

Just before riders get to the end of the straight they pass Cronk y Voddy crossroads, known to the early TT organisers as Ballakaighen Road, probably from the nearby farm of the same name. The crossroads are at the highest point on the Course between Ballacraine and Ramsey, with access to and from the course to east and west, plus spectating points and support facilities like parking, toilets and refreshments. For the past few years it has sometimes served as a 'Station' for a Travelling Marshal.

Retirement

There are few practice or race sessions when a rider is not forced to retire at Cronk y Voddy, where they will lean the bike against the bank at the crossroads and sit out the rest of the session. Some will be happy to chat to marshals or spectators, others will wander off to find a cup of tea. A top rider with another bike waiting for him at the Pits may try to persuade a spectator to return him to the start via back-roads if he dropped out early in the session, whilst others will just want to be left to their thoughts of what might have been had the lap gone right and what they are going to find has gone wrong when they eventually get the bike back to the garage.

Mountain Course Moments

It is told elsewhere in this book of how in the early 1930s, Stanley Woods was credited with being the first rider to establish a signalling station capable of giving him his position at a location other than the Pits. But Ruth Sheard tells in the book about her grandfather Tom Sheard - 'The Modest Manxman' - how in his winning ride in the 1923 Senior TT, Tom had "arranged with a friend that he should stand on a quiet part of the course and make certain signals with a handkerchief on a stick, so that I knew from him that I was leading".

That 1923 Senior TT was the first really wet race since the 1912 Junior. It was also very misty which made it harder to judge the approach of bends, particularly on the Mountain. In his post-race review, Graham Walker wrote that Manxman Sheard "was even rumoured to have a battalion of friends stationed at strategic braking points on the Mountain", to provide reinforcement to his local knowledge of the course.

◆ Most riders of the 1920s wore brown leather but there was no compulsion to wear leather until 1930. Geoff Davison told how in a desperate attempt to lighten the load for the little 250cc two-stroke Levis that he rode to victory in the 1922 Lightweight TT, his riding kit consisted of "tight-fitting cricket sweater, wash-leather gloves, drill breeches, stockings and dance pumps".

◆ Jimmy Simpson was a TT great of the 1920s and 1930s and although he only achieved one win, he was the first to set 60, 70 and 80 mph laps during a race. Many riders of the 1920s treated the early-morning practice sessions as an opportunity to put in a single lap before retiring to the Cadbury's cocoa tent for a chat and then returning to their hotels for breakfast. But as speeds increased, serious riders managed to cover ever more laps in practice. Jimmy claims that in 1931 he was the first to do 4 laps in a session. He did it as a fuel-consumption test, to prove that he only needed to stop once in a 7 lap race, when the custom was to stop twice.

◆ In 1933 'The Motor Cycle' told its readers in a preview of the TT that "Some additions to the course will be noted this year in the form of poles along the roadside to carry the overhead transmission cables under a scheme to electrify the Island. For the peace of mind of new competitors who are not too sure about their cornering abilities it can be said that these poles are set well back". The reality was that they were not always set back, and some could be found on the apex of bends.

◆ Everything was organised to run the MGP in early September 1939, even the Programmes were printed. But the worsening political situation in Europe and imminent outbreak of war, saw the races called off at the last moment.

Sidecars at the end of the undulating Cronk y Voddy straight position themselves to take its fast right-hand bend during late evening practice.

Mountain milestone

MOUNTAIN COURSE

CRONK Y VODDY ... THE ELEVENTH MILESTONE ... HANDLEY'S CORNER ... BALLASKYR ... BARREGARROW ... THIRTEENTH MILESTONE ... KIRK MICHAEL ... RHENCULLEN ... BISHOPSCOURT ... ALPINE ... BALLAUGH

CHAPTER FIVE

Cronk y Voddy to Ballaugh

Coming off the end of the Cronk y Voddy straight a vista of fields and villages stretches up the north-west coast of the Island, with Galloway in Scotland visible on a clear day. But riders are too busy to take that in, for all their attention is concentrated on the unfolding ribbon of road that falls away steeply towards the Eleventh Milestone.

Where is it?

With a name that suggests quite a precise location, there can be confusion as to the exact position of the Eleventh Milestone, for there are four markers that might be taken to represent the Eleventh Milestone and they are spread over a distance of about 400 yards. Surprisingly, the prominent old black and white milestone does not indicate the eleventh mile from anywhere on the course, because it records the distance between Castletown and Ramsey, not from Douglas. In true distance from the start, the eleventh mile probably lies between the first two nameboards, but in reality the Eleventh Milestone (usually called just the 'Eleventh') is generally regarded as a section of the course rather than an exact location. And what a sweeping and exhilarating section it is, for riders and spectators, or as three-times a TT winner Tom Herron put it "one of those super corners that if you get it right you feel like going back, wheeling the bike up the field and having a go again".

Bounded by banks and hedges on both sides, with a footpath along the right-hand one, the sweeping curves of the Eleventh run downhill and anyone who has ridden a road bike through here must have wished they could use all of the road like the racers do. They approach from Cronk y Voddy in top gear, with just enough time after monowheeling over a big hump on the way down to take a deep breath as the bike is cranked over for the first right hander and knocked back at least one gear. A dab on the brakes for some while leant over and then the second right requires the rider to almost clip the splayed kerb on the apex - multi TT winner of the 1990s, Jim Moodie, always hung his knee over the footpath - before the bike is hauled up to the vertical for a split second then over to the left to take the series of linked curves that follow.

Leant Over

The performance of a committed rider here has always been a joy to watch, whether on an early racer or one of today's top machines and the rider of a big bike will be continuously leaning over through this section for several hundred yards, apart from the moment of transition from right-hand to left-hand action, and all at something like 140 mph. What skill and courage they show!

When done with precision, a rider's passage through the Eleventh looks so smooth, but hitting the exact apex of the second right-hander can be difficult and getting it fractionally wrong can upset the line. Fortunately, if a rider does make a slight error on the apex, there is just enough room at the following left to right transition point to run a little deeper towards the old milestone and recover, although that can bring the rider very close to brushing the roadside bank and serves as a reminder to get it right next time. On the left-handers exiting the Eleventh, the really fast riders have to concentrate on not drifting too far into the gutter while seemingly cranked over for ever. They can be frustrated by slower riders on those long exit curves, for it is not safe to overtake until they hit the short straight leading down towards Handleys Corner.

Sidecars are also very fast through the Eleventh and at the moment when their cornering forces switch from right to left, passengers experience a moment of almost weightlessness. It is one that they take advantage of to shift position and handholds, and perhaps take a quick look back for outfits wanting to overtake. Driver and passenger will have a pre-arranged signal for such a circumstance, for the roads of the TT course are so narrow

Mountain milestone

Looking back to Cronk y Voddy in this very early view from Cronk Bane.

Riders have already come round one right-hander and are now faced with another right and long left through the 'Eleventh'

that overtaking is frequently only possible with the co-operation of the outfit being overtaken.

IT HAPPENED HERE . . .

An onlooker's report of the 1935 TT told how "the right and left are taken at full bore, riders using all the road, cutting both swerves to the finest limits. Evidence of this is to be seen in the Manx milestone opposite us - a very badly knocked stone structure which Gleave and Ovens did their utmost to utterly destroy". That remark was made after both riders cut things so close that they clipped the surround to the milestone during practice.

♦ A few years later, the bespectacled Senior TT winner Harold Daniell told of his satisfaction in taking the Eleventh "flat in top by using every inch of the road - but only just - the procedure is not recommended for beginners".

♦ Amongst the sweeping bends of the Eleventh is a spot known as Drinkwater's and it is marked by the white painted bridge parapet over a culverted stream that, like countless others, runs off the hills and under the course on its way to the sea. Few spectators who lean on this parapet realise that the spot is named in memory of the unfortunate Ben Drinkwater, who lost his life when he crashed his Norton there in the 1949 Junior TT. Ben was a good rider whose best result was third place in the 1947 Lightweight TT. He had indicated his intention to retire after the 1949 event, but planned to stay involved in racing and had already spoken to one or two young riders about sponsoring them. Sadly, it was not to be.

♦ During practice for the 1958 MGP three top runners were involved in a crash here. Seems that Ned Minihan decided to show Peter Middleton how quick he was by trying to go under him on the sweeping left exit bends. They touched, Peter fell and third man Colin Broughton had nowhere to go but into the roadside bank, breaking his shoulder and wrist, so putting him out of the race.

♦ In Mike Hailwood's comeback year of 1978, Phil Read was forced to press his 'works' Honda over its mechanical limits in his unsuccessful attempts to hold off Mike's Ducati in the Formula 1 race, and it was in a gateway just before the Eleventh that Phil finally bought his oil-smothered bike in to retire on the fifth lap. Then, before a marshal could reach him, he disappeared across the adjoining fields, still clad in helmet, boots and leathers.

♦ It was whilst sitting on the bank at the approach to the Eleventh at the 1983 TT that Steve Hislop's life underwent a major change. He told in his autobiography how Norman Brown came past so fast "he almost blew me off the banking", followed by Joey Dunlop of whom Steve said "I just couldn't believe that anything on earth could go as fast. I was completely blown away by the whole spectacle. In those split seconds that it took for Brown and Dunlop to hammer past me, my life had changed forever. I determined there and then that if I never did anything else in my meaningless life, I would try to round the TT circuit like those boys. My mind was totally made up". Steve had the ability to match his ambition and recorded 11 TT victories in ensuing years, the last of which was commemorated in the title of Duke's video of the 1994 TT, called 'The Eleventh Milestone'.

♦ The year 1994 was the one in which Bob Jackson was brought off at the Eleventh when passing a slower rider in the Supersport 400 race and finished entangled in the roadside fence, fortunately without injury. Despite demanding much skill and commitment to ride its high-speed curves, the Eleventh has seen relatively few accidents down the years. When Manxman Barry Wood fell here during the 2006 TT, he was the first rider to come off at the spot in the TT and MGP for almost 10 years.

♦ Shortly before roads were due to close for Friday's racing in the late 1990s, a double-decker bus broke down at the 11th Milestone and deposited oil on the bend. A heavy salvage lorry and road-cleaning vehicles had already been sent for when Travelling Marshal Jim Hunter arrived to evaluate the situation on behalf of the Clerk of the Course. After the bus had been towed away, and on what was a perfectly dry day, the bend here had to be wetted over a stretch of about one hundred and fifty yards as detergents

Clipping the apex of the right-hander with knee over the curb at the 'Eleventh'.

Spectators take their ease in the grounds of the White House as sidecar racers speed through Kirk Michael.

Mountain milestone

Tackling the twists and turns of Main Road, Kirk Michael.

race through on a high-performance machine and where even Joey Dunlop admitted that as a racer "it scares you". Tightly enclosed by buildings and footpaths, it has become covered in white lines, double yellow lines, street lights, manhole covers and road-signs, as it snakes and undulates to the right, to the right again, to the right yet again, to the left, to the right, to the left, to the right, to the left, all in just over half a mile. In that short distance it passes shops, bank, garage, church and houses, demanding the utmost concentration and bravery, for apart from the first right-hander and last left-hander, it is tackled at full throttle. Riders sometimes short-shift up through the gears after they accelerate out of Douglas Road corner, choosing spots in the village to change where they will least upset the stability of the bike, rather than by what it says on the rev counter, with top riders monowheeling down the dip in the middle of this village street. The whole stretch provides a terrific sensation of speed and with their passage often captured on video, the performance of the fastest riders is breathtaking as they skim projecting kerbs when seemingly drifting on the limits of adhesion.

John Surtees told that in the late 1950s "I always like rushing through the streets of Kirk Michael - that was fun, going as fast as you could through a little village with all the people looking out. But you had to be very careful of the wind between the houses, especially with the full streamlining we used before it was banned for 1958". On the point of streamlining, John had an ironic tale for journalist and former TT rider Alan Cathcart, revealing in an interview: "The TT is a full-fairing circuit because it's so fast, provided you knew what you were doing when you designed the streamlining. Guzzi and Gilera certainly did, because their bikes were quite safe, but ours were not! MV suffered from being aeronautical engineers, so we applied aircraft technology to motorcycles and got it all wrong". The FIM ruled out full streamlining from 1958, so bringing about the dolphin style used today. In really windy conditions on the Island, MV Agusta would use just a cockpit fairing, the thinking being that what they lost on top speed they gained in improved handling and safety.

Villagers are no longer permitted to bring a chair out from their cottages and sit on the pavement to watch riders streak past as they used to, but at the edge of the village riders pass the White House on the right which offers spectating facilities, and the Old Vicarage on the left which is home to the splendid A.R.E. collection of vintage motorcycles. Just before the Old Vicarage there is one of those TT course specialities that is difficult to avoid - a studded iron manhole cover located under trees and right on the racing line. Worrying enough in the dry, it is menacing when encountered with the throttle against the stop in the wet. Recently the highway authority gave the cover a coating of Shell Grip special surfacing, but riders would much prefer it to be set under the tarmac surface.

Until early-morning practice was abandoned in 2004, riders leaving Kirk Michael received the first of several warnings of dazzling low morning sun as they headed over the nine-mile stretch to Ramsey.

RHENCULLEN

Riders leave the built-up area of Kirk Michael at speed on a snaking road, with a blind rise leading to the fearsome Rhencullen. It is a spot whose Manx name translates to Holly Ridge, but that fails to create the tingle down the spine that TT fans get when they hear the word Rhencullen, for it is a place where the passage of the fast men is truly awesome. For years spectators crowded the banks and even the edges of the pavement here to witness one of the finest road-racing spectacles, as they were offered scenes of speed, skill and excitement that would lodge in their memories and be re-run with like-minded friends for years to come.

The Same Line, Every Time

Viewing a practice session here at the 1949 TT, 'The Motor Cycle' wrote "The two who appeared to be the fastest were Frith and Fry who nearly 'earholed' their way round. Bell, Daniell, Lockett, masters all, Graham, Frend,

Doran, Foster, Lyons - all were terrifically fast, and had a postage stamp been stuck on the road at any point on the bends over which, say, Frith's wheels had passed, the same stamp would have been crossed by the wheels of all the others mentioned. Is anything more thrilling to watch, than an ace at real speed on a course he knows?"

As speeds have risen, so Rhencullen has become ever more demanding for the riders of today's 200+ bhp machines. That does not prevent them trying their utmost, but it has forced the organisers to recognise that an accident here could have catastrophic results for watchers as well as the rider. While riders have to look after themselves, the only reasonable way of protecting spectators has been to ban them. This is another stretch that has become a prohibited area, resulting in the loss of a major viewing point on the Mountain Course.

The stretch that is today known as Rhencullen has four warning boards for riders and comprises a right-hand bend that they approach over a blind rise with the front wheel pawing the air as it continues to the right, followed by a left-hander (almost a double) that falls away downhill. That is a simplistic description of a demanding stretch between hedges and roadside cottages that has caught out many riders. One remembered incident here was the death of 'Archie' Birkin when riding a 500 McEvoy during practice for the 1927 TT and it led many people to call the spot Birkin's Bend, with some still calling the first bend by his name, although Rhencullen is now in more general use, even though there is no trace of the name on early maps of the area.

Roads Closed for Practice

At the time of Archie Birkin's death the roads of the Mountain Course were not closed to other traffic during practice. The sessions were all early morning ones and traffic was thin, but there were road users who went about their business at an early hour and Birkin met one of those, a fish van heading from Ballaugh to Peel to collect supplies. Many accounts say he collided with the van, but in his well-researched book 'Isle of Man TT & MGP Memorial 1907-2007', Paul Bradford says there was no collision (and the van did not stop), but Birkin was forced to swerve violently, lost control and fell from his machine.

Birkin's death resulted in the Road Closing Orders being extended to include practice sessions, something that had previously been asked for, but had been refused.

RIDERS VIEWS . . .

Graham Walker in 1935: " . . . you're into the right and left of Birkin's Bend. An easing of the throttle as you enter the bend to get your line and you're round at 95-100 mph".

◆ A few years later and Norton 'works' rider Harold Daniell must have been a bit quicker, for he wrote "I have done it flat, but don't like to".

◆ Twenty years later and Bob McIntyre (Gilera) was also calling the spot Birkin's and told "on the blind rise I hook back to fourth and sight by the same telegraph pole as everyone else, but the line through the long S-bend is governed by the need to be upright where the road drops away from the front wheel on the exit".

◆ Of the same era as Bob Mac and after the road had been widened here in the early 1950s, John Surtees called it Rhencullen and rated it "A difficult one but I used to love it. I treated it as a short circuit corner - clip all the sides all the way through in top gear. You had to lift the bike up smartish after the house though, because there's quite a jump immediately after".

◆ Most riders called the stretch Rhencullen from the late 1950s on, with Mike Hailwood saying he went down one 'cog' here on his 500 Honda

This classic TT image shows that spectators were allowed to crowd the banks and path at Rhencullen when David Jefferies wheelied through on his Suzuki.

Mountain milestone

John McGuinness copes with the changes of direction required at Rhencullen and with variations in the road surface where traffic has scrubbed off the chippings and left polished tar.

and Peter Williams classed it as one of only three places on the course that he used a marker. This was a telegraph pole that he sighted on to help get the right line for a kerb that could not be seen on the approach, so enabling him to take his Norton Commando through flat out.

♦ Joey Dunlop also used a course-side telegraph pole to aid his line here, but in the 1993 Ultra-Lightweight race where he was on target for a record-breaking 15th win, he recalled "At Rhencullen I always line up one of the telegraph poles. But I couldn't see the bottom of the pole with people standing around it waving. The crowd was going crazy. I know they were well-meaning but they put me off a bit". Joey's presence in an Island race always got the crowds waving, but the reason for their greater than usual enthusiasm was because victory in this one saw him better Mike Hailwood's total of 14 TT wins.

♦ Rhencullen was one of Steve Hislop's favourite spots where he deliberately pulled a big wheelie and put on a show for the crowds. Whilst Steve admitted showing-off a bit, his style gave a very quick passage, and it was one that he refined over time, saying over a particular aspect of the exit "it took me years to learn".

♦ Other observations on Rhencullen include the wise words to newcomers from Ray Knight: "Until you have worked it out completely, it could pay to go slow in, fast out. You will get faster with practice". Nick Jefferies made the point that it is "superb on a good bike, but petrifying on something that doesn't handle", whilst John McGuinness warns about the power of the latest Superbikes, telling that "the bike can get a real weave on here if you are not careful".

Flat in Top!

Rhencullen is not particularly wide, but amazing as it may seem it is taken flat out by the top Sidecar men. At the 2006 Sidecar TT it was Dave Molyneux and passenger Craig Hallam who were race favourites until disaster struck the super-fast pair during Thursday practice. Moly always made a big-effort to take the right-hand sweep of Douglas Road Corner on the entry to Kirk Michael as fast as possible, because it determined the speed he would achieve with the throttle against the stop over the next three miles to Ballaugh Bridge. On Thursday he made a good passage through the village and so set his Honda-powered machine to take Rhencullen flat-out. Travelling at 145 mph, the front of the outfit lifted slightly over the rise as usual, but then it

Dropping through the right-hand bends that lead on to the Bishopscourt Straight.

continued to lift, rose vertically on the rear wheel, turned over onto its back, slid down the road and burst into flames. Hallam was thrown clear, and despite being trapped underneath when it inverted, Molyneux also got out before it caught fire. Although it was a major disaster in one sense, both men realised how incredibly lucky they were to escape death.

BISHOPS COURT

A short downhill run through very solid walls, trees and banks completes the Rhencullen section, and riders then burst onto the Bishopscourt straight, unleashing all their power. In 1911 that was not much more than 5 bhp and took some coaxing, today it can be well over 200 pent-up 'horses' waiting to provide an instant response to a twist of the throttle. The straight actually starts with two right-hand kinks and the faster the rider the more they intrude, particularly as the last one has a slight crest that lightens the front wheel. Tree-lined for its entire length, the Bishopscourt section is one of the widest on the course, but if coming from bright sunlight at Rhencullen, it can seem like a dark green tunnel. At TT time the roadside bank on the right is covered with the flowers of wild garlic, the smell of which is picked up by passing riders, whatever their speed. The trees also have a use as Travelling Marshal 'Kipper' Killip found when sent to investigate why a batch of riders had 'gone missing' between Kirk Michael and Ballaugh in the 1960s. Thinking that there had been a major incident, 'Kipper' was relieved to see some thirty riders stopped under the trees, taking shelter from a localised rainstorm.

The name Bishopscourt comes from the imposing building on the left towards the end of the straight, which for centuries was home to the Bishop of Sodor and Mann, before being sold into private ownership. Opposite Bishopscourt is Bishopscourt Glen, which offers a pleasant walk and contains a few historical oddities.

Jock Taylor was to win TT races and the World Championship on three-wheels, but one of his nastiest racing moments came early in his Island racing career on the Bishopscourt straight. Thinking that this was one of the few chances he would get to check his engine temperature, Jock snatched a look at his gauge and the next instant found himself scraping alongside the bank for over 100 yards at 130 mph. That was a bit too close to the garlic for Jock's liking, so he repositioned the temperature gauge so that he could not see it, but his passenger could - a neat delegation of responsibility.

Mystery Crash

The year of 1978 is best remembered for Mike Hailwood's winning return to the TT. But whilst Mike was a TT veteran with vast knowledge of the Mountain Course and of the tactics required to race over it, the practice performances of American Pat Hennen raised high levels of expectation, particularly for the Senior race. In only his second year of Island racing, Hennen became the first man to lap in under 20 minutes when he rode his Suzuki around in 19 minutes 54 seconds, an average speed of 113.75 mph. Seemingly determined to carry his form through to the race, Hennen set a furious pace throughout the Senior TT. Such speed from a relative newcomer

John McGuinness is totally committed at 180 mph through the bends at the end of the Bishopscourt Straight.

was of concern to serious TT watchers, and on the last lap his ultra-rapid progress came to an end with a crash at Bishopscourt.

It was the experienced Tom Herron who led throughout that fateful 1978 Senior race and with the aid of five signallers he knew his exact race position and that of his nearest rivals, saying "I was always in control of my position on the Course", as he paced himself over the long six-lap and 226 mile race distance. On lap four Hennen moved into second place, riding closely on the road with Herron. With his signallers telling him the position, Herron knew through lap four that Hennen was coming and the exact moment that he arrived on his tail, but he also knew that the young American had to pass and then draw away from him by over twenty seconds to win - an extremely difficult task.

No one could blame Pat Hennen for trying, but maybe he should have realised that even if he got in front of Herron, the experienced Irishman was more than fast enough to just sit on his tail and not let him get away. It was an ideal situation for the race-leader and riding in such close proximity, Tom tried to indicate to Pat to hold station, but the American went down at Bishopscourt when just a few yards from Herron's back wheel. There was talk of a bird strike, or maybe a seized engine, also of him hitting the kerb while removing a tear-off visor, but no one really knew the reason for the crash that he was lucky to survive, but which ended his racing career.

Twists and Turns

The next mile and a half to Ballaugh Bridge tends to be referred to in rather vague terms by race fans as "the stretch after Bishopscourt", or as "the bends on the way to Ballaugh", for there are few common names for the twists and turns on a part of the course where the experienced rider can make time. As they reach the end of the Bishopscourt straight competitors are faced with the start of those bends, comprising sweeping left and right-handers that pass Bishopscourt Farm, with the course still running under an arch of overhanging trees and with prominent double white lines to be cranked over down the centre of the road.

This is a spot that receives varying degrees of commitment from riders, for with a high kerb bordering a footpath on the exit, there is no room for error. One man who always gives it his all is John McGuinness, but just read what is involved with today's all-powerful machines: "You come down that little hill from Rhencullen, get into top (on the straight), then it goes left and right. Well that right-hander, you're a complete passenger . . . sat on a moving torpedo that's doing its own thing and you just have to let it do it". John added "You're probably doing about 180, it's blind, you can't see through it".

Another man not keen to back the throttle here is Ian Lougher, but again it requires the utmost skill and bravery, for he tells "It drifts right across the road where you cross the ripples". He also recalls following Joey Dunlop through here in 2000 after it had been resurfaced, and how "It was a bit damp and Joey had a big moment here, with both feet down".

Lucky Men!

Leaving the Bishopscourt Farm right-hander, riders then pass a junction on the left that is known to the organisers as Orrisdale North. It was after an incident just before here in 1984 that Ray Swann was described in headlines as the "Luckiest Man on the Island on Friday". A less than precise report told that "he had his front brake seize on when going flat out at over 130 mph at Orrisdale". That was an unusual mechanical problem and in his subsequent fall, Ray was fortunate to get away with little more than "banging the wrist which was already plastered from an earlier mainland spill". Almost thirty years before, Geoff Duke hit a patch of molten tar here in 1955 that sent his Gilera into a full-lock slide, causing the normally unflappable Geoff to admit that the incident left him de-tuned for the next couple of miles. Less fortunate was the dashing young Ollie Linsdell, who fell here at the 2010 MGP and was badly injured.

Bishop's Dub

In the absence of recognised 'TT names' for the spectator-free stretch that follows Orrisdale, use is made here of a local name that refers to the boggy pond located on the left-hand side of the course. In winter this expands to become a small lake which is visited by wildfowl from Northern Europe.

Although poor road surfaces may have caused riders to temper their speeds on the twists after Orrisdale in the very early days, it has for many years been recognised as a fast stretch through to Ballaugh. It was widened in 1930 and by the mid-1930s Graham Walker wrote of "nearly two miles of perfect surface and full throttle, with one right sweep that can only just be negotiated (flat out)". That would have been Alpine, but the fact that Graham did not mention it by name suggests that it was not known as such at the time. However, before they reach Alpine, riders have those tricky rights and lefts, whose kerbs are painted black and white for several hundred yards to guide them between high hedges.

Today's riders crouch behind their screens and try to straight-line these three pairs of left and right twists, what one described as "getting tucked in for all the little wriggles to Ballaugh", but it is a challenging stretch at speed. Ray Knight warns learners to pay attention to their road-positioning on the entry to these bends, whilst reminding seasoned riders that "no slack should creep into the throttle wire, no matter what you are riding".

ALPINE

Whilst riders have sought to pass through the series of preceding bends at maximum speed, it is essential that they maintain the proper line, for the cumulative effect of errors in those bends can leave them severely off-line at Alpine's fast right-hander with potentially disastrous results, for the outside of the corner is lined with the stone wall of the property called Alpine House. In TT nomenclature the bend was for many years Alpine Cottage, although that was previously the name of the cottage located 150 yards back up the course and now named Iceman's.

The correct line at Alpine is a conventional one, coming from the left to almost clip the stout sod bank of the apex, that seems to jut into the road, and a good run through will set a rider up for a fast run to Ballaugh. Just over fifty years ago, John Surtees was top man at the TT riding the MV Agusta four, when the big-bike limit was 500cc. He told how "I could never do Alpine Cottage flat on the 500 MV. I got to the stage of easing it slightly and scrubbing off a little speed on entry to be able to crack it wide open for the exit. With the MV's high centre of gravity it was sometimes difficult to get it back on line if you went into a corner late, so my natural short-circuit style of going in early suited the bike more than perhaps the Manx Nortons. Full tanks only made this behaviour worse, so you had to go into the corner early even if it meant using up more road - it wasn't really a slide, more a controlled high-speed drift which tyre developments were beginning to permit".

It was the four cylinder Italian MV Agustas and Gileras of the 1950s that brought an end to the post-war victory run of the fine-handling British Manx Nortons in the Junior and Senior TTs, but they did it through power advantage that was partially offset by their inferior handling. Geoff Duke moved from riding Norton to Gilera for 1953 and Alpine showed him that there was a price to pay for the Gilera's superior power. Being his first TT with the Gilera, Geoff had many settings to experiment with during practice to get the bike into race trim, but one thing he did not try was how the four went with a full-tank containing 7 gallons of petrol. On the first lap of the Senior race, Geoff set the bike up to sweep through Alpine at 140 mph, but as he cranked it over the extra weight up high caused it to weave, taking him far over into the left-hand gutter on the exit, raising dust and loose chippings, and giving him a very nasty moment.

Ducati Mystery?

Clearly both the MV and Gilera of the 1950s found Alpine testing, but what was it about this spot that troubled Mike Hailwood and his Ducati when he returned to TT victory in 1978 after 11 years absence? John Watterson is Sports editor of the 'Manx Independent' and he watched the 1978 Formula 1 race here. Although unable to offer an explanation as to why, he recalls that on every lap Mike approached the right-hander at full throttle, "only for the ear-rumbling thunder of the Duke to go deadly silent as it free-wheeled coasted round the bend at more than 120 mph". Seemingly, Mike pulled in the clutch lever of the v-twin at Alpine on each lap, releasing it again as he ran quite wide on the corner. It was an action that is still unexplained.

The previous facility for spectators to watch at Alpine has long been withdrawn, and with it the custom for marshals' wives to sell refreshments from the Dave Featherstone memorial shelter, located on the approach to the bend.

Helicopter Rescue

A wind-sock in the field adjoining Alpine is there for the benefit of one of the two rescue helicopters that use it as a base during practice and race times at the TT and MGP, the other usually being stationed at Keppel Gate.

Back in the early 1960s the TT races were still the pinnacle of the motorcycle racing world and most riders wanted to take on the challenge of the Mountain Course. The racing game being what it is there were accidents to riders, sometimes in circumstances that would not be tolerated today. A major advance in providing support to those involved in accidents on the Mountain Course came at the 1963 TT, when Shell-Mex and BP provided funds for a helicopter to carry a doctor to an injured rider and the means to transfer him to hospital with a minimum of delay. The rescue helicopter service has developed down the years and now uses two 'choppers' during practice and race weeks. With each carrying medical equipment and able to take two casualties, the average time from call-out to delivering an injured rider to hospital is under 20 minutes and the provision of rescue helicopters has saved many lives.

The helicopters used at the TT are paid for from the Isle of Man Government's grant to the races, but the massive cost of those at the MGP has to be met by the organisers. This is why spectators are sometimes asked to contribute to the 'Helicopter Fund' at both the TT and MGP, for it takes a huge effort to raise the sum required to hire these life-savers. That fund is augmented by many money-raising efforts both on and off the Island outside race periods and they really are essential, for without helicopters there would be no MGP.

Benelli Bodge

Having taken Alpine with care, riders have a full-bore, slightly downhill, between-the-hedges run to Ballaugh, with the swiftly appearing left sweep of Brough Jairg straight-lined without easing, and the side-turning to Ballacobb passed with the throttle still against the stop. It was on this stretch that four-times a TT winner over the Clypse Course, Tarquinio Provini, had an accident during early morning practice in the mid-1960s that ended his racing career. Attributed at the time to Provini being dazzled by the morning sun, it was only 25 years later that 'Classic Bike' magazine revealed that Provini's Benelli had been subject to a mechanical bodge that resulted in a gearbox seizure.

If bodging a 'works' bike for the TT seems scarcely believable, how about the ruling introduced for the sidecars in 1965 that banned overtaking on the twists and turns to Ballaugh? Warning signs were erected at the side of the course and the penalty for breaking the new rule was exclusion.

Mountain Course Moments

When racing returned to the Island after the Second World War, it was the MGP of 1946 that was first to run and one reporter was moved to write at the start of practice: "the last seven years rolled away like the mist off Snaefell this morning and brought us back the scene we've waited for so long".

For the first time a Thursday afternoon practice session was used.

- ◆ Geoff Duke's high speeds at the 1951 TT had everyone talking and as editor of 'Motor Cycling' and a past TT winner, Graham Walker offered the opinion that if Duke had ridden his 'works' Norton over the same primitive unmade roads of the Mountain Course that 1911 winner Oliver Godfrey (Indian) rode at an average speed of 47.63 mph, he would have been pressed to average 62 mph - his actual race-winning speed was 93.83 mph in 1951. It would have been interesting to hear Graham's estimate of what Godfrey and his Indian might have averaged on the roads of 1951.

- ◆ Many riders took advantage of the publicity and fame they acquired from competing at the TT & MGP by setting up in business as motorcycle dealers. Victorious in 1911 and 1912, Oliver Godfrey and Frank Applebee did so in partnership, whilst in later years other top riders like Wal Handley, Alex Bennett, Jack 'CJ' Williams and Tommy Bullus, Artie Bell and Cromie McCandless, Bob Foster, Geoff Duke, John Surtees, Eric Oliver, Bill Lomas, Mike Hailwood and Rod Gould, Reg Armstrong, Bill Smith, Phil Read, Chris Vincent, Tommy Robb, Phillip McCallen, took the same route.

- ◆ The national press seized on the fact that Inge Stolle-Laforge became the first woman to compete at the TT, when she passengered Jaques Drion in 1954.

- ◆ Today the changing of disc-brake pads before a race is a fairly quick process carried out by a rider's mechanic, but in 1957 at the time of the first 100 mph lap, everyone used drum brakes and the process of relining was entrusted to a specialist, usually Ferodo who supplied brake linings to most of the entry. Prior to the 1957 Junior race, Ferodo mechanics worked through the night and between 9.00 pm and 5.15 am a total of 89 brake assemblies were relined and skimmed to suit individual drums. Similar support was given by other elements of the Trade, with Renolds usually rivetting 1,600 chains over a TT period and spark-plug company representatives carrying over 30 different grades of plug.

- ◆ The 1957 Golden Jubilee TT was attended by several competitors from the 1907 race and a report said "To survivors who raced in the infancy of our sport, the efficiency and complexity of present day engines must be almost frightening". That comment was made in the light of Bob McIntyre's double win on four-cylinder Gileras and the TT debut of the V-8 Moto Guzzi with an output of some 70 bhp.

- ◆ Prior to 1959 race numbers were allocated by ballot, meaning that a fast rider might have to start near the rear of the field and have the handicap of overtaking many slower riders. Conversely, his nearest rival might be lucky enough to gain a low number, start near the head of the field and be involved in far less overtaking. From 1959 a form of 'seeding' was introduced to reduce the effect of this unintentional handicap, and riders were allocated numbers based primarily on past performance. The top five seeds balloted for numbers 1 to 5.

Mountain milestone

BALLAUGH BRIDGE ... BALLACRYE ... QUARRY BENDS ... SULBY STRAIGHT ... SULBY BRIDGE

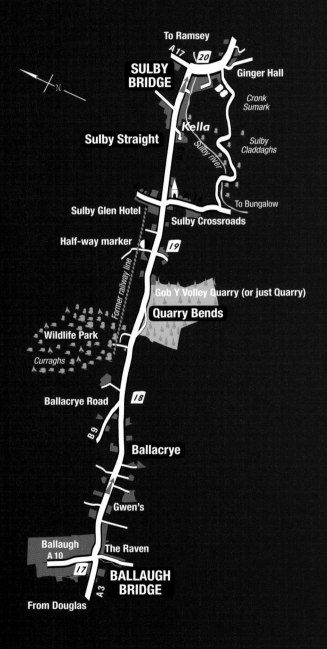

CHAPTER SIX

Ballaugh to Sulby Bridge

For 100 years riders have approached the village of Ballaugh with throttles wide-open and whilst a century ago they only needed to knock about 25 mph off their approach speeds before tackling the formidable hump of Ballaugh Bridge, today they must lose nearer 125 before leaping it at some 50 and more mph. Not everyone judges their braking correctly and many have gone over considerably faster than intended, thus providing spectators with plenty of excitement! Located 17 miles from the start, this is one of the most photographed spots on the Mountain Course and one that every spectator should see at least once, for the spectacle involved and for the fact that access and facilities are good, although room is limited.

The name Ballaugh is said to derive from the Manx Gaelic for homestead of the lake. No doubt there was a ford here before the bridge was built with its enclosing stone parapet walls allowing single-file passage by horse and cart. There was discussion about widening it in 1924 and in 1931 the owner on the right was approached about releasing land for widening. As he demanded £300, the Highways Board asked the owner of Bridge House on the left and he wanted £40. This was still deemed excessive and eventually £20 was paid, so allowing the bridge to be widened at a cost of about £250. In 1933 the stone parapet on the west side was replaced with metal railings, and the same treatment applied to the east side in later years, giving a little more road width and creating better sight-lines.

Whilst the hump is seen as the main barrier to a speedy passage by racers, the bridge also takes the road in a left-hand direction, and upon landing they are required to bank quite sharply to the right.

For the first forty years of racing, riders leaping the bridge knew that they also had to miss the projecting garden wall of The Raven pub that jutted out on the right. (It was known as The Railway Hotel until well into the 1950s.) Caution was also needed when accelerating out of the following right-hander, for the projecting window of a butcher's shop came perilously close as riders got carried out to the left. The shop has now gone and the pub wall was cut back prior to the 1954 TT, but it is still a relatively confined exit that needs treating with respect.

RIDERS VIEWS . . .

An unknown rider of 1930 spoke of taking the bridge at 55 mph on his rigid-framed and girder-forked bike, and Graham Walker would have been similarly mounted when he won a TT in the early 1930s and wrote a little later: "Ballaugh Bridge looms up. Hugging the right side of the approach you brake hard - harder - down into third - second - cut into the left railings, twitch right as you leap at fifty-five". That seems rather a fast passage for a rigid-framed bike, and one that would have given an exciting flight and stressful landing.

♦ Geoff Duke favoured bottom gear in the early 1950s, though as he only had four-speeds in his Norton gearbox, bottom would have been a somewhat higher gear than on today's bikes. He took a similar line close to the railings on the left, leaping from the crown and banking right immediately on landing.

♦ Always thinking and planning ahead, John Surtees took the bridge on the conventional line, whilst concentrating on a landing that let him "use every bit of power to get out of there as quickly as possible". John described the spot as "universally disliked by riders".

♦ With the need for fierce use of brakes on the approach, Steve Hislop spoke of "basically treating it like a short circuit hairpin - very hard braking, 5th, 4th, 3rd, 2nd and you're over".

♦ Ray Knight told other competitors how to tackle the bridge in some detail in his 'TT Riders Guide', but

Mountain milestone

This was Ballaugh Bridge in 1932. Ernie Nott is trying to get his Rudge straight before he jumps. Spectators on the front garden wall of The Raven show how it projected into the road until removed in 1954.

then admitted in print: "after 2,000 tries, I still don't get it right every time".

◆ With little suspension travel, sidecars have an unenviable time here, and because the approach offers a chance for drivers to try and outbrake each other, if they do not complete their moves it can get a bit crowded with six wheels or more on the bridge at the same time.

Front, back, or both?

Much has been spoken and written about the correct way to tackle Ballaugh Bridge, with the principal point of argument being whether a solo rider should land on the front-wheel, back-wheel, or on both wheels. It is one on which top riders have conflicting views. Not many favour front-wheel landings, although multi-winners Mick Grant and Steve Hislop said they did. Mick claimed that in doing so he was easing the strain on his bike's transmission and Steve agreed, adding "if you land smoothly on the front and get on the gas again, you carry more speed". While their front-wheel landings were seemingly deliberate, the same cannot be said for many others whose nose-down efforts are mostly involuntary.

Amongst those who aimed to complete their jumps of the bridge on the back-wheel were Bob McIntyre - who was prepared to sacrifice a little speed here to minimise stresses on the chain, gearbox, tyre, frame, fairing fixings, etc, plus other winners like Giacomo Agostini, Charlie Williams - who maintained a rear-wheel landing was "far better than a front wheel because it means you can transmit the power and get away quite safely".

A strong advocate of easing off slightly just before the hump to achieve a two-wheel landing was Geoff Duke. Many others used that method and a later top runner who also favoured fairly level flight was Joey Dunlop, although he admitted that he did not get it right every time.

Another question related to this tricky point is, how high? Again, the high-flights of a few riders whose brakes were not as good as they thought has not been from choice, but there have always been individuals who liked to put on a show and though some will say that time in the air is time lost, there is no doubt that the crowds like to see wheels off the ground. Perhaps the man who best 'rose to the occasion' offered by Ballaugh Bridge in recent years

This plaque at Ballaugh commemorates Karl Gall, who died after an accident in practice at the 1939 TT.

The point of no return.

Having been checked by medics, Les telephones his sponsor to explain matters.

with horse droppings, they had to slow their machines to some 20-25 mph with primitive dummy belt-rim brakes, while they also took one hand off the handlebars to operate the gear-change, worked via a lever mounted on the right side of the tank. It was still a time when using front brakes was considered a dangerous activity, (and it could be if they were used carelessly on loose surfaces), so some manufacturers did not bother to fit them and the primitive versions that were fitted were used sparingly.

As late as 1920 the Norton team still used the stirrup-type front brake inherited from the pedal-cycle, which Graham Walker warned was "only to be applied at the rider's peril". He ignored his own advice whilst seeking to reduce speed for the Bridge in the 1920 Senior race and the brake blocks turned into the spokes and fetched him off. It was the rear brake that provided early riders with the most retardation, and as hard application could eventually lock the back wheel, many approached the bridge in lurid slides.

Sulby Bridge was substantially widened and flattened at a cost of £1,500 just before the 1923 TT and the road surface was also tar-sprayed, so allowing riders a better run along the Sulby Straight and reducing the clouds of dust raised. However, things were far from perfect, for it was reported that the "new surface is inclined to be loose" and with the Senior TT of 1923 run in very wet conditions, fourth-place finisher Graham Walker said "The newly tarred sections of road, which we had blessed in the fair-weather practising, became death traps", presumably because exposed tar became slippery when wet.

Internal expanding drum brakes gained popularity in the early 1920s and they provided improved retardation to cope with growing speeds. Although initially very small, they were more effective than previous arrangements.

Recognising that changes were occurring in braking techniques, in its summary of the 1923 TT the magazine 'Motor Cycling' told its readers "the old idea that an effective front brake is dangerous is dead" and indeed, the Senior TT of that year was won by Manxman Tom Sheard riding a Douglas fitted with experimental disc brakes in which the disc was comprised of friction material onto which was depressed a wedge-shaped alloy shoe. Despite Tom's success, it was many years before disc brakes were seen again at the TT and a drum-braked machine won the Senior as late as 1974. But better brakes were certainly needed in those early years, for as seen in the figures quoted under 'Sulby Straight', approach speeds climbed to 100 mph in the early 1930s and 135 mph in the 1950s, since when they have continued upward, until today's riders are hitting over 190 mph and trusting everything to the disc brakes front and rear that have been developed to deal with such massive speeds. The enormous forces involved in slowing see front forks going on full compression, rear wheels wanting to leave the ground, tyres subject to extreme stress, and riders having to brace themselves from sliding over the front. In the mid 1950s Ferodo said that a rider braking hard for Sulby Bridge had to disperse 250,000 ft lbs of energy with his drum brakes, and today riders are travelling almost half as fast again as they hit their stoppers.

Mind the Drop

With the course taking such a sharp right turn at Sulby Bridge it is essential that riders get their braking right, because the road that goes off to the left is not a viable slip-road and there is no real straight-on route, except as a very last resort. For many years riders' straight-on view was of a hawthorn hedge and fence. That has now gone and the route ahead looks slightly more inviting, as it seems to offer an unobstructed run into a field. However, riders who have done an on-the-spot inspection will know that the level drops about three feet from the road to the field and even those with exceptional off-road riding skills would be hard pressed to stay on a race bike after leaving the road at speed in those circumstances.

An additional hazard for riders in the days of early-morning practice, was the low-level sun that shone straight into their eyes as they burst at speed from the tree cover several hundred yards from the bridge. At the 1937 MGP the organisers experimented with a 'sun barrier' to reduce its blinding effect. It was a canvas device on poles and for the first couple of sunny practice sessions it was said to have proved extremely useful. However, a change in the weather was accompanied by an overnight gale and by the next morning it was in shreds. For many years permanent course-side 'SUN' warning boards and marshals displaying 'SUN' flags were used, but the blinding dawn light is still believed to have been the cause of rider fatalities here.

RIDERS VIEWS . . .

John Surtees rode the four-cylinder MV Agusta to six victories over the Mountain Course between 1956-1960 and in his words "used to drift the rear wheel as a matter of course". Whilst that tactic was more often used on fast bends, he also told how "I used to turn the power on quite early in a bend like Sulby Bridge, because if the rear wheel stepped out unduly, then the bridge parapet on the left would stop it going too far". It was a time when the parapet on the exit still had no protective bales.

Peter Williams wrote of braking at the marker board 200 yards from the bridge on his 150 mph Norton twin, and Steve Hislop used the same marker when travelling 40 mph faster and indulging in "serious hard braking" as he went down through the gearbox from sixth to second for the turn.

The actual turn is treated as a simple right-hander with riders approaching on the left-hand side of the road, riding across to sweep past what is now a very low and flat kerb on the apex, before starting to feed in the power and running out wide to the left on the exit. Not too wide though, because despite the fact that the solid stonework of the bridge is now protected by bales, there still remains an unprotected low stone wall on the exit to keep ordinary traffic in bounds. For many years the stretch in front of that wall used to have a steep drainage camber, although it is less pronounced today. Speeds through here were 40-45 mph after the road was first surfaced in the 1920s and although Mike Hailwood still spoke of it as a 40 mph corner in the late 1960s, the removal of the shoulder-rubbing wall on the apex, plus further changes to road-surfacing, bikes and tyres have lifted the rate of progress.

TM Station

With its straight approach and strategic location, Sulby Bridge is a 'station' for one of the team of Travelling Marshals - known within the organisation as TMs - who assist at incidents if called to do so by Race Control from the tower at the Grandstand. 'Control' sometimes receives radioed reports from static marshals around the course who have noticed a bike with a loose fairing, or one that is suspected of leaking oil. That information will be passed to a Travelling Marshal 'station' like Sulby with a request that the rider be black-flagged. With the fairly long and straight approach to the Bridge, the rider will be expected to see and heed the black flag with orange centre that is displayed with a board showing his number. Together they mean he must stop and leave the course by the side-road on the left. There the bike will be inspected by the Travelling Marshal and the rider either allowed to proceed or be retired.

It may sound a straightforward task, but stopping a rider in the middle of a race can create a situation of high drama, for the adrenaline is pumping hard and emotions running high. The Travelling Marshal must keep a cool head, whilst showing understanding of the rider's agitation; make a thorough inspection of a very hot motorcycle and afterwards report his action to Race Control. If he allowed the rider to continue he will advise of the time he recorded the rider as stationary and that information will be passed to the time-keepers who, if directed to do so by 'Control', will credit the rider with the time lost in slowing, whilst stationary, and in accelerating back up to speed.

IT HAPPENED HERE . . .

A report from a practice session for the 1911 TT told that three machines - said to be Rudge-Whitworths - failed to take Sulby Bridge corner and were badly damaged and they "had to be brought home by train".

◆ First awarded in 1920, the Nesbit Shield (commemorating J.R. Nesbit a former Chairman of the ACU) was given to a rider at the TT "for triumphing over difficulties". In 1921 it went to G.W. Jones after he crashed his 250 New Hudson into the unprotected parapet of Sulby Bridge, was pitched into the Sulby River, (some reports say the bike also went into the river), climbed out, straightened his bent machine and went on to finish the race.

◆ Tommy Bullus told of his dramatic contact with Sulby Bridge in 1925 on a 'works' Panther, saying "On the very first morning of practice Oliver Langton and I arranged to do a couple of laps together. There was to be no racing and we arranged to travel at a reasonably fast speed in order to learn the course. We had only covered one inspection lap on the Sunday on a bike and sidecar. Starting from Ramsey where the team was based we accordingly rode at a very sedate speed over the Mountain, gradually increasing our pace but taking care not to come unstuck at any of the corners. When we arrived around Ballaugh, however, we were rapidly gaining confidence and, since this was the very first time we had been able to test the machines at any speed, we were anxious to give them at least one sustained full throttle burst and, after negotiating the humped bridge at Ballaugh, and the few fast bends through the Quarry, there was before us a long straight where we could really have a go.

There we were, two youngsters of 18 and 19 years, neither having ridden in any road race, let alone a TT before, flat out and heads well down racing along together in sheer youthful exuberance little guessing that this notorious and dangerous course was about to teach us a sharp lesson. Without noticing the warning boards we were suddenly in sight of the bridge at Sulby and realised that we were going far too fast to get round this very sharp right-hander, with brakes full on and the rear wheels locked solid we slid forward, Oliver on my left, both making for the gate which leads down to the river. But Oliver had the 'right of way' and opened the gate at about 50 miles per hour and shot down the steep slope into the river. I had no option but to take on the hard stone bridge parapet - however, not before I had got the machine into a gigantic dirt-track type slide, thus at least slowing a little but finally striking the stonework with a resounding crash and going over the top of the parapet onto soft earth mercifully without suffering the slightest scratch. Looking down at Oliver below neither of us could stop laughing for 10 minutes . . .".

◆ In the Junior TT of 1925, Len Horton (New Imperial) missed his braking point for the bridge whilst trying to attract the attention of a fellow rider to the fact that he had a bad petrol leak. Realising that he had left things too late and with no hope of getting round, Len opted to scrub off speed and slide into the side road on the left, only to collide with an Orange Cart and knock it clean over. Although he was able to jump to his feet and rejoin the race, when he returned to work at New Imperial's the following week, there was a letter from the Orange Cart attendant claiming for a new set of clothes to replace those she claimed were ruined in the crash. New Imperial paid up!

◆ A report of the 1946 MGP told how local Travelling Marshal Harry Craine misjudged the corner and "came a beautiful box of tacks", damaging bike and pride but, fortunately, not himself. Another Manxman who got away with getting it wrong here was Newcomer to the 1947 MGP Peter Crebbin, on a newly acquired Triumph twin. On only his second lap of practice, the cap from his oil-tank fell off somewhere on the western side of the course. Approaching Sulby Bridge flat-out in the company of two other competitors who he was determined to out-brake, he suddenly found himself sliding down the road pursued by the Triumph and those fellow competitors. After the dust had settled, the Triumph was found to have an extremely oily rear-tyre that had lost grip under the heavy braking. Peter also went on to become a Travelling Marshal.

◆ The 1959 TT had the first Formula 1 race for 350 and 500 machines. Terry Shepherd took a fine third place on his Manx Norton and afterwards reported that his only worrying moment had been at Sulby Bridge in the middle of the race. Seems he had been using a course-side marshal as a braking point, but on the fourth lap found, too late, that the marshal had moved nearer the bridge. In the vigorous cornering that ensued, he wore away a large portion of his right boot to expose his sock.

◆ Keith Heckles clouted the wall on the exit from the bridge in the 1959 MGP. Managing to stay on he pulled in at Ginger Hall. There he found he had ripped off his rear-brake pedal, so he returned to the bridge and retrieved it from the road. As he did so he heard a bunch of bikes approaching and, looking for an exit route, decided to leap over the roadside wall which was seemingly waist high to the few spectators standing behind it. What Keith did not know was that they were standing on a ledge some seven feet off the ground, and as he leapt over the wall he saw - too late - that he was taking a ten feet plunge into a field.

◆ In almost 100 TT races, Joey Dunlop is only recorded as falling off once and that was at Sulby Bridge in the 1986 Junior. Troubled with a badly fitting filler-cap, Joey was caught out by petrol on the rear tyre and spilled at relatively low speed. Although he picked himself up and raced on, he had damaged his exhaust and later retired at the pits.

◆ During practice for the 1989 TT, the very experienced Ray Knight (yes, he who wrote the 'TT Riders Guide' telling others how to ride the Course!) failed to slow sufficiently to take the bridge safely. Deciding to go straight-on, Ray was forced to leap from the higher road level to the lower field level and was thrown forward, thus causing painful impact with the large humped petrol tank. A considerate spectator contacted Manx Radio and soon after they played an excerpt from 'The Nutcracker', which they dedicated to Ray.

◆ The two hottest young riders at the 1990 TT were Steve Hislop and Carl Fogarty, and although they rode together in the official Honda team and shared a garage, Carl spent the entire practice period ignoring Steve and refusing to speak, clearly attempting to psyche him out prior to the racing. Come the opening Formula 1 race and 'Foggy' started 20 seconds behind 'Hizzy', but by lap 2 was on his tail and drew alongside on the Sulby Straight, for Steve had front brake problems and had lost time when he overshot Ballacraine. Both knew on the approach to the Bridge that it was to be an out-braking battle and whilst Carl recalls he "somehow got round the corner", Steve had to carry his braking far too deep and lost momentum. So, whilst Carl rode away to the race win, Steve pitted to change his front wheel and brake and then went out and smashed the outright lap record.

◆ The 2004 Production 1000 race started in excellent weather, but in the few minutes from the start until riders reached the western part of the course, low mist swept in from the sea along sections near Cronk y Voddy. As this meant that rescue helicopters would not have been able to see and use their landing areas in the case of an accident, the organisers were forced into the rare step of stopping the race and Sulby Bridge was the point at which they were red-flagged to a standstill.

At the 2010 TT a sizeable temporary grandstand appeared in the field on the left just before Sulby Bridge. With ever more restrictions on casual spectating, this may be the forerunner of similar structures around the Course.

Mountain Course Moments

Until well into the 1960s it was the custom to display the Tourist Trophies above the Leaderboard in front of the Grandstand on race days. It was a time when Trophies were still handed over to riders after their wins to take away and a post-TT report from 1928 told "the Senior Trophy was seen at Liverpool being carried in undignified manner by a porter along the landing stage".

After his victory at the 1939 Senior, Georg Meier took the Trophy home to Germany and, miraculously, it survived the war. Then, after his victory in the 1975 Senior TT, Mick Grant was allowed to take the Senior Trophy away with him and eventually put it on display in Denis Parkinson's Wakefield showroom. Whilst there in full view, the premises were burgled, money was taken from the safe, but this valuable chunk of silver was ignored.

Now the trophies are retained on the Island, kept in safe storage and generally handled with cotton gloves, as befits their muti-million pounds value.

◆ During the early 1960s the TT was still the top road race meeting in the world, with factories and riders making huge sacrifices to gain the glory

Tim Wood rounds Sulby Bridge on his way to victory in the 1913 Senior TT.

of a TT win. However, it was only in 1963 that, grudgingly, the organising ACU started to make a contribution to riders' expenses (£15) and prize-money had remained unchanged for many years. Later in the decade, ACU Chairman Norman Dixon proclaimed "You do not go to the TT to make money, for it is one occasion when, in a professional sport like motor cycle racing, the vast majority become amateurs and do it for the honour of taking part". Unfortunately, Dixon mis-judged the professionals of the day, for with the withdrawal of the Japanese factories from racing in the late 1960s and a general reduction in Trade support, riders had to tighten their belts and look for economies. They sought meetings that paid good start and prize-money, that were easy to get to and that were of short duration so that they did not incur expensive machine rebuilding costs. The TT fell short on all those criteria and as the ACU was to find in the early 1970s, tradition was not enough to maintain its top place, although a much increased contribution to riders' expenses and prize-money might have done. In 1977 it lost its World Championship status.

◆ A motorcycle is an assembly of many components, and the major elements of engine, gearbox and frame are supported by others such as carburettors, ignition, suspension units, etc, some of which the machine manufacturer may buy in. Any component can fail under race conditions and if it is an engine failure, a manufacturer will commonly seek to deflect criticism of his product by blaming it on something like an electrical problem. Ray Knight won the 500cc Production class of the 1968 TT and felt that his Triumph Daytona had gone so well that he would use it in the Senior race. A few laps into the Senior the Triumph cried enough and he was reported as having retired with ignition trouble. Ray was rather more honest about his 'ignition' problems, explaining "it must have been when the connecting rod cut the plug lead, because it was a blow up . . .".

◆ Sidecar grids at TTs of the 1960s and 1970s were made up of many 'specials' and the three-wheeler class developed a reputation as bodged oil-leakers that were hard-pressed to complete a few laps of a flat and undemanding short circuit, yet alone three laps and 113½ miles of the Mountain Course. The driver of one such outfit lost a megaphone from a Triumph twin during TT practice because of vibration and decided to run in the next session with open-pipes. To do this he had to fabricate securing brackets, and he described his bodge thus: "a stick of firewood, a four inch nail and a jubilee clip rendered them rigid" and to solve a problem with petrol-tank mountings he "used some old angle-iron found in a nearby garden". As the last practice loomed "the outfit was now festooned with tape, jubilee clips, rubber bands securing the primary chaincase, firewood and nails, rusty angle-iron; with the frame scorched mercilessly by the welding torch".

Mountain milestone

MOUNTAIN COURSE

SULBY BRIDGE ... GINGER HALL ... KERROMOAR ... GLEN DUFF ... GLENTRAMMAN ... CHURCHTOWN ... MILNTOWN ... SCHOOLHOUSE CORNER ... PARLIAMENT SQUARE ... CRUICKSHANKS ... MAY HILL ... RAMSEY HAIRPIN ...

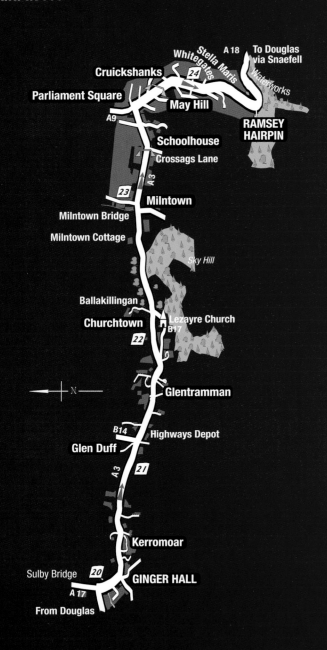

CHAPTER SEVEN

Sulby Bridge to Ramsey Hairpin

The stretch from Ginger Hall to Ramsey is considered to be the part of the course that has been least 'improved' over the past 100 years. Still relatively narrow, sinuous, bumpy and often heavily cambered, its 3½ miles of between-the-hedges going tests riders of modern bikes to the limits of their riding skills and puts them under heavy physical strain. It was no easier a century ago, for announcing after his first lap of practice in 1912 that "the roads were shocking, almost unrideable", local man Dougie Brown singled out the length from Ginger Hall to Ramsey as being the worst, a point acknowledged in an official report of the time that said "the long stretch of road between Sulby Bridge and Ramsey urgently needs a large sum to be expended upon it". Seemingly, that expenditure had not taken place before the 1913 races, for Highway Board members inspected it a few weeks earlier and considered it to be in a very bad state. The Surveyor General advised the Board that he proposed to lay 1,000 tons of foundation stone and macadam "directly after the holding of the motorcycle races" scheduled for the 4th and 6th June.

Improve or Else!

When the TT returned in 1920 after the First World War, Tom Loughborough, Secretary of the ACU, visited the Island in early May and felt that the course was in generally good condition, except for the same problematic stretch from Sulby to Ramsey. Its wretched state caused him to inform the authorities that unless the surface was improved the races would have to be abandoned. Something must have been done, as they did take place, although a constitutional spat that saw members of the House of Keys (Manx Parliament) decline to meet for the transaction of public business for several months in the Spring of 1920, almost caused them to be abandoned. No sitting of the Keys meant that the essential Road Closing Order was not passed and brought another warning of abandonment letter from the ACU to the Isle of Man on 4th May, for practice was due to start at the beginning of June, with the races on 15th and 17th June. Fortunately, normal Government business was resumed a few days later and a local newspaper welcomed the fact that the TT would take place, saying "these are the finest advertising medium which the Island possesses".

Despite the evident difficulty this stretch presented to riders, a document from the 1920 event shows that there were no official marshal points between Sulby Bridge and Ramsey during practice! On race days extra marshals were shown as provided on the stretch at Abbey Ville, Glentramman Road, Church Road, Milntown and Birchmore Avenue (probably Bircham Avenue).

Stanley's View

Talking later in the 1920s, Stanley Woods said how prior to 1924 the stretch from Ginger Hall to Ramsey "was loose and dusty, it was deeply pitted with pot-holes which were quite impossible to avoid, even at touring speeds". He also told that "horse-drawn vehicles still outnumbered motors and the number of nails which used to drop from horses shoes was enormous". In those days it was felt that a rider with a bit of weight and strength had an advantage here, whilst today's riders realise that they have to hang on and make the best of it, because trying to set up their suspensions to better deal with this stretch, would be counter-productive for the rest of the lap. But not everyone dislikes it, for when the talented Gary Johnson made his TT debut in 2007, he declared "My favourite bit was definitely from Ginger Hall to Ramsey. I loved that, probably because it was bumpy as hell and reminded me of my motocross days". Speaking some years earlier, Nick Jefferies described the stretch as "the most demanding in the world".

It is also a hugely demanding few miles for sidecars, with passengers very restricted as to how much they can move to help balance the outfits, or as experienced 'ballast' Dickie Gale put it recently "From Ginger Hall to

Mountain milestone

David Jefferies gives the railings a fairly wide berth as he sweeps through Ginger Hall on his V&M Yamaha in 2001.

Ramsey you do not passenger, you just hang on and cope with the bumps", while another said "we all take a deep breath leaving Sulby, hang on and breathe out at Ramsey".

GINGER HALL

At the start of the stretch is the Ginger Hall Hotel - known locally as 'The Ginger' - which is located on the left-hand bend that follows Sulby Bridge. Quite how it acquired its name is not known, although Leslie Quilliam's 'Gazetteer' says it is possibly derived from a corruption of the name of a nearby farm - Grangey. Like most licensed premises it is a magnet to race fans, particularly as the viewing is good and some movement is possible when roads are closed by riding under the shadow of Cronk Sumark with its legend of witchcraft ceremonies, and across the Claddaghs to Sulby Crossroads, or up via Tholt y Will to the Bungalow. Of course, in this strictly regimented age, spectating at Ginger Hall is not quite the relaxed business it used to be, when coaches brought hundreds of people out for the day to crowd the very edge of the course.

Riders accelerate up a short straight from Sulby Bridge to Ginger Hall, where from a position on the right they crank to the left and sweep past the apex where there used to be a stout hedge and bank, but which is now metal railings on top of a low stone wall. The bend then tightens, the camber on the exit is not favourable and riders exercise caution here, for there is only a narrow footpath on the outside to counter any mistakes. Many choose to use a higher gear than usual which allows the bike to pull them through at slightly lower revs, thus discouraging any rear wheel slides. Harold Daniell said of this spot "the inviting left sweep after the Ginger Hall Hotel is one of the most misleading sections of the course - as skid marks in queer places at the end of a day's racing will always show". For years a telegraph pole stood on the course side of the bank on the apex of the bend. This was before the railings

Bill Smith, B Randell and A Bergamonti in the 1969 Senior.

were installed and Geoff Duke's method here was to "cut the corner so close that it is necessary to lift the head to avoid striking the telegraph pole".

When railings were placed here they shielded the pole, but touching the railings was thought to have contributed to the heavy crash of Alex George during practice for the 1980 TT. They are now covered by sheets of plywood at race times.

It is not only spectators who appreciate the refreshments that are available here, for in the early 1960s a reporter for 'The TT Special' wrote "Ginger Hall is a delightful spot for a pressman who wishes to combine speed with relaxation. He can sit on the steps of the hotel armed with a notebook, pencil and glass of ale - which is what this particular pressman did".

Leaving the 'Ginger'

Modern bikes make easy work of the uphill run away from Ginger Hall, but in the past this was quite taxing, as the short but sharp climb often saw riders caught between gears in their three-speed boxes. Today sees the most powerful bikes cranked over to the left as they power away and then lifting their front wheels over the top of the rise, just as they need to crank right and drop downhill to Kerromoar. MGP veteran David Taylor advises "try to use a lot of engine braking around the Kerromoar section as using the brakes hard tends to soak up a lot of suspension movement which is better used dealing with the bumps". Those bumps are a warning to riders of the serious work they face over the next few miles, ones which Steve Hislop considered to be the most important part of the course to get right and thus gain time over the opposition.

KERROMOAR

This Manx name refers to 'large quarterland'. A non-racer would probably describe the course here as running downhill to a tightish left-hander, followed by a right and then an opening left; but former competitor Ray Knight, with over 25,000 Mountain Course racing miles under his wheels, sees the three bends in much more detail and this is what he says of Kerromoar: "Braking downhill and at the optimum point is, perhaps not too surprisingly, the way to get through here quickly. There is little to help judge braking points, but I use the telegraph pole set in the bank on the right-hand side, braking momentarily to set myself up. Then, heeling into the gutter and letting the brakes off, going hard left across the road and getting close to the chequered kerb, put the power back on and drive through hard. You can see through the first part of the section and are conscious of the kerb on the outside - it looks a couple of feet high, with a background of the inevitable flint wall. It is more than enough to curb the enthusiasm, but there is time to be made. It lies in driving hard through the right that immediately follows the left, thus getting a quicker run into what is to follow - a very fast section".

Ray is still only part way through this tricky sequence of bends and continues: "The right-hand portion is, or can be, quite quick enough to make you wonder if there is enough adhesion between your rear tyre and the tarmac and whether you should ease the throttle. You should be laid over far enough to provoke this thought if you are making the best use of your machine. Laid hard over, going right now, it is a left-hand exit from the section as you cross the sharp camber of the road, and at the angle you are forced to do so, it becomes something of a hump. The consequence of this is one more wheelie, but the danger is that you may not have the bike pointing in a straight line. When you regain contact with the tarmac, or the front wheel does, you could well be in for a mighty weave and wobble". Ray's words really do give an experienced 'from the saddle' view of how these bends should be ridden, although Carl Fogarty's feelings about the spot were somewhat blunter, with "I hate Kerromoar and the two bends after - it's really bumpy and a struggle to get the bike upright".

Like many places on the course, the bends of Kerromoar were once much narrower and tighter until the first two were widened in 1954. It has always been important to get them right to maximise speed on the stretch that follows, something emphasised by many riders down the years, with Graham Walker also reminding fellow competitors in 1935 that "when flat out in top again, you pass the lady in the bath-chair - don't forget to wave - she will bring you luck".

Incidents!

Kerromoar itself is not really a spectator point, nor are the ensuing few miles, although some enthusiasts find themselves holes in the hedges. But

A wide spread of tyremarks on the road suggest that the line from Sulby Bridge to Ginger Hall is not too critical.

Jake de Rosier on his Indian, after his stop in the 1911 Senior race.

though there have been few watchers, the Kerromoar bends have seen their share of incidents. One which had consequences that are difficult to believe involved American Jake de Rosier and his Indian during the 1911 Senior race. A reporter stationed at Ramsey wrote "Word filtered through that de Rosier had come to grief at Kerromoar, and, at 1.17, when the redoubtable American arrived (at Ramsey), there was a rush to learn what had happened. De Rosier's casualties comprised the trifling details of a lost footrest, a broken tappet rod, a broken throttle control, and a bandaged chin. He remained over 15 minutes in hospital, and though he made gallant attempts afterwards to make up lost ground . . .".

Jake's gallant attempts saw him finish the race, but he was later disqualified for having borrowed tools on the course to adjust his valve gear. But did he really spend 15 minutes in hospital in the middle of a Senior TT race? Whatever, and despite his disqualification, the first three places in the 1911 Senior still went to riders of two-speed, chain-driven Indians; thus giving a lesson on transmissions to British makers.

George Tottey was leading the 1923 Junior until Kerromoar when, in his words: "I ran into a lot of loose stuff in the gutter and came off". He straightened a badly bent bike and rode on to finish eighth. Someone else to get into trouble here was a man riding his first MGP in 1952, Alan Rutherford, and he tells how his racing dreams were shattered for the year when his Velocette "clipped the kerb and the bike hit a telegraph pole to be written off. I was OK apart from a few scratches. The marshals told me that they had been betting on which lap I would come off, being on a totally wrong line". Alan's riding must have improved in later years, as he took two 4th places in the Senior MGP - he also named his house 'Kerromoar'.

Top Swiss Florian Camathias managed to crash here during the 1962 Sidecar race, but it was his passenger Horst Burkhardt who came off worse, breaking his leg and suffering other injuries. In another three-wheeled incident at the spot, the marshals who struggled to lift and carry Howard Langham away on a stretcher, suggested that he might lose a bit of weight before next TT if he had thoughts of repeating the performance.

Very fortunate to damage only his ankle and a finger in a 1970s crash here was four-time TT winner John Williams. Just how fortunate can be judged by the fact that as he physically heaved his bike over for a change of direction and rolled on the power, the twist-grip rubber came off in his hand. Rubbers are usually wired in place, but such an incident shows the heavy demands of the Mountain Course.

GLEN DUFF

The one mile 'straight' from Kerromoar to Glentramman is, in Steve Hislop's words "a bit of a twister but basically it's a straight". It is certainly twisty enough to test the nerves of those who are not totally sure which way the road goes through its mostly tree-lined tunnel; it also makes a slight descent and is the bumpiest section of the whole course - Guy Martin describing it as "a different track altogether", where riders try to carry their weight on the footrests. The bump factor is one that lesser riders allow to govern their speeds, but which the top men say they cannot afford to do - although most have to. A 1920s publication mentions a marshalling point of Abbey Ville after Kerromoar, and an Abbey Ville and Abbey Hill can still be found here. Half way along is Glen Duff (Black Valley) which is identified by little more than a Highways Depot and former Quarry that provided stone for the roads, but riders are barely aware of these, for they are literally hanging on, "mere passengers" as one put it, with other comments being "going so quick on the bumps you can't focus properly" and "snaking off the crest at Glen Duff at max".

The ordinary road-rider can have little concept of what is involved in retaining control at near maximum speeds at such a hugely bumpy location, for the bike is fighting to take control and the rider has to use every ounce of skill to keep it on the road. But although this is acknowledged as perhaps the worst stretch of the course, the racing pressures of accelerating, braking, cornering, coping with the bumps, damp patches, trying to read signals and ride to a race strategy, etc, etc, are unrelenting over the whole of today's 37¾ mile lap, and apart from pit-stops - which bring pressures of their own - it continues over the full 226 miles race distance. With the huge power available at the twist-grip, there is simply no let-up today, mental or physical.

Nearly a Disaster!

Riding on the edge sometimes results in going over it and John McGuinness recalls "There's a bit, slightly left and right by the council yard (Glen Duff Highways Depot). It looks nothing . . . but my 600 went lock-to-lock, big time. Smashed the clocks, bent the fairing stays, I ran over my feet and broke the brake lever. It took the wind right out of my sails. I bet I'm the slowest guy through there now". Subsequent 130 mph laps on his big bike showed that John was still more than on the pace, and it was another multi-TT winner, Phillip McCallen, who earlier explained the need to ride on the edge with "You simply can't afford to ride well within your limits or you're going to be left behind".

After the Battle

Of his epic struggle with Carl Fogarty in the record-breaking six-lap 1992 Senior TT, Steve Hislop recalled how he was forced to back the throttle on his rotary-engined Norton here to avoid "being spat through the hedge". Each time he did so he thought to himself "I bet Foggy's pulling away from me now on this bit". The two combatants met for dinner on the evening of the race, Steve to celebrate his win and Carl to ponder defeat. At one stage in their verbal re-run of the race, it was Carl who revealed that he also had to ease the throttle of his Yamaha at Glen Duff, adding "I thought at the time, I'll be losing out here to Steve"!

Speed Restrictions

It was after reflection on their all-out efforts in the 1992 Senior TT that Carl and Steve called for a reduction in upper engine size for solo races and asked for a capacity limit of 600cc to be introduced. Both felt that the big bikes were too fast for the Mountain Course, but not all top riders agreed and now 1,000cc (actually 1010cc four and 1200cc twin-cylinder) engines are allowed.

The call to curb speeds at the TT was nothing new. Bikes have been considered too fast for the course at intervals down the years, with the organisers, the FIM, riders, etc, muttering about reducing or restricting engine capacities. But growth in speeds has been accompanied by course improvements, plus better brakes, tyres and suspension that make bikes more manageable, while manufacturers like the fact that engine sizes used in today's race bikes reflect those used in the road machines they are trying to sell.

GLENTRAMMAN

The very last thing a rider will be thinking of on the ultra-fast passage past Glen Duff and into Glentramman, is the fact that the name translates as Elder Tree Valley. But, just as riders have a hard time of it here, so Manx people suffered in a very different way in years gone by as they eked out an existence through the cold winter months, living in a far different world to today's TT racers and visiting fans. The elder tree has a strong place in Manx folklore and most cottages grew one. In early May the outsides of buildings

CHAPTER EIGHT

Ramsey Hairpin to the Bungalow

The Mountain

A few introductory words about the upland stretch of the TT course that riders are about to tackle will not go amiss at this stage, for while the 24 lowland miles from Douglas to Ramsey puts many demands on riders and creates its share of mechanical failures, it is the Mountain stretch after Ramsey that has proved the real bike-breaker during its 100 years of use. Indeed, it is fair to say that the many machine failures on the Mountain, often of race leaders, have broken the hearts of countless riders, designers, manufacturers, sponsors and trade reps, whilst also bringing much disappointment to race fans who have seen their favourites' chances of success snatched away by the demands of this notorious climb.

Going Up

Competitors are virtually at sea-level as they come through the centre of Ramsey and start the upgrade to Ramsey Hairpin, but it is after the Hairpin that the gradient of the Mountain Road steepens and then climbs almost continuously for the next seven miles up North Barrule and across the side of Snaefell (Snow Mountain), until it reaches its highest point of 1398 feet at Hailwood's Height, located just before Brandywell and beneath the slopes of Mullagh Ouyr (Brown Summit). That may not sound particularly high and today's bikes appear to make relatively easy work of it, but it is unquestionably seven miles of prolonged extra engine-loading.

Early machines found it very hard going over the rutted and loose-surfaced upward track that passed for the Mountain Road, for with their small number of gears, general lack of power, tendency to lose some of that power through belt-slip and over-heating, this climb was a real challenge and one that had to be tackled six times (later seven) in a race.

The Mountain section of the course is barren, lonely and offers little succour to riders who break-down or fall. Marshals were miles apart in the earliest years and a rider forced to stop could not expect to find shelter, whilst anyone who fell and was injured might well lie in the road until spotted by the next rider.

Even today's marshals are widely spaced on the Mountain and further problems are created by the weather, for as riders climb the wind increases, temperature drops and there is the risk of low cloud and moisture. Whilst those factors have served to increase riders' discomfort, particularly before the advent of fairings in the 1950s, bikes are also affected because a headwind reduces speed, further increases engine loading, can prevent top gear being pulled and sees more fuel consumed. Variations in height and temperature used to affect carburation, particularly on the two-strokes that formed part of the entry from the TT's earliest years, but while today's fuel-injected bikes automatically compensate for such variations, their consumption figures and speeds are still affected by changes in wind direction.

Poor Conditions

For many years the steep climb allied to poor road conditions deterred local people from using the Mountain Road. Little use meant little maintenance and a local paper reported of a practice session at the 1914 TT "more than one rider was held up on the Mountain where the mud was said to baffle description". What vehicular use there was in those early days tended to be horse and cart or lorries hauling timber and peat, with the occasional charabanc carrying tourists in summer, but as some of those vehicles were fitted with solid tyres, they tended to cut up the road surface. Such were the Highway Board's concerns that they banned 'heavy' vehicles (about 2½ ton upwards) from the Mountain Road from June 1922 until August 1938, and in the early 1920s when the surface was still unsealed macadam, they closed it to all traffic on the Sunday prior to race-week so they could

Mountain milestone

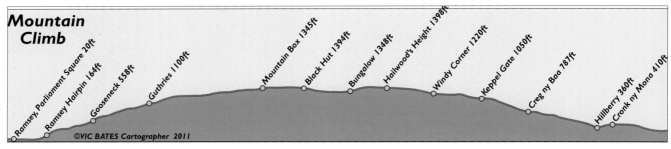

The ascent and descent of the Mountain is shown in this cross-section view from Ramsey to Hillberry.

sweep and prepare it for racing.

Writing in 1938, Stanley Woods said of conditions in his TT debut year of 1922: "From just above the hairpin there was no true surface at all - it was just a stony mountain track . . . even at its widest point the track was far narrower than it is now at its narrowest".

As tar-spraying and tarmacadam treatments spread across the Island's roads, the Mountain section was improved in stages from the 1920s, but it remained a mostly narrow single-track road, there was no proper edging and grass encroached on the surface from the sides. Also, road treatments of the time did not retain chippings as well as they do today and the edges were thick with loose material, a problem made worse by the fact that there were no mechanical road-sweepers to remove it. The effect was to reduce the useable road width, making passing a difficult and risky business, for getting off line could be distinctly dangerous.

In their pre-TT assessments, the motorcycle magazines of the 1920s always reported on the state of the road surfaces, with 'The Motor Cycle' saying in June 1926 that lots of loose chippings nullified the effect of improvements made by easing the radius of corners. It was a view shared by riders who "all complained strongly of the amount of loose chippings spread over the tar". The chippings of the time were noticeably larger than those used today and apart from the obvious danger they presented when ridden at speed, they also put competitors out of races by causing mechanical damage.

Better Roads mean Greater Speeds

Howard Davies (HRD) won the 1925 Senior at record-breaking speed and afterwards said "The Course was very much improved between 1924 and 1925. A lot of work has been done on the Mountain stretch. The surface was tarred almost all the way round and there was a noticeable absence of stones". Lap record speeds in 1922 and 1923 were fractionally under 60 mph, but with surfacing of the Mountain section starting in 1923-24, the outright lap record climbed to 63.75 in 1924 (an increase of just over 6%), and to 68.97 mph (an addition of a further 8%) in 1925. To put that in perspective, a lap speed of 130 mph would need to increase to over 145 mph in two years to match the mid-1920s increase brought about mostly by course improvements. These continued through the 1920s and 1930s and a report from 1937 confirmed that "the Mountain Road is being improved each

Seeking to maintain speed for the climb that follows, star rider of the early 1920s Bert Le Vack gives his New Imperial too much throttle on the loose surface of Ramsey Hairpin.

R Farrer James raises the dust as he climbs away from the Hairpin on his NSU in the 1914 Senior TT.

winter, and its condition today can hardly be compared with that prevailing in the early years of the TT".

For much of the last century the Isle of Man relied on the seasonal tourist and fishing industries to employ much of its population in summer, but winter could see many such people out of work. The Highways Board was charged with finding winter work for the unemployed and it would save major projects for the winter months, taking on as many as 700 men as casual labour in the 1920s. It may not have been the first choice for someone employed as a hotel waiter in the summer to spend the winter as a labourer on a road-widening scheme up on the Mountain, but they had little choice.

Mountain Gates and Fencing

Until 1934 there were several gates across the Mountain Road to keep the sheep on the high ground, and much of the road was un-fenced. It is part of Island racing legend that the first rider in morning practice had to open those gates. That may have been true for the earliest races, but it is called into question by the procedural regulations for 1922 which specified that the flagmen at "East Snaefell Gate, Bungalow Gate (Brandywell) and Keppel Gate" were "responsible for seeing that these gates are securely fixed open during practising, and the flagman on Keppel Gate must inform the Starting Point by telephone not later than 5.00am that all these gates are open". One slight problem with the latter requirement was that there was no telephone at Keppel Gate in 1922, so he probably used the one at Creg ny Baa.

There was also mention that "special arrangements will be made by the police with the shepherds on the mountain to ensure that the course is clear of sheep". This involved financial inducements to keep flocks away from the road during TT. Inevitably, some shepherds sought to take advantage of the races and the ACU complained of their excessive charges in 1932. But as riders found, despite shepherds charging for their services, there was no guarantee that they could keep stock off the road. Then 'The Motor Cycle'

greeted the news of the fencing of the Mountain Road and removal of the gates in 1934 with "the risk of sheep and other stock straying on to the course will be entirely eliminated", but those were the words of a journalist sitting at his desk in London and they did not reach the ears of the agile Manx Mountain sheep, so riders still met an occasional stray beast.

One unforeseen problem associated with the new course-side fencing was of riders complaining that at speed the new white concrete posts "flash by with a disagreeable dazzling effect". It was a time when all practice sessions were early morning ones and it was the low morning sun that created a 'strobing' effect from the posts as riders headed across the Mountain to Douglas.

Maintaining the Mountain Road has been, and still is, a difficult task. Winter rainfalls are heavy and although the main watercourses running off the mountains are now culverted under the road, there are still occasional slippages of waterlogged soil and rocks from the mountain-sides that can bring thousands of tons of it down to destroy the road. Such winter happenings are invisible to race fans, for the Highway Authority always ensures the road is restored before TT time.

Coming Down

The Mountain climb is rightly regarded as the big challenge to bikes and riders, but the demands of the descent should not be underestimated, for there are places where brakes and handling are tested to their limits, and others where engines can be over-revved if the bike is under-geared and the throttle is not backed off.

The ascent and descent of the Mountain takes riders from Ramsey to Hillberry, a distance of over 12 miles, and whilst they will pass marshals' posts along the way, from the time they leave the last house in Ramsey until they come to Kate's Cottage on the descent, they have not passed any human habitation in the 10 miles covered.

Mountain milestone

Harry Harris heads for Tower Bends during his ride to second place in the 1923 Junior race before he lines up to take the Gooseneck

WATERWORKS

Having got a bit ahead of ourselves with that overview of the Mountain section, it is time to return to Ramsey where riders leave the Hairpin under full power and today's bikes hit their rev limiters in the intermediate gears, as they accelerate up the tree-lined stretch that follows. The distinct right-hand curve part of the way up was not a problem to early machines, for riders would push the throttle lever open to its maximum coming away from the hairpin and just wait for the curve to come to them, or as one early rider put it "I got down a little flatter and nearly pushed my throttle lever off the handlebars in my attempts to make it open more fully".

Ray Knight advises present day riders to think ahead and shape their exit from the Hairpin to get the best line up the hill and thus minimise the effect of this bend that is followed by a drifting left that Canon Stenning was calling the "Little Gooseneck" in 1963. Riders are aware that it can be damp under the trees coming away from the Hairpin, but back at the 1948 MGP the Thursday afternoon practice was only able to go ahead after swift clearing of a landslide that blocked the road here.

Soon after comes Waterworks which comprises two right-hand bends, with the second being tighter than the first; the name of the corner coming from the nearby Ballure reservoir and associated activities.

Early riders were still plodding slowly upwards at the Waterworks, but by the 1930s they were going well in second and by the late 1950s John Surtees was saying "two nice challenging corners, the bike would be really singing again and you'd be recapturing your rhythm after that lousy hairpin". Steve Hislop spoke of approach speeds of 130 mph to Waterworks in the early 1990s, making them a real challenge to slow for and get right. And riders must be right, not least because there is a kerb and wall on the left with a nasty drop over it. Such is the power of today's machines that even 600cc sidecars are getting up into fourth gear on the climb after Ramsey hairpin, then dropping one gear for the first Waterworks right-hander and down another for the second corner, before powering away past the twenty-five mile marker towards Tower Bends.

Waterworks has seen its share of fallers, as riders are reluctant to sacrifice speed so early on the Mountain climb and because the second bend is frequently found to be tighter than remembered. Such fallers include leaderboard men such as Ernie Lyons (1947), Mike Kelly (1963), Charlie Williams (1976) and, surprisingly, past-master Sammy Miller, who leapt the road-side wall on his Ducati during a Parade Lap in 2008. Another was Phillip McCallen, who fell during the 1995 Junior when trying to take the race lead from Iain Duffus. Phillip passed Iain on the road on the third lap but Iain held his own pace, saying "McCallen was riding so hard - much harder than I wanted to - he was hairy". On the last lap Iain rounded the second Waterworks bend to find bits of McCallen's Honda all over the road and Phillip crawling off the track, his mission to recover from a poor first half of the race at an end.

The roadside footpath out of Ramsey ends just after Waterworks and there are no paths and virtually no kerbs until past Hillberry, as the road runs between a mixture of grassed and rocky banks, sometimes with exposed drainage channels at the edges.

TOWER BENDS

Often vaguely described by competitors and spectators as the curves leading to the Gooseneck, the Tower Bends comprise quite a sharp right-hander followed by two lefts. There then follows another right, then a left just before the Gooseneck, all under the shadow of North Barrule.

These bends get their name from the prominent stone tower in a nearby field that was erected by the people of Ramsey to commemorate a royal visit by Queen Victoria and Prince Albert in 1847.

Like many groups of bends on the course, the sequence here needs working on to get the best line and speed. They are not just a link between Waterworks and the Gooseneck, for they are a demanding section in themselves that established stars like Ian Lougher stress should be attacked, so as to gain time over those who are not applying themselves fully; although the current bumpiness of the final left-hander is not welcomed by riders trying to set themselves up for the almost immediate right-hander of the Gooseneck, whilst some find the jagged walls inhibiting.

GOOSENECK

With the road snaking in from the left, curving sharply to the right, then curving away uphill to the left, it is not difficult to see how this corner got its name.

Riders have now climbed 600 feet and in the early days of the TT they were still going slowly at what was quite a narrow spot.

So slowly were early riders travelling and so difficult was this right-hand turn over the poor surfaces of the day, that winner of the 1914 Senior, Cyril Pullin, said afterwards "how I hated the Gooseneck with its tiresome foot-slogging". It was a point repeated by 1922 Lightweight winner Geoff Davison who told how he also used leg power to help his Levis round here on the way to victory. A year after and a pre-TT report from 1923 told that the basic macadam road had received attention between the Gooseneck and the Bungalow, but how the top surface remained loose. It went on "but a few good heavy showers, assisted by a steam-roller will make all the difference". The Gooseneck was substantially widened and slightly banked in 1939, which must have improved matters for riders, who now make an approach on the left, take a late right sweep across to the climbing apex, and use much of the road on the exit as they strive to maintain revs and momentum for the continuing climb that follows.

Most riders take the Gooseneck in their stride, but Charlie Williams was careful not to try too hard because "I feel it would be so easy to lose it". Steve Hislop actually disliked it, as he always thought about the potential for slippery fuel droppings on the inside of the bend from normal traffic. This was part of the reason that he avoided cranking over too hard and too tight here, thus he used all of the road coming out and in doing so gave spectators an extra close-up thrill on the exit; something that he was aware of.

IT HAPPENED HERE . . .

Howard Davies won the 1921 and 1925 Senior TTs, but when in with a chance of winning the 1926 Senior he took a heavy tumble here. He remounted and continued, but later retired with engine trouble.

◆ At the 1949 TT the 350cc AJS 7R was still a relatively new racing bike in its post-war form, but showing the potential of a model that went on to offer great service to a generation of racers in over-the-counter form, the 'Ajay' held the lead in the Junior in the hands of Bill Doran until the last lap, when he retired here with gearbox trouble. It was a year when there were 100 entries in the Junior spread over just three makes - 27 Norton, 35 Velocette, 38 AJS.

◆ The exit from the Gooseneck is a favourite place for riders to receive signals. Master of the 125s and much more, Joey Dunlop claimed that this was the only spot on the course that he lifted his head from under the screen when riding the Ultra-Lightweight bike, and that was just to read his signal. Mind you, Joey's signalling crew sometimes kept the cold at bay with liberal amounts of warming liquor and on more than one occasion Joey was shown his board upside down.

◆ Mick Grant earned his living as a racer, with wife Carol an active part of the team who occasionally signalled here. Mick told how if he was not going too well, Carol would dispense with the board and, as he put it, "if there's a fair amount of finger and thumb rubbing meaning we need the cash, I know that things are not good and that I must go quicker".

◆ Riders sometimes turn on the style here to entertain spectators and none did it better than Graeme Crosby, for in his TT debut year of 1979, he wowed them by pulling monster wheelies on his big Moriwaki Kawasaki fitted with high bars and just a cockpit fairing. Although 'Croz' admitted that he struggled to learn the circuit in his first year, spectators loved his performance and he did take a fourth place.

◆ Many spectators make regular pilgrimage to this popular vantage point, for it is one where riders used to be almost in touching distance. The thrill spectators gained from such close proximity was understandable, but in 2007 the risk they were taking was judged unacceptable and restrictions were imposed from that year's MGP. Whilst they are still a lot closer to the action than they would be at a short circuit, they are now forced to sit higher up the bank.

The Gooseneck remains a popular spectating spot and is accessible from the course before the roads close, but parking can be awkward and access

Manxman Dan Kneen takes an almost shoulder-touching line at Tower Bends in 2010, on the way to the Gooseneck.

Mountain milestone

difficult unless plenty of time is allowed. There is also a single-track road that allows passage to the Gooseneck from the Maughold to Laxey Road.

Onwards and Upwards

Having got the bike vertical coming out of the Gooseneck, riders pour on the power as the road continues to climb into a long, long, left curve. With over 200 bhp available to them, today's top riders revel in the acceleration here, but John McGuinness tells how full power is now just a bit too much for that left curve and he rolls the throttle a fraction. Such power demands maximum concentration and there is no time to admire the scenery. But just like riders of earlier days, current competitors become aware of the increasing exposure to bare mountainside and buffeting winds from here on, whilst also gaining an early indication of whether they have their gearing right to suit the wind direction.

JOEY'S

A little over half a mile from the Gooseneck, the next identifiable point on the course goes by just a single name, but no one is in any doubt as to who it represents - Joey Dunlop MBE, OBE, who lost his life racing in Estonia in 2000. The spot was previously known as the Twenty-sixth milestone and the authorities considered it appropriate that this should be the place on the course at which to remember the twenty-six times TT winner.

Well into the Mountain climb now, the road is open with not a tree or house to be seen and the feeling of leaving civilisation behind and heading for the wilds increases with every upward yard covered. Many riders revel in these open conditions where they can now see through the bends, although Joey was never a fan of the Mountain, preferring instead the between-the-hedges going from Ballacraine to Ramsey.

Joey's is a fast right-hander followed by another climbing right and then

David Jefferies uses all of the road on the exit from the Gooseneck

the road curves left on the way to Guthrie's. Graham Walker told in the 1930s that the right-hander could be taken flat in third, but added the warning "but you must hold on tight and you may have to use the grass"; whilst writing in 1990, Ray Knight described these bends as "easy enough on a moderately-powered machine". But bikes have moved on in the last 20 years to the tune of some 50 extra bhp, something that turns "easy" bends into much more challenging affairs. And of course, as has always been the case, the more power a rider has available, the more bends there seem to be.

Repercussions

The Centenary TT of 2007 enjoyed a happy atmosphere both on and off the track. Then, just as the organisers were beginning to think how well things had gone, a major incident at Joey's in the last race of the week brought the death of a rider, two spectators and also serious injuries to two marshals.

The repercussions were wide-ranging, involving the Coroner's office, police, insurance implications, official resignations, etc. For the TT and MGP that had evolved gradually over many years to meet changing conditions, the immediate actions demanded changed forever the relatively relaxed attitude to spectators that had existed for a century.

That is not to say that the organisers were previously unaware of the dangers to spectators, for as racing speeds had increased on the Mountain course, then year on year they had sensibly restricted viewing from positions they no longer considered to be safe. But this had been an occasional action and although often not well received by those affected, many of whom had watched from the same point for years, it was generally accepted as 'progress'. However, post 2007 TT, the situation changed dramatically. The edges of the Mountain Course used to be an almost 37¾ mile grandstand, but overnight literally miles of it were turned into prohibited areas and this affected almost everyone's race viewing. To make things worse, many of the bans on spectating were imposed at the most exciting viewing spots.

Alternative Lappers

On a lighter note, the challenge of the Mountain course has attracted many other forms of competition over its 37¾ miles. Walkers, runners and pedal cyclists are just a few of those who have sought to cover them in the shortest time, whilst those seeking to raise money for charity have pushed various wheeled contraptions over it with rather less urgency. However, by this point 26 miles into a lap and with over 5 miles of climbing road still to cover, many such lappers must have questioned why they agreed to take on the challenge.

Ariel caught the TT atmosphere in its advertising, showing Travelling Marshal Bertie Rowell on one of its 'Square Four' models above Ramsey in 1948. Riders have now climbed 600 feet and in the early days of the TT they were still going slowly at what was quite a narrow spot.

A rider's view of Joey's, where 26th mile marker is passed..

Major Repairs

In 1989 the stretch from the Gooseneck to Guthrie's Memorial was described as "billiard-table smooth". But complacency is a bad thing where the Mountain is concerned and over the winter of 1991-92 the nearly three-mile stretch from Joey's to the Mountain Box had to be closed for major repairs to the road foundation, which had been damaged by continuous water run-off from the Mountain. The road banks were also cut back in places and reinforced to prevent slippage, whilst water run-off was better channelled.

GUTHRIE'S

Riders see the sequence of left-hand bends on the approach here differently, with some regarding them as a big double left, whilst Mike Hailwood, Steve Hislop and John McGuinness claim there are three. Twice winner Tom Herron thought the same but found it a difficult spot, adding "the only person I have seen go through there properly is Mike Hailwood (in 1978). I learnt it that night and forgot it again the next day". Whatever the difficulty, competitors do need to be clear in their minds of the layout, for at today's speeds they form a tricky approach, being followed by a right, left, right at Guthrie's. Both the approach and Guthrie's itself offer the potential to gain or lose substantial time, and John McGuinness tells how it took him years to get the correct approach and how Guthrie's itself has seen him have a few slides whilst seeking to maintain maximum momentum.

What is today called Guthrie's should perhaps be known as Guthrie's Memorial, but it was earlier called the Cutting, before being renamed in honour of six-times TT winner Jimmy Guthrie in the late 1930s, when it required bottom gear on a four-speed Norton of the type that he rode through here so many times.

The plaque on the cairn tells that Jimmy was killed racing in Germany, and it was erected here because this was the spot where he retired during his final TT race, the 1937 Senior. He was a respected rider whose reputation was built on action rather than words, and when 'The Motor Cycle' invited subscriptions to a fund in Jimmy's memory it raised £1,500. That was a substantial sum, being sufficient to pay for the memorial that looks across to his native Scotland and for beds in Nobles Hospital.

The road steepens on the exit from Guthrie's and there is still a long way to go on the Mountain climb, meaning that engines have to be revved to their limits in each gear to recover and maintain the best speed. It is here that riders cross the first of several white-painted stone bridges that occur over the next couple of miles and which come with conspicuous black/yellow/black visibility stripes. It used to be first or second gear at about 45 mph at the tightest point of the Cutting in the 1930s, with Mike Hailwood talking of a pretty quick 60 mph in the 1960s and it being only slightly faster now.

Three-wheels

Sidecar outfits are just as reluctant to yield speed here and they have had their share of incidents at Guthrie's. There is only room to recount a couple, but back in 1960 three-wheelers returned to race over the Mountain Course for the first time since 1925. They had spent the period 1954-1959 contesting the Sidecar TT over the Clypse Course, but for 1960 it was back to the Mountain, with top British driver and ex-World Champion, Eric Oliver, returning after a period away.

Eric had a huge reputation as a sidecar-driver in the 1950s, but the rise to supremacy of BMW power left him struggling with his Norton. However, with earlier knowledge gained from racing solos over the Mountain Course, he was expected to offer a serious threat to the BMW boys in a planned return at the 1960 TT. In only the second lap of practice, Eric and passenger Stan Dibben were just leaving Guthries when they suffered a fracture of the steering-head, lost control and crashed, with bike and riders going part-way down the mountain-side. Walter Radcliffe was one of two marshals in the area who attended the incident and he tells how one of the other race favourites, Florian Camathias, pulled up at the accident, momentarily surveyed the wreckage, gave his BMW a handful of revs and sped away from the scene, knowing that a major threat had been removed.

In the second Sidecar race of 1987, Dick Greasley's passenger Stu Atkinson fell from the chair at Guthrie's and was whisked to hospital by helicopter. Stopped nearby with terminal mechanical problems were Des Founds and Gary Irlam. As there was nothing wrong with Greasley's machine, he 'borrowed' passenger Irlam and rode back to the Paddock, saving himself a chilly wait on the mountainside.

The Mountain is a desolate spot to get a puncture!

Early in his TT career, Cameron Donald tackles the testing bends of the Mountain climb.

Bizarre

It was up here that an unusual incident occurred just before roads closed for Wednesday evening's practice at the 1999 MGP. A van driving down towards Ramsey had a brake disc shatter and the hot metal fragments set fire to the heather and gorse at the road-side. Fire engines were called, but they were handicapped by lack of water and the practice session was abandoned.

First Lady Marshal?

Soon after Guthrie's, riders are faced with a climbing left-hander that takes them past the twenty-seventh milestone. It was here that Miss May Kneale, a trained nurse, did marshalling duties from the late 1920s, thus establishing a claim to have been the first lady marshal. It can be a bleak spot and getting anyone to marshal the Mountain in the days before personal transport became common-place must have been difficult, for people with their own vehicles were few and those who wanted to marshal on the Mountain at 4.30 am were even fewer!

Bob McIntyre told how he made a big effort and held just over 100 mph through the long left-hander here when he set the first 100 mph lap in the 1957 Senior TT and after the twenty-seventh milestone comes a demanding right-hander that Bob reckoned he was summoning all his courage to take at 120 mph, as he flashed over another of those striped bridges and on to the Mountain Mile. Known to some riders as Milligan's Bridge it is hardly any wider than sixty years ago, yet John McGuinness told of getting through it "just about flat-out" when setting his 130 mph lap in 2007 and earlier star Mick Grant considered it "one of the most important corners on the circuit".

MOUNTAIN MILE

It is nearer one and a half miles of upward drag to the next significant landmark at the Mountain Box, which for much of the run can be seen silhouetted against the skyline and from which riders are seen as moving specks in the wild Manx landscape. Throughout TT history the 'Mountain Mile' has been renowned for taking all the power that can be thrown at it, and for wrecking the engines of many who have done so. It is by no means straight, and when it was loosely surfaced and of only single-track width, riders probably had to ease the throttle in places, just as most do on the fastest bikes today.

It is sometimes said that this stretch offers riders the opportunity for a breather, but in the first few years of racing on the Mountain Course, those riders who were really trying would be fighting the road surface all the way. When it was tar-sprayed and chipped in the mid 1920s they just might have had chance to draw breath, but as speeds on their unstreamlined machines climbed to over 100 mph, it was the buffeting of the wind they often had to fight; although if it happened to be coming from behind, they would bless it for the extra 10 mph it gave them. The streamlining introduced in the 1950s eased things for riders in some circumstances, but the full form of streamlining initially used could be a handful in cross winds at the 145 mph

Joey Dunlop nears the end of the Mountain Mile and prepares to tackle the Mountain Box.

achieved up here by 500cc MV Agustas and Gileras in the mid 1950s. Today it is a case of tucking everything in, coping with unrelenting power, barely being aware of passing the three-quarter way marker as the concrete fence posts flash by close to the right-hand verge, and summoning the courage to tackle the several right-hand kinks with the minimum of easing; these usually coinciding with the crossing of more striped bridges.

IT HAPPENED HERE . . .

Harold Daniell told how in the late 1930s he hit 115 mph in top gear along the Mountain Mile. But he also told that a headwind would force him to hold on to third. Today, many-times Sidecar TT winner Dave Molyneux tells that he gears his bikes to hit just over 15,000 rpm on the stretch to Ramsey, but how with the throttle hard against the stop they pull just 13,000 on the Mountain Mile. The effect of the Mountain is also felt by today's solos, for although top riders average over 140 mph from the Grandstand to Ramsey, the climb of Snaefell results in just over 130 mph laps.

◆ Geoff Duke's Norton engine seized here during his first Island race at the 1948 MGP, when a split oil-tank allowed the essential lubricant to gradually drain away. Leaving the bike leaning against the bank, Geoff walked to the Mountain Box to report his retirement. There he gave the marshal his number and received the reply "What a pity, you were leading the race!"

◆ In his book 'Japanese Riders in the Isle of Man', Ralph Crellin tells how Suzuki had a three-man 'works' team entered for the 50cc TT in 1967, which it wanted Yoshimi Katayama to win. That was fine in theory, but in the early stages Katayama had to stop to change plugs and Ralph says "Teammates Hans-Georg Anscheidt and Stuart Graham rode as slowly as they could to let him catch up. Catching the dawdling pair at Ramsey, he took the lead up the Mountain, but, looking back at his teamsters, he drifted into the ditch alongside the Mountain Mile and crashed, retiring unhurt, leaving Graham to win from a slowing Anscheidt".

◆ Lonely at the best of times, this stretch seemed even lonelier to Geoff Rushbrook and Steven Leitch, when in the second Sidecar race of 1985 their Yamaha outfit caught fire and they had to stand and watch it burn out as fire-extinguishers were brought from many hundreds of yards away. The fire created dense smoke and flames for the full width of the road, that other outfits drove through.

Testing

With its relative freedom from traffic, the Mountain Road was frequently used by competitors for unofficial testing of race bikes from the mid 1920s, with the Mountain Mile being particularly popular.

Norton also used the Mountain Road for testing outside TT and MGP periods, for over the winter of 1949/50 they bought examples of their racers with the existing 'garden-gate' frame and a new 'featherbed' frame, to try in as near as they could get to race conditions. Under the benevolent eye of the local constabulary, Geoff Duke and Artie Bell put the bikes through their paces in a handling test that the featherbed won.

A post-TT feature in the motorcycle magazines for many years was the testing over the Mountain of race winning bikes by trusted journalists from 'The Motor Cycle' and 'Motor Cycling', usually early in the morning on the day after the race. They would then write of their experiences for the benefit of their huge world-wide readership. For many years, those two magazines were the principal source of information on the TT and MGP for race fans. Whilst readers in Australia and New Zealand might have to wait over six weeks for their copies to come by sea, in Britain they were generally available on Thursday of each week. But at TT time in the 1950s, both magazines made special efforts to get them flown in and on sale in the Isle of Man at breakfast

Rider's view of the approach to the Mountain Box.

time on Tuesday morning of race-week. That was just some 16 hours after Monday's Lightweight race had finished and the covers were overprinted 'Special Air Edition'.

MOUNTAIN BOX

Quite when this name was established is not known, but it reflects the fact that the race organisers had a telephone provided here in the early years of the races. It was a place that for the first twenty and more years of the TT had a gate across the road, known as the East Mountain Gate. Today there is a substantial stone shelter used by marshals, that also has an emergency telephone.

Although the expression Mountain Box is now recognised as referring just to this location, in early post-war years it was sometimes called the First Mountain Box, with Brandywell being called the Second Mountain Box.

The Mountain Mile finishes with a climbing double left-hander at the Mountain Box, a spot where Ray Knight stresses "late braking and the correct peel-off point are particularly critical here", adding that the pace at which it will be taken depends upon the riders "quotient of bravery". It was 65 mph in second in the mid 1930s and after the road was widened in 1955 it was a bit faster in third, while for the big Production racers of the early 1970s it was faster still in fourth. Steve Hislop described this as "one of my favourites - it's beautiful", whilst warning that the wind could be gusty as you came out of the lee of North Barrule. Steve went into it in fourth and came out in fifth, as does John McGuinness who sees it as four left-handers which "through time and experience" he takes as one.

IT HAPPENED HERE . . .

Mention has been made of the scarcity of marshals over the Mountain section in earlier days, but even if marshals were present at an incident, there was often little they could do for a fallen rider. Seemingly, there were none just past the Mountain Box during the 1927 Lightweight race, for when John 'Jack' Cooke came off his DOT on the second lap at 11.30 am, he lay unconscious in the road until fellow competitor Leonard Higson stopped and pulled him to the edge. Higson then rode on and told spectators a few hundred yards away of the incident before continuing a further two miles to the Bungalow and telling the marshals there. In his admirable book 'Isle of Man TT & MGP Memorial 1907 - 2007', author Paul Bradford recounts poor Jack Cooke's fate, saying "A group of spectators including two (spectating)

Short Circuit?

Forty years later the topic of an alternative course was raised again when the difficult 1970s saw the TT lose world championship status and brought well-meaning attempts to create a short circuit on the Island to GP standards. The Island's Tourist Board set up a Short Circuit Sub-committee in the early 1980s, and well-known racers like Geoff Duke and Phil Read lent their names to schemes.

Despite the several attempts to find alternative courses, to most fans the TT remains firmly linked to the Mountain Course, and their annual trip is a form of pilgrimage, with one enthusiast writing: "to lack this urge . . . to go to the TT races . . . is, I suggest, to lack an essential quality as a motorcycle enthusiast".

Too Easy

All the concerns about the difficulties of the Mountain Course have been matched at regular intervals by claims that changes and improvements were making it too easy, with 'The Motor Cycle' claiming in 1927 "that if more road improvements are carried out the value of the races will be seriously decreased". By 1938 'Motor Cycling' was asking "is the Highways Board wise in attempting year by year, to make the TT course resemble a race-track?". The easing of bends, smoothing of bumps, widening of roads, laying of ever grippier surfacings have all contributed to making things easier, but today's riders would hardly wish to race over the course in its pre-war state. Also, visitors who focus just on the racing should recognise that as the course is improved year on year, most changes are actually aimed at improving conditions for day to day Manx traffic.

Travelling Marshals

Mention has been made of a fatality during the 1934 Lightweight TT that was run in poor conditions. This occurred on the Mountain and involved experienced runner Syd Crabtree, whose crash went undetected until the organisers realised he was missing and sent marshals out to search in the fog.

That a crashed rider could lie undetected was unacceptable and the Crabtree incident of 1934 caused the ACU to take a detailed look at its organisation for 1935 - particularly in respect of racing in bad weather - and was a major factor in the establishment of the Travelling Marshals service.

BUNGALOW BRIDGE

Soon after the Verandah comes yet another wide open, fast-as-you-dare bend that sweeps to the left and carries the official nameboard Bungalow Bridge. The spot is quite widely known as the Stonebridge, with some calling it Stonebridges, whilst others use the term Graham Memorial, after the construction of a shelter here in 1955 in memory of TT-winner Les Graham.

The unusual roof-line of the Graham Memorial is seen here in Rob Temple's photo, as riders leave the Verandah and crank over for Bungalow Bridge.

The bend is somewhat featureless and used to be a much narrower second gear affair, but today its open nature tempts riders to go to their limits and some have come unstuck. One earlier rider who was unlucky to do so was 'Esso' Gunnarsson back in 1963. In the last TT practice session 'Esso' cranked his Norton in here, discovered that the bend was soaked in oil, and man and machine skated up the road. It was the days of open-face helmets and he suffered deep facial wounds that required 36 stitches. Despite that, he was on the line for the start of the Junior race and though he recalled "every bump in the road was like a stab in the back", he finished a courageous

Rob Temple's photo captures the roofline of the Graham Memorial during evening practice.

fourteenth. There have been many other 'offs' here and when Gary Thrush crashed in 1988, conditions were too misty for the helicopter to land, so practice was stopped whilst an ambulance drove along the course from Ramsey and took him to hospital.

It was just after Bungalow Bridge that Graham Walker hit a sheep with his Norton during his first TT in 1920. Travelling at close to 60 mph, Graham later recalled "I was lucky to get off with a black eye and a dislike for mutton which has lasted ever since".

The stretch of road that follows, comprises three linked rights that have been described as "another Verandah in miniature". These are known to some as Bungalow Bends and again the aim of riders is to take them as one and come out on the correct line for the approach to the Bungalow.

After the death of TT-winner Bob McIntyre while racing at Oulton Park in 1962, a timber box for a flag-marshal was provided on the approach to the Bungalow, for Bob had often remarked how tough conditions were for marshals on the Mountain. The 'McIntyre Box' was later moved further back from the corner, but as a timber structure that was put out just for the races, it failed to survive the handling it received during its repeated moves.

Mountain Course Moments

It needs a particular mental state to be a passenger on a racing sidecar outfit and entrust your life to another person, whilst it also makes huge

Mountain milestone

Phil Read is the filling in the sandwich as three riders tackle Bungalow Bridge.

physical demands to race for three laps and 113 miles of a modern TT on a thinly-padded, flat metal platform, with a couple of handholds, virtually no suspension, noise, heat, vibration, acceleration that can see you sliding off the back, and braking at points like Governor's Bridge that can easily pitch you out of the front. During practice for the 1983 Sidecar TT, three passengers felt they had reached their limits and withdrew their services. Two of the drivers found replacements, but prospective race winner Trevor Ireson could not locate someone of the right calibre, so was forced out of the races.

♦ The Ultra-Lightweight TT for 125cc machines was reintroduced in 1989 and it was great fun for spectators to see them in practice sessions where they sometimes got out with bigger machines, for what a 125 lost on the straights was more than made up for on the corners by determined and experienced riders, particularly on the twisty bits. Robert Dunlop described riding one as "you just sit there with it buzzing at 13,000 rpm and wait for the corners to come". Despite the fact that power output of 1,000cc race bikes was probably three times that of the 125s, when their race was dropped from the Mountain Course after 2004 the 125 lap record was just a fraction short of 110 mph.

♦ For the past 40 years the majority of racing bikes have transmitted power from the engine to the gearbox by the way of geared primary drives, but when gearboxes were first adopted, the primary transmission of power

A BSA Gold Star of the 1950s with exposed primary chain. Note the carpet felt fixed to the lower frame rail to catch excess lubricant from the drip-feed oiler.

1990 but the MGP retained starting in pairs until 2011, when it moved to single starts.

♦ Sponsorship - where would modern racers be without someone subsidising their racing? Of course, not everything is what it seems in the world of race support and big stickers are not necessarily accompanied by big money. 'Motorcycle News' carried a report in 1993 that read "American oil giant Redline is backing Chris Petty on an RC30 Honda in this year's Formula One and Senior TTs". Whilst racing friends were no doubt envious of his seeming good fortune in having landed support from an "American oil giant", privateer Chris had to point out that "The deal was actually nine litres of oil"!

♦ The centenary of the first running of the TT (over the St Johns or Short Course) was celebrated in 2007 and John McGuinness marked it with the first 130 mph lap of the Mountain Course. There were celebrations to mark the centenary, including a re-enactment of the first race, with TT celebrities like Mick Grant, Nick Jefferies and Guy Martin joining members of the Vintage Motor Cycle Club in riding a lap of the St Johns course.

Riders line up at St Johns in 2007, in preparation for the start of the re-enactment of the first Isle of Man TT of 1907.

was done by way of a chain, called the primary chain (to distinguish it from the secondary chain that took drive from the gearbox to the rear wheel).

On a racing bike the primary chain ran exposed to air with little more than a guard over the top run. This was as much to allow cooling air to the clutch as to the chain, but meant that unlike the enclosed 'oil-bath' primary chain set-up used on most road bikes, the exposed chain needed a regular drip feed of oil, for informed sources said that the primary chain performed 35 million articulations between its links during a TT race. Without lubrication it would barely last a lap.

♦ The TT went over to starting riders singly at 10 second intervals in

Mountain milestone

MOUNTAIN COURSE

BUNGALOW ... HAILWOOD'S HEIGHT ... BRANDYWELL ... DUKE'S ... WINDY CORNER ... THIRTY-THIRD MILESTONE ... KEPPEL GATE ... KATE'S COTTAGE ... CREG NY BAA

The Bungalow to Creg ny Baa

Giacomo Agostini begins the descent of the Mountain on his MV Agusta.

away at number 115 in a race in which the 1,000s ran with the 350s. Going into an early lead on time, he was soon amongst the slower 350s, which had started first. However with a 40 seconds advantage over the second place man, his race came to an end at Brandywell when he shaped to take a slower rider on the outside, but the rider moved over and forced him to the edge of the road where his tyres lost grip in loose material and he was off. Richard spent the rest of the TT (and his honeymoon) in hospital. That was Richard's only TT appearance, but nearly sixty years later his son David Madsen Mygdal is a veteran of many races over the Mountain Course.

Long Distance

By 1930 the Mountain Course was famous throughout the motorcycling world. Looking to benefit from the prestige associated with the name, the Dunelt concern ran a long-distance reliability test over its 37¾ miles in March of the year. It proved to be an endurance test of both men and machine as the 7 riders endured snow, ice and fog to cover 350 laps (13,000) miles on a stock 500cc model over a period of 16 days; a performance rewarded with the award of the prestigious Maudes Trophy from the ACU.

The Descent

From 2009 Brandywell became a 'Station' for the Travelling Marshal who used to be based at the Bungalow. This high-spot marks the start of the Mountain descent and there is a myth that anyone who breaks down here can coast the six miles to the finish in Douglas. That is not so, even though Nick Jefferies once described the effect of the change from uphill to downhill as "Pass Brandywell and collect an extra 200cc". It is a fact that after slogging

Brandywell's left-hander. It is mostly downhill from here to the finish.

A member of the Dunelt Maudes Trophy team of 1930 at Brandywell.

up the Mountain for over seven miles, riders over the past century have sought to take advantage of this downhill section by building up to maximum speeds and taking the bends on the limit. The results have been predictable, with plenty of spills and blown engines between Brandywell and the finish.

DUKE'S

The course continues to curve away from Brandywell in left and right-hand sweeps as riders gather speed whilst almost continuously cranked over, before reaching three descending left-hand bends that for many years were known as the Thirty-second Milestone. With a drainage channel on the inside and formerly a line of concrete fence posts 'protecting' a drop to the Baldwin Valley on the outside, this is a challenging stretch where a knowledgeable rider can gain time. One man who rated the Thirty-second amongst his favourite locations was five times international TT winner in the 1950s Geoff Duke and some fifty years after his great achievements, he was honoured with the Thirty-second being formally renamed Duke's in 2003.

Some wondered why it took so long to add Geoff Duke's name to the Mountain Course, for his Island successes were impressive with a win in the Clubman's TT, a win in the MGP and five international TT victories. A late starter in racing due to the war, he only had 11 TT rides in all, and when two second places and a fourth are added to his wins, it makes a fine record. Geoff was also World Champion on six occasions and has always been a great ambassador for the TT and the Island, where he eventually made his home. He surely earned the honour of having Duke's named after him - a spot that some choose to call Duke's Bends.

But not all Geoff's memories of this stretch are pleasant, for he recalled riding in close company with Ray Amm (Norton) in the 1953 Senior TT when, experiencing gear selection problems on his Gilera, he also had concerns about Ray's lurid riding style. He explained perhaps the worst moment was at the Thirty-second "when Ray drifted to the outside edge of the road and on to the grass . . . convinced that Ray had had it, I backed off, only to discover that this was all in a day's work for the Rhodesian, who stayed upright and pressed on with undiminished fury".

In his 'TT Riders Guide' to the Mountain Course, Ray Knight gives a good description of how to tackle this spot, saying "The Thirty-second consists of three left-handers, which can just about be blended into one - at least, the top men will do so; there is time to be made here. Getting this one right means staying out wide for the first part, passing very close to the apex of the second, and running out wide on the exit. The run down to Windy Corner will be so much quicker when you get this one right".

Vic Willoughby was a staff writer for 'The Motor Cycle' and a good racer in the early post-war period. Vic considered the Thirty-second to be one of the most satisfactorily demanding points on the course and writing nostalgically after his retirement from racing, he told how he would "no longer have to summon courage at the 32nd to 'ignore' the first downhill left-hander of the three in order to find the line that takes you through them all on full bore without risking an airborne trip into the valley below", that being the picturesque green valley of East Baldwin.

Today, Dave Molyneux tells how the sidecar aces take it "in top gear, not quite flat all the way through".

WINDY CORNER

It is an open downhill run of a few hundred yards from Duke's to the wide sweeping right-hander that is today's Windy Corner. With the Baldwin valley on the right and the drop away to Laxey to the left, the road runs above them at a height of 1,220 feet and in a more mountainous region the spot might be termed a Col. It is often - though not always as some claim - subject to wind that is funnelled through its modest heights with sufficient force to affect a bike's handling.

The road here has grown in width from the single-track of early days, to one that was described as being "eased to a great extent" in 1922, to a wider version that was said in the early 1930s to have "curiously deceptive cambers" and that is now a broad, well-surfaced road that is positively cambered. Unfortunately, it also carries a mass of ugly traffic warning signs that do nothing for the beauty of the wild Manx landscape.

The former line of concrete posts that partially masked the apex some 50 years ago are now gone and the bank on the outside of the exit has been lowered, although there is a stone wall some way back from the road. From the 1930s to the 1960s Windy Corner was seen by racers as a second gear corner taken at about 55-60 mph, but that speed has risen considerably, with Tom Herron telling of using third on a 750 in the mid-1970s and it now being taken in fourth. The last batch of widening and improvements were carried out in early 2006 as a road safety measure for ordinary users. Like most such improvements it made things faster for the racers, and perhaps it was the positive cambering that made sidecar ace Dave Molyneux estimate that the work would yield a saving of a substantial four seconds a lap. However, after a couple of laps of practice that year, solo star John McGuinness complained that he was catching his knee-sliders on cats eyes, not on the centre line but those located on the apex of the revised bend. It was perhaps unreasonable to expect the highway engineers to have thought of that one!

By the time of his knee-slider incident in 2006, John had been riding the TT for 10 years and as well as having to learn course changes each year, he claimed that he was still learning the complete course, hoping to shave seconds off his lap, which had reached an average speed of 129 mph.

Misty Again

Due to its height, Windy Corner is subject to the mist that can bedevil the Mountain Course. Today's racers are not expected to ride in conditions of heavy mist, but they were in earlier days. In only its second year of use (1912), Harrison Watson was groping his way across the Mountain in poor conditions and failed to see Windy Corner in time to get round. Running up the bank and breaking the front forks of his Humber, he was unable to ride it and so hitched a lift back to Douglas on the rear carrier of another competitor's Triumph.

Tommy Robb came across mist at Windy during morning practice for the 1960 TT, which created a visibility problem made worse because he had

Overlooked by Carn Gerjoil, Windy Corner sweeps to the right and away towards the Thirty-third Milestone.

forgotten to treat his goggles against internal misting. Tommy was stretching the throttle wire on his 500 Matchless on the downhill stretch after the Thirty-second, when he found the corner upon him sooner than anticipated. His attempt to get round failed, and hitting the roadside wall he was thrown from the bike and broke his neck. Tommy shared a ward at the local Nobles Hospital with fellow racers Ernst Degner, Eric Oliver and Mitsuoh Itoh. However, that did not last for long, for Itoh cut the plaster off his broken leg and 'escaped' via a toilet window, whilst Tommy was transferred to a hospital in Belfast. Though his injury virtually immobilised him for three months, it did not prevent him returning to racing, earning 'works' rides with Honda and Yamaha in the 1960s, then crowning his TT career by winning the Lightweight 125 race in 1973.

Tempting

It does not have to be misty for riders to be caught out by Windy Corner for its open right-hand sweep tempts them to push hard here and Steve Hislop observed "You always think you could go in much later on the brakes and 10 mph faster". Unfortunately, although the roadside wall is now some way from the road on the exit, the uneven ground in front of it means that anyone who runs wide is still likely to fall. For some years there was a gravel trap for ordinary traffic here.

This breezy spot has also seen its share of machine problems and one man to suffer was Jim Whalley in the 1921 Junior. Fighting for the lead with Eric Williams (AJS) and Howard Davies (AJS), Whalley was ahead on a couple of occasions on his Blackburne-engined Massey Arran, until on the last lap he collected a front-wheel puncture at Windy and crashed heavily. Bleeding from various places, he remounted and rode his bike in on the rim with it stuck in second gear and with the exhaust-pipe scraping the road. His gallant effort secured fifth place, whilst Eric Williams took the win by 4 minutes from Howard Davies, who came in second after losing 12 minutes in mending a puncture.

A few other incidents of note here include one involving Jimmy Simpson in the early 1920s, when his engine cut just after Brandywell and Jimmy whipped the clutch in and started to coast. Undecided as to whether he should stop or coast as far as possible, Jimmy told that as he approached Windy a spectator held a bottle of beer out to him, bringing the reaction "I just could not pass it". Jimmy stopped, sampled the beer, then looked over his Norton. He changed the plug, checked compression, sparks and petrol, before discovering a blocked main jet. Getting back into the race, he still finished eighth.

Over forty years later there was a classic battle between Mike Hailwood (Honda) and Giacomo Agostini (MV Agusta) in the 1967 Senior TT, during which the lead changing several times, before Ago's chain broke near Windy towards the end of the race, robbing him of possible victory. Ago was the only rider in the race not using Renold chain (his was Italian Regina), however he was a rider who never missed an opportunity to show off to the crowds and his multiple jumps during a race at places like Ago's Leap, coming over the top of Creg Wyllies and at Ballaugh Bridge, would certainly have put extra strain on his MV's chain.

As well as his involvement in that 1967 duel with Ago, Mike Hailwood featured in another incident here during the 1978 Senior when he ran out of petrol, borrowed some to get to the finish and was announced as having completed the race within Silver Replica time. Though others must have known about the situation, it was Mike who did the right thing and told the organisers that he had broken the rules, thus bringing about his exclusion.

Nearly a Slip-up

On the last lap of his winning ride in the 1998 Formula 1 race when he

The view from inside Windy Corner looking out.

was perhaps beginning to think about holding the trophy aloft at post-race celebrations, Ian Simpson pushed his Honda a little too hard through Windy, the front wheel tucked in as he cranked over and he thought the race was lost. Fortunately he was able to get the bike back in shape by using his knee to push it up again and rode on to victory.

Experienced riders know the value of getting Windy Corner right because the correct exit will see them achieve the fastest speed along the next stretch of road, but it is easy to get sucked in by its wide open sweep, lose line and momentum, and so pay for it in time lost on the run to the Thirty-third milestone that follows.

Spectating can be worthwhile here on the bend, though facilities are nil. An unmade road, known to some as Noble's Park Road - nothing to do with Noble's Park near the Grandstand - runs eastwards from the outside of Windy Corner to Glen Roy and is occasionally used by riders of off-road bikes. It used to be in better condition and was recognised as an ambulance access point to the course in the 1930s, if a rather rough one. In the Highway's Board minutes of the 1920s, Windy Corner was officially referred to as Noble's Corner.

A reminder that the TT course is subject to the vagaries of public road use came on the Friday afternoon of practice week in 1991. A five bike pile up at Windy Corner on open roads saw three catch fire and inflict damage to the road surface on the racing line, which meant that emergency repairs were required before it could be used by the racers.

Widening of the Mountain Road between Windy Corner and Keppel Gate was carried out in early 1933 when it was claimed to have been given a non-skid surface, and further major repairs brought weeks of road closure to normal traffic over the winter of 1991-92. In past years the material required for such work would have been obtained from Slieau Lhost quarry that stands adjacent to the road just after Windy Corner. Nowadays it is likely to be hauled from miles away, for you no longer use any old stone for road-making. Everything has to be to prescribed specifications, with even the chippings used in maintenance meeting skid-resistance values.

THIRTY-THIRD MILESTONE

It is a little over half a mile of slightly descending and curving road between Windy Corner and the Thirty-third Milestone. Like the rest of the Mountain section there are no footpaths here, but the side of Carn Gerjoil climbs away from the left of the road, while the green slopes of Baldwin fall sharply away on the right. Back in 1929 this stretch was said to be narrow and bumpy, whilst today it is one where some of the old concrete fence posts remain at the edge of the course and are used by competitors as marker points for braking.

Incorporating two left-hand curves before the tricky un-named right-hander that riders must get correct on the approach to the Thirty-third, this run from Windy Corner gets full throttle. For early riders with limited power, wringing every possible mile per hour out of their bikes was vital and it was on this sort of exposed stretch of road that those on naked bikes adjusted their riding styles to try and minimise wind resistance. Getting their heads down and tucking legs and elbows in was part of the process, and a report on Sid Gleave's ride in winning the 1933 Lightweight told how he "scientifically utilises hedges and walls as shelter from the wind".

Having achieved their best possible speeds on the approach to the Thirty-third, riders have always sought to surrender as little as possible to its two sweeping left-hand bends. That usually meant trying to find a line that allowed them to be ridden as one, whilst also picking a route that avoided the worst of the bumps. In early years most could maintain good speed here even though it was quite narrow, and writing of the late 1930s, Freddie Frith recalled that it was "very nearly flat out" through the Thirty-third on his Norton. However, given its demanding nature, the high speeds achieved and the fact that riders are on the down slope of the Mountain road heading for home, it is hardly surprising that the spot has seen its share of incidents, although it was not until 1931 that the organisers erected one of their warning boards here, and then a report from 1933 says it was "altered and widened to make it considerably less acute".

RIDERS VIEWS . . .

Whilst Freddie Frith's view was that in the late 1930s the Thirty-third was "very nearly flat out", by the early 1950s Geoff Duke reckoned "full bore in third gear at 115 mph" - which, in speed terms, probably amounted to about the same as Freddie's assessment.

Peter Williams rode various bikes at the TT, but of riding his Norton Commando in the early 1970s he told of the importance of being on the right line not only for the corner immediately faced, but also to gain the best positioning to take the ones ahead, saying "if you come out of the Thirty-third right you save yourself a couple of seconds going into Keppel Gate".

Twenty years later, Steve Hislop had appreciably more power to play with and spoke of having to work really hard to get the correct line at the right-hander before the Thirty-third. Knocking it back to fifth, Steve then treated the two bends of the Thirty-third as "one big corner - in fifth gear you are probably doing 150 to 160 mph around there", or as Ray Knight put it: "quite simply, a brave man's corner"!

Though lap speeds get ever faster, Ian Hutchinson said of the Mountain stretch recently, "the bikes are so fast and twitchy, it only takes a little gust to shove you way off line", whilst John McGuinness says he uses "posts on the outside to get a peeling-off point" for the Thirty-third, although it is something in his subconscious that tells him which is the correct one to use, because John generally tends to ride by feel (as he puts it), rather than by use of markers.

IT HAPPENED HERE . . .

A TT report of 1928 said that as a result of accidents that had occurred at the Thirty-third, a detachment of St John Ambulance men from Laxey instituted a service of regular attendance at the point. All practice sessions were early-morning ones and it continued: "These sportsmen thus imposed upon themselves a four-mile climb up the mountain paths from sea level to 1,000 feet at 4.00 am each day".

◆ With less marshals on duty back in 1939 than today, crowd control

Stan Basnett was on hand to catch Trevor Nation (5) and Robert Dunlop (4) on their John Player Nortons at the Thirty-third in the 1990 Formula 1 race.

was more difficult. Les Higgins told how on his last lap of the Junior he rounded the Thirty-third at speed to find spectators walking in the road. They had obviously seen enough racing for the day and were making for home - but how selfish.

◆ New machines were in short supply when racing restarted after the Second World War in 1946, but one man who did have new bikes at that year's MGP was Peter Aitchison. The Norton factory had left a stock of spares on the Island after the 1939 TT ready for use at the 1939 MGP (that was never held). Those spares remained intact and their availability contributed to the construction of several new racing motorcycles, with Peter, who first rode the MGP in 1937, able to purchase two. He asked Francis Beart to prepare them but, sadly, after taking a fine second place in the Junior race, Peter crashed at the Thirty-third during wet and misty conditions in the Senior and was killed. Peter's parents later paid for the course to be widened here in his memory.

Nigel Seymour Smith also contested that 1946 Senior MGP, but in contrast to Peter Aitchison and his two new Manx Nortons, he rode his ten-year old Vincent HRD TT Replica.

Nigel told how his riding kit consisted of "a Cromwell helmet, bought in 1939 for thirty-shillings from Lewis's, racing leather jacket and breeches from Marble Arch Supplies, price two pounds ten shillings, army despatch rider lace-up boots" and a body-belt to cope with rough roads. Unfortunately, Nigel fell on the first lap at The Bungalow damaging his knee and spent the rest of the race awaiting an ambulance ride back to Douglas. He recalled "At the end of the race I had the sad experience when the ambulance that picked me up at the Bungalow, went on to pick up Peter's body. I remember as I sat next to the driver, looking through the window behind my head and seeing the still form, covered by a blanket, with a blue crash helmet resting on his chest like some Viking warrior".

◆ Many riders have cast the bike away at the Thirty-third and survived, amongst them being Maurice Quincey who had his 'works' Norton seize solid with a broken con-rod during the 1955 Senior. Then, in the 1969 Production race, Graham Penney was holding third place on the first Kawasaki Mach 3 to be seen in Island racing. Living up to its reputation of being fast, it also proved that reports of its evil handling were true as it threw Graham through the fence here.

◆ It is not only solos who seek to maintain their hard-earned speed and sidecars have had their share of incidents at the Thirty-third. Seasoned sidecar competitor Roy Hanks was in his twenty-second year of riding the TT and had passed the spot hundreds of times before, but on the first lap of the 1988 Sidecar 'A' race it caught him out. Honest enough to admit it was his fault, Roy explained "We had already caught the outfit that ended up winning the race (Mick Boddice and Chas Birks); there were three outfits all together (approaching the Thirty-third) and I remember thinking this would look spectacular if I passed them all at once at Creg ny Baa and, while I was thinking all of this they all went left and I went straight on. There were a lot of theories that it was windy or that I was trying some special tyres but it was simply that I showed the course a little bit of contempt - I wasn't paying 100% attention - and it flew off the Mountain and landed some 300 yards down it". Typically, despite sustaining serious injuries, Roy was back at the following year's TT and is still racing. An interesting point arising from that incident is that several years later, top Sidecar racer Dave Saville said "At the Thirty-third I can't help thinking about the time Roy Hanks flew over the top . . . in a way I suppose it's a good thing, it keeps me steady".

◆ The Thirty-third is a spot where riders must be fully committed if they are to feel satisfied with their riding of it, and in consequence there are many who have had nasty moments there. One was Bill Smith, who had a clip-on snap during the 1968 Lightweight 250 race and still managed to ride

This rider is dwarfed by the scenery as he exits the Thirty-third's left-hand bends and prepares for the right at Keppel Gate . . .

. . . into the right-hander

. . . dealing with the left-hander.

Mountain milestone

Jack Sheard (OEC-Blackburne) tackles the exit from Keppel Gate in the 1925 Lightweight TT.

into sixth place. Charlie Williams suffered the same fate here in 1981, at a moment when he was well cranked over. He also got away with it.

Winner of the 1987 Junior was Eddie Laycock (EMC) who survived what he called "an enormous last lap slide at the Thirty-third that really scared me". Although tempted to ease the pace, Eddie knew that Brian Reid was snapping at his heels, so he kept the power on and won by 5.6 seconds.

◆ One incident that has gone into TT history featured Trevor Nation on his black and gold John Player rotary-engined Norton in the 1990 Formula 1 race. He was travelling close behind fellow Norton 'works' teamster Robert Dunlop, when Robert missed a gear on the approach to the Thirty-third and slowed slightly. Looking to take advantage of the situation, Trevor moved out to overtake. By then they were upon the corner and, aided by a gust of wind, Trevor ran wider than intended. Indeed, he ran so wide that he went off the road surface onto the grass at the edge where he knocked down several plastic marker posts, threw up grass and muck on to Robert's screen, and felt that he was about to go over the edge at 130+ mph. Fighting for control of the Norton, he managed to get it back onto the road and continue the race, albeit with a smashed screen and broken fairing. Watchers at the scene dubbed it the Great Escape!

The road at the Thirty-third has received further widening down the years and what are now almost lay-byes on the outside of the bend, give riders who overdo it a little extra width and a slight chance to recover.

KEPPEL GATE

Keppel Gate is really a stretched 'S' bend, starting with a right-hand sweep, followed by a short straight that is useful for knocking off speed between the two, for the second bend is a much sharper left-hander (in fact, almost two left-handers on a road bike). In times past, some people called the right-hander Clark's Corner after Ronald Clark spilled from his Levis here during the 1921 Junior, recovering to win the Lightweight class. Soon after the second bend the road begins to steepen its fall towards Kate's Cottage.

With just four speeds available to them, riders of the late 1930s used second through here, changing up into third on the exit where it was very bumpy, a fact confirmed by Vic Willoughby in the early 1950s, who wrote of "the ghastly undulations between Keppel Gate and Kate's Cottage". Some 40 years later riders used second or third and short-shifted up as they exited, whilst today some riders wheelie off the slight summit as the road falls away towards Kate's.

White Lines

In Joey Dunlop's years of Island racing he had to cope with ever more white lines on the course which had to be crossed while cranked over. The bends at Keppel Gate call for serious angles of lean and the white lines here were amongst a number that, although assured that they were applied in a non-slip paint, always made him feel a bit twitchy.

Joey was not the only one to be affected and an edition of 'TT Times' in 1990 said: "The extensive use of white lines on the TT course, more particularly on the Mountain section, has attracted criticism". Highways Engineer and former racer Steve Woodward tried to alleviate riders concerns by pointing out "The Mountain Road is one of the busiest on the Island and has to be marked in accordance with modern safety standards for all vehicular traffic in all weathers. The paint used for these lines gives a better skid resistance than the normal road surface. It is mixed with a high percentage of calcined flint and the barest minimum of ballatoni beads which give the reflectivity in bad weather".

But the problem did not go away, and after almost crashing when his Honda slid on the white lines at the Verandah during early morning practice in 1995, Joey's subsequent comments persuaded the Highways Board to take a grinder to the shiny white bits down the centre of the road.

Over a decade later and close of practice at the 2007 MGP saw many complaints about the slipperiness of newly painted white lines on the

Mountain. Riders old and new spoke of slides and the press described it as the "paddock's hot topic . . . that needed to be addressed".

It was back in 1925 when the Island's road authorities first experimented with "white painted lines in the centre of the road on corners". Seemingly they were a success and most bends were treated over the next couple of years. Then in 1935 a yellow 'fog' line was put down the middle of the road over the Mountain from the Gooseneck to Creg ny Baa, to provide a guide in misty conditions. It was a broken line on the straight that was made solid on corners. To further help at times of poor visibility, in the 1930s 'John Bull' supplemented its normal milestones with additional markers every quarter mile from the Bungalow to Creg ny Baa.

By 1965 a broad white line was painted along the outside edge of the Mountain Road from the Black Hut to the Bungalow to give riders something extra to see in the mist that so often bedevilled the upper reaches.

Today's riders still have double white lines at places like Keppel Gate, but they also have the advantage of Shell Grip special surfacing through some of those bends. Keppel Gate is also one of several locations on the course that have permanent wet patches in the road created by underground water, something that riders need to be aware of, although they actually present little problem.

Spectators

Spectating was a far more casual affair in early years than it is today, with marshals and 'Prohibited' notices few and far between on the Mountain. One spectator recollected watching a late 1920s Senior with: ". . . the weather was well-nigh perfect . . . a friend and I watched the race from various points of the course from Windy Corner down to Creg ny Baa. The marshals were very tolerant, and one could wander about almost as one chose. I distinctly remember being on Alec Bennett's line at the Keppel Gate right-hander and having to take rapid evasive action". Somewhat better controlled in the 1950s, spectators packed the banks between Keppel Gate and Kate's.

Spectating areas are now more restricted, but for those who want both a fairly close-up view of the racing, plus a long view of approaching riders, watching at Keppel Gate can be rewarding, despite it being exposed and largely lacking in facilities. That has always been the case, as AJS-mounted Bert Haddock found when he fell and sustained serious injury here during early-morning practice for the 1920 Junior race. A message was conveyed to the marshals some three-quarters of a mile away at Creg ny Baa and they used their telephone to summon the doctor from the start. After he had made his own way to Keppel Gate and examined the rider, the doctor then requisitioned the horse-ambulance from Douglas Hospital (presumably via the telephone at Creg ny Baa). In words of the time "The ambulance was sent with two horses at breakneck speed, and succeeded in reaching there in what must have constituted a record time for this old ambulance". No doubt the gallant horses did their best, but faced with a constant uphill gradient from Douglas, their efforts would hardly have stood comparison with the response-times demanded of today's high-speed medical rescue services.

Things were not a lot better in the early 1930s, for when C. Atkins crashed his Norton here during practice for the 1934 TT a report in 'The TT Special' told "Within a few minutes the wires were humming with instructions from Mr Huggett (ACU) to Mr W Mullin the Chief Local Marshal. Dr Pantin, and the ambulance were sent out from the Start, and as they had to drive against the course from Cronk ny Mona, the marshals at Hillberry, Creg ny Baa and the Bungalow were instructed to flag oncoming competitors until Atkins had been removed". In a broad hint that riders' response to flagging had been less than satisfactory, they were later reminded that a waved blue flag meant go slowly, a stationary blue meant that a rider was to keep to his left, while "if a yellow flag is displayed, competitors must stop".

The marshalling situation remained inadequate here, for in Tuesday morning's practice at the 1950 TT, Alan Westfield fell from his Triumph and lay unconscious in the road until next man through, Johnny Lockett, found him. Lockett then had to ride on to Creg ny Baa and ask the marshals there to provide assistance at far off Keppel Gate.

Just Where Was Keppel Gate?

In the early days of the TT the stretch from soon after the Thirty-third Milestone through to Kate's Cottage and beyond was all known as Keppel Gate (sometimes locals called it just the Keppel - even Kepple in some official documents), with Kate's Cottage not being a recognised name, for the property there was known as the Shepherd's House. It was probably the occupant - "Mr Maclaren of Kepple Gate" - who just prior to the 1911 TT wrote to the Highways Board complaining of motorists and others leaving the Mountain gates open, and of the excessive speed of motor cars on the Mountain Road.

There have been two locations for gates in this area, one between Keppel's bends and an earlier one at Kate's Cottage. But despite all the emphasis on gates, Canon Stenning claims the place name here actually has nothing to do with gates, being derived from the combination of two Scandinavian words: kapl gata (or kapalla gatta), which meant road to the summit, or the horse road, although he also felt that it could have meant champion or hero's mountain - take your pick!

Up to 1922, Keppel Gate was located opposite what is now called Kate's Cottage, but it was then relocated to a point between the bends at what is now known as Keppel Gate.

The original gate was a single one but the new arrangement of 1922 was of double gates with a central post in the middle of the road. Reports of the time tell of that post still being in position when practice was due to start in 1923, then a bunch of riders removed the post and delivered it to the Highways Board office. What looks like the central post-hole can be seen in the above picture, with variations in the road surface suggesting that only one gate was kept open for normal traffic.

In 1934 the Keppel Gate was removed altogether. This was good news for riders and ordinary road users, but a local paper pointed out "the news will not be so welcome to the little band of juveniles who used to trek up to the gate at the weekend, and obtain a small harvest of coppers for their services in opening and closing the gate for traffic as it came along".

It was the fencing of the Mountain Road that allowed the removal of the gates, but it took some years before the local sheep accepted this restraint on their roaming, and there were still incidents of riders encountering sheep on the course. However, whilst there is no record of a competitor being killed through colliding with a sheep in the years before the road was fenced, sadly, there were fatalities from riders hitting the concrete posts that lined the Mountain section of the course after the fence was erected.

More Incidents

Earlier mention was made of the bends here claiming their share of fallers, and one was Harry Reed in 1912. Like many riders of the time, Harry supplemented the inadequate brakes of his DOT machine by use of an exhaust valve lifter when slowing for corners. This had the effect of cutting engine power, but unbeknown to Harry his valve lifter pin had broken and an account of the time said "as he dashed down the Mountain at a mile a minute, he raised the lever at Keppel Gate only to find it inoperative. He made direct for the wall and was only saved by the turf at the roadside. He had six more spills at corners owing to the failure of this pin", but still finished in fourteenth place. Charlie Dodson was another Keppel victim en-route to winning the rain-lashed 1928 Senior race. On the sixth lap Charlie's eye was

Mountain milestone

Steve Hislop leaves Keppel Gate for Kate's Cottage in 1990.

distracted by someone waving a Programme and he crashed into the bank, splitting his crash helmet from top to bottom at the back. Remounting, he was still slightly dazed at the finish from the battering he had taken from the weather and the course, so it took a little time to register that he and his Sunbeam had won the race, a feat they repeated in 1929.

A potentially worse incident was avoided by the experienced eyes of local Travelling Marshal George Costain (MGP winner in 1954), when on duty one morning in the early 1960s. Coming through Keppel Gate in thick fog, George spotted a couple of fresh gouges in the road surface. Stopping to investigate, he found that a rider and machine had left the course and disappeared from the sight of other riders, but there were no marshals in the vicinity. Fortunately, the rider was only shaken. Checking that he was OK, George rode on to the next group of marshals and ensured that they gave attention to man and machine. Although most Travelling Marshals of the 1960s used sporting roadsters provided by the race organisers (who borrowed them from the factories), George always had the loan of a fully-faired Manx Norton from racing sponsor Reg Dearden to carry out his duties in the six years that he did the job.

One incident at which marshals were present but, thankfully, were not required, occurred during the 1992 Junior race. Top runners Steve Hislop and Brian Reid fought for the lead throughout the race on their 250 Yamahas. On the last lap Steve ran wide and glanced off the bank here and the time lost may have cost him the race, for Brian went on to win by 3.2 seconds, at a speed of 115.13 mph. It is difficult to tell whether Steve was more incident-prone than other riders, or whether his mistakes were better recorded, but he certainly left his mark at a number of spots on the course.

One of the two helicopters used for the 'Flying Doctor' rescue service is usually based at Keppel Gate.

KATE'S COTTAGE

Usually abbreviated to just Kate's, this has been an evocative name to generations of TT enthusiasts, as its small white form sitting high up on the mountain-side is one of the first points on the Mountain Course that can be positively identified by race fans and riders as they approach the Island by sea. For those who know where to look, the sight of Kate's guarantees to lift their spirits and allow them to put day to day cares from their minds, as they contemplate another holiday dedicated to racing and its associated excitement.

Whether there was ever a Kate at the cottage is doubtful and the name seems to have come from the attribution of the name Tait's (or Tate's), which was quoted by a radio commentator to reflect the owner in the 1930s. But while Mr Tait has long gone, the imaginary Kate lives for race fans in the name of this Manx cottage sitting above the drop to Creg ny Baa, with its sweeping vista of the east of the Island and across to the hills of Cumbria.

Kate's has been a marshalling point where those putting in long hours of duty have often received hospitality from the owners of the cottage. John and Marie Webber from England were typical of many who gave up their holidays to marshal in the 1960s and for this they were 'rewarded' with stations on the Mountain section between Brandywell and Kate's Cottage. Marie remembers enduring some particularly bad weather up there, but says "Whatever the weather, much the best spot to marshal was at Kate's Cottage, so long as the householder, a retired hospital matron, happened to take to you. If all was sweetness and light, a full-scale luncheon would be served in the cottage garage - tablecloth, china and all".

It was mentioned earlier that here, 34 miles from the start, was the location of the original Keppel Gate until 1922. As photographs show, it was an iron one that was normally kept closed to retain the sheep on their

Riders plunge down from Keppel Gate towards the tight left-hander of Kate's Cottage.

Frank Applebee (Scott) corners on the limit over the loose-surfaced road at Kate's Cottage during the 1912 TT, with the original Keppel Gate in the background.

mountain pastures. Of single farm-gate width, it was propped open with a piece of rock while practice or racing was in progress.

Danger

Tackling this narrow, loose-surfaced hazard whilst rapidly gathering speed for the downhill run to Creg ny Baa was not for the faint-hearted. A report from 1914 said "The gate at the Keppel is a horrible place, and I never wish to see a rider take it again like Alfie Alexander did. He took it at terrific speed - far faster than any other rider", whilst in 1921 riders complained that the underlying rock was beginning to show through the road surface, something the authorities promised to deal with by blasting.

The whole Mountain Course was a dangerous one to race over in 1911 and 100 years later the danger found in every one of its 37¾ miles is still evident for all to see, not least from the toll it continues to take of riders' lives. Whether the risks are greater for the super-skilled stars or for the average rider is not clear, but they are ever-present. Ian Hutchinson's words when interviewed by Chris Moss before his five wins at the 2010 TT give a rider's view that is probably often thought, but rarely spoken. Ian said: " . . . when you're coming down to the Creg on the last lap on the Friday it's a relief to see the sea and know you're going back home on it".

Hole in a Hedge

This downhill left-hand sweep of the course past Kate's Cottage has been subject to some 'improvements' by the highway authority down the years, but it is not a great deal wider than it was in the 1920s; moving one rider to describe the spot as looking like "a hole in a hedge" when approached at

Looking back from Kate's Cottage towards Keppel Gate in the early 1950s.

Mountain milestone

This is how spectators sat on the exit from Kate's Cottage in the early 1950s, when Geoff Duke estimated his speed here as 115 mph!

speed. Judging the pace through here so as not to get too close to the right-hand bank on the exit is important. It was reckoned a 95 mph spot in the 1930s "using every inch of the road, including the gutters", but by the early 1950s Geoff Duke rated it as third gear and 115 mph, adding "Aim to come out of the corner about one foot to the right of the yellow centre line. Only with superb road-holding is it advisable to attempt speeds that carry you on to the adverse camber near the right bank".

Today riders are tight to the left-hand bank coming into Kate's and use all of the road on the exit - where spectating is now forbidden. It is another corner to build up speed with experience, for being too brave, too soon, can have nasty consequences, but a quick passage through Kate's - being wary of the often wet patch in the road - sees riders ready for the all-out plunge and change into top on the downhill run to Creg ny Baa that follows.

A report from 1931 told that the section of road from Shepherd's Hut (probably the Black Hut) to Creg ny Baa Hotel was reconstructed, tar-grout macadam surfaced, camber reduced and width of roadway materially increased.

It was not until the mid 1930s that rev counters came into general use on racing bikes. Prior to that, riders judged engine speed by ear and they needed to be particularly careful on this drop not to over-rev the engine. Some top riders at the 1937 TT felt that from the information yielded by their new rev counters, the drop to Creg ny Baa was the place at which they achieved their highest speeds on the Mountain Course. On the last lap of a race, the spectators crowding these banks would wave the fast men on to the chequered flag, moving Stanley Woods to say that their actions: "tempted one to go faster than is safely possible".

Mountain Course Moments

Some riders start TT fortnight as though they had never been away. They do not seem to need laps to build up to race speed, for they are on the pace immediately. In 1991 Steve Hislop's average speed on his Honda RVF on his first lap of practice was 121.54 mph and that was just 1 mph slower than the lap record that he set the previous year. Steve's lap was probably done in an attempt to psyche the opposition, (for a certain Carl Fogarty was similarly mounted). Huge talent is needed to be so quick, so soon, and in a taste of what was to come, Ian Hutchinson's first practice lap at his five-win TT of 2010 was 129mph.

♦ Few women ride the TT, but the two entered in 1995 had an unusual shared experience when, after a brief rain shower in Monday evening's practice, Sandra Barnett slid off at Quarter Bridge and Kate Parkinson did exactly the same half a mile up the road at Braddan Bridge. Both escaped with bruises.

♦ The year 2000 was Joey Dunlop's last at the TT. At the age of 48 he won an incredible three races and took his tally of TT wins to twenty-six. In his last Island race, riding what turned out to be his last TT lap, he set his fastest ever lap time around the Mountain Course to average 123.87 mph.

♦ In 2001 the United Kingdom was hit by widespread outbreaks of foot and mouth disease. Striving to prevent the disease spreading to the

Before the Mountain Road was tar-sprayed, this rider summons all his courage to hold his bike flat out on the loose-surfaced drop to Creg ny Baa.

An entry in the 2009 TTXGP.

An alternative stopping method.

Island and destroying the livelihoods of its farming community, the Manx Government reluctantly cancelled the TT and MGP races. The disease did not reach the Island, but the cancellation brought home to many local people just what the races meant to the Isle of Man.

◆ Early morning practice was a wearing business for riders, officials, spectators and the Manx public who lived within earshot of the course. It did offer a unique experience that was enjoyed by some, but the majority were glad when it was dropped prior to the 2004 TT. There were more changes to the event in 2005 when 125, 250 and 400s were dropped from the Mountain Course.

◆ A much-publicised 'Green TT' was introduced for 2009. Officially called the TTXGP and claiming zero emissions, it was run on the morning of Senior race day. After a number of withdrawals, a dozen riders eventually set off on the single lap race. In a performance with which he impressed some of the cynics by hitting 100 mph on the Sulby Straight and lapping at 87 mph, Rob Barber brought his 'Team Agni' bike home in first place.

Just as in the first TT race of 1907 when competitors received a fixed amount of fuel and had to ride with petrol economy in mind, so the 'electrically powered' machines of 2009 had to be ridden with a thought to battery life, with some competitors riding the stretch to Ramsey at less than maximum speed, in order to conserve enough power to propel them up and over the Mountain.

◆ Carolynn Sells made history when she rode to victory in the 2009 Lightweight MGP and so became the first woman to win a race over the Mountain Course.

Carolynn Sells is all concentration as she approaches Douglas Road Corner, Kirk Michael.

Mountain milestone

MOUNTAIN COURSE

CREG NY BAA ... BRANDISH ... HILLBERRY ... CRONK NY MONA ... SIGNPOST CORNER ... BEDSTEAD ... THE NOOK ... GOVERNOR'S BRIDGE ... FINISH

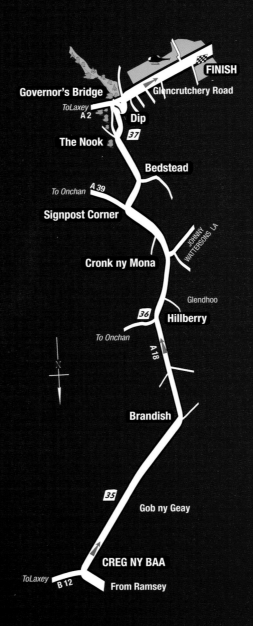

CHAPTER TEN

Creg Ny Baa to the Finish

The Manx name Creg ny Baa translates into English as Rock (or Hill) of the Cows but few of the countless thousands of TT enthusiasts who have watched the races from the spot will care about that, for in respect of Creg ny Baa they all seem naturally fluent in Manx. Even for those on the other side of the world who have never made it to Mona's Isle, the name rolls off the tongue with easy familiarity, and there are even race fans that use it as a house name to remind them of their enthusiasm for the Island and, no doubt, to intrigue the neighbours and postman.

Part of the reason for this spot's popularity is its fairly easy access from Douglas. Far enough out in the early days to make it feel like an adventure for day-trippers and those without transport; shuttle services by charabancs, taxis and then coaches, started from Douglas at 7.00 am on race-days, bringing crowds of people to spectate.

In 1931 there were said to be 118 charabancs at the Creg on a race day, whilst a police-sergeant told how there were 5,000 spectators present in 1936 and numbers at the 1949 Senior were estimated at 8,000. Many of those would have come from the 14,600 people who crossed to the Island on Thursday night as foot passengers from Liverpool and Fleetwood on excursions to see the Senior; others would have been staying the week and would have left their hotels and boarding houses with packed lunches in brown paper bags, ready for a day of excitement. With plenty of road in view in both directions, some remained at the corner, perhaps taking a seat in one of the small grandstands, while others wandered uphill towards Kate's Cottage or down towards Brandish Corner to watch from the heather and gorse-clad roadside banks. For many years a footbridge across the course on the exit from Creg ny Baa allowed spectators to change the side they viewed from.

Ever-present for 100 years of TT racing has been the hostelry on the bend, that was variously known as the Keppel Hotel and Creg ny Baa Hotel but now calls itself just 'Creg ny Baa'. It has had many landlords and mixed trading fortunes over the past century and of one landlord in the late 1930s it was told in regard to early morning practice, "he would open up for hot drinks at 5.00 am, if he felt inclined". It now offers corporate hospitality on race-days with prime viewing, whilst still catering for the spectating masses as it has done for a century.

Telling the World

A popular spot for a BBC radio commentator and for professional photographers in the past, the Creg usually carried advertising banners on the front of the building and on the roadside banks. Today they even have banners on the roof that appear in shots taken by helicopters filming from overhead. Such banners are at risk of getting blown away from this exposed location, for Creg ny Baa is situated 800 feet above sea level, roughly half-way down the Mountain descent.

Fried Brakes

This acute right-angled bend comes after the straight but stepped half-mile drop from Kate's Cottage, which the brave take with the throttle hard against the stop. Braking at the end of the high-speed downhill run has always been a challenge, particularly as the approach to the bend has no easy features to serve as braking reference points, although distance indicator boards are of some help to today's riders. With competitors needing to turn sharp right but with the laws of physics dictating that their bikes should go straight on, braking systems are tested to the limit here and many is the rider who has over-taxed his 'stoppers' at this difficult to judge corner; a place which could be even harder to get right in the early days, when often shrouded in mist. Once described as "the sharpest pull-up on the Course", it presented particular difficulties for riders of those early machines, for as they sought to aid the slowing process by changing down, they had to take a hand off the bars

Mountain milestone

Spectators crowd the banks on the drop to Creg ny Baa in the 1930s, as Alec Bennett leads Syd Crabtree in the 1929 Junior TT. Both were on Velocettes.

to use the tank mounted gear lever that was always mounted on the right-hand side. Many developed a technique of putting the left hand over the tank to use the lever, thus allowing them to keep the front brake applied with the right. That would have been after the growing adoption of front-brakes approaching the mid-1920s, for prior to that Paddy Johnston said of 1922: "the front brake in those days was more or less an ornament", a point confirmed in a report from the 1921 Senior race, when at the finish it was noted that the black enamel on the front wheel rim of winner Howard Davies AJS was unmarked by any action from the stirrup front-brake. He explained that he never used the front brake during the entire race.

Can you imagine trying to stop on just a puny back-brake when flat out down to the Creg? Paddy Johnston had to in 1922, but in one practice session as he went to brake, he found that he had lost his back-brake rod on the Mountain. It was only by quickly getting down through the box from third (top) to bottom, that he managed to struggle round.

Many riders have lost control and dropped their machines here and others have been on the verge of doing so before scrambling round in untidy fashion. It was as difficult to slow from 75 mph over the loose surfaced roads of the early 1920s as it is from over twice that speed today, and before the road was tar-sprayed and chipped in the mid-1920s, Stanley Woods recalled how a bunch of Scottish lads brought brooms and swept the course on the approach to the corner, for the benefit of riders.

'Bolsters'

There was originally a wall around the outside of the bend before it met the roadside bank of the field next to the hotel. Quite when the organisers started putting protection on it is not known, but there are references from the early 1920s of riders hitting what were variously described as bolsters,

mattresses and sandbags, whilst a report from 1929 talks of "the dreaded Creg where the walls are mattressed in case a rider misjudges the corner or has trouble with his brakes". By then the protection was provided by the Leicester Rubber Company and carried its trademark 'John Bull' advertising. The hotel itself has received growing amounts of protection down the years and nowadays it is enclosed in a veritable steel cage to safeguard spectators, with an outer layer of air-fencing for the benefit of riders.

Creg ny Baa crammed with coaches, cars and bikes in the early 1970s. There is no longer a footbridge allowing access to the inside of the course.

RIDERS VIEWS . . .

In the early 1920s the Creg was said to be a 35 mph bend, and 30 years later Geoff Duke was still saying "Drop to about 35 mph and when the left bank begins to fall away, swing sharply to the right to make a 90 degree turn, using the clutch. Accelerate out of the corner at 45/50 mph, almost brushing the right-hand wall with your shoulder". In the late 1950s, when setting the first 100 mph lap, Bob McIntyre spoke of having to pin his 500cc Gilera down to 50 mph and reinforcing the point about the need for hard braking here, an account of the time said of the Gilera "the brakes dissipate almost enough energy to cause a local heat wave". The widespread adoption of fairings in the 1950s made bikes 'slipperier', so putting more demand on brakes.

Although this right-hand corner is taken by using a conventional approach on the left and sweeping across to the apex, many riders go in deep and take a squared off line rather than riding through on a constant radius. Ten years after Bob Mac it was 50 mph and second gear for Mike Hailwood and his big Honda, after "dropping like a stone" from Kate's.

Many of today's bikes aviate on the undulations coming down to the Creg, with Steve Hislop saying how he would choose his moment to hook into top gear to coincide with one of these wheel lifts because "it stops the bike from coming up too much". Steve also said of the corner, "you can try and be too brave on the brakes and just end up in a mess". Nowadays virtually every bike will be squirming vigorously at the back end as the rear goes light under heavy braking and weight is transferred to the front. With top riders conscious of the crowds, some will want to scratch round the bend and put on a show, but multi-winner John McGuinness warns "it's an easy corner to fall off on".

Sidecars also have to resort to heavy braking here and both driver and passenger must brace themselves to avoid being thrown too far forward, as they drop down through the gears into second and the passenger shifts his weight as they drift through.

Changes

The route of the Mountain Road near the Creg changed in the years before the first TT and the corner that was created has also been subject to alteration down the years. Originally having a high stone bank and fence at the apex, this has gradually been lowered. After the 1938 TT it was reported "the Creg as we know it will probably be no more in 1939, for during the coming winter it is proposed to eliminate the corner and convert it into a wide

A classic view from Creg ny Baa looking back up the course to Kate's Cottage.

Mountain milestone

sweeping bend, cutting across the land on the inside of the corner". That did not happen, but in recent times the bank was cut back to provide a better sightline for ordinary traffic, although the kerb line remained near its original position.

Towards the end of the 1930s the bank on the left side of the approach to the Creg was cut back quite considerably just before it led off to the road for Laxey. It was believed that this would allow a better approach line for riders and faster passage through the corner, but the new strip of road had a looser surface than the existing, so riders tended to keep to the original line. Like many parts of the Mountain Course, even normal traffic use brings about change and early in practice for the 1939 TT, word quickly went around among riders that the Creg "had grown a new bump" and that the following straight was "anything but smooth".

When the Clypse Course was brought into use in 1954, this was the point where riders using the Clypse (on which they approached Creg ny Baa from Brandish) left the Mountain Course and headed over the back road in the direction of Laxey. That same road is now available as an escape route for riders who overshoot the corner and it allows spectators to come and go from the Creg whilst racing is in progress.

Creg ny Baa may be immensely popular for thousands of spectators on a sunny race-day, but during a cold early morning in 1953, G Clark (Vincent) has the corner to himself while practising for the 1,000cc Clubman's TT.

IT HAPPENED HERE . . .

Norton won the twin-cylinder race at the first TT in 1907, but experienced a barren period over the next few years. However, hopes were high in 1911 when the company's owner 'Pa' Norton and Percy Brewster entered on 500cc singles, but 'Pa' was soon out of the race when his engine seized at Ramsey and Brewster's ride came to an end when his new Armstrong three-speed hub locked up on the approach to the Creg.

◆ In the 1913 Senior race a report told how A.J. McDonagh was in second place, but as he passed Russian rider Kremleff "his machine turned over, and he sustained a broken collar bone". The report went on "His injury did not necessitate his removal to hospital".

◆ By 1925 the road surface was tarred which meant faster speeds on the approach. Of one it was told "Probably the most miraculous escape of the whole race took place at Creg ny Baa . . . A. M. Harvey descended from Keppel Gate at a terrific speed, tried to negotiate the corner too fast and failed and was thrown from his machine a distance of several yards. He hit the wall squarely, was dazed, and fortunately rolled with the force of the impact. The machine turned somersault in the air and hit the wall at the exact spot as the rider. Had Harvey not rolled, it would have meant instant death".

◆ Later to win TT races, Tim Hunt was on his way to victory in the Junior class of the 1928 Amateur race when he suffered valve-gear failure on the drop to Creg ny Baa. Such was his lead that coasting and pushing to the finish would almost certainly have seen him take a podium place; but Tim elected to retire to the bar of the 'Keppel Hotel'.

◆ Another rider who went on to TT wins at record-breaking speeds, Harold Daniell told how in his first Island race at the 1930 MGP he hit the sandbags on the outside of the bend due to over-confidence. It was an incident that taught him an early lesson on the need to respect the course.

◆ In the 1933 Lightweight TT, former winner Henry Tyrell Smith almost lost his chance of winning a replica when marshals at the Creg inexplicably opened the course to normal traffic before he had finished his last lap. The race organisers eventually credited Henry with a seventh and last lap time that equated to that set on his sixth lap. Five other riders were affected by the early road opening.

◆ Huge crowds at the 1936 Senior TT were entertained by stars like Stanley Woods (Velocette) and Jimmy Guthrie (Norton), but it was Freddie Frith who had them on their feet, for he approached the corner far too fast whilst seeking to overtake 'Ginger' Wood and in the words of a contemporary report "he banked over at an alarming angle, so much so that the back wheel slid from under him and the Norton went into a waltz, but superb riding soon avoided what seemed a certain crash".

◆ Despite attempts to protect it, the Creg ny Baa Hotel has felt the impact of several crashes. In 1939 W.N. Webb failed to slow his Excelsior sufficiently, hit it and broke a leg. Then during opening practice for the 1964 MGP, Michael Bennett drove straight into the front door. After he had gathered his wits and broken Norton, he admitted that he had forgotten the corner.

◆ With the wonderful confidence shown by press photographers as they crouch at the side of the track in places prohibited to ordinary mortals, 'Motor Cycling's Alf Long found at the 1955 TT that he was in the precise spot that Dick Thomson's out of control Matchless was determined to occupy after Dick dropped it. Long's big plate camera was smashed in the ensuing melee, but he escaped injury.

◆ Top Yamaha riders Bill Ivy and Phil Read were required to ride to team orders in 1968, and Read was down to win the Lightweight 125 race. Bill was on top form (he won the Lightweight 250 at record-breaking speed) and, perhaps making a point to Yamaha, he led the 125 race on his four-cylinder two-stroke, before stopping at the Creg on the last lap, ostensibly to check his position, but in reality to let Read make up time. Read duly went on to win, but said in later years of the much-publicised incident "I think I'd have preferred it if he had won".

◆ Mike Hailwood rode his last Island race in 1979, finishing in second place in the Classic TT after a hard-fought battle with winner Alex George. The margin between the two at the finish was just three seconds but the result could easily have gone the other way, for Alex told how on the fifth lap he was so physically and mentally tired that he "started to go to pieces and made a mess of the Creg, scattering spectators . . .". However, history tells that he pulled himself together and held on for a memorable win.

◆ Graeme Crosby was a Newcomer to the TT in 1979 and soon learned an important lesson, saying "The main thing I learned is that if you do break down it must be near a pub". During that first year he managed, with some difficulty, to reach the Creg after his machine holed a piston. Continuing with the pub theme, 'Croz' later described a TT race as "just like a quick blast home from the pub with your mates" - something of an over-simplification from the three-time TT winning Kiwi.

Creg Ny Baa to the Finish

George O'Dell shows the style of a TT winner and World Champion.

through the portion called Gob ny Geay (Mouth of the Wind), known to some as Sunny Orchard. This spot, which can live up to its name in being windy, is also known by some riders as the Cutting, with the high banks creating an impression that one described as "like going through a pipe". It is a spot where 50 years ago a really committed rider would take the kink flat in top at about 135 mph, whilst others feathered the throttle. The road would have had a tar and coarse stone chip surface, and by 1980 speeds had risen to just over 160 mph. Today the road is perhaps smoother, but it still receives regular application of tar and chippings and now top runners muster the courage to hold it flat in top gear at approximately 190 mph. All this is on a stretch of road without footpaths that is cambered at the edges and runs between unyielding high banks. It must seem worryingly narrow at speed and even more so when overtaking, or when riding in a group.

In the early 1980s Mick Grant felt that this stretch shared with The Highlander the title of fastest part of the course, with the wind direction sometimes being the determining factor. Mick tells that he tended to set his gearing to gain maximum benefit on the lower part of the course before Ramsey, so he sometimes had to back off a touch on this stretch to avoid over-revving, and in his long Island racing career Nick Jefferies learned that "if you're having to roll the throttle slightly to avoid over-revving before braking for Brandish, then overall your gearing is about right. Easing off may cost you a second here, but over the Mountain you'll have clawed back much more".

◆ Norman Brown led the confusingly named Senior Classic race of 1983 which was for modern bikes, but after becoming the first man to lap at 116 mph, he ran out of petrol at the Creg and had to sit out the rest of the race as Rob McElnea went on to win.

◆ Starting his Island racing on a solo but then taking nine victories in the Formula 2 sidecar class between 1985 and 1990, Dave Saville was an experienced Mountain Course competitor. Popular with spectators and fellow competitors, Dave raced a solo at the Classic MGP in the early 1990s, but a heavy fall at Creg ny Baa turned his two-wheel ride into a tragedy that ended his racing career and left him in a wheelchair.

BRANDISH

Forced to brake fiercely for Creg ny Baa, riders like to be as smooth as possible rounding the corner so that they come out at the fastest possible speed and gain maximum benefit from the one mile stretch to Brandish Corner that follows.

Fastest?

Reckoned by many to be the fastest part of the course, the road here falls downhill (with a slight rise in the middle) and has a noticeable right-hand kink

Geoff Duke's line through Brandish in 1950.

Incidents

The long run to Brandish passing the 35 mile marker should be straightforward, but like every other stretch of the Mountain Course it has seen its share of incidents. Racer turned journalist Vic Willoughby told how 1953 was the year in which privately-owned Nortons suffered many failures of the rear suspension piston rods due to the relatively flimsy bolted attachment of the sub-frame to the main featherbed frame, and how in one exciting incident Cyril Julian had both rods break whilst riding the straight between Creg ny Baa and Brandish. Cyril got away with what was a very nasty moment and Vic Willoughby's own race ended at Sarah's Cottage on lap six when one of his suspension rods snapped, sending both he and his Norton clattering into the gutter.

With their BMW hitting maximum speed on the drop to Brandish in the 1961 sidecar race, Claude Lambert with wife Marie-Louise passengering, could do little as the outfit gradually began to weave down the road. Eventually it hit the bank, turned over and threw driver and passenger into the road, killing Marie-Louise and leaving Claude with a broken leg and other injuries. Travelling Marshal Dennis Craine soon arrived to render assistance at what was a highly distressing scene and then showing the incredible spirit that so many motorcycle racers possess, Claude returned to the TT the following year with a new passenger and took fourth place.

Winner of the 1977 Sidecar TT, George O'Dell summarised the Creg ny Baa to Brandish section with the words "The drop down to Brandish is the place to give you the sh . . s. You know you're going fast there. As you turn right from the Creg it drops away and there's a bit of an adverse camber. The bike always feels it's travelling in the air at that part especially with the wind getting under you". Known for his vigorous racing style, George continued "I treat Brandish with the utmost caution. If you hit that bank you'd bury yourself in the field".

Maximum speed has always been the target on this so inviting stretch of road, but though it is a place to put the chin on the tank and tuck everything in, riders must remain aware of the looming Brandish Corner that, until recently, bore sharp left between tightly enclosing roadside banks. The spot has caused many falls and probably the best known is that of Walter

Mountain milestone

Brandish Corner in 2006.

Brandish Corner after alterations prior to the 2007 TT.

Brandish, after whom the bend was named when he fell from his Triumph in 1923 and broke a leg. Second in the previous year's Senior race, Walter entered too fast, got carried out to the right on the exit, clouted the bank and parted company with his machine.

Quite how Walter's spill was dealt with is not known, because the organiser's list of "points on the course where Officials will be stationed" for 1923 makes no mention of a flagman, marshal or policeman doing duty anywhere between Creg ny Ba and Hillberry.

The road must have been at its narrowest when Walter came off because in 1931 the corner was improved and slightly widened by having the ditch filled in. As a result, experienced riders Arthur Simcock and Charlie Dodson ventured the opinion that it could then be taken about 8 mph faster.

Braking

The organisers must have been worried about riders mis-judging Brandish in the 1950s, for in 'History of the Clubman's TT Races' it tells that in 1952 "a red warning line had been painted across the road on the approach to Brandish Corner". Back in those days brakes had probably not returned to 100% efficiency for Brandish after having been brought to a state of overheating for the big downhill stopping job at Creg ny Baa, so riders would generally sit up while braking, to create more wind resistance and so help their tired brakes.

Despite the speed of approach to Brandish having increased considerably down the years, brakes have also improved and Charlie Williams drew the interesting observation from three members of the Jeffries clan of TT winners, Tony, Nick and David, that the braking point for Brandish was virtually the same in the early 2000s as it had been when Tony first rode the Mountain course in 1968.

Whoops!

Walter Brandish's spill has been repeated by several TT winners at this point and by many lesser lights. Amongst the former was Derek Minter, who fell and injured his arm while practising on a Gilera that had been taken out of mothballs for a comeback at the 1966 TT. Phil Read was also on the TT comeback trail when he rode a 'works' Honda to victory in 1977, but when engaged on a bit of unofficial open roads practising before the last race of the week, he spilled here and broke his collar-bone. Then during practice in 1982 Alex George approached at top speed, left his braking too late, came off his Honda, broke knuckles, and put himself out of the event.

One TT winner lucky to escape a spill here was Carl Fogarty in 1990. He recalled the incident in his auto-biography 'Foggy' with "I also rode in the 600cc race, which I didn't enjoy, especially when the bike spewed out oil approaching Brandish, one of the fastest parts of the circuit. With the back wheel sliding round, it was a miracle that I kept the bike upright to avoid hitting the banking head on". Another fortunate escape was by multi-Sidecar TT winner Mick Boddice at the 1982 TT. The first Sidecar race was run in sweltering conditions and Mick was leading on the last lap as he approached Brandish, only to run out of fuel. Pulling in to retire, Mick stepped off the bike and fainted.

Joey Dunlop always managed to stay on board here, but admitted in the mid 1980s "To be honest, Brandish scares me. There is no room and it is so quick".

Trouble for Travelling Marshals

This is a bend that is even tricky enough to catch out Travelling Marshals like Keith Trubshaw and Tony Duncan - and they were not racing! It is not unknown for Travelling Marshals to fall, but this pair set a new record when they were involved in a double spill here in 2006. Brandish had seen an earlier crash by a Travelling Marshal when Mike Kneen lost the front end of his Honda CBX750 towards the end of practice for the 1985 MGP. The Honda was badly damaged but Mike got away lightly and was quickly back on duty on a spare bike. How he managed to write-off a second official Honda on open roads just before the first race is another story. Suffice to say it was not his fault.

Widening

The inside bank at Brandish was removed in post-war years but the road width stayed much the same. In place of the bank was erected a line of white-painted metal railings to improve visibility, but it remained a tricky spot and it was not just the racers who mis-judged the corner, for many ordinary road users took it too fast and hit the roadside bank, when travelling in both directions. Concerned at the way that it so often featured in its accident statistics, the Island's Department of Transport effected changes here before the 2007 TT, to make it safer for ordinary traffic.

TT traditionalists were not impressed, for the radius was greatly eased and much of the left-hand bank taken away, with the troublesome bank on the outside also removed for some distance. Gone for all time was the demanding corner where 50 years ago, a late peel-off allowed it to be taken in second gear at an exciting 60 mph.

There was much speculation before the 2007 TT as to just how fast the corner could now be taken. John McGuinness reckoned that it had turned a "50 mph corner into a 150 mph one", that it would "reduce average lap times by three seconds" and that it was "the easiest corner on the track now". Because it was so smooth and open, John felt it would be top gear on a 600 but said "I'll knock it back one and roll it on the big one".

Just before leaving Brandish, a couple of "did you know?" items get a mention. Prior to the corner being given the name of Brandish it was known as O'Donnells, a name without any apparent TT connections. A report from 1913 tells how "a bump just before O'Donnell's had, before the end of the race become a somewhat dangerous beach, which was productive of several hair-raising leaps and wobbles". It is a spot that has also been referred to as Upper Hillberry.

HILLBERRY

Having dealt with Creg ny Baa and Brandish, riders are faced with yet another corner that has a straight downhill approach, so giving them time to summon up courage to tackle it at their personal maximum speeds. It was John McGuinness who said in the book 'TT 100' by Mick Duckworth, "I never used to be a lover of Hillberry corner, the trouble is you can see it coming for far too long". Certainly Hillberry is a place where a rider with plenty of 'bottle' will gain over the more cautious. It has always been a very fast run between the roadside banks from Brandish, with sod hedges eventually changing to stone walls which riders keep close to on the left and that spectators lean on.

It is at this point that riders can focus on Hillberry's right-hand bend disappearing in an uphill sweep under trees, with initial unyielding vertical banks on the exit easing to give only slightly less daunting sloping grass ones.

Close to the action - spectators lean on the course-side wall at Hillberry.

Jimmy Guthrie (Norton) sweeps through Hillberry on his way to victory in the 1934 Junior TT. In later years the bank and wall on the inside was cut back, making the corner measurably faster.

Mountain milestone

Committed to Hillberry's right-hander.

This is truly a corner for the brave, and many courageous riders have experienced 'moments' here.

It was former TT winner Jimmy Simpson who, asked about the degree of commitment that riders put into such challenging corners, claimed with particular reference to the epic 1930s battles between Wal Handley and Jimmy Guthrie "the race of the moment was the one thing that mattered - there was no thought of the future, no count of the risk". Such commitment and bravery has been evident at Hillberry over the past century of racing, providing generations of spectators with high-speed images of their favourite riders to tuck away in their memory banks.

Exciting Moments

Jimmy Alexander competed in the TT from 1910 to 1914 and recalled that in 1911 he brought top American rider Jake de Rosier over to the Island a week before official practice started, to carry out course-learning. Seemingly fearless, Jake had several spills in that first week and although he normally raced in woollen tights, he soon swapped them for leather trousers. Come official practice and Jake was trying hard when he hit loose stones and lost control of his Indian here, the bike going on to hit the bank and throw him into the dusty road. It was a spill in which Jake acquired yet more cuts and bruises and in which the Indian was badly bent.

Another man who travelled a long way to race at the TT was Russian 'Boris' Kremleff. Going well on his belt-driven Rudge in 1913, he crashed here just two miles from the finish of his race.

When Alec Bennett became the last man to win a Senior TT on a side-valve bike in 1922, the speed at which he tackled Hillberry on his Sunbeam drew praise from spectators, except on the last lap, when in words of the time "he gave everyone a fright by swinging out after rounding the bend at a good mile a minute, and scraping by".

No one ever doubted the commitment of Stanley Woods whilst racing and during his winning ride in the 1923 Junior race over still largely unsurfaced roads, he was said to have taken Hillberry in one huge steady drift "throwing up a bow-wave of dirt and stone from his wheels. All through this giant slide at perhaps 60 mph he was rock steady, master of his machine and himself". It was while spectating at Hillberry two years earlier that Stanley had felt inspired to race in the TT, something he first did in 1922, finishing 5th in the Junior.

Later in 1923, just past the corner was the scene of a sad incident when local rider Thomas Brew was killed during practice for the September 'Amateur' race. Brew's sister was standing in the lane to Glendhoo just after Hillberry, out of sight as she thought, but he saw her, took one hand off the handlebars to wave, got into a speed wobble and crashed. On race day the flags at the Grandstand were lowered to half-mast and Brew's number on the scoreboard was covered with black ribbon.

The short-lived sidecar TT races of 1923-1925 put enormous stresses on the cobbled together outfits of the day, for most were not originally designed to be raced. Frames, forks, wheels, tyres, etc, were pushed to their limits and George Rowley who was a passenger to Jimmy Simpson in the 1925 race, spoke of catching glimpses of the front inner tube as they went flexing through Hillberry. Freddie Dixon also found it taxing on the Douglas-powered banking sidecar outfit that he rode to victory in 1923, because on the last lap one of the bike's frame tubes broke here. A report of the incident said "Bike and sidecar then went into a mutually opposed bank, leaning upon each other so steeply that the nearside handlebar fouled the chair body".

Fast Corner

By 1924 Hillberry was considered to be among the fastest corners on the course and with bikes doing 80 mph on the approach, several top riders of the day were asked to estimate their speeds through the corner. Local man Tom Sheard said 50-55 mph, George Dance 55 mph, Graham Walker 55 mph, Freddie Dixon 55-60 mph and Alec Bennett 60 mph. By 1932 the spot was being described with seeming precision as an 80 mph corner, although virtually no race bikes were fitted with speedometer or rev counter at the time.

Controversy arose after the 1936 Junior TT when Norton star Jimmy Guthrie was excluded from the race after an incident at Hillberry. He stopped here on the fifth lap to refit his drive chain, but some claimed that he then infringed the rules by receiving outside assistance to restart. Flagged off the course at Ramsey, he was stopped there for some 7 minutes before restarting and finishing the race. Norton Motors Ltd protested at his exclusion and after deliberation by the stewards, Jimmy was reinstated as a finisher.

'Ginger' Wood had an exciting moment here when in his words "in 1932 I endeavoured to modify Hillberry Corner . . . took it too fine. Took also, with my right handlebar-grip, the entire Dunlop advertising banner. Ricocheting over to the other bank, there ensued a fight for possession, which I lost". Then of the 1937 Lightweight race it was said "The best first lap thrill is provided by 'Ginger' Wood . . . he simply tore round the bend. The road is barely wide enough for him . . . every inch of space . . . a fine attempt at hedge slicing", then adding "The Surveyor-General witnessed it; perhaps he can do something about it for Ginger". It was not actually a hedge that he 'sliced' but a stone bank on the inside of the bend with grass and bushes growing from it, and it was one that Freddie Frith clipped on the last lap of his winning ride in the 1937 Senior, when he was put slightly off line by spectators leaning over the stone wall on the entry to the bend and waving their programmes. Fortunately he got away with it.

This inside bank determined speed through Hillberry and before it was cut back in 1948 riders took it revving hard in second gear, but its realignment added speed and meant third gear.

Drifting

Geoff Duke was one of the stars of the early 1950s who was brave enough to tackle Hillberry on the limit, aided by the confidence-inspiring handling of his Featherbed Norton. He estimated his speed at 115 mph and although he could take it in top, he preferred third so that he maintained maximum revs and drive on the uphill stretch to Cronk y Mona that followed. Although easing of the inside bank had certainly raised speeds, he always kept a little way out from the apex to avoid the bumps and camber there, and also tried to avoid running wide on the exit, where the camber fell away. Geoff knew that in the early 1950s he slid the rear-wheel in controlled fashion on fast sweeps, but he expressed concern when well-informed sources suggested that he was actually drifting both wheels at spots like Hillberry.

With his square jutting chin, Bob McIntyre was the personification of the determined racer and one who battled through his share of adversity, earning the admiration of race fans in the process. His first Island race was the 1952 Clubman's TT on a 350cc BSA Gold Star. On the second lap of practice he fancied he could take Hillberry flat out in top gear. Perhaps he could have done so on a proper racing Goldie, but he forgot that his Clubman's model still had the centre-stand fitted. This grounded, lifted the rear-wheel and off he came at some 90 mph. Severely knocked about by the spill, Bob was threatened with having his bike taken away from him by his employer, Sam Cooper of Troon, but he had learned his lesson and kept his wheels upright in his following nine-year TT career that included three wins and other podium places.

Many riders have frightened themselves by getting carried into the steeply cambered left-hand gutter on the exit from Hillberry, and whilst some have got away with it, others have gone down. Before he became a TT winner, Jack Findlay fell from his Norton in practice for the 1961 TT and the results were evident for over 130 yards of torn banking and scarred road surface. Jack damaged an arm but was quickly back racing. Later that year, Ned Minihan had a scare during the second lap of his winning ride in the Senior MGP. A report of the time tells: "As he zipped through the 100 mph right-hander he found Bernard Hunter (Norton) rebounding off the grass bank on the exit. Hunter swerved violently to the right. The startled Ned dived through a gap on the left - and was clear".

It was in the 750cc class of the massed-start Production race of 1970, that Peter Williams (Norton Commando) spent most of the race clawing back the lead of Malcolm Uphill (Triumph Trident). Catching and passing him at Creg ny Baa on the last lap, Peter's hopes of victory were dashed when going through Hillberry the Norton coughed - he was low on fuel - and Uphill passed him to win by 1.6 seconds. Despite this disappointing incident, Hillberry suited the fine-handling Norton and Peter said "I like Hillberry. I just have a picture of it and I roll it off to steady myself and just literally aim for this tree on the right-hand side of the road. I always miss it but it puts me right for going up the hill . . . must be 120 mph". It was Peter's choice to hold top (because it suited him and the Norton), but nowadays most riders take it in fourth or fifth gear.

Joey Dunlop did not like Hillberry, rating it as "dangerous", though watching his progress through there you would hardly have suspected it. It was the uncertain conditions created by the trees overhanging the course here that concerned the twenty-six times TT winner. They are fine specimens of beech that have probably been there since the first TT, but as they have grown their influence on road conditions has increased.

Despite the way that the camber falls away towards the left-hand bank on the exit from Hillberry, today's committed riders use all the road, leaving little margin for error. It is an impressive sight!

Oh, No!!

Many riders have run out of fuel, either on the Mountain or during its descent and races have been lost through the lack of it. With its downhill approach causing fuel to run to the front of the tank, Hillberry has had its share of low fuel scares. Brian Reid had been threatening to win a TT in the early 1980s, but had yet to cross the finishing line in first place. Things were looking good in the 1985 Lightweight 250, but then at Hillberry Brian brought his EMC to a halt with an empty tank and lost another chance of a win. His pit crew had been careful to fill the tank to the brim at his earlier pit-stop because he had spluttered in to refuel completely empty, but pressure in the second half of the race from Joey Dunlop meant that Brian had to break the lap record in a freshening wind and there was just not enough of the vital essence to get him to the line. Brian eventually gained his first TT win in the Formula 2 race of 1986.

One rider with a TT win to his name and a host of Silver Replicas, used his long experience of TT racing to save himself a push to the finish after running out of petrol on the rise out of Hillberry on the last lap of a race. Coasting to a halt with his tank empty, he spotted a bike of the same type as his leaning against the bank 50 yards further on where its unfortunate rider had retired. Running up to the bike our anonymous rider quickly undid the fixing strap, trotted down the road with the tank under his arm and fitted it to his bike. A hurried shout to a nearby marshal to look after his empty tank and our canny rider was away again to finish yet another TT.

Not so fortunate was a rider whose Laverda run out of petrol here during an evening practice in the 1970s. Unable to borrow any on the spot, he hitched a lift via side-roads back to the Paddock. Grabbing a can of petrol, he then managed to get a lift back to Hillberry, refuelled the Laverda and was credited by the time-keepers with completing a lap in just over one hour.

Both solo and sidecar riders stress the need to take Hillberry in a gear that gives the best drive and control for the uphill stretch of Hillberry Road and the bends of Cronk ny Mona that follow. Top sidecar driver Dave Molyneux, a man not known for releasing the tension on the throttle wire a fraction if he could avoid it, explained: "Hillberry - fast, but people kid themselves it's faster than it is . . . kick it down one and nail it. You need fifth for the drive up the hill". In an indication of a passenger's influence in sidecar racing, Dave Saville told how he used to take Hillberry flat in top gear on his two-stroke Yamaha, but how this took him too far out to the left to get the best drive up the hill. Talking about it with passenger Nick Roche, it was agreed that Nick would position himself differently on the 'chair' and the affect of this allowed Dave to come out in the middle of the road and so achieve better drive.

More Restrictions

Spectators have enjoyed a good view of racing at Hillberry down the years, for there is a small grandstand on the bend, plus limited seating on the bank, but this is another spectator area to suffer from the widespread restrictions introduced in 2008. Even private householders were advised to impose those restrictions on their course-side gardens where they had long invited spectators to view the races. Vague talk, even threats related to insurance, liabilities, claims, etc, made them feel they had to comply. Not all were happywith the situation.

Access to and from Hillberry is available by a small back road from Onchan during racing, although the best viewing positions are taken early on race-days. For the ßregulars at this spot, sitting waiting for the first rider in a race to come into view as a distant dot rounding Brandish and then rushing down towards this daunting right-hander, offers a Mountain Course experience that never palls. As the rider passes there is a combined intake of breath from those watching, for the speed of passage never ceases to amaze.

CRONK NY MONA

The uphill swoop out of Hillberry takes riders through a left kink and then from a line on the right-hand side of the road, onto the further left sweep of Cronk ny Mona (Hill of the Turf or Hill of the Peat Turbary). Exhilarating for riders when things are going well, for Mick Grant spoke of taking the second left in a two-wheel drift on a 350 Yamaha in the early 1970s between banks crowded with spectators and John McGuinness said in 2007 "I love Cronk ny

Mountain milestone

Ken Bills (Vincent) is well into the climb towards Cronk ny Mona in the 1950 Senior TT.

Mona, it's one of my favourite parts of the course", it is also a place where things have been known to go badly wrong. Francis Beart was an engineer of repute who prepared and entered many machines in post-war TT and MGP races with considerable success but, as ever in racing, tragedy was never far away. Beart entered Ivor Arber on his 350 and 500 Nortons in the 1952 MGP but early in practice Arber failed to take Hillberry on the 500 and was killed. Thinking to pack up and go home, Beart was persuaded to make his 350 available to Ken James. On only his third lap of practice on the bike, James crashed at Cronk ny Mona and was also killed. The double tragedy saw Beart withdraw from providing bikes for Island racing, although he continued to prepare engines for Formula 3 cars that often used Manx Norton power. He did return to bikes and entered MGP winners of the early 1960s like Peter Darvill, Clive Brown and Joe Dunphy, on machines that were renowned for their thorough preparation and that ran in his apple-green colours.

Hartle Off!

In 1968 John Hartle was brought back into the MV Agusta team at the TT to support Giacomo Agostini. Unfortunately a fall from a Production Triumph at Windy Corner left him with injuries which prevented him racing the Junior MV, but he was fit enough to ride in Friday's Senior. Whilst John was fit to race, the big MV was not, for it developed several false neutrals in its gearbox and these contributed to his downfall at Cronk ny Mona. Seeking to reduce his gear-changing he came through in a higher gear than usual but the bike started to wag its head, went off line and hit the kerb, fetching John off at high speed - fortunately without major injury. It was a time when even good runners received only a minimal contribution towards their expenses, and when jokingly asked why he had chosen two out of the way places like Windy Corner and Cronk ny Mona to fall, John's reply was "For only £15 start money I wasn't going to do it in front of the Grandstand".

The rise to Cronk ny Mona is the point where those competitors who have nursed ailing machines over the Mountain, and especially those who have been coasting with failed engines, realise just how steep a climb it is. One of the unluckiest men here was Les Graham mounted on an AJS Porcupine in the 1949 Senior TT. Speeding up to Cronk ny Mona in the lead on the last lap, the magneto armature sheared, the engine lost all power, and Les pushed in to a disappointing 10th place. Some consolation for the TT failure came later in the year when he was crowned as the first 500cc World Champion.

On the Saturday of practice week for the 1956 MGP, the Highways Department was forced to bring in a steam roller and gang of workmen to iron out bumps in the road surface at Cronk ny Mona that had caused riders several anxious moments. Something similar had happened in 1952 when the junction here served as the point where competitors in a car race joined the Mountain Course from the Willaston direction to head towards Signpost. For them it was almost a hairpin bend and the copious rubber deposits they laid just before the two-wheelers were due to race over it, required considerable remedial work to return it to race fitness.

It was lack of fuel that robbed Iain Duffus of victory at Cronk ny Mona on the last lap of the 1998 Production race while leading on his Honda, but worse was to follow when he fell (not from a bike) in the Paddock the same year and broke a leg.

With housing development coming to the area in the 1990s, the outside of the bend was widened in conjunction with road improvements and a marshals' shelter was provided in memory of Manx road racers Andy Bassett, Paul Rome and Phil Hogg, who were all killed in Island racing. Phil is also remembered by the Hogg Motorsport Association which, with a pool of unpaid volunteers, provides rescue ambulances and supporting services to a variety of Isle of Man speed events.

Change of Route

It should be remembered that in the first four years of TT racing over the Mountain Course (1911-1914), riders turned right at Cronk ny Mona and headed via Willaston to the finishing line at the top of Bray Hill. That road off is known as Johnny Watterson's Lane and with the nearby Scollag Road they provide access from inside the course for spectators to reach other viewing points.

How Long is the Mountain Course?

There are many who consider that the straightening of bends over a period of 100 years must have reduced the distance around the Mountain Course. But has it? Well, here are a few 'facts' to consider and help determine whether riders are still covering a full 37¾ miles lap.

From 1911 to 1914 the lap length was given as 37½ miles, although the method of measurement is unknown. The distance covered increased in 1920 when the course was extended from Cronk ny Mona to include Signpost Corner, Governor's Bridge and Glencrutchery Road, with the organiser's 'General Conditions for the International 1920 Auto-Cycle Tourist Trophy Races' describing it as 37 miles and 6 furlongs (37¾ miles), a figure that seemed generally accepted. But perhaps not everyone accepted it, for in its issue of 12th April 1923, 'The Motor Cycle' said "The TT Course is to be measured accurately: to do this it will be necessary to employ a revolution counter attached to the sidecar wheel of an outfit driven slowly". The figure that resulted was 37 miles 1299 yards, and as ¾ mile = 1,320 yards, they obviously thought it to be 21 yards short of a full 37¾ miles. But did the sidecar take a central line or a racing line, and is a three-wheeler's racing line the same as a two-wheeler's?

From 1923 the organisers used a figure of 37 miles 1,300 yards and a speed/distance table was published based upon this, quoting an equivalent lap distance of 37.739 miles. After another official measure in 1937 it was declared to be 37 miles 1291 yards or 37.733 miles and the ACU used the figure of 37 miles 1290⅔ yards to calculate lap times in the pre-war period, which they said was 37.7333 miles, with the race regulations for 1937 settling for 37 miles 1,290 yards.

Presumably there was more agitation over lap distance in the early 1950s, for assisted by TT winner Fergus Anderson, 'The Motor Cycle' measured it in 1954 using a Smith's Instruments wheel-revolution counter, and came up with a figure of 37.449 miles which equates to 37 miles and 790 yards. So was it really some 500 yards shorter than in pre-war days?

It was the turn of 'Motorcycle News' to have a measure before the 1978

TT. The method used was not declared, but its result of 37.195 miles (37 miles 343 yards) was. It seemed perfectly happy with its figure which showed that getting on for 1,000 yards had disappeared since 1920, although the race organisers seemed equally content to persevere with calculating lap times based on 37.733 miles, as it had done for many years.

The next record of the course being measured comes from the centenary year of 2007 when Alex Coward, a veteran cyclist and official course measurer for Manx cycle trials, was asked to ride a lap and establish a figure. The line he took was generally one yard out from the left-hand kerb, with just the occasional cutting of a corner. After several hours of pedalling, Alex came up with a distance of 37.7224 miles, which is just 17½ yards (52.8 feet was the actual figure quoted) short of the official 37.73 miles given in the current regulations for the race, although the distance is now shown as 60.70km (37.73 miles).

Perhaps in this age of on-bike telemetry and data-logging we should be able to get a precise figure of distance covered, but then the power put down by modern bikes sees them wheel-spinning under full-throttle, even in the dry (and with traction control), and they also spend more time than in the past with their wheels off the ground over bumps. So how do those factors affect recorded lap distances? And do all riders cover the same distance, for the 'racing line' of tyre marks on bends can be six to eight feet wide, which begs the question, do the fastest riders take the tightest line (and thus the shortest lap distance) or does their greater speed through corners mean they take the widest? And so the questions go on, but with so many variables involved in establishing lap distance, it really is doubtful that a figure which satisfies all of the people all of the time will ever be achieved.

SIGNPOST CORNER

After Cronk ny Mona comes Signpost Corner and with about one mile to go to the Grandstand and Pits, it is a spot where riders can lose concentration as they think about an impending refuelling stop or the end of the race. With those additional tensions it is hardly surprising that there have been an above average number of 'incidents' at this deceptive downhill bend.

The most popular word employed by riders to describe the approach to Signpost from Cronk ny Mona has always been "tricky" and this is because they make a fast approach over a blind summit some 200 yards before, and run downhill under heavy braking to make the quite sharp, dropping right-turn where the signpost (now a signboard) points to Douglas. Many of today's riders feather the throttle slightly before coming over the brow, so that braking is better controlled. That was a policy adopted by Geoff Duke almost 60 years ago, who also told "on these slow bends like Signpost, it is better to go easily and concentrate on getting out quickly". Whether it was related to stresses set up by the fierce braking here is not known, but back in 1921 Bert Kershaw had the exciting experience of having his handlebars break whilst slowing for the corner during practice. Then, on the third lap of the race, he suffered a repeat of the problem, at exactly the same spot.

Much rubber is laid on the road approaching this corner under heavy braking and riders who get Signpost wrong will usually take to the slip road that runs off to Onchan. Conditions must have been particularly slippery in 1949, for a report from the Saturday practice tells that 1 in 3 of the riders out that morning had to make use of the slip-road. There they tackled the frustrating and time-consuming process of doing a multi-point turn in the road, a task not aided by a racing bike's limited steering lock. Others, even

Fast lady Maria Costello cranks into Signpost Corner.

Mountain milestone

NSU riders inspect the course in 1954 when Signpost Corner was undergoing alterations.

This is Signpost Corner in 1927 looking down the narrow road towards Bedstead...

though not travelling on the right line or at the right speed, succumb to the temptation to try and get round the bend and often run wide and hit the outside bank. One of many to do so was Phil Haslam during his winning ride in the 1973 Junior MGP in which he became the first rider to lap at over 100 mph in the Manx. Phil got away without falling and those who do fall generally avoid injury, but any spill here results in the deposit on the road of petrol, oil, cooling fluids and earth from the roadside bank, resulting in a job for marshals and the slowing of other competitors under yellow flags, so losing them precious time.

Though it is such a simple turn, mistakes made at Signpost have cost riders TT race wins and good placings. In the 1922 Junior, Bert Le Vack crashed his New Imperial here on the 4th lap whilst leading, handing victory to Manxman Tom Sheard (AJS) and in 1953 Walter Zeller, riding the first BMW solo entry in the TT since 1939, crashed out of the race at this reasonably straight-forward right-hander, despite having made a pre-race visit to the Island to put in 150 course-learning laps on a road bike. In an incident that was not one of rider error, Michael Rutter was brought off here by oil on the rear tyre when leading the 1997 Senior race. Disappointed to have lost the chance of winning, Michael spotted the silver lining to this particular cloud of disappointment when he observed that he was glad the oil had caused his downfall at Signpost and not at the ultra-fast Hillberry that he had rocketed through half a mile before.

Flagmen

In 1920 some formality was brought to the important position of Flagman, when the Dunlop Tyre Company contracted 26 men to duty for the entire practice and race period, paying them a fee but making their services available to the organisers without charge.

Allocated white coats, sashes and hats, they were officially termed Flag Signallers and became known within the race organisation as 'Dunlops', for their sashes and hat-bands carried the company name. However, although allowed to display such advertising during practice, initially they were not allowed to wear sash and hat-band on race days.

Issued with just a red flag, they were told "When a motorcycle approaches you, look around your corner, and, if you see that the road is clear, keep your flag furled by your side. Should you see the road obstructed at all near your corner, vigorously wave the Red Flag to indicate Stop". On some corners the Flagman would be on his own, but his instructions said "Should an accident happen you must not go to help but remain at your post waving your Red Flag to prevent other motorcycles rushing into danger".

With today's requirements for marshals to be stationed so that they are within line of sight of each other, it is difficult to imagine how the organisers managed to control things in 1920 with just 26 points manned during practice, supplementing those with marshals and policemen to have some 60 locations covered on race days.

Changes

The layout of Signpost Corner has been subject to change down the years, although it still retains its main characteristic of being a sharp, dropping right turn. In 1923 it received improvements that saw it eased and widened, but still with a high stone wall on the apex. Further changes took place in early 1954, with some of the work relating to the fact that Signpost Corner became the point where Sidecars and Lightweights using the Clypse Course (1954-1959) rejoined the Mountain Course from the Onchan direction, for the run to the finish. The appearance was much changed, with the inside wall removed to be replaced by railings, and 'The Motor Cycle' claimed that "considerably higher speeds were possible", whilst 'Motor Cycling' felt that allied to improved road surfacing between Cronk ny Mona and the finish, riders were able to save several seconds on their lap times.

With the advent of housing development in the area, in 1993 a mini-roundabout was formed to control day to day traffic at Signpost Corner and further traffic management works were carried out before the 2011 TT, but the racing line remained the same.

The Light is on at Signpost Corner

The scoreboards facing the Grandstand at the start and finish area are fitted with a lamp over each riders' number which is illuminated towards the end of a lap, so alerting pit-crews, spectators, commentator, etc, to a rider's approach for refuelling or the end of the race. That early-warning facility was first introduced in 1921 and was activated when a rider reached Governor's Bridge, but as speeds rose the place for the warning to be given was moved back along the course and it became a stock phrase of commentators to announce a rider's end of lap progress with "his light is on at Signpost Corner", an expression that entered into the language of TT fans and beyond. Whether he was a fan of the TT is not known, but celebrated author of the 'Flashman' series of books, George MacDonald, moved to live on the Isle of Man in 1969. He published his autobiography a few years ago at the advanced age of 80 and, perhaps in anticipation of his death two years later, called it 'The Light's on at Signpost Corner'. Although Signpost remained in use as the end of lap warning point for many years, it was eventually moved further back up the course to its present location of Cronk ny Mona.

Countless bikes have overshot the turn at Signpost, but Australian Jim

Scaysbrook tells of a four-wheeled episode here. Jim was a racing friend of Mike Hailwood and accompanied him on his return to the TT in 1978. He recalled how they used a Rover 3500 that had been loaned to Mike and did several laps in the company of the then course record holder Mick Grant, with the idea of helping Mike get back up to speed. One problem was that Mike had difficulty in believing some of the speeds Mick was telling him, though history shows that, come the racing, he overcame that little problem. Jim told in 'Classic Racer' how "We reeled off a few laps in the Rover with Mike getting a little more animated behind the wheel with each passing mile, until during one plunge down from Kate's Cottage we only just scraped around Creg ny Baa and both Jeff Sayle and I (from the back seat) said we could smell something burning. Mike was enjoying himself too much to even bother to answer, but arriving at Signpost, the brake pedal went straight to the floor. The cornering attempt was hastily aborted and we shot straight through the intersection and managed to stop with the handbrake some distance up the road. The rear brakes were on fire, the front smouldering wrecks".

The last word on Signpost goes to Joey Dunlop, a man who rode it hundreds of times on bikes of varying capacity. Talking in the mid-1980s he said "Signpost is a crucial bend that has to be tackled correctly. The line is most important in order to line up the following Bedstead Corner".

BEDSTEAD

Whilst Signpost Corner developed a justified reputation for riders overshooting, in 'History of the Clubman's TT Races' it is told how during opening practice in 1952 no rider actually took to the slip road at Signpost but instead, for reasons that were not explained, at Bedstead, the corner that followed "no fewer than 14 were glad to find the gate open" into a track that led off.

The run from Signpost Corner to Bedstead and The Nook was previously narrow and heavily-cambered at the sides between high banks. Bedstead is a left-hand bend, but the steep cambers deterred riders from using the far right-hand side of the road and so they approached it nearer the centre, taking it in second gear at about 60 mph in the 1950s and notching third on the downhill run to The Nook. Today this stretch does allow riders to approach Bedstead on the right with a short burst of acceleration and braking over a road that is barely recognisable from the one used 50 years before in terms of width, surfacing, cambers, addition of footpaths, street lighting columns, copious white lines, etc.

How So Named?

The surroundings to the road have also changed with the coming of housing development, and no longer are gaps in an adjoining field filled with old bedsteads, something that seems to be generally accepted as the reason why the corner acquired its name. Indeed, writing in 1963, the reliable Canon Stenning said "until the last few years the field boundary was decorated with old iron bedsteads". However, in 1936 a writer to 'The TT Special' claimed that "it is so called because of the ambulance bell-tent with the bedstead outside placed in the grounds opposite the corner" and editor Geoff Davison added "this . . . is the most accurate description of Bedstead Corner". The writer also mentioned "I believe it is sometimes called Rectory Corner" and for a short while the racing fraternity is said to have called it Bennett's Corner - reason unknown.

THE NOOK

Wise riders know that they are not going to win a race by their performance over this last part of the lap and so caution is the word for riding

All the farmland seen in this aerial view of Signpost Corner in the early 1970s is now covered in housing.

the stretch through Bedstead to The Nook and on to Governor's Bridge. However, not every rider is wise or cautious and because its downhill nature means that they have been able to pour on all available power on the short straight after Bedstead, many have been unable to lose sufficient speed to negotiate the unforgiving and slower-than-remembered corner at The Nook, despite heavy braking. It really is an awkward corner to try and race around, with Harold Daniell calling it "a nasty bumpy affair", whilst others have been even less complimentary. What the first Japanese rider to race at the TT, Kenzo Tada, said when he crashed here during practice in 1930 would probably have been unintelligible to bystanders, but what they would have noticed was that he was close on the tail of Charlie Dodson, the Senior TT winner of the previous two years, and whilst the experienced Charlie sailed round on his Sunbeam, Velocette-mounted Kenzo was caught out, probably by trying to go too fast, too soon.

Austin Munks was an MGP winner who told how early in practice for the 1934 MGP he got involved in a spirited ride with 'Crasher' White (later to take five TT podium places). After swapping places for a lap and a half, 'Crasher' was about 50 yards in the lead as they approached the blind corner of The Nook. Austin rushed round in pursuit and, in his words, "to my surprise and horror I saw 'Crasher' yards up the bank on the far side of the road, tearing chunks of grass and earth up with his footrests. I couldn't do anything about it, having got the model well laid over, and I suddenly realised that 'Crasher' looked like falling on top of me. By sheer luck I just got under him before he stalled and decided to come to earth again". Riding on to the finish of the lap, Austin was surprised when 'Crasher' drew up alongside him, his only concern being that his excursion had cost him the chance to be first to the finishing line.

A few years later at the 1937 MGP, Sub-Lieutenant H.A. Kidd managed to crash his Velocette at what was then a 45 mph bend. His was the only Velo in the Senior Manx that year and although this off-duty naval officer escaped injury, it seems that he was fortunate, for it was said "his crash helmet stood him in good stead".

To become a member of a 'works' team like MV Agusta was the height of success in the 1950s. The talented Terry Shepherd achieved that coveted status in 1957, but in Wednesday morning TT practice he fell just after The Nook and broke two fingers on his left-hand. The injury kept him out of racing for several months and prevented him from proving his worth to MV, with the result that he was back riding Nortons in 1958.

Many others have attempted to ride the banks at The Nook and fallen to earth, including stars such as Phillip McCallen who came off here in 1990 when his bike slipped into neutral while changing down, thus robbing him of

Mountain milestone

Ryan Farquhar commits his Kawasaki to The Nook at the 2010 TT.

the engine-braking that riders rely on in such situations. The road is particularly narrow for the sidecars, and multiple class winner Dave Saville rated the corner as "frightening" on three-wheels.

GOVERNOR'S BRIDGE

As riders exit The Nook, a spot referred to in the early 1920s as Okell's House, they pass the grounds of the Governor of the Island's residence on the left, itself formerly called Bemahague, and need to take a line that puts them in a favourable position to tackle the short run containing two narrow, tricky left-handers leading down under over-hanging trees to Governor's Bridge. When first brought into use in 1920, this was described as "an awful right-handed hairpin, which will be the worst corner on the whole course". Though it is the slowest point of the course, it has the greatest number of fallers. Fortunately, at this sharp right-hand hairpin it is usually more a matter of falling over than falling off, but riders have always come to grief at Governor's, for a lot of things come into play here, including the heavy braking used on the approach, the transition from high to low speed, the sharpness and camber of the bend, plus the effect of making the turn with tight steering dampers and limited steering-lock as the road falls away into the dip.

So many riders have fallen here, but perhaps one of the most unfortunate was Eric 'Mouse' McPherson who travelled 13,000 miles to do so at the 1948 TT. Spending 4 weeks on a steamer to reach Marseilles, he then travelled overland to the Island, arriving almost a week before practice started to familiarise himself with the course. Going out in the first official practice, 'Mouse' completed a lap in 32 minutes but on his second lap he took a simple fall at Governor's Bridge, damaged his pelvis and put himself out of the race.

Countless photographs have been taken at Governor's Bridge over the past century, but the one on the previous page of Bob McIntyre and his Gilera on the way to victory in the 1957 Golden Jubilee Senior TT is a true classic, for it also shows him setting the first 100 mph lap of the Mountain Course.

The steep downhill and tree-lined run from The Nook is shown in the photograph, as is the sharp drop on the turn under the gaze of massed spectators on the inside. Bob would have been down into bottom gear on this very tricky right-hander, coping with the effect of limited steering-lock on the Gilera, a nasty selection of cambers and a road surface lubricated with oil droppings.

Today's TT images from this spot are much changed, for not only are spectators no longer allowed on the inside of the bend, but even the white-washed stone pillar is not in its original position. A few years ago a careless lorry driver demolished this historic monument (well it is a historic monument to TT fans!) and it was rebuilt a couple of yards further back. Thereafter the original line was maintained with bales, but now there are permanent kerbs.

Arrows used to be painted on the road to guide riders into the part known as Governor's Dip that follows the turn, but presumably that stopped when the Clypse Course was brought into use in 1954, for although those riding the Clypse used the same approach to Governor's Bridge from the direction of The Nook, they did not ride the dip but turned right on to the higher 'main' road - a much more comfortable route.

So why do riders using the Mountain Course use the Dip? Well, in 1920 when the course was extended (riders previously turned right at Cronk ny Mona) the main road looped into the Dip and was the only one available. Then in 1923 the road that takes current traffic was built to bridge the glen at a better position, but use of the Dip was maintained for racing.

This busy road junction has had its share of highway improvements, the most recent being in 2008 when the original narrow stretch from The Nook to Governor's Bridge was by-passed for ordinary traffic with a new road and a

Creg Ny Baa to the Finish

The steep downhill and tree-lined run from The Nook is shown in the photograph, as is the sharp drop on the turn under the gaze of massed spectators on the inside. Bob McIntyre would have been down into bottom gear on this very tricky right-hander, coping with the effect of limited steering-lock on the Gilera, a nasty selection of cambers and a road surface lubricated with oil droppings.

revamped roundabout created, the original road being retained for the races.

The often dark and dank Governor's Dip is not the most popular part of the course. In the past Stanley Woods referred to a Lily Pool here and it was not quite so heavily overhung with trees as it is today.

The dip is now closed to ordinary traffic, so requires much sweeping and cleaning before racing, but it often remains damp. It is a place to be ridden with care, for as Geoff Duke explained "There is nothing to be gained by trying to win the race here; a split second gained is not worth the risk of falling off". Peter Williams put it even more emphatically with "I just trickle round there. Can't gain anything. I hate the place", whilst sidecar man Dave Molyneux describes it as "the sh . . . y part, it's always slippery and so bumpy (on the exit) it throws you wide onto Glencrutchery Road".

Back in 1952 the motorcycling press reported in critical tones of a distinct change of road surface where riders emerged from the dip onto Glencrutchery Road. What they claimed to be the "non-skid" surface that riders had experienced over the Mountain section maintained by the Highways and Transport Board, was followed at this point by something less grippy that came under the control of Douglas Corporation.

Today's six-speed race bikes are quite flexible at slow speeds, but in the days of three-speed gearboxes through the 1920s and highly-geared four speeds of the 1950s, the clutch was often pulled in for the actual rounding of Governor's Bridge, then slipped whilst riding through the Dip before gradually being fed in with increasing revs as the rider moved from the left to the right side of the road, gathered speed and changed up for the exit from the Dip to burst onto the Glencrutchery Road, where there has always been a hump to catch the unwary and carry them out wide to where the camber gets quite steeply adverse near the far gutter. Of course, 'bursting out' was reserved for those who still retained full power, because for those with mechanical problems or who were empty of fuel, the climb out was yet another moment of truth on the long journey to the chequered flag.

It is a Long Push

So close to the finish, the stretch of Glencrutchery Road that leads from Governor's Dip to the Grandstand takes but a few seconds to cover under full power, but it has been the scene of much racing drama and despair, because for those forced to push in it seems to take an eternity for the finishing line to appear.

One extraordinary race that culminated with a push along the Glencrutchery Road was the 1920 Junior TT. Cyril Williams was riding a 'works' AJS that, along with his team-mates' similar models had been much

Crowds watch a sidecar crew at Governor's Bridge in 1924.

Stanley Woods (Moto Guzzi) rides through the dip at Govenor's Bridge on his way to victory in the 1935 Senior TT.

Mountain milestone

Eric Twemlow leaves Governor's Bridge Dip for Glencrutchery Road in the 1929 Junior race, although he was later to retire his D.O.T. machine, just like the unfortunate rider pushing number 7.

faster than anything else in practice, so it was no surprise when Cyril and the other AJS riders were soon heading the race. It was run over 5 laps (187½ miles), would last for over 4½ hours and in those days it required an element of controlled riding to have a chance of staying the distance. However, experienced TT followers watched in alarm as the AJS team raced each other for victory at the head of the field. So hot was the pace that several of them dropped out and suddenly Cyril was the only AJS runner left with a chance of victory. Then further disaster struck when his bike gave out at Keppel Gate on the last lap. Coasting where possible, pushing where necessary, Cyril gave of his all over the last four miles. Fighting off attempts by well-wishers to help him push the bike along the Glencrutchery Road (outside assistance was not allowed), he staggered over the line to retain first place.

Fortunate

Another performance of note was recorded here by Jimmy Simpson (Norton) in the 1931 Senior TT. On the third lap of the race Jimmy became the first man to set an 80 mph lap (80.82), and it was achieved despite running out of fuel after leaving Governor's Dip, thus causing him to coast for several hundred yards along the Glencrutchery Road at diminishing speed, before crossing the line to complete his record-breaking lap and pulling into his pit to refuel. Clearly there was an element of good fortune involved for Jimmy and so there was back in 1912 when the twin-cylinder Douglas of Junior winner Harry Bashall snapped one of its connecting rods as it crossed the finishing line. Stanley Woods win in the 1935 Lightweight 250 on a Moto Guzzi was also blessed by good fortune, for his little single's valve-gear gave way as he crossed the line, or as a report of the post-race strip-down put it, "the Guzzi revealed unmistakable symptoms of a distaste for excess revs". The 500 on which Stanley won the Senior race two days later shared many of the head and valve-gear components of the 250, and maybe this explains why he made a slow start before throwing caution to the winds on the last lap, revving it into the danger area and snatching victory from the unfortunate Jimmy Guthrie, who (because of the interval starting system) had finished his race some 14 minutes before Stanley and was already being feted as the winner.

Lady luck smiled on another Irishman on Friday 13th June 1952, when Reg Armstrong had the primary chain of his Norton snap just as he was taking the chequered flag to win the Senior. Dave Molyneux was also 'lucky' when taking victory in the second Sidecar race in 2005. Having set a record-breaking pace, Dave had a rear-wheel bearing break up as he came onto the Glencrutchery Road and just managed to coax his outfit to the line. Mick Boddice was not so fortunate with his outfit in 1985, for having led the entire three laps (113 miles), he was within a few hundred yards of the flag when the drive chain jumped its sprockets and the outfit slowed dramatically, letting Dave Hallam through to snatch victory.

Races Lost

Both TT and MGP races have seen heroic sights as riders who have run out of petrol or suffered machine failure in the closing stages of a long and tiring race, have sought to push their stricken machines the length of the Glencrutchery Road to the finishing line, causing many to arrive in a state of exhaustion. Among those who have been leading a race on the last lap but then been forced to push in have been George Bower (1923 Amateur Race), John 'Crasher' White (1932 Junior MGP), George Brown (1948 Senior Clubman's TT), Charlie Salt (1948 Senior MGP), Les Graham (1949 Senior TT), Sammy Miller (1957 Lightweight 250 TT), Tom Thorp (1959 Junior MGP), Ken Huggett (1971 Senior MGP), John Williams (1976 Senior TT), Roy Richardson (1999 Lightweight Classic MGP). All lost their coveted race lead, and though some still finished in the first ten, others who had been pushing from far out were passed by forty or fifty riders. The above 'stars' have shared the pushing experience with many lesser lights, for whom the will to finish was just as strong. Most of them would have had the time while pushing (but hardly the inclination) to reflect on the fact that the word Glencrutchery is said to be from an earlier form of Glen Cruggery, meaning the Harper's Glen.

In 'Joey Dunlop - His Authorised Biography', author Mac McDiarmid tells of an unwanted mid-race push along the Glencrutchery Road performed by Joey. It was during the notorious 1981 Classic TT when Honda, having had a Stewards decision go against them earlier in the week, turned out their three man team of bikes and riders for the Classic TT in black paintwork and leathers as a silent protest. Riders Joey Dunlop, Ron Haslam and Alex George were mere pawns in the political game Honda were playing, but Joey did his best to give his employers the win they sought by running at record-breaking speed in the early stages of the race. Unfortunately, such was his pace that on the third lap, just a couple of miles before he was due to refuel, his Honda ran out of petrol. Joey explained "the petrol ran out at Cronk ny Mona and I free-wheeled to Governor's, then began the push. The bike is so big . . . I only made it to the pits by taking it one telegraph pole at a time, then having a little rest". Refuelled, Joey set off at a furious pace, only to have the cam chain break at the Gooseneck.

GLENCRUTCHERY ROAD

The majority of riders are still under engine power at Governor's, and for them the curving right-hand uphill exit from the dip to the finishing line is just under half a mile of road that starts with an upgrade then levels out. Care is needed not to apply too much throttle and get carried wide on exiting the Dip, but the ensuing stretch should receive straightforward full throttle treatment, although there were times in the late 1930s when hot weather caused problems with the surface, a report from 1939 saying "the sun had played havoc with the tar . . . it lay there in a thick, sticky mass . . . riders warned".

Account must be taken of the different situations under which competitors are expected to ride the Glencrutchery Road. As examples, an evening's practice at the TT will usually be divided into two sessions, one for solos and the other for sidecars, and at the end of the first session riders must be prepared to be red-flagged and stopped prior to the usual finishing line, because competitors in the other session will be starting from the pits area and so the usual return road after the finishing line is not available. At the MGP it is usually Classics, Newcomers and Ultra Lightweights who go out in one session and Senior, Junior and Lightweight in the other. Obviously, the situation will differ during a race, but even then two different methods apply to riding this stretch. In earlier days riders would refuel after 3 or occasionally 4 laps, but today it is usually after 2 laps. Accordingly, those intending to refuel need to position themselves on the Glencrutchery Road so that they can peel off left for pit-lane.

In the first year of use of this stretch of road (1920), those riders going through on a non-stop lap would have been hitting some 65-70 mph as they passed the Grandstand over a fairly narrow road that was substantially widened in 1936. Speeds have increased and today the top men are doing over 170 mph, which means it is essential that they keep over to the right hand side of the road to avoid competitors who are stopping and those who are leaving pit-lane and rejoining the course after a pit stop.

The Ultimate!

Then there is the ultimate Glencrutchery Road scenario, the one that riders dream of and that has them racing straight down the middle to take the chequered flag at the end of a race, with the crowd in the Grandstand on their feet and cheering. It is the winner's privilege to see the flag being waved for him, just as he will have seen course-side spectators enthusiastically waving their Programmes during most of the last lap. Those finishing further down the field are usually greeted with less enthusiasm and their lot is to see the chequered flag held stationary. Nevertheless, every finisher will experience a sense of achievement, and probably relief; with those who have done well feeling incredibly elated, whilst others who have not met their targets and expectations, either through machine problems or their own riding, often feeling completely deflated.

Completing a Race

A race is completed when a rider passes over the white finishing line painted across the Glencrutchery Road, after completing the requisite number

Stanley Woods manages a modest smile after winning the 1933 Junior TT.

of laps. As well as the waving of a flag (not chequered in the early days), riders were shown a large STOP sign. The finishing line is located at the start of the Grandstand in front of what are still called the Timekeepers Boxes, where for nearly 100 years riders were manually timed, although today it is done by signal from a transponder fitted to the bike, via pick-up points here and elsewhere on the course.

Unlike a massed-start race where all riders cross the finishing line within about half a minute, the combination of interval starting at the TT and MGP where the last man gets away perhaps 15 minutes after the first, the difference in racing ability between competitors that can amount to several minutes per lap, and unforeseen problems that may slow a rider; wide gaps (often exceeding a lap) open up between front runners in a race and those

This is just one of the twenty-six occasions that Joey Dunlop took the chequered flag as victor of a race for a Tourist Trophy.

Mountain milestone

The three top men at the record-breaking 2007 Senior TT. Left: Guy Martin, 2nd, centre: John McGuinness, 1st, right: Ian Hutchinson, 3rd.

Undoubtedly the star of the 2010 TT with five wins in the week, Ian Hutchinson poses with the Tourist Trophy.

bringing up the rear. Organisers have always been reluctant to flag off later runners until they have completed all their laps, but there have been instances where individual riders have been so far back that they have circulated in glorious isolation when everyone else has finished, riding almost a lap with the sole attention of hundreds of marshals and the entire race organisation devoted to their efforts.

With the races run to a timetable and the organisers wanting to hand the roads of the Mountain Course back to the Manx people without unreasonable delay, the regulations now specify how the finish of a race will be handled, saying: "Riders who have completed the designated number of laps for the race will be shown a chequered flag by an official at the finish line, at track level. Riders who cross the finishing line without completing the designated number of laps but after the leading rider on the road has been shown the chequered flag will be permitted to proceed on a further lap but after a race has been won and where in the opinion of the Clerk of the Course it is no longer possible to qualify for a cash award or replica, he will order all competitors to stop as each crosses the finishing line, irrespective of the number of laps completed. Red lights will then be shown prior to the return road gate and a marshal showing a red flag will stand at the return road gate. To be counted as a finisher in the race and to be included in the race results a rider must complete the full designated number of laps for the race and cross the finishing line within a time decided by the Clerk of the Course".

Confusion can occur with competitors who have completed their allotted laps being flagged off whilst others are going straight through, for, surprising as it may seem, riders are not always certain which lap they are on. In the seven-lap 1938 Junior race, Stanley Woods (Velocette) won from Ted Mellors (Velocette), but Ted did not see the finishing flag and rode on to almost complete an eighth lap, eventually being stopped some 34 miles later at Creg ny Baa.

The regulations covering the finish of a race contain an extra proviso that

Time for the boat trip home after another year's racing.

"The rider must be in contact with his/her machine". J. Malcolm Muir came home to win the 1931 Senior MGP by 10 minutes, but such was his exuberance that he slammed his brakes on so hard that he took an embarrassing tumble on damp roads, right in front of the crowded Grandstand. Fortunately he was in company with his bike as they crossed the line.

After the Race

There is rather more hustle and bustle about post-race procedures today than there used to be. Well before the last man has been flagged off, the first three will have been directed along the return road into the Winners Enclosure where they will receive enthusiastic welcomes from families, pit-crews and media. It is a place where emotions spill out and a few quick interviews are performed.

Then riders are ushered to the garlanding rostrum in front of the Grandstand, where dignitaries are gathered, anthems are played, the appropriate Tourist Trophy is available, champagne is sprayed, photographs taken and the crowds applaud. From there it is off to the media centre, where they are questioned for the benefit of radio, TV, newspaper and magazine representatives, who then prepare accounts of the race that will go around the world.

An intense fortnight of racing activity over the Mountain Course customarily finishes on a Friday with the race for the Senior Tourist Trophy (or Senior MGP) and whilst the top finishers in the race are busy giving their versions of it to the media, tail-enders are being flagged off, the Grandstand is emptying, the Roads Open car is nearing completion of its post-race lap, traders are thinking about closing down their stalls at the back of the Grandstand and as has happened for a century, the unique 37¾ mile Mountain Course is about to be handed back to the people of the Isle of Man, for what its roads were created for - their normal day to day use.

Ordinary roads returned to ordinary traffic.

INDEX

Agostini, G. 40, 41, 78, 79, 106, 126, 159, 161
Amm, R. 43, 78, 160
Anderson, F. 71, 91
Anstey, B. 109
Applebee, F. 167
Archibald, A. 56
Armstrong, R. 188

Baldwin, M. 126
Bashall, H. 14, 188
Barrington, M. 89
Bennett, A. 58, 178
Boddice, M. 92, 105, 176, 188
Brandish, W. 25, 175

Casey, M. 145
Collier, C. 11, 68, 71
Crabtree, S. 149, 173
Crosby, G. 35, 39, 139, 174
Crowe, N. 53
Cummins, C. 18, 124, 127

Dahne, H. 157
Daniell, H. 54, 57, 59, 60, 72, 75, 94, 97, 174
Davenport, L. 53, 57
Davies, H. 26, 136, 172
Davison, G. 81, 139
Dixon, F. 178
Dodson, C. 38, 65, 165
Donald, C. 91
Doran, W. 72
Duff, M. 58
Duffus, I. 138, 180
Duke, G. 15, 33, 42, 43, 50, 57, 68, 73, 78, 88, 90, 101, 103, 144, 160, 173, 178
Dunlop, J. 16, 18, 32, 38, 40, 44, 48, 51, 52, 62, 69, 73, 74, 98, 100, 106, 114, 139, 140, 143, 157, 164, 168, 178, 188, 189
Dunlop, R. 16, 65, 92, 107, 150, 164

Farquhar, R. 86, 184
Fath, H. 58
Findlay, J. 179
Fisher, R 19, 129
Fogarty, C. 20, 114, 119, 120, 176
Fowler, R. 11
Frith, F. 20, 69, 174

George, A. 20, 63, 80, 145, 174
Godfrey, O. 132
Graham, L. 40, 149, 180

Graham, S. 73, 144
Grant, M. 16, 20, 35, 38, 72, 73, 78, 114, 139, 147, 179
Guthrie, J. 43, 48, 141, 148, 177, 178

Hailwood, M 38, 51, 58, 72, 75, 76, 78, 79, 84, 88, 97, 98, 99, 101, 109, 111, 114, 122, 123, 129, 133, 135, 141, 145, 155, 157, 159, 161, 173, 175, 183
Handley, W. 15, 41, 53, 86, 88
Hanks, R. 69, 110, 163
Hartle, J. 145, 180
Haslam, R. 35, 40, 105
Herron, T. 56, 79, 83, 100
Hislop, S. 17, 18, 20, 32, 38, 50-52, 54, 57, 60, 63, 72-74, 76, 84, 93, 104, 109, 110, 120, 144, 161, 162, 166, 173
Holden, R. 76
Hunt, T. 20, 42, 174
Hutchinson, I. 4, 42, 43, 54, 76, 86, 109, 162, 167, 190

Ireson, T. 80
Itoh, M. 156

Jefferies, D. 17, 38, 40, 56, 71, 97, 111, 118, 140
Jefferies, N. 4, 59, 62, 78, 92, 98, 117, 121, 175
Jefferies, T. 27, 47

Knight, R. 17, 47, 54, 60, 75, 87, 92, 103, 114, 119, 140, 144, 160

Laycock, E. 164
Leach, D. 145
Lougher, I. 51, 54, 90, 156
Lomas, W. 16

McCallen, P. 55, 91, 109, 138
McElnea, R. 32
McGuinness, J. 28, 34, 36, 38, 53, 56, 62, 79, 86, 89, 98-100, 107, 108, 120, 131, 140, 144, 176, 179, 190
McIntyre, R. 38, 59, 75, 90, 97, 110, 149, 178, 185
Martin, G. 17, 32, 38, 54, 90, 146, 190
Meier, G. 78, 114
Minter, D. 16, 93, 176
Molyneux, D. 39, 54, 85, 90, 98, 108, 121, 144
Moodie, J. 71

Nott, E. 104

O'Dell, G. 61, 69, 121, 156, 175

Phillip, A. 57
Pickrell, R. 16
Plater, S. 15, 17

Provini, T. 101
Pullin, C. 25

Quayle, R. 17, 70, 158
Read, P. 20, 65, 84, 150, 174
Redman, J. 39
Reid, B. 88, 106, 179
Robb, T. 160
Rutter, M. 182
Rutter, T. 27

Saville, D. 72, 76, 124, 174
Seeley, C. 17, 39
Sheard, T. 35, 45, 81
Simpson, I. 161
Simpson, J. 39, 60, 75, 81, 96, 161, 188
Steinhausen, R. 110
Surtees, J. 17, 54, 58, 59, 73, 76, 79, 80, 90, 96, 100, 105, 113, 129, 138, 146

Taylor, J. 20, 39, 90, 110
Tenni, O. 156
Tonkin, S. 88
Tyrell Smith, H. 76, 174

Uphill, M. 35

Walker, G. 35, 36, 58, 75, 78, 81, 97, 110, 112, 119, 121, 149
Walker, M. 20
Williams, Charlie 40, 75, 90, 110, 128
Williams, Cyril 185
Williams, E. 25, 27, 92
Williams, J. 120
Williams, P. 20, 89, 90, 113, 122, 158, 162
Woodland, B. 91
Woods, S. 35, 59, 63, 70, 72, 77, 111, 117, 156, 172, 178, 187, 189